Proceedings of the 16th International Meshing Roundtable

Michael L. Brewer · David Marcum (Eds.)

Proceedings of the 16th International Meshing Roundtable

 Springer

Michael L. Brewer
Computational Modeling
Sciences Department
Sandia National Laboratories
P. O. Box 5800
Albuquerque, NM 87185
USA
mbrewer@sandia.gov

David Marcum
Mississippi State University/SimCenter
2 Research Blvd., MS 9627
Starkville, MS 3919
USA
Marcum@simcenter.msstate.edu

ISBN 978-3-642-43466-2 Springer Berlin Heidelberg New York

Springer is a part of Springer Science+Business Media
springer.com
© Springer-Verlag Berlin Heidelberg 2008
Softcover re-print of the Hardcover 1st edition 2008

Typesetting: by the authors and Integra using a Springer LATEX macro package

Cover design: WMX Design, Heidelberg

Printed on acid-free paper   SPIN: 12049370   5 4 3 2 1 0

Preface

The papers in this volume were selected for presentation at the 16th International Meshing Roundtable (IMR), held October 14–17, 2007 in Seattle, Washington, USA. The conference was started by Sandia National Laboratories in 1992 as a small meeting of organizations striving to establish a common focus for research and development in the field of mesh generation. Now after 16 consecutive years, the International Meshing Roundtable has become recognized as an international focal point annually attended by researchers and developers from dozens of countries around the world.

The 16th International Meshing Roundtable consists of technical presentations from contributed papers, keynote and invited talks, short course presentations, and a poster session and competition. The Program Committee would like to express its appreciation to all who participate to make the IMR a successful and enriching experience.

The papers in these proceedings were selected from among 41 submissions by the Program Committee. Based on input from peer reviews, the committee selected these papers for their perceived quality, originality, and appropriateness to the theme of the International Meshing Roundtable. We would like to thank all who submitted papers. We would also like to thank the colleagues who provided reviews of the submitted papers. The names of the reviewers are acknowledged in the following pages.

We extend special thanks to Lynn Washburn, Bernadette Watts, and Jacqueline Finley for their time and effort to make the 16th IMR another outstanding conference.

16th IMR Program Committee
August 2007

Reviewers

Name	Affiliation
Ahn, Hyung Taek	Los Alamos Nat. Labs, USA
Alliez, Pierre	Inria, Sophia-Antipolis
Ascoli, Ed	Pratt & Whitney Rocketdyne, USA
Benzley, Steven	Brigham Young Univ., USA
Borden, Micheal	Sandia Nat. Labs, USA
Branch, John	Univ. Nacional de Colombia, CO
Brewer, Michael	Sandia Nat. Labs, USA
Burgreen, Greg	Mississippi State U., USA
Cabello, Jean	U.G.S., USA
Clark, Brett	Sandia Nat. Labs, USA
Dey, Tamal	Ohio State Univ., USA
Frey, Pascal	Laboratorie Jacques Louis Lions. UPMC, France
Gable, Carl	Los Alamos Nat. Labs, USA
Garimella, Rao	Los Alamos Nat. Labs, USA
Helenbrook, Brian	Clarkson Univ., USA
Ito, Yasushi	U. of Alabama, Birmingham, USA
Knupp, Patrick	Sandia Nat. Labs, USA
Kraftcheck, Jason	U. of Wisconsin, USA
Ledoux, Franck	CEA/DAM, USA
Lipnikov, Konstantin	Los Alamos Nat. Labs, USA
Loubere, Raphael	University of Toulouse, France
Marcum, David	Mississippi State U., USA
Mezentsev, Dr. Andrey	CeBeNetwork Engineering & IT, Bremen, Germany
Michal, Todd	Boeing, USA
Mukherjee, Nilanjan	U.G.S., USA
Owen, Steve	Sandia Nat. Labs, USA
Patel, Paresh	MSC Software, USA
Pav, Steven	U. of California, San Diego, USA
Pebay, Philippe	Sandia Nat. Labs, USA
Quadros, William	Algor, USA
Rajagopalan, Krishnakumar	PDC, USA
Schneiders, Robert	MAGMA Giessereitechnologie GmbH, Germany
Simpson, Bruce	U. of Waterloo, CA
Shepherd, Jason	Sandia Nat. Labs, USA
Shewchuk, Jonathan	U. of California, Berkley, USA
Shih, Alan	U. of Alabama, Birmingham, USA
Shontz, Suzanne	Pennsylvania State Univ., USA
Staten, Matthew	Sandia Nat. Labs, USA
Steinbrenner, John	Point Wise, USA
Thompson, David	Mississippi State U., USA
Thompson, David	Sandia Nat. Labs, USA
Unger, Alper	University of Florida, USA
Vassilevski, Yuri	Russian Academy of Sciences, Russia
Xiangmin, Jiao	Georgia Institute of Technology, USA
Yamakawa, Soji	Carnegie Mellon U., USA
Zhang, Yongjie	U. of Texas, Austin, USA

16th IMR Conference Organization

Committee:

David Marcum (Co-Chair)
Mississippi State Univeristy/SimCenter, Starkville, MS
Marcum@simcenter.msstate.edu

Michael Brewer (Co-Chair)
Sandia National Labs, Albuquerque, NM
mbrewer@sandia.gov

Edward Ascoli
Pratt & Whitney Rocketdyne, Canoga Park, CA
edward.ascoli@pwr.utc.com

Rao Garimella
Los Alamos National Laboratory, Los Alamos, NM
rao@lanl.gov

Todd Michal
The Boeing Company, St. Louis, MO
todd.r.michal@boeing.com

Bruce Simpson
UW, Dept. of Computer Science, Waterloo, Canada
rbsimpson@uwaterloo.ca

Joseph Walsh
Simmetrix, Inc., Clifton Park, NY
jwalsh@simmetrix.com

Soji Yamakawa
Carnegie Mellon University, Pittsburgh, PA
soji@andrew.cmu.edu

Coordinators:

Lynn Janik Washburn
Sandia National Labs, Albuquerque, NM
lajanik@sandia.gov

Jacqueline A. Finley
Sandia National Labs, Albuquerque, NM
jafinle@sandia.gov

Web Designer:

Bernadette Watts
Sandia National Labs, Albuquerque, NM
bmwatts@sandia.gov

Web Site: **http://www.imr.sandia.gov**

Table of Contents

Session 5B Applications and Software

Session 1A

Tetrahedral Meshing 1

1A.1

Aggressive Tetrahedral Mesh Improvement

Bryan Matthew Klingner and Jonathan Richard Shewchuk

University of California at Berkeley

Summary. We present a tetrahedral mesh improvement schedule that usually creates meshes whose worst tetrahedra have a level of quality substantially better than those produced by any previous method for tetrahedral mesh generation or "mesh clean-up." Our goal is to aggressively optimize the worst tetrahedra, with speed a secondary consideration. Mesh optimization methods often get stuck in bad local optima (poor-quality meshes) because their repertoire of mesh transformations is weak. We employ a broader palette of operations than any previous mesh improvement software. Alongside the best traditional topological and smoothing operations, we introduce a topological transformation that inserts a new vertex (sometimes deleting others at the same time). We describe a schedule for applying and composing these operations that rarely gets stuck in a bad optimum. We demonstrate that all three techniques—smoothing, vertex insertion, and traditional transformations—are substantially more effective than any two alone. Our implementation usually improves meshes so that all dihedral angles are between $31°$ and $149°$, or (with a different objective function) between $23°$ and $136°$.

1 Introduction

Industrial applications of finite element and finite volume methods using unstructured tetrahedral meshes typically begin with a geometric model, from which a mesh is created using advancing front, Delaunay, or octree methods. Often, the next step is to use heuristic mesh improvement methods (also known as *mesh clean-up*) that take an existing mesh and try to improve the quality of its elements (tetrahedra). The "quality" of an element is usually expressed as a number that estimates its good or bad effects on interpolation error, discretization error, and stiffness matrix conditioning. The quality of a mesh is largely dictated by its worst elements. Mesh improvement software can turn a good mesh into an even better one, but existing tools are inconsistent in their ability to rescue meshes handicapped by bad elements. In this paper, we demonstrate that tetrahedral mesh improvement methods can push the worst elements in a mesh to levels of quality not attained by any previous technique, and can do so consistently.

There are two popular mesh improvement methods. *Smoothing* is the act of moving one or more mesh vertices to improve the quality of the elements

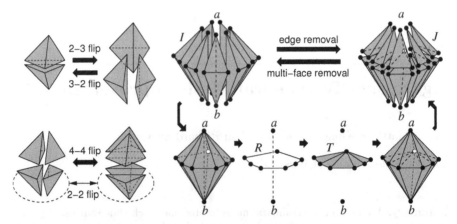

Fig. 1. Examples of topological transformations.

adjoining them. Smoothing does not change the topology (connectivity) of the mesh. *Topological transformations* are operations that remove elements from a mesh and replace them with a different set of elements occupying the same space, changing the topological structure of the mesh in the process. Smoothing lies largely in the domain of numerical optimization, and topological transformations in the domain of combinatorial optimization. The two techniques are most effective when used in concert.

Topological transformations are usually *local*, meaning that only a small number of elements are changed, removed, or introduced by a single operation. Figure 1 illustrates several examples, including 2-3 flips, 3-2 flips, 4-4 flips, and 2-2 flips. The numbers denote the number of tetrahedra removed and created, respectively. 2-2 flips occur only on mesh boundaries, and cause an edge flip on the surface.

Smoothing and topological transformations are usually used as operations in a *hill-climbing* method for optimizing the quality of a mesh. An *objective function* maps each possible mesh to a numerical value (or sequence of values) that describes the "quality" of the mesh. A hill-climbing method considers applying an operation to a specific site in the mesh. If the quality of the changed mesh will be greater than that of the original mesh, the operation is applied; then the hill-climbing method searches for another operation that will improve the new mesh. Operations that do not improve the value of the objective function are not applied. Thus, the final mesh cannot be worse than the input mesh. Hill climbing stops when no operation can achieve further improvement (the mesh is locally optimal), or when further optimization promises too little gain for too much expenditure of time.

In our opinion, the best research to date on tetrahedral mesh improvement is the work of Freitag and Ollivier-Gooch [11], who combine optimization-based smoothing with several topological transformations, including 2-3 flips, 3-2 flips, and an operation sometimes called *edge removal*. They report the performance on a variety of meshes of several schedules for applying these operations, show that their best schedule eliminates most poorly shaped tetrahedra, and offer empirical recommendations about what makes some schedules better than others.

Although we have long felt that the paper by Freitag and Ollivier-Gooch is a model of excellent meshing research, we also suspected that yet better results were possible through a more aggressive set of operations. Delaunay mesh generation algorithms achieve good results by inserting new vertices [12], often boosting the smallest dihedral angle to 19° or more [22]. But no "mesh clean-up" paper we know of uses transformations that add new vertices to the mesh—a strange omission. No doubt this oversight stems partly from the desire not to increase the size (number of elements) of a mesh.

We show here that vertex-creating transformations make it possible to achieve levels of mesh quality that, to the best of our knowledge, are unprecedented. Given meshes whose vertices are somewhat regularly spaced (as every competent tetrahedral mesh generator produces), our implementation usually improves them so that no dihedral angle is smaller than 31° or larger than 149°. It sometimes achieves extreme angles better than 40° or (with a different objective function) 120°. No previous software we know of for tetrahedral mesh generation or mesh improvement achieves angles of even 22° or 155° with much consistency.

As a combinatorial optimization problem, mesh improvement is not well behaved. The *search space* is the set of all possible meshes of a fixed geometric domain. A transformation (including smoothing) is an operation that transforms one mesh of the domain to another. These operations give the search space structure, by dictating what meshes are immediately reachable from another mesh. The objective function, which maps the search space to quality scores, has many local optima (meshes whose scores cannot be improved by any single transformation at hand), and it is unlikely that any mesh improvement algorithm will ever find the *global optimum*—the best possible mesh of the domain—or even come close.

However, our goal is merely to find a local optimum whose tetrahedra are all excellent. Intuitively, a powerful enough repertoire of operations ought to "smooth out" the objective function, thereby ensuring that few poor local optima exist. The question is, what is a powerful enough repertoire?

In this paper, we take the operations explored by Freitag and Ollivier-Gooch and enrich them considerably, adding the following.

- A topological transformation that inserts a new vertex (usually into a bad tetrahedron). This operation is not unlike Delaunay vertex insertion, but it is designed to optimize the worst new tetrahedron instead of enforcing the Delaunay property. Sometimes it deletes vertices as well.
- Smoothing of vertices constrained to lie on the boundary of the mesh.
- Edge removal for edges on the boundary of the mesh.
- The *multi-face removal* operation of de Cougny and Shephard [7].

- Compound operations that combine several other operations in the hope of getting over a valley in the objective function and finding a better peak. If unsuccessful, these operations are rolled back.

These additions are motivated by two observations: to repair a tetrahedralization it is often necessary to repair the boundary triangulation; and inserting a new vertex often breaks resistance that cannot be broken by topological operations that do not change the set of vertices in a mesh. We have implemented and tested schedules that use these operations, and we investigate the consequences of turning different operations on or off.

The main goal of this paper is to answer the question, "How high can we drive the quality of a tetrahedral mesh, assuming that speed is not the highest priority?" Questions like "How quickly can we consistently fix a mesh so all its dihedral angles are between, say, 20° and 160°?" are important too. But we think it is hard to do justice to both questions in one short paper, and studying the former question first will make it easier to answer the latter question in future work.

2 Mesh Quality

The success of the finite element method depends on the shapes of the tetrahedra. Large dihedral angles (near 180°) cause large interpolation errors and rob the numerical simulation of its accuracy [14, 17, 24], and small dihedral angles render the stiffness matrices associated with the finite element method fatally ill-conditioned [2, 24]. Although anisotropic tetrahedra with extreme angles are desirable and necessary in some contexts, such as aerodynamics, we restrict our attention here to isotropic simulations, which are the norm in mechanics, heat transfer, and electromagnetics. Often, a single bad tetrahedron can spoil a simulation. For example, a large dihedral angle can engender a huge spurious strain in the discretized solution of a mechanical system. Therefore, our top priority is to produce a mesh in which the worst tetrahedra are as good as possible.

Most mesh improvement programs encapsulate the quality of a tetrahedron t as a single numerical *quality measure* $q(t)$. Many such quality measures are available [9, 24]. All the mesh operations we use are flexible enough to accommodate almost every measure in the literature. We assume each measure is "normalized" so that a larger value of $q(t)$ indicates a better tetrahedron, and $q(t)$ is positive if t has the correct topological orientation, zero if t is degenerate, and negative if t is "inverted" (meaning that there is a wrinkle in the fabric of the mesh). We assume no input mesh has inverted tetrahedra; all our operations will keep it that way.

We tried four quality measures in our implementation.

- The minimum sine of a tetrahedron's six dihedral angles, or the *minimum sine measure* for short. This measure penalizes both small and large dihedral angles, and Freitag and Ollivier-Gooch [11] find it to be the most effective measure they considered. It also has the advantage that dihedral angles are intuitive.

- The *biased minimum sine measure*, which is like the minimum sine measure, but if a dihedral angle is obtuse, we multiply its sine by 0.7 (before choosing the minimum). This allows us to attack large angles much more aggressively without much sacrifice in improving the small angles.
- The *volume-length measure*, suggested by Parthasarathy, Graichen, and Hathaway [19] and denoted V/ℓ_{rms}^3, is the signed volume of a tetrahedron divided by the cube of its root-mean-squared edge length. We multiply it by $6\sqrt{2}$ so that the highest quality is one, the measure of an equilateral tetrahedron.
- The *radius ratio*, suggested by Cavendish, Field, and Frey [5], is the radius of a tetrahedron's inscribed sphere divided by the radius of its circumscribing sphere. We multiply it by 3 so that the highest quality is one, the measure of an equilateral tetrahedron. We experimented with this measure because of its popularity, but we found that it is inferior to the volume-length measure in mesh optimization, even when the goal is to optimize the radius ratio. So we will revisit it only once—in Section 5 where we demonstrate this fact.

The first two measures do not penalize some tetrahedra that are considered bad by the last two measures. For example, an extremely skinny, needle-shaped tetrahedron can have excellent dihedral angles, whereas its skinniness is recognized by the volume-length measure and the radius ratio. There is evidence that a skinny tetrahedron with good dihedral angles is harmless, hurting neither discretization error nor conditioning [24]; its worst crime is to waste vertices, because its accuracy is inversely proportional to the length of its longest edge, not its shortest. Moreover, such a tetrahedron is indispensable at the tip of a needle-shaped domain. Readers not convinced by this argument will find the volume-length measure invaluable.

We need to extend quality measures from individual tetrahedra to whole meshes. The worst tetrahedra in a mesh have far more influence than the average tetrahedra, so the objective function we optimize is the *quality vector*: a vector listing the quality of each tetrahedron, ordered from worst to best. Two meshes' quality vectors are compared *lexicographically* (akin to alphabetical order) so that, for instance, an improvement in the second-worst tetrahedron improves the overall objective function even if the worst tetrahedron is not changed. A nice property of the quality vector is that if an operation replaces a small subset of tetrahedra in a mesh with new ones, we only need to compare the quality vectors of the submeshes constituting the changed tetrahedra (before and after the operation). If the submesh improves, the quality vector of the whole mesh improves. Our software never needs to compute the quality vector of an entire mesh.

3 The Fundamental Tools: Mesh Operations

Here we describe the mesh transformation operations that form the core of our mesh improvement program. Simultaneously, we survey the previous work in mesh improvement.

3.1 Smoothing

The most famous smoothing technique is *Laplacian smoothing*, in which a vertex is moved to the centroid of the vertices to which it is connected [13]. Typically, Laplacian smoothing is applied to each mesh vertex in sequence, and several passes of smoothing are done, where each "pass" moves every vertex once. Laplacian smoothing is popular and somewhat effective for triangular meshes, but for tetrahedral meshes it is much less reliable, and often produces poor tetrahedra.

Better smoothing algorithms are based on numerical optimization [20, 4]. Early algorithms define a smooth objective function that summarizes the quality of a group of elements (e.g. the sum of squares of the qualities of all the tetrahedra adjoining a vertex), and use a numerical optimization algorithm such as steepest descent or Newton's method to move a vertex to the optimal location. Freitag, Jones, and Plassman [10] propose a more sophisticated *nonsmooth* optimization algorithm, which makes it possible to optimize the worst tetrahedron in a group—for instance, to maximize the minimum angle among the tetrahedra that share a specified vertex. A nonsmooth optimization algorithm is needed because the objective function—the minimum quality among several tetrahedra—is *not* a smooth function of the vertex coordinates; the gradient of this function is discontinuous wherever the identity of the worst tetrahedron in the group changes. Freitag and Ollivier-Gooch [11] had great success with this algorithm, and we use it essentially unchanged (though we have our own implementation).

Whereas Freitag and Ollivier-Gooch only smooth vertices in the interior of a mesh, we also implemented constrained smoothing of boundary vertices. If the boundary triangles adjoining a vertex appear (within some tolerance) to lie on a common plane, our smoother assumes that the vertex can be smoothed within that plane. Similarly, we identify vertices that can be moved along an edge of the domain without changing its shape. However, we did not implement constrained smoothing for curved domain boundaries, so some of our meshes do not benefit from boundary smoothing.

We always use what Freitag and Ollivier-Gooch call *smart smoothing*: if a smoothing operation does not improve the minimum quality among the tetrahedra changed by the operation, then the operation is not done. Thus, the quality vector of the mesh never gets worse.

3.2 Edge Removal

Edge removal, proposed by Brière de l'Isle and George [3], is a topological transformation that removes a single edge from the mesh, along with all the tetrahedra that include it. (The name is slightly misleading, because edge removal can create new edges while removing the old one. Freitag and Ollivier-Gooch refer to edge removal as "edge swapping," but we prefer the earlier name.) It includes the 3-2 and 4-4 flips, but also includes other transformations that remove edges shared by any number of tetrahedra. In general, edge removal replaces m tetrahedra with $2m - 4$; Figure 1 (right) illustrates replacing seven tetrahedra with ten. De Cougny and Shephard [7] and Freitag and Ollivier-Gooch [11] have shown dramatic evidence for its effectiveness, especially in combination with other mesh improvement operations.

Let ab be an edge in the interior of the mesh with vertices a and b. Let I be the set of tetrahedra that include ab. Each tetrahedron in I has an edge opposite ab. Let R be the set of these edges. (R is known as the *link* of ab.) R forms a (non-planar) polygon in three-dimensional space, as illustrated. An edge removal transformation constructs a triangulation T of R, and creates a set of new tetrahedra $J = \bigcup_{t \in T} \{\text{conv}(\{a\} \cup t), \text{conv}(\{b\} \cup t)\}$, as illustrated, which replace the tetrahedra in I.

The chief algorithmic problem is to find the triangulation T of R that maximizes the quality of the worst tetrahedron in J. We solve this problem with a dynamic programming algorithm of Klincsek [16], which was invented long before anyone studied edge removal. (Klincsek's algorithm solves a general class of problems in optimal triangulation. Neither Brière de l'Isle and George nor Freitag and Ollivier-Gooch appear to have been aware of it.) The algorithm runs in $O(m^3)$ time, but m is never large enough for its speed to be an impairment.

3.3 Multi-Face Removal

Multi-face removal is the inverse of edge removal, and includes the 2-3 and 4-4 flips. An m-face removal replaces $2m$ tetrahedra with $m + 2$. It has been neglected in the literature; so far as we know, it has appeared only in an unpublished manuscript of de Cougny and Shephard [7], who present evidence that multi-face removal is effective for mesh improvement.

Multi-face removal, like edge removal, revolves around two chosen vertices a and b. Given a mesh, say that a triangular face f is *sandwiched* between a and b if the two tetrahedra that include f are $\text{conv}(\{a\} \cup f)$ and $\text{conv}(\{b\} \cup f)$. For example, in Figure 1, the faces of T are sandwiched between a and b in the mesh J. An m-face removal operation singles out m of those sandwiched faces, and replaces the tetrahedra that adjoin them, as illustrated. (An m-face removal actually removes $3m - 2$ faces, but only m of them are sandwiched between a and b.)

Our software uses multi-face removal by singling out a particular internal face f it would like to remove. Let a and b be the apex vertices of the

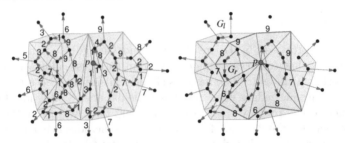

Fig. 2. Vertex insertion as graph cut optimization. In this example, the smallest cut has weight 6. The weights of the cut edges are the qualities of the new elements.

two tetrahedra adjoining f. The optimal multi-face removal operation does not necessarily remove *all* the faces sandwiched between a and b. We use the algorithm of Shewchuk [23] to find the optimal multi-face removal operation for f (and to determine whether *any* multi-face removal operation can remove f without creating inverted tetrahedra), in time linear in the number of sandwiched faces.

3.4 Vertex Insertion

Our main innovation in this paper is to show that mesh improvement is far more effective with the inclusion of transformations that introduce new vertices. We use an algorithm similar to Delaunay vertex insertion: we choose a location p to insert a new vertex and a set I of tetrahedra to delete, such that p lies in, and can "see" all of, the star-shaped polyhedral cavity $C = \bigcup_{t \in I} t$. We fill the cavity with a set of new tetrahedra $J = \{\text{conv}(\{p\} \cup f) : f \text{ is a face}$ of $C\}$. Choosing the position of p is a black art; see Section 4.2 for how we choose it. To choose I, we solve this combinatorial problem: given a point p, which tetrahedra should we delete to maximize the quality of the worst new tetrahedron?

Our algorithm views the mesh as a graph M with one node for each tetrahedron, as depicted in Figure 2. For simplicity, we identify nodes of the graph with the tetrahedra they represent. M contains a directed edge (v, w) if the tetrahedron v shares a triangular face with the tetrahedron w, and v occludes w from p's point of view. The edge (v, w) reflects the geometric constraint that w can only be included in the set I (i.e., in the cavity C) if v is included— that is, the cavity must be star-shaped from p's perspective. (If p is coplanar with the triangular face that v and w share, we direct the edge arbitrarily.) Although M can have cycles, they are rare, so we adopt some nomenclature from trees: if $(v, w) \in M$ then w is a *child* of v and v is a *parent* of w. Any tetrahedron that contains p is a *root* of M. Usually there is just one root tetrahedron, but sometimes we insert a new vertex on a boundary edge of the domain, in which case all the tetrahedra sharing that edge are roots. If a vertex is inserted at p, all the roots must be deleted.

Our algorithm for finding an optimal cavity computes a cut in M that induces a cavity in the mesh. It begins by constructing the subgraph G of M whose nodes are the roots of M and all the tetrahedra that are reachable in M from the roots by a directed path of length six or less. We select G this way because we do not want to search the entire graph M for a cut, and we find that in practice, tetrahedra further away from the root rarely participate in the optimal cavity. We find that G typically has 5–100 tetrahedra. For each triangular face that belongs to only one tetrahedron in G, we add a "ghost node" to G to act as a neighboring tetrahedron. Then, every leaf of G is a ghost node, as Figure 2 shows.

The tetrahedra in G, except the leaves, are candidates for deletion. For each edge $(v, w) \in G$, let f be the triangular face shared by the tetrahedra v and w. Our algorithm labels (v, w) with the quality of the tetrahedron $\mathrm{conv}(p \cup f)$—the tetrahedron that will be created if v is deleted but w survives.

The problem is to partition G into two subgraphs, G_r and G_l, such that G_r contains the root tetrahedra and G_l contains the leaves, as illustrated in Figure 2. The deleted tetrahedra I will be the nodes of G_r, and the surviving tetrahedra will be the nodes of G_l. Because the cavity $C = \bigcup_{t \in I} t$ must be star-shaped from p's perspective (to prevent the creation of inverted tetrahedra), no tetrahedron in G_l may be a parent of any tetrahedron in G_r. Our goal is to find the partition that satisfies this constraint and maximizes the smallest edge cut (because that edge determines the worst new tetrahedron).

The algorithm in Figure 3 computes this optimal cut. (We omit the proof.) The algorithm iterates through the edges of G, from worst quality to best, and greedily ensures that each edge will not be cut, if that assurance does not contradict previous assurances. Upon termination, the tetrahedra labeled "cavity" become the set I of tetrahedra to be deleted, and the set J of tetrahedra to be created are determined by the triangular faces of the cavity C, which are recorded by the ninth line of pseudocode. In practice, I typically comprises 5–15 tetrahedra. After an initial $O(|G| \log |G|)$-time sorting step, the rest of the algorithm runs in $O(|G|)$ time.

Sometimes, a vertex insertion operation deletes one or more of the other vertices, as in Figure 2. When a tetrahedron with three parents is deleted, the vertex it shares with all three parents is deleted too. Thus, our vertex insertion operation sometimes reduces the number of vertices in the mesh.

3.5 Composite Operations

Mesh improvement methods often get stuck in local optima that are far from the global optimum. Joe [15] suggests that this problem can be ameliorated by composing multiple basic operations to form new operations. These composite operations sometimes get a hill-climbing optimizer across what was formerly a valley in the objective function, thereby leading the way to a better local optimum.

We have found that vertex insertion, as described in Section 3.4, rarely improves the quality vector of the mesh immediately, but it is frequently effective

```
Sort edges of G from smallest to largest quality.
H ⟸ a graph with the same vertices as G but
    no edges (yet). (H need not be stored as a
    separate graph; let each edge of G have a bit
    that indicates whether it is in H too.)
All vertices of G are initially unlabeled.
Label every root of G "cavity."
Label every leaf of G "anti-cavity."
for each directed edge (v, w) of G (in sorted order)
    if v is labeled "cavity"
        if w is labeled "anti-cavity"
            Record (v, w), which determines a new
                tetrahedron in J.
        else if w is unlabeled
            CAVITY(w)
    else if v is unlabeled
        if w is labeled "anti-cavity"
            ANTICAVITY(v)
        else { w is unlabeled }
            Add (v, w) to H.
```

```
CAVITY(w)
    Label w "cavity."
    for each unlabeled parent p of w in G
        CAVITY(p)
    for each unlabeled child c of w in H
        CAVITY(c)


ANTICAVITY(v)
    Label v "anti-cavity."
    for each unlabeled child c of v in G
        ANTICAVITY(c)
    for each unlabeled parent p of v in H
        ANTICAVITY(p)
```

Fig. 3. Algorithm for computing the cavity that optimizes the new tetrahedra when a new vertex is inserted. Upon completion, the tetrahedra to be deleted are labeled "cavity."

if traditional smoothing and transformations follow. To create an operation that composes vertex insertion with subsequent operations, we implemented a rollback mechanism that allows us to attempt a sequence of transformations, then reverse all the changes if the final mesh is not better than the initial one.

The ATTEMPTINSERT pseudocode in Figure 4 shows how we follow vertex insertion with smoothing and topological transformations, then decide whether to roll back the insertion. Immediately after inserting the new vertex (as described in Section 3.4), we smooth it (by optimization), then we run passes of topological transformations and smoothing on the tetrahedra adjoining it in an attempt to improve them. (See the pseudocode for details.) Finally, we compare the quality vector of all the new and changed (smoothed) tetrahedra with the quality vector of the deleted tetrahedra (the set I). If the mesh has not improved, we roll it back to the instant before we inserted the new vertex.

Even though the algorithm in Figure 3 is technically optimal, we have learned by experiment that *composite* vertex insertion is more effective if we bias the algorithm to prefer larger cavities than it would normally choose. To encode this bias, our implementation multiplies edge weights by 1.0, 1.4, 1.8, or 2.1 if they are a distance of zero, one, two, or greater than two from the nearest root, respectively. These weights sometimes cause worse-than-optimal tetrahedra to be created, but these are often improved by subsequent operations. In the end, the vertex insertion operation is only accepted (not rolled back) if the *unbiased* quality vector improves.

4 Scheduling the Operations

4.1 Examples from Previous Work

Joe's algorithm [15] checks each face of the mesh to see if any of his transformations (including composite transformations) will improve the local tetrahedra. It performs passes over the entire mesh (checking each face), and terminates when a pass makes no changes. His experiments show that he can eliminate most, but not all, tetrahedra with radius ratios below 0.3. (In our experiments, we eliminated all tetrahedra with radius ratios below 0.51 by optimizing the objective V/ℓ_{rms}^3.)

Freitag and Ollivier-Gooch's schedule [11] begins with a pass of 2-3 flips that enforce the Delaunay in-sphere criterion (testing each interior face of the mesh once), then a pass of 2-3 flips that optimize the minimum sine measure, then a pass of edge removal operations that optimize the minimum sine, then two passes of optimization-based smoothing. Next, a procedure that targets only the worst tetrahedra in the mesh attempts to remove them with 2-3 flips and edge removal operations. Two more passes of smoothing complete the schedule. For many of their meshes, they obtain dihedral angles bounded between about 12° and 160°, but these results are not consistent across all their test meshes. Dihedral angles less than 1° occasionally survive, and in more examples dihedral angles under 10° survive.

Edelsbrunner and Guoy [8] demonstrate that that a theoretically motivated technique called *sliver exudation* [6], which uses sequences of 2-3 and 3-2 flips to remove poor tetrahedra from meshes, usually removes most of the bad tetrahedra from a mesh, but rarely all. Again, dihedral angles less than 1° sometimes survive, and in most of their examples a few dihedral angles less than 5° remain.

Alliez, Cohen-Steiner, Yvinec, and Desbrun [1] propose a variational meshing algorithm that alternates between optimization-based smoothing (using a smooth objective function) and computing a new Delaunay triangulation from scratch. This algorithm generates meshes that have only a small number of tetrahedra under 10° or over 160°, but it does not eliminate all mediocre tetrahedra, especially on the boundary. (See the STGALLEN and STAYPUFT input meshes in Section 5.) Note that variational meshing is a standalone mesh generation algorithm, and cannot be used as a mesh improvement algorithm, because the mesh it generates does not conform to a specified triangulated boundary.

4.2 Our Mesh Improvement Schedule

Pseudocode for our mesh improvement implementation appears in Figure 4. Like all such schedules, ours is heuristic and evolved through trial and error.

We find that prescribing a fixed number of improvement passes, as Freitag and Ollivier-Gooch do, is too inflexible, and we get better results by adapting our schedule to the mesh at hand. We begin with a pass of optimization-based

MESHAGGRESSION(M)   { M is a mesh }
    Smooth each vertex of M.
    TOPOLOGICALPASS(M)
    $failed \Leftarrow 0$
    while $failed < 3$
        $Q \Leftarrow$ list of quality indicators for M.
        Smooth each vertex of M.
        if M is sufficiently better than Q
            $failed \Leftarrow 0$
        else
            TOPOLOGICALPASS(M)
            if M is sufficiently better than Q
                $failed \Leftarrow 0$
            else
                if $failed = 1$ { desperation pass }
                    $L \Leftarrow$ list of tets with a dihedral
                            $< 40°$ or $> 140°$.
                else $L \Leftarrow$ list of the worst 3.5% of
                            tets in M.
                INSERTIONPASS(M, L)
                if M is sufficiently better than Q
                    $failed \Leftarrow 0$
                else $failed \Leftarrow failed + 1$

INSERTIONPASS(M, L)   { L is a list of tets }
    for each tetrahedron $t \in L$ that still exists
        for each face f of t on the mesh
                boundary (if any)
            $p \Leftarrow$ point at barycenter of f
            if ATTEMPTINSERT(M, p)
                Restart outer loop on next tet.
        $p \Leftarrow$ point at barycenter of t
        if ATTEMPTINSERT(M, p)
            Restart outer loop on next tet.
        for each edge e of t on the mesh
                boundary (if any)
            $p \Leftarrow$ point at midpoint of e
            if ATTEMPTINSERT(M, p)
                Restart outer loop on next tet.

TOPOLOGICALPASS(L)   { L is a list of tets }
    for each tetrahedron $t \in L$ that still exists
        for each edge e of t (if t still exists)
            Attempt to remove edge e.
        for each face f of t (if t still exists)
            Attempt to remove face f (multi-face
                    or 2-3 or 2-2 flip).
    return the surviving tetrahedra of L and
            all the new tetrahedra
            created by this call.

ATTEMPTINSERT(M, p)   { New vertex at p }
    $I \Leftarrow$ deleted tetrahedra, computed as
                discussed in Section 3.4
    $q \Leftarrow$ quality of the worst tetrahedron in I
    Replace I with the new tetrahedra J (see
                Section 3.4 and Figure 3).
    $attempts \Leftarrow 8$
    repeat
        Smooth p.
        { In next line, the cavity may expand }
        $J \Leftarrow$ TOPOLOGICALPASS(J)
        $attempts \Leftarrow attempts - 1$
    while $attempts > 0$ and some topological
                change occurred
    $K \Leftarrow J \cup$ the tetrahedra in M that share
                a vertex with J
    repeat
        $q' \Leftarrow$ quality of the worst tet in J
        Smooth each vertex of J.
        $q'' \Leftarrow$ quality of the worst tet in K
        $attempts \Leftarrow attempts - 1$
    while $attempts > 0$ and $q'' > q'$
    if $q'' > q$
        return true.
    Roll back all changes since the beginning
                of this procedure call.
    return false.

Fig. 4. Our mesh improvement schedule.

smoothing (smoothing each vertex once) and a pass of topological transformations (leaving out vertex insertion), because these are always fruitful. Our topological pass visits each tetrahedron once, and searches for a transformation that will eliminate it and improve the quality vector of the changed tetrahedra (and therefore of the entire mesh). If no edge removal operation succeeds in removing a particular tetrahedron, we try multi-face removal on each face of that tetrahedron that lies in the mesh interior. However, in Section 5 we test the effect of disabling multi-face removal but still allowing faces to be removed by 2-3 flips (which are easier to implement). If an interior face has an edge on the mesh boundary, we also test the possibility that a 2-2 flip on that edge might improve the mesh.

Our implementation then performs passes over the mesh until three subsequent passes fail to make sufficient progress. We gauge progress using a small list of *quality indicators*: the quality of the worst tetrahedron in the entire mesh, and seven *thresholded means*. A mean with threshold d is computed by calculating the quality of each tetrahedron in the mesh, reducing to d any quality greater than d, then taking the average. The purpose of a thresholded mean is to measure progress in the lowest-quality tetrahedra while ignoring changes in the high-quality tetrahedra. For the minimum sine measure, we compute thresholded means with thresholds $\sin 1°$, $\sin 5°$, $\sin 10°$, $\sin 15°$, $\sin 25°$, $\sin 35°$, and $\sin 45°$. (Each tetrahedron is scored according to its worst dihedral angle; we do *not* compute thresholded means of *all* dihedral angles.) A mesh is deemed to be sufficiently improved (to justify more passes) if at least one of the thresholded means increases by at least 0.0001, or if the quality of the worst tetrahedron increases at all.

Each pass begins with smoothing. If smoothing does not make adequate progress, a topological pass follows. If progress is still insufficient, we resort to vertex insertion, which is less desirable than the other operations both because it often increases the size of the mesh and because our compound vertex insertion operation is slow. Vertex insertion is usually aimed at the worst 3.5% of tetrahedra in the mesh, but our implementation never gives up without trying at least one "desperation pass" that attempts to insert a vertex in every tetrahedron that has an angle less than $40°$ or greater than $140°$.

Most mesh generation algorithms create their worst tetrahedra on the boundary of the mesh, and boundary tetrahedra are the hardest to repair. Thus, when our vertex insertion pass targets a tetrahedron on the boundary of the mesh, it always tries to insert a vertex at the barycenter of the boundary face(s) first. For tetrahedra where that fails, and tetrahedra not on the boundary, we try the tetrahedron barycenter next. We also try the midpoints of tetrahedron edges that lie on the boundary, but we try them last, because (for reasons we don't understand) we obtain better meshes that way.

5 Results and Discussion

We tested our schedule on a dozen meshes.

- CUBE1K and CUBE10K are high-quality meshes of a cube generated by NETGEN [21].
- TFIRE is a high-quality mesh of a tangentially-fired boiler, created by Carl Ollivier-Gooch's GRUMMP software.
- TIRE, RAND1 and RAND2 come courtesy of Freitag and Ollivier-Gooch, who used them to evaluate their mesh improvement algorithms. TIRE is a tire incinerator. RAND1 and RAND2 are *lazy triangulations*, generated by inserting randomly located vertices into a cube, one by one. Each vertex was inserted by splitting one or more tetrahedra into multiple tetrahedra. (Unlike in Delaunay insertion, no flips took place.) The random meshes have horrible quality.
- DRAGON and COW are medium-quality meshes with curved boundaries, generated by isosurface stuffing [18]. The curvature prevents us from smoothing the original boundary vertices.

- STGALLEN and STAYPUFT are medium- to low-quality meshes with curved boundaries, generated by two different implementations of variational tetrahedral meshing [1], courtesy of Pierre Alliez and Adam Bargteil, respectively.
- HOUSE and P are Delaunay meshes generated by Pyramid [22] configured so the vertices are nicely spaced, but no effort is made to eliminate sliver tetrahedra.

Tables 1 and 2 show these meshes before and after improvement with the MESHAGGRESSION schedule in Figure 4. We tested the minimum sine measure (upper right corner of each box), the biased minimum sine measure (lower right), and the volume-length measure V/ℓ^3_{rms} (lower left) as objectives. (We omit meshes optimized for the radius ratio objective, which was not competitive with the volume-length measure, even as measured by the radius ratios of the tetrahedra.)

Our main observation is that the dihedral angles improved to between 31° and 149° for the minimum sine objective, between 25° and 142° for the biased minimum sine objective, and between 23° and 136° for the volume-length measure. Even the pathological meshes RAND1 and RAND2 end with excellent quality. These numbers put our implementation far ahead of any other tetrahedral mesh algorithm we have seen reported. Freitag and Ollivier-Gooch reported angle bounds as good as 13.67° and 156.14° for TIRE, versus our 28.13° and 125.45°; as good as 15.01° and 159.96° for RAND1, versus our 36.95° and 119.89°; and as good as 10.58° and 164.09° for RAND2, versus our 34.05° and 126.61°.

Of course, to obtain such high quality takes time. Meshes that begin with high quality take a modest amount of time to improve. RAND1 and RAND2 take disproportionately longer—both because our implementation tries to hold back vertex insertions until they prove to be necessary, and because the composite vertex insertion operation is slow, often accounting for about 90% of the running time. Of that 90%, about one third is spent in the basic vertex insertion operation, one third in smoothing the cavity, and one third in topological transformations in the cavity. It seems impossible to predict which quality measure will run faster on a given mesh, and the differences in running times are erratic.

No mesh increased in size by more than 41%, and some meshes shrank (because the vertex insertion operation can also delete vertices).

Table 3 shows the effects of turning features on or off. The top half of the page explores the question of which features are most important to have if the programmer's time is limited. We try all combinations of three operations: optimization-based vertex smoothing in the interior of the mesh (but not on mesh boundaries); vertex insertion in the interior of the mesh (but not on boundaries); and edge removal (but no other topological transformations). Smoothing proves to be the most indispensable; substantial progress is almost impossible without it. Vertex insertion is the second-most powerful operation. We were surprised to see that it alone can substantially improve some meshes, even though most vertex insertion operations fail when neither smoothing nor

Table 1. Twelve meshes before and after improvement (continued in Table 2). In each box, the upper left mesh is the input, the upper right mesh is optimized for the minimum sine measure, the lower right mesh is optimized for the biased minimum sine measure, and the lower left mesh is optimized for V/ℓ_{rms}^3. Running times are given for a Mac Pro with a 2.66 GHz Intel Xeon processor. Red tetrahedra have dihedral angles under 5° or over 175°, orange have angles under 15° or over 165°, yellow have angles under 30° or over 150°, green have angles under 40° or over 140°, and better tetrahedra do not appear. Histograms show the distributions of dihedral angles, and the minimum and maximum angles, in each mesh. Histograms are normalized so the tallest bar always has the same height; absolute numbers of tetrahedra cannot be compared between histograms.

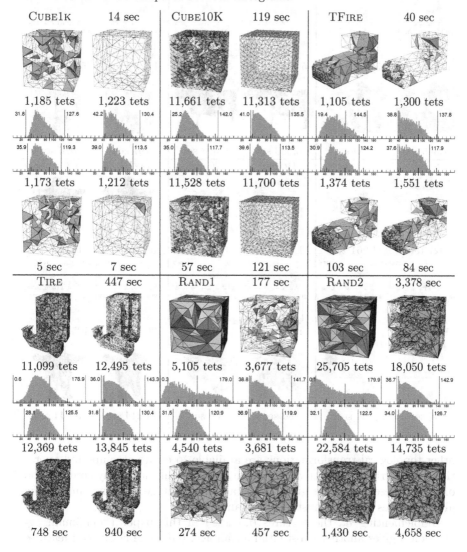

Table 2. Continuation of Table 1. Red histogram bars should have their heights multiplied by 20 to account for the fact that in the semi-regular meshes DRAGON and COW, angles of 45°, 60°, and 90° occur with high frequency.

other topological transformations are available to create a compound operation (as described in Section 3.5). Edge removal ranks last. Any combination of two of these operations gives a substantial advantage over one, and having all three gives another substantial advantage.

Implementing *all* the features discussed in this paper ("maximum aggression") gives another substantial advantage, but these additional features

Table 3. Histograms showing the dihedral angle distribution, and minimum and maximum dihedral angles, for several meshes optimized with selected features turned on (upper half of page) or off (lower half). The objective was to maximize the biased minimum sine measure. Multiply the heights of the red histogram bars by 20. "Maximum aggression" has all features turned on.

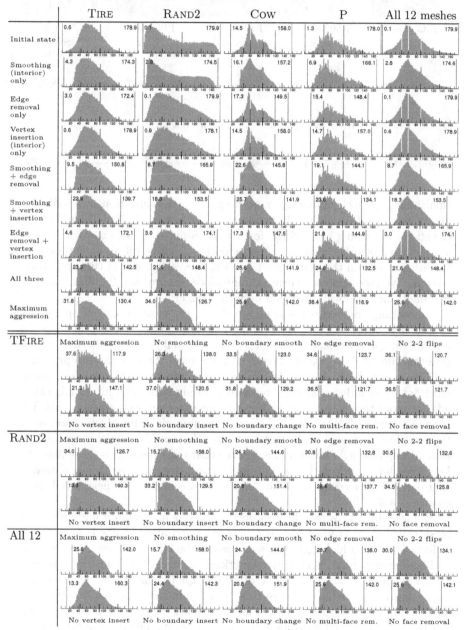

Table 4. A stretched input mesh and four output meshes optimized with different quality measures as objective functions. The histograms tabulate, from top to bottom, dihedral angles, radius ratios (times 3), and $6\sqrt{2}V/\ell_{\mathrm{rms}}^3$.

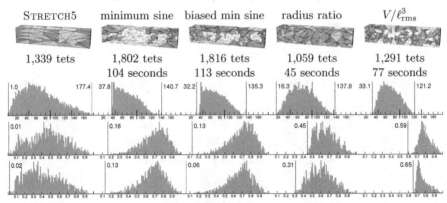

STRETCH5	minimum sine	biased min sine	radius ratio	V/ℓ_{rms}^3
1,339 tets	1,802 tets	1,816 tets	1,059 tets	1,291 tets
	104 seconds	113 seconds	45 seconds	77 seconds

(multi-face removal, boundary smoothing, boundary insertion) are individually responsible for only small increments. The bottom half of Table 3 shows the effects of turning just a single feature off. Some of the switches listed there are inclusive of other switches. "No smoothing" turns off all smoothing—in the interior and on the boundary. Likewise, "no vertex insert" turns off all insertion. "No face removal" turns off multi-face removal and 2-3 flips, whereas "no multi-face removal" turns off only the former.

Smoothing and vertex insertion are clearly the most disastrous operations to lose. The effort to extend smoothing and vertex insertion so that they can operate on the boundary of the mesh was also well rewarded. Besides vertex insertion, no single topological operation is crucial if the others are present.

The papers by Freitag and Ollivier-Gooch and by de Cougny and Shephard both concluded that edge removal is rarely successful for an edge that adjoins more than seven tetrahedra; but our experience contradicts this. We see many successful removals of edges adjoining eight tetrahedra, and even the occasional removal of eleven tetrahedra. (Klincsek's algorithm makes this easy to implement.) However, we observe that edge removal is most likely to be successful for edges that adjoin four tetrahedra, and multi-face removals that remove two faces predominate, so the most common beneficial topological change is a 4-4 flip.

Table 4 illustrates the effect of optimizing our four quality measures on a mesh called STRETCH5, which is CUBE1K scaled along one axis by a factor of five. This mesh highlights the weakness of the minimum sine measure and its biased counterpart as objective functions—namely, they sometimes permit skinny tetrahedra to survive. The other two quality measures are better for improving the distribution of vertices and producing "rounder" tetrahedra. The minimum sine objective is best for optimizing the smallest dihedral angle, but the volume-length measure is the best all-around objective of the four. It even optimizes the radius ratio better than the radius ratio does! (We suspect that the radius ratio behaves poorly as an objective function because of the instability of the circumscribing radius as a function of vertex position.) In our twelve-mesh test suite, the volume-length objective always improved the worst radius ratio to at least 0.51, whereas the radius ratio objective left behind many worse tetrahedra, the worst having a radius ratio of 0.30.

6 Conclusions

We see two important avenues for future work. First, our mesh improvement implementation assumes that the spacing of the vertices in the input mesh is already correct. A better code would take as input a *spacing function* that dictates how large the tetrahedra should be in different regions of the mesh, and insert or delete vertices accordingly. Second, algorithms and schedules that achieve results similar to ours in much less time would be welcome. Our composite vertex insertion operation accounts for most of the running time, so a more sophisticated vertex insertion algorithm might improve the speed dramatically.

Because we can produce meshes that usually have far better quality than those produced by any previous algorithm for mesh improvement or mesh generation, even when given pathological inputs, we suggest that algorithms traditionally considered "mesh improvement" might become standalone mesh generators. If the barrier of speed can be overcome, the need to write separate programs for mesh generation and mesh improvement might someday disappear.

Acknowledgments. We thank Pierre Alliez, Adam Bargteil, Moshe Mahler, and Carl Ollivier-Gooch for meshes and geometric models. Pixie rendered our meshes. This work was supported by the National Science Foundation under Awards CCF-0430065 and CCF-0635381, and by an Alfred P. Sloan Research Fellowship.

References

1. Pierre Alliez, David Cohen-Steiner, Mariette Yvinec, and Mathieu Desbrun. *Variational Tetrahedral Meshing.* ACM Transactions on Graphics **24**:617–625, 2005. Special issue on Proceedings of SIGGRAPH 2005.
2. Randolph E. Bank and L. Ridgway Scott. *On the Conditioning of Finite Element Equations with Highly Refined Meshes.* SIAM Journal on Numerical Analysis **26**(6):1383–1394, December 1989.
3. E. Brière de l'Isle and Paul-Louis George. *Optimization of Tetrahedral Meshes.* Modeling, Mesh Generation, and Adaptive Numerical Methods for Partial Differential Equations, IMA Volumes in Mathematics and its Applications, volume 75, pages 97–128. 1995.

4. Scott A. Canann, Michael Stephenson, and Ted Blacker. *Optismoothing: An Optimization-Driven Approach to Mesh Smoothing*. Finite Elements in Analysis and Design **13**:185–190, 1993.

5. James C. Cavendish, David A. Field, and William H. Frey. *An Approach to Automatic Three-Dimensional Finite Element Mesh Generation*. International Journal for Numerical Methods in Engineering **21**(2):329–347, February 1985.

6. Siu-Wing Cheng, Tamal Krishna Dey, Herbert Edelsbrunner, Michael A. Facello, and Shang-Hua Teng. *Sliver Exudation*. Journal of the ACM **47**(5):883–904, September 2000.

7. Hugues L. de Cougny and Mark S. Shephard. *Refinement, Derefinement, and Optimization of Tetrahedral Geometric Triangulations in Three Dimensions*. Unpublished manuscript, 1995.

8. Herbert Edelsbrunner and Damrong Guoy. *An Experimental Study of Sliver Exudation*. Tenth International Meshing Roundtable (Newport Beach, California), pages 307–316, October 2001.

9. David A. Field. *Qualitative Measures for Initial Meshes*. International Journal for Numerical Methods in Engineering **47**:887–906, 2000.

10. Lori A. Freitag, Mark Jones, and Paul Plassman. *An Efficient Parallel Algorithm for Mesh Smoothing*. Fourth International Meshing Roundtable (Albuquerque, New Mexico), pages 47–58, October 1995.

11. Lori A. Freitag and Carl Ollivier-Gooch. *Tetrahedral Mesh Improvement Using Swapping and Smoothing*. International Journal for Numerical Methods in Engineering **40**(21):3979–4002, November 1997.

12. William H. Frey. *Selective Refinement: A New Strategy for Automatic Node Placement in Graded Triangular Meshes*. International Journal for Numerical Methods in Engineering **24**(11):2183–2200, November 1987.

13. L. R. Hermann. *Laplacian-Isoparametric Grid Generation Scheme*. Journal of the Engineering Mechanics Division of the American Society of Civil Engineers **102**:749–756, October 1976.

14. P. Jamet. *Estimations d'Erreur pour des Élements Finis Droits Presque Dégénérés*. RAIRO Analyse Numérique **10**:43–61, 1976.

15. Barry Joe. *Construction of Three-Dimensional Improved-Quality Triangulations Using Local Transformations*. SIAM Journal on Scientific Computing **16**(6):1292–1307, November 1995.

16. G. T. Klincsek. *Minimal Triangulations of Polygonal Domains*. Annals of Discrete Mathematics **9**:121–123, 1980.

17. Michal Křížek. *On the Maximum Angle Condition for Linear Tetrahedral Elements*. SIAM Journal on Numerical Analysis **29**(2):513–520, April 1992.

18. François Labelle and Jonathan Richard Shewchuk. *Isosurface Stuffing: Fast Tetrahedral Meshes with Good Dihedral Angles*. ACM Transactions on Graphics **26**(3), August 2007. Special issue on Proceedings of SIGGRAPH 2007.

19. V. N. Parthasarathy, C. M. Graichen, and A. F. Hathaway. *A Comparison of Tetrahedron Quality Measures*. Finite Elements in Analysis and Design **15**(3):255–261, January 1994.

20. V. N. Parthasarathy and Srinivas Kodiyalam. *A Constrained Optimization Approach to Finite Element Mesh Smoothing*. Finite Elements in Analysis and Design **9**:309–320, 1991.

21. Joachim Schöberl. *NETGEN: An Advancing Front 2D/3D-Mesh Generator Based on Abstract Rules*. Computing and Visualization in Science **1**(1):41–52, July 2007.

22. Jonathan Richard Shewchuk. *Tetrahedral Mesh Generation by Delaunay Refinement*. Proceedings of the Fourteenth Annual Symposium on Computational Geometry (Minneapolis, Minnesota), pages 86–95, June 1998.

23. _____. *Two Discrete Optimization Algorithms for the Topological Improvement of Tetrahedral Meshes*. Unpublished manuscript at http://www.cs.cmu.edu/~jrs/jrspapers.html, 2002.

24. _____. *What Is a Good Linear Element? Interpolation, Conditioning, and Quality Measures*. Eleventh International Meshing Roundtable (Ithaca, New York), pages 115–126, September 2002.

Three-dimensional Semi-generalized Point Placement Method for Delaunay Mesh Refinement

Andrey N. Chernikov and Nikos P. Chrisochoides

Department of Computer Science
College of William and Mary
Williamsburg, VA 23185
{ancher,nikos}@cs.wm.edu

Summary. A number of approaches have been suggested for the selection of the positions of Steiner points in Delaunay mesh refinement. In particular, one can define an entire region (called picking region or selection disk) inside the circumscribed sphere of a poor quality element such that any point can be chosen for insertion from this region. The two main results which accompany most of the point selection schemes, including those based on regions, are the proof of termination of the algorithm and the proof of good gradation of the elements in the final mesh. In this paper we show that in order to satisfy only the termination requirement, one can use larger selection disks and benefit from the additional flexibility in choosing the Steiner points. However, if one needs to keep the theoretical guarantees on good grading then the size of the selection disk needs to be smaller. We introduce two types of selection disks to satisfy each of these two goals and prove the corresponding results on termination and good grading first in two dimensions and then in three dimensions using the radius-edge ratio as a measure of element quality. We call the point placement method semi-generalized because the selection disks are defined only for mesh entities of the highest dimension (triangles in two dimensions and tetrahedra in three dimensions); we plan to extend these ideas to lower-dimensional entities in the future work. We implemented the use of both two- and three-dimensional selection disks into the available Delaunay refinement libraries and present one example (out of many choices) of a point placement method; to the best of our knowledge, this is the first implementation of Delaunay refinement with point insertion at any point of the selection disks (picking regions).

1 Introduction

Delaunay mesh generation algorithms start with the construction of the initial mesh, which conforms to the input geometry, and then refine this mesh until the element quality constraints are met. The general idea of Delaunay refinement is to insert additional (Steiner) points inside the circumdisks of poor quality elements, which causes these elements to be destroyed, until they are gradually eliminated and replaced by better quality elements. Traditionally, Steiner points are selected at the circumcenters of poor quality elements [3, 6, 14, 16, 17]. Ruppert [14] and Shewchuk [17] proved that if the radius-edge upper bound $\bar{\rho}$ is greater than or equal to $\sqrt{2}$ in two dimensions and $\bar{\rho} \geq 2$ in three dimensions, then Delaunay refinement with circumcenters terminates. If, furthermore, the inequalities are strict, then good grading can also be proven both in two and in three dimensions. In two dimensions, in addition to good grading, one can also prove size-optimality which refers to the fact that the number of triangles in the resulting mesh is within a constant multiple of the smallest possible number of triangles satisfying the input constraints.

Recently, Üngör [20] introduced a new type of Steiner points called *off-centers*. The idea is based on the observation that sometimes the elements created as a result of inserting circumcenters of skinny triangles are also skinny and require further refinement. This technique combines the advantages of advancing front methods, which can produce very well-shaped elements in practice, and Delaunay methods, which offer theoretical guarantees. Üngör showed that the use of off-centers allows to significantly decrease the size of the final mesh in practice. While eliminating additional point insertions, this strategy creates triangles with the longest possible edges, such that one can prove the termination of the algorithm and still produce a graded mesh.

Chew [4] chooses Steiner points randomly inside a *picking sphere* to avoid slivers. The meshes in [4] have constant density; therefore, Chew proves the termination, but not the good grading.

Li and Teng [9, 10] extended the work in [4] by defining a picking sphere with a radius which is a constant multiple of the circumradius of the element. They use two different rules for eliminating the elements with large radius-edge ratio and for eliminating the slivers. In particular, in [9] the rules are defined as follows: "Add the circumcenter c_τ of any d-simplex with a large $\rho(\tau)$" and "For a sliver-simplex τ, add a good point $p \in \mathcal{P}(\tau)$", where $\rho(\tau)$ is the radius-edge ratio, $\mathcal{P}(\tau)$ is the picking region of simplex τ, and the good point is found by a constant number of random probes. The authors in [9] prove that their algorithm terminates and produces a well graded mesh with good radius-edge ratio and without slivers.

In the present work, we define Type-2 selection disks similarly to the picking region in [9]. We extend the proofs in [9] to show that any point (not only the circumcenter) from the selection disk (picking region) can be used to eliminate the elements with large radius-edge ratios. We do not address the problem of sliver elimination, however, our work can be used in conjunction with the sliver removal procedure from [9] such that the Delaunay refinement algorithm can choose any points from the selection disks (picking regions) throughout both the stage of the construction of a good radius-edge ratio mesh ("almost good mesh" [9]) and the stage of sliver removal. Intuitively, the requirement of good grading has to be more restrictive than the requirement of termination only, and, therefore, the definitions of the selection disk has to be different to satisfy each of these goals. The present paper improves upon our previous result [2] by decoupling the definitions of the selection disk used for the proof of termination (Type-1) and the selection disk used for the proof of good grading (Type-2). As can be seen further in the paper, the selection disk of Type-2 is always inside the selection disk of Type-1 of the same element, and as the radius-edge ratio ρ of an element approaches the upper bound $\bar{\rho}$, the Type-2 disk approaches the Type-1 disk.

The traditional proofs of termination and size optimality of Delaunay refinement algorithms [14, 17] explore the relationships between the insertion radius of a point and that of its parent. Stated shortly, the *insertion radius* of point p is the length of the shortest edge connected to p immediately after p is inserted into the mesh, and the *parent* is the vertex which is "responsible" for the insertion of p [17]. The proofs in [14, 17] rely on the assumption that the insertion radius of a Steiner point is equal to the circumradius of the poor quality element. This assumption holds when the Steiner point is chosen at the circumcenter of the element, since by Delaunay property the circumdisk of the element is empty, and, hence, there is no vertex closer to the circumcenter than its vertices. However, the above assumption does not hold if we pick an arbitrary point inside the selection disk of the element. For example, in Figure 2(right) the Steiner point p_i within the selection disk can be closer to the mesh vertex p_n than to any of the vertices p_k, p_l, p_m which define the skinny triangle. Therefore, we need to extend the existing theory in the new context, i.e., Steiner points can be inserted anywhere within the selection disks of poor quality elements.

One of the important applications of the flexibility offered by the use of selection disks is in conforming the mesh to the boundary between different materials. The advantages are especially pronounced in medical imaging, when the boundaries between different tissues are blurred, see Figure 1(left). In this case, after the image is segmented, instead of a clear separation, we have a boundary zone of some none-negligible width, see Figure 1(right). Then the goal of the mesh generation step is to avoid creating edges that would intersect the boundary, which can be achieved by inserting Steiner points inside the boundary zone. Another use for the selection disks would be to put vertices on a point lattice to reduce the occurrence of slivers, see [8].

In [2] we presented an example of a two-dimensional optimization-based method which allows to decrease the size of the final mesh in practice. Here we present a preliminary evaluation of an analogous three-dimensional method. The three-dimensional results are not yet as good as two-dimensional and require additional development.

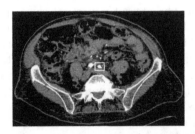 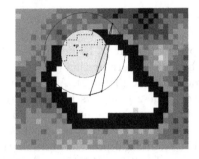

Fig. 1. The use of the selection disk for the construction of boundary conformal meshes. **(Left)** An MRI scan showing a cross-section of a body. **(Right)** A zoom-in of the selected area containing an artery: the inside is white, the outside has different shades of gray and the black zone is an approximate boundary between these regions. The standard Delaunay refinement algorithm would insert the circumcenter c. However, in order to construct a mesh which conforms to the boundary, another point (p) would be a better choice.

2 Delaunay Refinement Background

We assume that the input domain Ω is described by a Planar Straight Line Graph (PSLG) in two dimensions, or a Planar Linear Complex (PLC) in three dimensions [14–17]. A PSLG (PLC) $\mathcal{X}$ consists of a set of vertices, a set of straight line segments, and (in three dimensions) a set of planar facets. Each element of $\mathcal{X}$ is considered *constrained* and must be preserved during the construction of the mesh, although it can be subdivided into smaller elements. The vertices of $\mathcal{X}$ must be a subset of the final set of vertices in the mesh.

Let the mesh $\mathcal{M}_{\mathcal{X}}$ for the given PSLG (PLC) $\mathcal{X}$ consist of a set $V = \{p_i\}$ of vertices and a set $T = \{t_i\}$ of elements which connect vertices from V. The elements are either triangles in two dimensions or tetrahedra in three dimensions. We will denote the triangle with vertices p_u, p_v, and p_w as $\triangle(p_u p_v p_w)$ and the tetrahedron with vertices p_k, p_l, p_m, and p_n as $\tau(p_k p_l p_m p_n)$. We will use the symbol $e(p_i p_j)$ to represent the edge of the mesh which connects points p_i and p_j.

As a measure of the quality of elements we will use the circumradius-to-shortest edge ratio specified by an upper bound $\bar{\rho}$, which in two dimensions is equivalent to a lower bound on a minimal angle [11, 15] since for a triangle with the circumradius-to-shortest edge ratio ρ and minimal angle A, $\rho = 1/(2 \sin A)$. We will denote the circumradius-to-shortest edge ratio of element t as $\rho(t)$.

Let us call the open disk corresponding to a triangle's circumscribed circle or to a tetrahedron's circumscribed sphere its *circumdisk*. We will use symbols $\bigcirc(t)$ and $r(t)$ to represent the circumdisk and the circumradius of t, respectively. A mesh is said to satisfy the *Delaunay property* if the circumdisk of every element does not contain any of the mesh vertices [5].

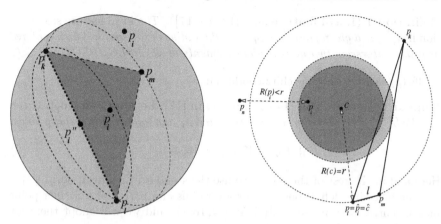

Fig. 2. (Left) Encroachment in three dimensions. **(Right)** Selection disks of Type-1 (light shade) and Type-2 (dark shade) of the skinny triangle $\triangle(p_k p_l p_m)$.

We will extensively use the notion of *cavity* [7] which is the set of elements in the mesh whose circumdisks include a given point p. We will denote $\mathcal{C}_{\mathcal{M}}(p)$ to be the cavity of p with respect to mesh $\mathcal{M}$ and $\partial \mathcal{C}_{\mathcal{M}}(p)$ to be the set of boundary edges in two dimensions (or triangles in three dimensions) of the cavity, i.e., the edges or the triangles which belong to only one element in $\mathcal{C}_{\mathcal{M}}(p)$. When $\mathcal{M}$ is clear from the context, we will omit the subscript. For our analysis, we will use the Bowyer-Watson (B-W) point insertion algorithm [1, 22], which can be written shortly as follows:

$$
\begin{aligned}
V^{n+1} &\leftarrow V^n \cup \{p\}, \\
T^{n+1} &\leftarrow T^n \setminus \mathcal{C}_{\mathcal{M}^n}(p) \cup \{(p\xi) \mid \xi \in \partial \mathcal{C}_{\mathcal{M}^n}(p)\},
\end{aligned}
\tag{1}
$$

where ξ is an edge in two dimensions or a triangle in three dimensions, while $\mathcal{M}^{n+1} = (V^{n+1}, T^{n+1})$ and $\mathcal{M}^n = (V^n, T^n)$ represent the mesh before and after the insertion of p, respectively.

In order not to create skinny elements close to the constrained segments and faces, sequential Delaunay algorithms observe special *encroachment* rules [14–17]. In particular, if a Steiner point p is considered for insertion but it lies within the open equatorial disk of a constrained subfacet f, p is not inserted but the circumcenter of f is inserted instead. Similarly, if p is inside the open diametral circle of a constrained subsegment s, then the midpoint of s is inserted instead. Consider the example in Figure 2(left). The new point p_i is inside the three-dimensional equatorial disk of a constrained face $\triangle(p_k p_l p_m)$. In this case, p_i is rejected and the algorithm attempts to insert the circumcenter p'_i of $\triangle(p_k p_l p_m)$. If p'_i does not encroach upon any constrained segments, it is inserted into the mesh. If, however, it encroaches upon a constrained segment, which is $e(p_k p_l)$ in our example, p'_i is also rejected and the midpoint p''_i of the constrained edge is inserted.

Definition 1 (Local feature size [14, 16, 17]). *The* local feature size function lfs (p) *for a given point p is equal to the radius of the smallest disk centered at p that intersects two non-incident elements of the PSLG (PLC).*

lfs (p) satisfies the Lipschitz condition:

Lemma 1 (Lemma 1 in [14], Lemma 2 in [17], Lemma 2 in [16]). *Given any PSLG (PLC) and any two points p_i and p_j, the following inequality holds:*

$$\text{lfs} (p_i) \leq \text{lfs} (p_j) + \|p_i - p_j\|. \tag{2}$$

Here and in the rest of the paper we use the standard Euclidean norm $\| \cdot \|$.

The following definitions of insertion radius and parent of a Steiner point play a central role in the analysis in [14, 16, 17] and we will adopt them for our analysis, too.

Definition 2 (Insertion radius [16, 17]). *The* insertion radius $R(p)$ *of point p is the length of the shortest edge which would be connected to p if p is inserted into the mesh, immediately after it is inserted. If p is an input vertex, then $R(p)$ is the Euclidean distance between p and the nearest input vertex visible from p.*

Here a vertex is called visible from another vertex if the straight line segment connecting both vertices does not intersect any of the constrained segments.

Remark 1. If p is a midpoint of an encroached subsegment or subfacet, then $R(p)$ is the distance between p and the nearest encroaching mesh vertex; if the encroaching mesh vertex was rejected for insertion, then $R(p)$ is the radius of the diametral circle of the subsegment or of the equatorial sphere of the subfacet [16, 17].

Remark 2. As shown in [16, 17], if p is an input vertex, then $R(p) \geq \text{lfs}(p)$. Indeed, from the definition of lfs (p), the second feature (in addition to p) which intersects the disk centered at p is either a constrained segment, a constrained facet, or the nearest input vertex visible from p.

The following definition of a parent vertex generalizes the corresponding definition in [16, 17]. In our analysis, even though the child is not necessarily the circumcenter, the parent is still defined to be the same vertex.

Definition 3 (Parent of a Steiner point). *The parent $\hat{p}$ of point p is the vertex which is defined as follows: (i) If p is an input vertex, it has no parent. (ii) If p is a midpoint of an encroached subsegment or subfacet, then $\hat{p}$ is the encroaching point (possibly rejected for insertion). (iii) If p is inserted inside the circumdisk of a poor quality element (triangle or tetrahedron), $\hat{p}$ is the most recently inserted vertex of the shortest edge of this element.*

The quantity $D(p)$ is defined as the ratio of lfs (p) over $R(p)$ [16, 17]:

$$D(p) = \frac{\text{lfs}(p)}{R(p)}. \tag{3}$$

It reflects the density of vertices near p at the time p is inserted, weighted by the local feature size. To achieve good mesh grading we would like this density to be as small as possible. If the density is bounded from above by a constant, the mesh is said to have a *good grading* property.

3 Two-Dimensional Generalized Delaunay Refinement

In this section we introduce two types of selection disks which can be used for the insertion of Steiner points in two dimensions. The two-dimensional results are given here for completeness and prepare the mindset for the three-dimensional results in the following section. The Type-1 selection disk corresponds to the definition of the selection disk we used in [2] which we briefly review. Then we define the Type-2 disk and prove good grading and size-optimality.

Definition 4 (Selection disk of Type-1 in 2D). *If t is a poor quality triangle with circumcenter c, shortest edge length l, circumradius r, and circumradius-to-shortest edge ratio $\rho = r/l > \bar{\rho} \geq \sqrt{2}$, then the Type-1 selection disk for the insertion of a Steiner point that would eliminate t is the open disk with center c and radius $r - \sqrt{2}l$.*

For example, in Figure 2(right), $e(p_l p_m)$ is the shortest edge of a skinny triangle $\triangle(p_k p_l p_m)$ and c is its circumcenter. The selection disk of Type-1 is the lightly shaded disk with center c and radius $r(\triangle(p_k p_l p_m)) - \sqrt{2}\|p_l - p_m\|$.

Remark 3. As $\rho(\triangle(p_k p_l p_m))$ approaches $\sqrt{2}$, the Type-1 selection disk shrinks to the circumcenter c of the triangle. If, furthermore, $\rho(\triangle(p_k p_l p_m)) \leq \sqrt{2}$, the selection disk vanishes, which coincides with the fact that the triangle $\triangle(p_k p_l p_m)$ cannot be considered skinny.

In [2] we proved that any point inside the Type-1 selection disk of a triangle can be chosen for the elimination of the triangle, and that the generalized Delaunay refinement algorithm which chooses Steiner points inside Type-1 selection disks terminates.

Using the definition of the selection disk, in [2] we suggested an example of an optimization-based method which allows to improve the size of the mesh by up to 20% over the circumcenter insertion method and up to 5% over the off-center insertion method, for small values of the minimal angle bound. The underlying idea of our method is that, by choosing a point within the selection disk we can vary the set of triangles in its cavity, we simultaneously minimize the number of deleted good quality triangles and maximize the number of deleted poor quality triangles.

3.1 Proof of Good Grading and Size Optimality with Selection Disks of Type II

Definition 5 (Selection disk of Type-2 in 2D). *If t is a poor quality triangle with circumcenter c, shortest edge length l, circumradius r, and circumradius-to-shortest edge ratio $\rho = r/l > \bar{\rho} \geq \sqrt{2}$, then the* Type-2 *selection disk for the insertion of a Steiner point that would eliminate t is the open disk with center c and radius $r(1 - \frac{\sqrt{2}}{\bar{\rho}})$.*

For example, in Figure 2(right), $e\,(p_l p_m)$ is the shortest edge of a skinny triangle $\triangle\,(p_k p_l p_m)$ and c is its circumcenter. The Type-2 selection disk for this triangle is the darkly shaded disk with center c and radius $r\,(\triangle\,(p_k p_l p_m))\,(1 - \frac{\sqrt{2}}{\bar{\rho}})$.

Remark 4. Note from Definitions 4 and 5 that the radius of the Type-2 selection disk is always smaller than the radius of the Type-1 selection disk of the same skinny triangle because $r(1 - \frac{\sqrt{2}}{\bar{\rho}}) = r - \frac{\rho}{\bar{\rho}}\sqrt{2}l$ and $\rho > \bar{\rho}$.

Remark 5. As $\bar{\rho}$ approaches $\sqrt{2}$ the radius of the Type-2 selection disk approaches zero, which means that the selection disk shrinks to the circumcenter point.

As can be seen further below, the price which we pay for the gain in the flexibility in choosing points is the increase of the constants which bound the size of the mesh. To classify the Delaunay refinement algorithms with respect to the theoretical bounds on mesh grading we need the following definition.

Definition 6 (δ-graded Delaunay refinement algorithm in 2D). *If for every triangle t with circumcenter c, circumradius r, shortest edge length l, and circumradius-to-shortest edge length ratio $\rho = r/l > \bar{\rho} \geq \sqrt{2}$ a Delaunay refinement algorithm selects a Steiner point p_i within the Type-2 selection disk such that $\|p_i - c\| < r(1 - \delta)$, where*

$$\frac{\sqrt{2}}{\bar{\rho}} \leq \delta \leq 1,$$

we say that this Delaunay refinement algorithm is δ-graded.

Lemma 2. *If p_i is a vertex of the mesh produced by a δ-graded Delaunay refinement algorithm then the following inequality holds:*

$$R\,(p_i) \geq C \cdot R\,(\hat{p}_i), \tag{4}$$

where we distinguish among the following cases:

(i) $C = \delta\bar{\rho}$ if p_i is a Steiner point chosen within the Type-2 selection disk of a skinny triangle;

Otherwise, let p_i be the midpoint of subsegment s. Then

(ii) $C = \frac{1}{\sqrt{2}}$ if $\hat{p}_i$ is a Steiner point which encroaches upon s, chosen within the Type-2 selection disk of a skinny triangle;

(iii) $C = \frac{1}{2\cos\alpha}$ if p_i and $\hat{p}_i$ lie on incident subsegments separated by an angle of α (with $\hat{p}_i$ encroaching upon s), where $45° \leq \alpha \leq 90°$;

(iv) $C = \sin\alpha$ if p_i and $\hat{p}_i$ lie on incident segments separated by an angle of $\alpha \leq 45°$.

If p_i is an input vertex or if p_i and $\hat{p}_i$ are on non-incident features, then

$$R(p_i) \geq \mathrm{lfs}(p_i).$$

Proof. We need to present a new proof only for case (i) since the proof for case (ii) is the same as in the corresponding lemma for the Type-1 selection disk [2], and the proofs for all other cases are independent of the choice of the point within the selection disk and are given in [17].

Case (i) By the definition of a parent vertex, $\hat{p}_i$ is the most recently inserted endpoint of the shortest edge of the triangle; without loss of generality let $\hat{p}_i = p_l$ and $e(p_l p_m)$ be the shortest edge of the skinny triangle $\triangle(p_k p_l p_m)$, see Figure 2(right). If $e(p_l p_m)$ was the shortest edge among the edges incident upon p_l at the time p_l was inserted into the mesh, then $\|p_l - p_m\| = R(p_l)$ by the definition of the insertion radius; otherwise, $\|p_l - p_m\| \geq R(p_l)$. In either case,

$$\|p_l - p_m\| \geq R(p_l). \tag{5}$$

Then

$$
\begin{aligned}
R(p_i) &> \delta r && \text{(from Delaunay property and Definition 6)} \\
&= \delta\rho\|p_l - p_m\| && \text{(since } \rho = \tfrac{r}{\|p_l - p_m\|} \text{)} \\
&> \delta\bar{\rho}\|p_l - p_m\| && \text{(since } \rho > \bar{\rho} \text{)} \\
&\geq \delta\bar{\rho}R(p_l) && \text{(from (5)).}
\end{aligned}
$$

Hence, $R(p_i) > \delta\bar{\rho}R(\hat{p}_i)$; choose $C = \delta\bar{\rho}$. □

Remark 6. We have proven Inequality (4) only for the case which involves the "free" Steiner points, i.e., the points which are chosen from selection disks and which do not lie on constrained segments. The proofs for the other types of Steiner points do not involve the definition of the selection disk and are applicable without change from [17].

Lemma 3. *If p is a vertex of the mesh produced by a δ-graded Delaunay refinement algorithm and C is the constant specified by Lemma 2, then the following inequality holds:*

$$D(p) \leq \frac{2-\delta}{\delta} + \frac{D(\hat{p})}{C}. \tag{6}$$

Proof. If p is inside a Type-2 selection disk of a skinny triangle with circumradius r, then

$$\|p - \hat{p}\| < 2r - \delta r \quad \text{(from the definition of } \hat{p} \text{ and Def. 6)}$$
$$= (2 - \delta)r$$
$$= \frac{2-\delta}{\delta}\delta r$$
$$< \frac{2-\delta}{\delta}R(p) \quad \text{(from Delaunay property and Def. 6).}$$

If p is an input vertex or lies on an encroached segment, then

$$\|p - \hat{p}\| \leq R(p) \quad \text{(by definitions of } \hat{p} \text{ and } R(p)\text{)}$$
$$\leq \frac{2-\delta}{\delta}R(p) \quad \text{(since from Def. 6, } \delta \leq 1\text{).}$$

In all cases,

$$\|p - \hat{p}\| \leq \frac{2-\delta}{\delta}R(p). \tag{7}$$

Then

$$\text{lfs}(p) \leq \text{lfs}(\hat{p}) + \|p - \hat{p}\| \quad \text{(from Lemma 1)}$$
$$\leq \text{lfs}(\hat{p}) + \frac{2-\delta}{\delta}R(p) \quad \text{(from (7))}$$
$$= D(\hat{p})R(\hat{p}) + \frac{2-\delta}{\delta}R(p) \quad \text{(from (3))}$$
$$\leq D(\hat{p})\frac{R(p)}{C} + \frac{2-\delta}{\delta}R(p) \quad \text{(from Lemma 2).}$$

The result follows from the division of both sides by $R(p)$. □

Lemma 4 (Extension of Lemma 7 in [17] and Lemma 2 in [14]). *Suppose that $\bar{\rho} > \sqrt{2}$ and the smallest angle in the input PSLG is strictly greater than $60°$. There exist fixed constants C_T and C_S such that, for any vertex p inserted (or considered for insertion and rejected) by a δ-graded Delaunay refinement algorithm, $D(p) \leq C_T$, and for any vertex p inserted at the midpoint of an encroached subsegment, $D(p) \leq C_S$. Hence, the insertion radius of a vertex has a lower bound proportional to its local feature size.*

Proof. The proof is by induction and is similar to the proof of Lemma 7 in [17]. The base case covers the input vertices, and the inductive step covers the other two types of vertices: free vertices and subsegment midpoints.

Base case: The lemma is true if p is an input vertex, because in this case, by Remark 2, $D(p) = \text{lfs}(p)/R(p) \leq 1$.

Inductive hypothesis: Assume that the lemma is true for $\hat{p}$, i.e., $D(\hat{p}) \leq \max\{C_T, C_S\}$.

Inductive step: There are two cases:

(i) If p is in the Type-2 selection disk of a skinny triangle, then

$$D(p) \leq \frac{2-\delta}{\delta} + \frac{D(\hat{p})}{C} \quad \text{(from Lemma 3)}$$
$$= \frac{2-\delta}{\delta} + \frac{D(\hat{p})}{\delta\bar{\rho}} \quad \text{(from Lemma 2)}$$
$$\leq \frac{2-\delta}{\delta} + \frac{\max\{C_T, C_S\}}{\delta\bar{\rho}} \quad \text{(by the inductive hypothesis).}$$

It follows that one can prove that $D(p) \leq C_T$ if C_T is chosen so that

$$\frac{2-\delta}{\delta} + \frac{\max\{C_T, C_S\}}{\delta\bar{\rho}} \leq C_T. \tag{8}$$

*(ii)*If p is the midpoint of a subsegment s such that $\hat{p}$ encroaches upon s, then we have 3 sub-cases:

(ii-a) If $\hat{p}$ is an input vertex, then the disk centered at p and touching $\hat{p}$ has radius less than the radius of the diametral disk of s and therefore lfs $(p) < R(p)$. Thus, $D(p) < 1$ and the lemma holds.

(ii-b) If $\hat{p}$ is a rejected point from the Type-2 selection disk of a skinny triangle or lies on a segment not incident to s, then

$$
\begin{aligned}
D(p) &\leq \tfrac{2-\delta}{\delta} + \tfrac{D(\hat{p})}{C} &&\text{(from Lemma 3)} \\
&= \tfrac{2-\delta}{\delta} + \sqrt{2}D(\hat{p}) &&\text{(from Lemma 2)} \\
&\leq \tfrac{2-\delta}{\delta} + \sqrt{2}C_T &&\text{(by the inductive hypothesis).}
\end{aligned}
$$

(ii-c) If $\hat{p}$ lies on a segment incident to s, then

$$
\begin{aligned}
D(p) &\leq \tfrac{2-\delta}{\delta} + \tfrac{D(\hat{p})}{C} &&\text{(from Lemma 3)} \\
&= \tfrac{2-\delta}{\delta} + 2\cos\alpha D(\hat{p}) &&\text{(from Lemma 2)} \\
&\leq \tfrac{2-\delta}{\delta} + 2C_S\cos\alpha &&\text{(by the inductive hypothesis).}
\end{aligned}
$$

It follows that one can prove that $D(p) \leq C_S$ if C_S is chosen so that both of the following relations (9) and (10) are satisfied:

$$\frac{2-\delta}{\delta} + \sqrt{2}C_T \leq C_S, \tag{9}$$

$$\frac{2-\delta}{\delta} + 2C_S\cos\alpha \leq C_S. \tag{10}$$

If $\delta\bar{\rho} > \sqrt{2}$, relations (8) and (9) can be simultaneously satisfied by choosing

$$C_T = \frac{(2-\delta)(\bar{\rho}+\delta)}{\delta\bar{\rho} - \sqrt{2}}, \quad \text{and} \quad C_S = \frac{(2-\delta)\bar{\rho}(1+\sqrt{2})}{\delta\bar{\rho} - \sqrt{2}}.$$

If the smallest input angle $\alpha_{\min} > 60°$, relations (8) and (10) can be simultaneously satisfied by choosing

$$C_T = \frac{2-\delta}{\delta} + \frac{C_S}{\delta\bar{\rho}}, \quad \text{and} \quad C_S = \frac{2-\delta}{\delta(1 - 2\cos\alpha_{\min})}.$$

□

The analysis in Lemma 4 assumes that all angles in the input PSLG are greater than 60°. Although such geometries are rare, in practice a modification of the algorithm with a concentric circular shell splitting [14, 17] allows to guarantee the termination of the algorithm even though the small angles adjacent to the segments of the input PSLG cannot be improved.

Theorem 1 (Theorem 8 in [17], Theorem 1 in [14]). *For any vertex p of the output mesh, the distance to its nearest neighbor is at least $\frac{\text{lfs}(p)}{C_S+1}$.*

The proof in [17] relies only on Lemmata 1 and 4 here and, therefore, holds for the arbitrary points chosen within selection disks of skinny triangles.

Theorem 1 is used in the proof of the following theorem.

Theorem 2 (Theorem 10 in [17], Theorem 14 in [12], Theorem 3 in [14]). *Let $\text{lfs}_{\mathcal{M}}(p)$ be the local feature size at p with respect to a mesh $\mathcal{M}$ (treating $\mathcal{M}$ as a PSLG), whereas $\text{lfs}(p)$ remains the local feature size at p with respect to the input PSLG. Suppose a mesh $\mathcal{M}$ with smallest angle θ has the property that there is some constant $k_1 \geq 1$, such that for every point p, $k_1\text{lfs}_{\mathcal{M}}(p) \geq \text{lfs}(p)$. Then the cardinality of $\mathcal{M}$ is less than k_2 times the cardinality of any other mesh of the input PSLG with smallest angle θ, where $k_2 = \mathcal{O}\left(k_1^2/\theta\right)$.*

Smaller values of δ offer more flexibility to a δ-graded Delaunay refinement algorithm in choosing Steiner points. However, from Lemma 4 it follows that as $\delta\bar{p}$ approach $\sqrt{2}$, C_T and C_S approach infinity, which leads to the worsening of the good grading guarantees. Therefore, along with satisfying application-specific requirements, the point insertion schemes should try to place Steiner points as close to circumcenters as possible.

4 Three-Dimensional Generalized Delaunay Refinement

In this section we introduce again two types of selection disks which can be used for the insertion of Steiner points. First, we prove the termination of a Delaunay refinement algorithm with the Type-1 selection disks. Then we give an example of an optimization based strategy for the insertion of Steiner points from the Type-1 selection disks. Finally, we introduce the Type-2 selection disk (which is always inside the Type-1 selection disk of the same skinny tetrahedron) and prove the good grading. As in [16], the proofs require that all incident segments and faces in the input geometry are separated by angles of at least $90°$. According to our own experience and that of other authors [16, 18, 19], the Delaunay refinement algorithm works for much smaller angles in practice, although skinny tetrahedra adjacent to the small input angles cannot be removed. To prevent the algorithm from creating edges of ever diminishing length (for details, see the proofs below and the dataflow diagram in Figure 4), one can compute the insertion radii of candidate Steiner points explicitly and forbid the insertion of points which lead to the introduction of very short edges [16].

4.1 Proof of Termination with Selection Disks of Type I

Definition 7 (Selection disk of Type-1 in 3D). *If τ is a poor quality tetrahedron with circumcenter c, shortest edge length l, circumradius r, and* *circumradius-to-shortest edge ratio* $\rho = r/l > \bar{\rho} \geq 2$, *then the* selection disk of Type-1 *for the insertion of a Steiner point that would eliminate* τ *is the open disk with center* c *and radius* $r - 2l$.

Following [16], the analysis below requires that the Delaunay refinement algorithm prioritize the splitting of encroached faces. In particular, when a Steiner point p encroaches upon several constrained faces, the encroached face which contains the projection of p is split. The projection of a point p onto a plane is the point in the plane which is closest to p. This requirement allows to achieve better bounds on the circumradius-to-shortest edge ratios in the final mesh.

Lemma 5 (Projection Lemma [16]). *Let* f *be a subfacet of the Delaunay triangulated facet* F. *Suppose that* f *is encroached upon by some vertex* p, *but* p *does not encroach upon any subsegment of* F. *Then* $\mathrm{proj}_F(p)$ *lies in the facet* F, *and* p *encroaches upon a subfacet of* F *that contains* $\mathrm{proj}_F(p)$.

Now we can prove the following lemma which establishes the relationship between the insertion radius of a point and its parent.

Lemma 6. *If* p_i *is a vertex of the mesh produced by a Delaunay refinement algorithm which chooses points within Type-1 selection disks of tetrahedra with circumradius-to-shortest edge ratios greater than* $\bar{\rho} \geq 2$, *then the following inequality holds:*

$$R(p_i) \geq C \cdot R(\hat{p}_i), \tag{11}$$

where we distinguish among the following cases:

(i) $C = 2$ *if* p_i *is a Steiner point chosen within the Type-1 selection disk of a skinny tetrahedron;*

(ii) $C = \frac{1}{\sqrt{2}}$ *if* p_i *is a circumcenter of an encroached constrained face;*

(iii) $C = \frac{1}{\sqrt{2}}$ *if* p_i *is a midpoint of an encroached constrained segment.*

If p_i *is an input vertex or if* p_i *and* $\hat{p}_i$ *are on non-incident features, then*

$$R(p_i) \geq \mathrm{lfs}(p_i).$$

Proof. We need to prove only cases (i) and (ii) since the proofs of all other cases are independent of the choice of the point within the selection disk and are given in [16].

Case (i) Without the loss of generality, let $e(p_l p_m)$ be the shortest edge of the skinny tetrahedron $\tau(p_k p_l p_m p_n)$ and $\hat{p}_i = p_l$. Then

$$
\begin{aligned}
R(p_i) &> 2\|p_l - p_m\| && \text{(from Delaunay property and Definition 7)} \\
&\geq 2R(p_l) && \text{(from Definition 2 of insertion radius)} \\
&= 2R(\hat{p}_i);
\end{aligned}
$$

choose $C = 2$.

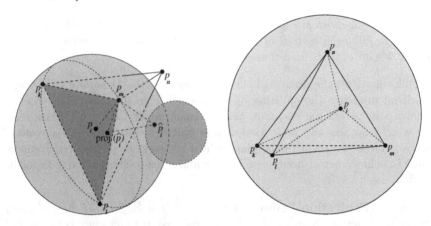

Fig. 3. **(Left)** An illustration of the relationship between the insertion radii of Steiner points, in the case of encroachment in three dimensions. **(Right)** A tetrahedron with high circumradius-to-shortest edge ratio.

Case (ii) Consider Figure 3(left). Let $\hat{p}_i$ be inside the selection disk (the smaller shaded circle) of some skinny tetrahedron (not shown), such that $\hat{p}_i$ encroaches upon the constrained face $\triangle\,(p_k p_l p_m)$, and p_i is the circumcenter of $\triangle\,(p_k p_l p_m)$. According to the projection requirement, let $\triangle\,(p_k p_l p_m)$ include $\mathrm{proj}_F\,(\hat{p}_i)$, where F is the facet of the input PLC containing $\triangle\,(p_k p_l p_m)$. Without the loss of generality, suppose p_m is the point closest to $\mathrm{proj}_F\,(\hat{p}_i)$ among the vertices of $\triangle\,(p_k p_l p_m)$. Because $\mathrm{proj}_F\,(\hat{p}_i)$ lies inside the triangle $\triangle\,(p_k p_l p_m)$, it cannot be father away from p_m than the circumcenter of $\triangle\,(p_k p_l p_m)$, i.e.,

$$\|\mathrm{proj}_F\,(\hat{p}_i) - p_m\| \leq r\,(\triangle\,(p_k p_l p_m))\,. \tag{12}$$

Furthermore, because $\hat{p}_i$ is inside the equatorial disk of $\triangle\,(p_k p_l p_m)$,

$$\|\mathrm{proj}_F\,(\hat{p}_i) - \hat{p}_i\| < r\,(\triangle\,(p_k p_l p_m))\,. \tag{13}$$

From (12) and (13), as well as the fact that the triangle with vertices $\hat{p}_i$, $\mathrm{proj}_F\,(\hat{p}_i)$, and p_m has a right angle at $\mathrm{proj}_F\,(\hat{p}_i)$, we have:

$$\|\hat{p}_i - p_m\| < \sqrt{2}r\,(\triangle\,(p_k p_l p_m))\,. \tag{14}$$

Since p_m has to be in the mesh by the time $\hat{p}_i$ is considered for insertion, $R\,(\hat{p}_i) \leq \|\hat{p}_i - p_m\|$. Then, using (14), we obtain:

$$R\,(\hat{p}_i) < \sqrt{2}r\,(\triangle\,(p_k p_l p_m)) = \sqrt{2}R\,(p_i)\,;$$

choose $C = \frac{1}{\sqrt{2}}$. $\square$

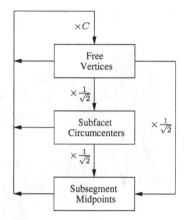

Fig. 4. Flow diagram from [16] illustrating the relationship between the insertion radius of a vertex and that of its parent in three dimensions. If no cycle has a product smaller then one, the algorithm will terminate. Input vertices are not shown since they do not participate in cycles. In [16] the constant $C = \bar{\rho} \geq 2$. In our case, with the use of Type-1 selection disks $C = 2$, and with the use of Type-2 selection disks $C = \delta\bar{\rho} \geq 2$.

Figure 4 shows the relationship between the insertion radii of mesh vertices and the insertion radii of their parents. We can see that if Inequality (11) is satisfied then no new edge will be created whose length is smaller than half of the length of some existing edge and the algorithm will eventually terminate because it will run out of places to insert new vertices.

4.2 An Example of a Point Selection Strategy

In two dimensions, by selecting the new Steiner point, we can construct only one new triangle which will be incident upon the shortest edge of the existing skinny triangle and which will have the required circumradius-to-shortest edge ratio. The three-dimensional case, however, is complicated by the fact that several new tetrahedra may be incident upon the shortest edge of the existing skinny tetrahedron. The example in Figure 3(right) shows a skinny tetrahedron $\tau\,(p_k p_l p_m p_n)$ with two new tetrahedra $\tau\,(p_i p_k p_l p_n)$ and $\tau\,(p_i p_k p_l p_m)$ that are incident upon the shortest edge $e\,(p_k p_l)$. By having to deal with multiple incident tetrahedra we face a multi-constrained optimization problem which to the best of our knowledge has not been analyzed with respect to the existence of the optimal solution and the construction of this solution.

A heuristic approach was suggested by Üngör [21] who proposed two types of off-center points in three dimensions. Based on the experimental data, he observes the following: "Use of both types of off-centers or the use of TYPE II off-centers alone outperforms the use of TYPE I off-centers alone, which in turn outperforms the use of circumcenters." If a is the midpoint of the shortest edge $e\,(p_k p_l)$ of the tetrahedron and c is its circumcenter, than the TYPE II

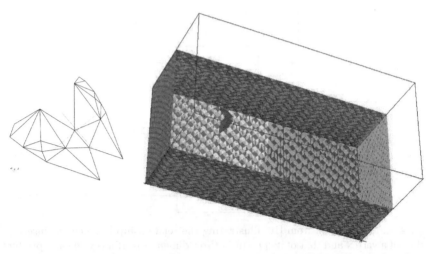

Fig. 5. (Left) A wireframe model of a flying bat. **(Right)** The bat inside a box.

Table 1. The number of tetrahedra produced with the use of different point selection methods for three models in three dimensions. The first method (CC) always inserts circumcenters of skinny tetrahedra, the second method (OC) always inserts off-centers, and the third method (OPT) is optimization based. The minimal values in each cell are hilighted.

Model	Point position					$\bar{\rho}$				
		2.0	4.0	6.0	8.0	10.0	12.0	14.0	16.0	18.0
Points in cube	CC	1126	560	527	482	457	429	432	417	417
	OC	**777**	**298**	190	184	236	**157**	180	181	**169**
	OPT	1723	1619	**174**	**174**	**174**	174	**174**	**174**	174
Rocket	CC	1017	629	566	**567**	**562**	**542**	**542**	**542**	**542**
	OC	1060	729	**540**	673	679	660	660	660	660
	OPT	**937**	**610**	628	678	598	**542**	**542**	**542**	**542**
Bat	CC	**24552**	**16561**	**15427**	**15226**	**15083**	**14923**	**14921**	**14923**	**14894**
	OC	24985	21019	20781	20820	19764	19247	21058	17816	18301
	OPT	24628	17599	15970	15533	15267	15053	15074	15084	14939

off-center b is computed in [21] on the segment $\mathcal{L}(ac)$ such that $|\mathcal{L}(ab)| = \alpha_3 \left(\bar{\rho} + \sqrt{\bar{\rho}^2 - 1/4} \right)$, where α_3 is the perturbation factor. From experimental evidence in [21] the author suggests that a good choice for α_3 is 0.6. The insertion of TYPE II off-centers was implemented by Hang Si in Tetgen version 1.4.0 along with the circumcenter insertion method [18, 19]. We also added the implementation of an optimization-based point insertion method which with every insertion of a Steiner point within a Type-1 selection disk of a skinny tetrahedron minimizes the signed difference between the number of deleted good quality tetrahedra and the number of deleted poor quality tetrahedra. As opposed to the two-dimensional case, we did not restrict the position of the

Steiner point to a specific arc, but instead sampled 1,000 points uniformly in spherical coordinates of the Type-1 selection disk and chose the best one. For our experiments, we used the following three geometries. The "Points in cube" model is the unit cube with two additional points close to its center at 10^{-4} distance from each other. The "Rocket" model is the example PLC supplied with the Tetgen distribution. The "Bat" model [13] is shown in Figure 5. Table 1 summarizes the results of our experiments. From our point of view, these results do not provide a basis for conclusions, and further research is required.

4.3 Proof of Good Grading with Selection Disks of Type II

Definition 8 (Selection disk of Type-2 in 3D). *If τ is a poor quality tetrahedron with circumcenter c, shortest edge length l, circumradius r, and circumradius-to-shortest edge ratio $\rho = r/l > \bar{\rho} \geq 2$, then the Type-2 selection disk for the insertion of a Steiner point that would eliminate τ is the open disk with center c and radius $r(1 - \frac{2}{\rho})$.*

Remark 7. Note from definitions 7 and 8 that the radius of the Type-2 selection disk is always smaller than the radius of the Type-1 selection disk of the same skinny tetrahedron because $r(1 - \frac{2}{\rho}) = r - \frac{\rho}{\rho}2l$ and $\rho > \bar{\rho}$.

Definition 9 (δ-graded Delaunay refinement algorithm in 3D). *If for every tetrahedron τ with circumcenter c, circumradius r, shortest edge length l, and circumradius-to-shortest edge length ratio $\rho = r/l > \bar{\rho} \geq 2$ a Delaunay refinement algorithm selects a Steiner point p_i within the Type-2 selection disk such that $\|p_i - c\| < r(1 - \delta)$, where $2/\bar{\rho} \leq \delta \leq 1$, we say that this Delaunay refinement algorithm is δ-graded.*

Lemma 7. *If p_i is a vertex of the mesh produced by a δ-graded Delaunay refinement algorithm then the following inequality holds:*

$$R(p_i) \geq C \cdot R(\hat{p}_i), \tag{15}$$

where we distinguish among the following cases:

(i) $C = \delta\bar{\rho}$ if p_i is a Steiner point chosen within the Type-2 selection disk of a skinny tetrahedron;
(ii) $C = \frac{1}{\sqrt{2}}$ if p_i is a circumcenter of an encroached constrained face;
(iii) $C = \frac{1}{\sqrt{2}}$ if p_i is a midpoint of an encroached constrained segment.

If p_i is an input vertex or if p_i and $\hat{p}_i$ are on non-incident features, then

$$R(p_i) \geq \text{lfs}(p_i).$$

Proof. We need to present a new proof only for case (i) since the proof for case (ii) is the same as in Lemma 6, and the proofs for all other cases are independent of the choice of the point within the selection disk and are given in [16].

Case (i) Without the loss of generality, let $e\,(p_l p_m)$ be the shortest edge of the skinny tetrahedron $\tau\,(p_k p_l p_m p_n)$ and $\hat{p}_i = p_l$. Then

$$
\begin{aligned}
R\,(p_i) &> \delta r\,(\tau\,(p_k p_l p_m p_n)) && \text{(from Delaunay property and Def. 9)}\\
&= \delta\rho(\tau\,(p_k p_l p_m p_n))\|p_l - p_m\| \\
&> \delta\bar{\rho}\|p_l - p_m\| && \text{(since the tetrahedron is skinny)} \\
&\geq \delta\bar{\rho}R\,(\hat{p}_i)\,;
\end{aligned}
$$

choose $C = \delta\bar{\rho}$. $\square$

Lemma 8 (Extension of Lemma 3 to 3D). *If p is a vertex of the mesh produced by a δ-graded Delaunay refinement algorithm and C is the constant specified by Lemma 7, then the following inequality holds:*

$$
D\,(p) \leq \frac{2 - \delta}{\delta} + \frac{D\,(\hat{p})}{C}. \tag{16}
$$

Proof. The proof is formally equivalent to the proof of Lemma 3. $\square$

Lemma 9 (Extension of Theorem 5 in [16]). *Suppose that $\bar{\rho} > 2$ and the input PLC satisfies the projection condition. Then there exist fixed constants $C_T \geq 1$, $C_F \geq 1$, and $C_S \geq 1$ such that, for any vertex p inserted or rejected by a δ-graded Delaunay refinement algorithm, the following relations hold:*

(i) $D\,(p) \leq C_T$ if p is chosen from the selection disk of a skinny tetrahedron;
(ii) $D\,(p) \leq C_F$ if p is the circumcenter of an encroached constrained face;
(iii) $D\,(p) \leq C_S$ if p is the midpoint of an encroached constrained segment.

Therefore, the insertion radius of every vertex has a lower bound proportional to its local feature size.

Proof. The proof is by induction and is similar to the proof of Lemma 7 in [16]. The base case covers the input vertices, and the inductive step covers the other three types of vertices: free vertices, the circumcenters of constrained faces, and the midpoints of constrained subsegments. The constants are chosen as follows:

$$
C_T = C_0 \frac{C_1 + 1 + \sqrt{2}}{C_1 - 2}, \quad C_F = C_0 \frac{(1 + \sqrt{2})C_1 + \sqrt{2}}{C_1 - 2}, \quad C_S = C_0 \frac{(3 + \sqrt{2})C_1}{C_1 - 2},
$$

where $C_0 = (2 - \delta)/\delta$ and $C_1 = \delta\bar{\rho}$. As δ approaches $2/\bar{\rho}$, the constants approach infinity. $\square$

Theorem 3 (Theorem 6 in [16]). *For any vertex p of the output mesh, the distance to its nearest neighbor is at least $\frac{\mathrm{lfs}(p)}{C_S + 1}$.*

The proof of this theorem does not depend on the specific position of the Steiner point inside the selection disk. Hence, the theorem holds in the context of three-dimensional generalized Delaunay refinement.

5 Conclusions

In this paper we presented the first to our knowledge decoupled definitions of two types of selection disks along with the corresponding proofs of termination and size optimality, both in two and in three dimensions. The primary goal which we pursued while developing the selection disk ideas was the design of parallel Delaunay refinement algorithms. A number of techniques have been developed by multiple authors for the selection of Steiner points for sequential Delaunay refinement. The development of the corresponding parallel algorithm for each of the sequential point selection techniques would have been cumbersome and time consuming. Instead, we have shown that there exists an entire region for the sequential selection of Steiner points. Therefore, if a parallel algorithm framework is designed with the assumption that the points are chosen from this region, all sequential point placement techniques can be used as a black box.

The additional flexibility in choosing the Steiner points can be used to satisfy a number of diverse mesh improvement goals. For example, one can construct the difference of the selection disk with the forbidden regions which create slivers to find the region which does not create slivers. Then, instead of using randomized point selection for sliver elimination like in [4, 9, 10], a deterministic insertion of a single point is sufficient. As another example, if a selection disk intersects a boundary between different materials, one would like to insert the point along the boundary with the goal of constructing a boundary conformal mesh.

Delaunay refinement algorithms can also use selection disks for the splitting of constrained segments and faces; we plan to address it in the future work along with the complete generalization of point insertion in all dimensions.

Acknowledgments

We thank Gary Miller for helpful references and George Karniadakis for the bat model. This work was supported (in part) by the following NSF grants: ACI-0312980, CNS-0521381, as well as a John Simon Guggenheim Award. We also thank the anonymous reviewers for insightful comments and constructive suggestions.

References

1. A. Bowyer. Computing Dirichlet tesselations. *Computer Journal*, 24:162–166, 1981.
2. A. N. Chernikov and N. P. Chrisochoides. Generalized Delaunay mesh refinement: From scalar to parallel. In *Proceedings of the 15th International Meshing Roundtable*, pages 563–580, Birmingham, AL, Sept. 2006. Springer.
3. L. P. Chew. Guaranteed quality mesh generation for curved surfaces. In *Proceedings of the 9th ACM Symposium on Computational Geometry*, pages 274–280, San Diego, CA, 1993.

4. L. P. Chew. Guaranteed-quality Delaunay meshing in 3D. In *Proceedings of the 13th ACM Symposium on Computational Geometry*, pages 391–393, Nice, France, 1997.
5. B. N. Delaunay. Sur la sphere vide. *Izvestia Akademia Nauk SSSR, VII Seria, Otdelenie Mataematicheskii i Estestvennyka Nauk*, 7:793–800, 1934.
6. W. H. Frey. Selective refinement: A new strategy for automatic node placement in graded triangular meshes. *International Journal for Numerical Methods in Engineering*, 24(11):2183–2200, 1987.
7. P.-L. George and H. Borouchaki. *Delaunay Triangulation and Meshing. Application to Finite Elements*. HERMES, 1998.
8. F. Labelle. Sliver removal by lattice refinement. In *Proceedings of the 21nd Symposium on Computational Geometry*, pages 347–356, New York, NY, USA, 2006. ACM Press.
9. X.-Y. Li. Generating well-shaped d-dimensional Delaunay meshes. *Theoretical Computer Science*, 296(1):145–165, 2003.
10. X.-Y. Li and S.-H. Teng. Generating well-shaped Delaunay meshes in 3D. In *Proceedings of the 12th annual ACM-SIAM symposium on Discrete algorithms*, pages 28–37, Washington, D.C., 2001.
11. G. L. Miller, D. Talmor, S.-H. Teng, and N. Walkington. A Delaunay based numerical method for three dimensions: Generation, formulation, and partition. In *Proceedings of the 27th Annual ACM Symposium on Theory of Computing*, pages 683–692, Las Vegas, NV, May 1995.
12. S. A. Mitchell. Cardinality bounds for triangulations with bounded minimum angle. In *Proceedings of the 6th Canadian Conference on Computational Geometry*, pages 326–331, Saskatoon, Saskatchewan, Canada, Aug. 1994.
13. I. Pivkin, E. Hueso, R. Weinstein, D. Laidlaw, S. Swartz, and G. Karniadakis. Simulation and visualization of air flow around bat wings during flight. In *Proceedings of the International Conference on Computational Science*, pages 689–694, Atlanta, GA, 2005.
14. J. Ruppert. A Delaunay refinement algorithm for quality 2-dimensional mesh generation. *Journal of Algorithms*, 18(3):548–585, 1995.
15. J. R. Shewchuk. *Delaunay Refinement Mesh Generation*. PhD thesis, Carnegie Mellon University, 1997.
16. J. R. Shewchuk. Tetrahedral mesh generation by Delaunay refinement. In *Proceedings of the 14th ACM Symposium on Computational Geometry*, pages 86–95, Minneapolis, MN, 1998.
17. J. R. Shewchuk. Delaunay refinement algorithms for triangular mesh generation. *Computational Geometry: Theory and Applications*, 22(1–3):21–74, May 2002.
18. H. Si. On refinement of constrained Delaunay tetrahedralizations. In *Proceedings of the 15th International Meshing Roundtable*, pages 509–528, Birmingham, AL, Sept. 2006. Springer.
19. H. Si and K. Gaertner. Meshing piecewise linear complexes by constrained Delaunay tetrahedralizations. In *Proceedings of the 14th International Meshing Roundtable*, pages 147–163, San Diego, CA, Sept. 2005. Springer.
20. A. Üngör. Off-centers: A new type of Steiner points for computing size-optimal guaranteed-quality Delaunay triangulations. In *Proceedings of LATIN*, pages 152–161, Buenos Aires, Argentina, Apr. 2004.
21. A. Üngör. Quality triangulations made smaller. In *Proceedings of the European Workshop on Computational Geometry*, Technische Universiteit Eindhoven, Mar. 2005. Accessed online at http://www.win.tue.nl/EWCG2005/Proceedings/2.pdf on April 28, 2007.
22. D. F. Watson. Computing the n-dimensional Delaunay tesselation with application to Voronoi polytopes. *Computer Journal*, 24:167–172, 1981.

1A.3

SVR: Practical Engineering of a Fast 3D Meshing Algorithm*

Umut A. Acar[1], Benoît Hudson[2], Gary L. Miller[2], and Todd Phillips[2]

[1] Toyota Technological Institute
[2] Carnegie Mellon University

Summary. The recent Sparse Voronoi Refinement (SVR) Algorithm for mesh generation has the fastest theoretical bounds for runtime and memory usage. We present a robust practical software implementation of the SVR for meshing a piecewise linear complex in 3 dimensions. Our software is competitive in runtime with state of the art freely available packages on generic inputs, and on pathological worse cases inputs, we show SVR indeed leverages its theoretical guarantees to produce vastly superior runtime and memory usage. The theoretical algorithm description of SVR leaves open several data structure design options, especially with regard to point location strategies. We show that proper strategic choices can greatly effect constant factors involved in runtime.

1 Introduction

At last year's IMR conference we introduced a new meshing algorithm, Sparse Voronoi Refinement (SVR), which provided the typical guarantees for theoretical meshing algorithms, along with an unusual one that the algorithm ran in near-linear time [HMP06]. The goal in designing SVR was to create a meshing algorithm that was similar in implementation and style to many widely used meshing algorithms, but with the added benefit of very strong worst-case bounds on the runtime complexity and space usage. An additional achievement of SVR is that the algorithm can work in any fixed dimension d. More recently, we proved that the algorithm can be run in parallel, and we showed that the the Li-Teng sliver removal algorithm could easily be incorporated into the SVR framework [LT01, HMP07].

The main goals of the present work are twofold. First, to show preliminary results on a new implementation of the sequential version of this algorithm. Second, to discuss new data structures to empirically improve the run time of the point location parts of the algorithm, which may be of more generally applicable use. We focus on point location because both in theory and practice, the dominant cost of SVR and most other algorithms for meshing is the point location cost.

* This work was supported in part by the National Science Foundation under grants CCR-0122581 and Intel faculty gift.

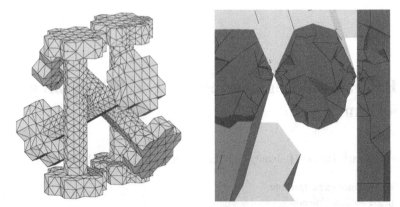

Fig. 1. Left: A radius/edge quality 2.0 mesh of an assemblage of four hexagonal dumbbells meshed by our software. Each dumbbell is described as a set of facets defining the two ends and the connecting rod. The output is a set of tetrahedra that fill space and the barbells while resolving the input. For visual effect we removed those tetrahedra in a standard postprocessing step. **Right:** detail on the nearest approach of three of the dumbbells. Notice that the tetrahedra grade smoothly away from the pinch point.

We compare SVR with two related codes: Pyramid by Shewchuk [She05a, She98], which is available by request to him; and TetGen by Si [Si07, Si06], which is available online. Our implementation of SVR compares very favorably to them, generating meshes that are of similar quality and size in less time. Furthermore, on pathological examples, prior codes run out of memory even at small input sizes whereas SVR sees no difficulty. Our implementation is available at http://www.sparse-meshing.com, free for the research community.

SVR produces a quality conforming mesh that is size-optimal in the number of vertices [HMP06]. One important concern in quality meshing is how we define the quality of tetrahedron. We use two separate measures of element quality in the algorithm and code. Both rely on the *circumball* of a tetrahedron, the smallest ball containing the tetrahedron's vertices. We denote it's radius (the *circumradius*) as R. The first quality measure we use is the radius-edge ratio of a tetrahedron: we we compare the circumradius R versus the shortest edge e of the tet. In a good-radius-edge tetrahedron, R/e must be less than some value ρ. In three dimensions, this metric is somewhat lacking, as it can admit poorly-shaped *slivers*. Radius-edge is still useful however, since in a good radius-edge mesh, every vertex has bounded degree [MTTW99].

The other quality measure is the radius/radius ratio: we compare the circumradius R to the radius r of the largest ball *inscribed* by the simplex. The ratio R/r of a good quality element must be less than some value σ. This quality criterion does not admit slivers, and is thus the one desired for output.

In the finite element method, an element with good radius/radius ratio is known to be numerically good under standard assumptions. The reason we use both measures is that is it not well understood how to avoid creating slivers during the algorithm (though we do scour them from the final output). The SVR algorithm provably never creates excessively bad radius-edge simplices even in the intermediate stages of the algorithm.

Time and Space Usage: Our code takes as input a Piecewise Linear Complex (PLC) [MTTW99]. Let n be the total number of input features (vertices, segments, polygons, etc). Let L/s be the *spread* of the input, i.e. the ratio of the diameter of the input space to the smallest pairwise distance between two disjoint features of the PLC.

SVR has worst case runtime bounded by $O(n \log L/s + m)$, where m is the number of output vertices. This runtime bound is a vast improvement over prior meshing algorithms for three and higher dimensions. For almost all interesting inputs, this bound is equivalent to $O(n \log n + m)$, which is optimal (using a sorting lower bound). SVR also has optimal output-sensitive memory usage $O(m)$, which means that even on pathological inputs it can process moderately large inputs entirely in memory.

2 Related Work

There have been several different approaches to the meshing problem. The idea of generating a mesh whose size is within a constant factor of optimal was first considered by Bern, Epstein, and Gilbert [BEG94] using a quadtree approach. A 3D extension was given by Mitchell and Vavasis [MV00], who later released an implementation under the name of QMG [Vav00].

Chew introduced a 2D Delaunay refinement algorithm [Che89] and showed termination. The quality of the initial triangulation was improved by adding the circumcenters of poor quality triangles as extra vertices. This produced a mesh with no small angles, but inserted many more new vertices than necessary. Ruppert [Rup95] extended this idea of adding circumcenters for 2D meshing to produce a mesh that was within a constant factor in size from the optimal and also handled line segments as input features. Shewchuk implemented Ruppert's algorithm in the very popular Triangle code [She05b], which has since been extended with various enhancements and remains actively maintained.

The extension of Ruppert's algorithm to 3D has been ongoing research. Some methods assume that that Ruppert's local feature size function is given [MTTW99]. Others refine a bad radius-edge ratio mesh directly [She98, MPW02]. These methods by themselves do not give quality meshes because they include slivers; a large number of other techniques have been concocted that aim to eliminate slivers while only slightly (at most linearly) increasing the output size. A 3D version of Ruppert's algorithm in conjunction with a sliver-eliminating post-process produces a quality, optimal-sized mesh. Shewchuk has implemented his higher-dimensional version of Ruppert's algorithm in Pyramid [She05a], but it has not yet achieved an official release and remains in alpha stage.

Of the many other algorithms for Delaunay refinement in 3D that have been proposed [She98, MPW02, CP03, CD03], none except SVR have eluded nontrivial runtime analysis. Trivial runtime bounds such as $O(m^3)$ be found in most cases. Simple examples can usually give bad worst-case performance for naive implementations of

SIMPLIFIED-SVR
1: **while** Volume Mesh is not Conforming **do**
2:      REFINE a non-conforming tet
3:      **while** Volume Mesh is not of Quality ρ **do**
4:          REFINE a poor-quality tet
5:      **end while**
6: **end while**

REFINE
7: **if** FINDWARPPOINT **then**
8:      Insert a vertex from some nearby feature
9: **else if** FINDENCROACHED **then**
10:     Recursively REFINE all encroached lower-dimensional feature meshes
11: **else**
12:     Add a circumcenter as a new vertex in this Mesh
13: **end if**

Fig. 2. A very loose description of a much simplified SVR. Most of the runtime work is spent on point location in evaluating lines 7 and 9 (See Section 8). More detailed pseudo-code and the full algorithm can be found in [HMP06].

these algorithms. As mentioned, they will all suffer from intermediate size $\Omega(n^2)$ in the worst case.

2.1 Optimal Time Meshing Algorithms

Finding refinement algorithms that have provably good run times has also been of interest. Spielman, Teng, and Üngör [STÜ07] proved that Ruppert's and Shewchuk's algorithms can be made to run in $O(\lg^2 L/s)$ parallel steps. They did not, however, prove a work bound. Miller [Mil04] provided the first sub-quadratic time bound in 2D with a sequential work bound of $O((n \lg \Gamma + m) \lg m)$, where Γ is a localized version of L/s (in particular, $\Gamma \leq L/s$). In addition to an $O(n \lg L/s + m)$ sequential runtime, the SVR algorithm has been shown to be parallelizable to run in $O(\lg L/s)$ iterations with the same work [HMP07].

3 Overview of SVR

For a detailed description of SVR, refer to [HMP06]. Herein, we briefly review SVR for the purpose of describing the implementation needs of the algorithm. Coarse pseudocode for SVR is described Figure 2. The input is given as a piecewise linear complex (polygonal faces, segments, and vertices), that we will refer to as a set of *features*. Constraints on the input are described further in Section 4. Over the life of the algorithm, SVR maintains a separate mesh for each feature. A three dimensional volume mesh is maintained covering the entire domain.

In SVR, these feature meshes begin very coarsely, and initially do not conform to the input at all. The meshes are then gradually refined for two reasons: to maintain quality, and to eventually conform. Refinement is done by attempting to insert the circumcenter of a poor quality tetrahedron or of a non-conforming tetrahedron. But because the mesh does not yet conform, we may instead choose to WARP, and insert a vertex from a nearby input feature instead of the circumcenter. We do this to make sure that new points are not created to close to the features or close to input points. Additionally, before the volume mesh can insert near a feature, we may first need to refine the feature itself. This is accomplished by protecting the feature with *protective balls* that cannot be entered by the volume mesh. If the volume mesh desires to insert a circumcenter that *encroaches* upon a lower-dimensional protected ball, it first *yields* to the lower-dimensional feature, and allows the latter to refine itself. After the lower-dimensional feature has refined, the volume mesh may warp to the point that was inserted into the lower-dimensional mesh. Facet meshes operate similarly with respect to their bounding segments: if a facet mesh wishes to eliminate a bad triangle (because it has poor quality, or because the volume mesh encroached upon it), it attempts to insert the circumcenter of the triangle, possibly yielding to encroached boundary segments and/or warping to points not yet resolved in the facet mesh. A putative new circumcenter may encroach on many protective balls simultaneously; if so, for good runtime we must make sure all the affected features refine themselves.

4 Input Format

The input format is that of Shewchuk's Pyramid, which is the obvious extension of the Triangle format to the third dimension: An input file starts with a header listing how many inputs vertices there are, then lists their coordinates. If there are segments to conform to, it then lists their number and describes each as a pair of vertex IDs. Finally, if there are polygons, it lists their number and describes each as a list of segment IDs. Holes and concavities are automatically found and need not be mentioned in the input.

For SVR to properly produce a quality mesh, users must somewhat restrict their input. As usual, our mesher needs clean inputs: the code will report an error if a polygon is not watertight, or if two polygons intersect each other geometrically but no intersection is mentioned in the input. More importantly, the algorithm sometimes also fails if two input polygons intersect at an angle less than 90°, even on clean input: the problem is that refinement on one of the faces may encroach on the other, which may loop back to encroach on the former. We can tell the mesher not to refine an element smaller than some minimum size, which will return us a mesh, albeit with possibly some bad elements. It remains active research to solve these issues in a more principled but practical manner.

Our runtime proofs also require additional properties of the polygonal features: (1) polygons must be defined by only a constant-bounded number of points, (2) the initial triangulation of each polygon must be of good quality, (3) the initial triangulation must not self-encroach. Violating any of these requirements impinges upon the runtime properties, but the code will still properly return an optimal-size, good-quality mesh. Polygons with n_i vertices on their boundary currently take time $O(n_i^2)$ to preprocess; it would not be hard to reduce this to $O(n_i \lg n_i)$ time, though it seems unnecessary for all inputs we have on hand since generally n_i is on the order of 4–20. The second and third condition are repaired automatically by adding more points to the boundary and interior when creating the initial triangulation; this causes the code to run only logarithmically slower than optimal.

Constant	Constraints	Default Value
Output Radius-Edge, ρ	$\rho > 2$	2.0
Output Radius-Radius, σ	$\sigma \gg 3$	not set
Sliver Growth, B	$B > 2$	3.0
Perturbed Insertion, δ	$0 < \delta < 1$	0.1
Yielding Ratio, k	$0 < k < 1$	0.9

Fig. 3. Table of constants for SVR.

5 Algorithm Constants

Several constants for SVR are left to be chosen by the user. Figure 3 gives an overview of these constants and their defaults in our implementation.

The first constant, ρ, determines the radius-edge quality bound on every element in the final output mesh. For most numerical methods using the mesh, ρ will have a strong effect on the quality and runtime of the method used, with a lower ρ (better shaped elements) being preferred. Of course, driving ρ lower will necessarily increase the number of elements output by any meshing software, SVR included.

Many numerical methods also require the elimination of slivers [ELM+00]. We have extended SVR to eliminate extreme slivers using the Li Teng sliver removal algorithm [LT01]. The algorithm randomly perturbs circumcenter insertions by a factor of δ to ensure that any new slivers created must be larger by a factor of B, then recursively works on the larger slivers. Eventually no larger slivers can be created, and so the mesh is sliver-free. The constant σ gives an upper bound on the radius/radius ratio of any output tetrahedron. Like all quality bounds, tightening σ (reducing it) will increase output size. The published proofs [LT01, Li03] suggest the settings of δ and B listed above; unfortunately, for σ they yield a gargantuan number. Therefore, by default the sliver-removal code is off. However, the proof is known to be very loose, and in practice we can eliminate much worse slivers than is provable. See the experiments in Section 9.

5.1 Warping parameter

One of the more interesting constants to specify for SVR is the *warping parameter*, k. This parameter controls the behavior of the warp operation: when we decide to split a simplex s, we will warp to a point within distance $k \cdot R(s)$ of the center of s if there is one. A higher value of k will more aggressively look for a vertex to warp to, which intuitively would help reduce the output mesh size. However, a higher value of k also worsens the intermediate quality bound, which allows high-degree vertices and thus hurts runtime. This section discusses the tradeoff between k, runtime, and output size, and aims to find a reasonable default setting for k.

First, in our original paper, we proved that if $\rho k^3 > 2$, our algorithm will terminate with the aforementioned guarantees. This somewhat limits the user's choice of parameter settings. However, it is easy to show that we can internally use a ρ' that satisfies the requirement, and, once all points have been inserted into the mesh (and therefore k is rendered irrelevant), we can improve the mesh quality up to ρ. In other words, the user may ask for any k, and any $\rho > 2$.

Under this framework, we can then ask about the optimal setting of k in terms of mesh output size, or in terms of runtime, leaving ρ fixed. We ran an experiment on the point-cloud refinement code, tracking runtime and output size versus k. See Figure 4. The main finding is that in practice, the output size is indeed quite strongly affected by the value of k, whereas runtime is much less affected until k becomes very close to 1. Based on these results, we set the default k value to 0.9 as being a reasonable tradeoff between output size and runtime.

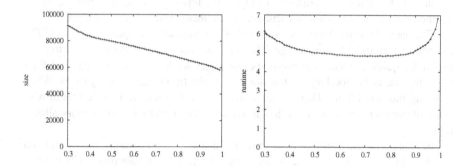

Fig. 4. *Left:* Number of vertices SVR outputs for the Stanford bunny, versus k. *Right:* Runtime of SVR on the Stanford bunny in arbitrary units, versus k. Notice the tradeoff between mesh size and runtime. Other natural and pathological examples showed similar tradeoffs.

Another experiment we ran was to see how to best choose the point to warp to. When there are many choices, it may be that one is better than another. In particular, we expected that it would be best to choose the point closest to the prospective Delaunay center among all available points to warp to; the intuition is that this is, in a sense, the "median" and we are, in a sense, sorting. Whether this is a good idea depends on the order in which valid warp points will be seen. In early implementations, the difference was substantial. However, in the current implementation, with the Voronoi shelling approach described in Section 8.4, the first point we find is far from any vertex, making it a good choice. Therefore, we currently report the first point we can safely warp to.

6 Implementation in C++

We implemented our algorithm in C++, in the hopes of being most accessible to the community, but still highly efficient. Almost all of the code is general-dimensional; the exception is in the numerical predicates, where we use Shewchuk's specific-dimensional code [She97]. While the code is written in a general dimensional style, we use the C++ template mechanism to choose the dimension at compile-time. That is, an SVR<2> is a mesher in two dimensions, while an SVR<3> is in three dimensions. This allows the compiler to unroll loops, perform constant propagation on the dimension term, and otherwise optimize the code to achieve almost no overhead over a specific-dimensional code. The data structures are accessible via an API; we expect this will allow more easily experimenting with post-processes and variations on our algorithm.

7 Mesh Data Structures

Using any reasonable data structures, SVR will be faster than its competitors on pathological examples. However, to also be useful on more normal inputs, we must be somewhat careful. We implemented SVR in a way that sharply separated the implementation into four modules, so that we can easily change various pieces one by one: (1) basic data structures and I/O, (2) the topological mesh structure; (3) the geometry code and Delaunay structures; (4) the mesh refinement algorithm.

The basic data structures are essentially C++ Standard Templates Library (STL) structures optimized for our usage. The STL and the Boost libraries are extremely useful for quickly generating prototype code, but we have seen that many of their structures are either too large or too slow for production usage in many cases. When profiling indicated that STL libraries were limiting factors, we replaced them with some of our own versions, in particular to have greater control over memory allocation.

For the topological structure, we currently use a standard pointer-based simplicial complex (4 vertex pointers and 4 neighbour pointers per tetrahedron). Client code can optionally attach additional data using a template argument; the type of vertices is itself also specified by a template argument. Simplices are reference-counted to avoid complications and memory bugs when simplices are destroyed yet remain in the SVR work stack. Access is typically via a generic depth-first-search routine.

The API has changed little as we tried various types of structures; we therefore predict that, with relatively little effort, we could scale our implementation to much larger sizes than tested here simply by replacing the underlying structure with a compressed mesh structure [Bla05]. The topological structure is also largely compatible with the CGAL structures; one could write a thin adapter class to view one structure as the other. We chose to write our own structure in order to have full control over the implementation.

The Delaunay triangulation structure is a thin veneer over the simplicial complex. SVR computes the circumcenter and circumradius of every simplex (to test for quality), and does so repeatedly: simple empirical evidence shows about 10 times per simplex in 3d. Therefore, we compute these values only once, and cache them as data attached to each simplex. An LRU cache can easily be substituted for relatively little runtime overhead to reduce the per-simplex memory usage; this will become critical if we use a compressed mesh structure.

The geometric code is largely drawn from Shewchuk's notes on geometric calculations [She99] and from his public-domain library [She97, NBH01]. We use LAPACK for some calculations, but have found that for very low-dimensional operations, which dominate our geometric requirements, the cost of marshaling and unmarshaling data between our code and LAPACK dwarfs the cost of the numerical calculations, and it is thus advisable to write them by hand.

The point location structures of the mesher are the most performance-critical ones; we describe them separately in great detail in Section 8.

8 Point Location Data Structure

A point location data for SVR provides a good deal of room for variation in implementation. SVR requires that every time a vertex is considered for addition into the volume mesh, SVR locally searches for any lower dimensional feature meshes, so that the volume mesh might conform to these instead. Additionally, SVR searches for lower dimensional features that may need refinement before it can proceed with this volume mesh insertion. Essentially, a correspondence must be maintained between the volume mesh and the feature meshes because they do not conform to one another. In the following section, we describe implementing such a correspondence.

Recall that SVR runs in $O(n \lg L/s + m)$ time. The $n \lg L/s$ term is driven by the cost of two operations: finding a point to warp to, and testing for encroachment. We conflate these operations and call them both *point location*. The m term also includes these operations, plus some other work. Therefore, point location is the leading term in the asymptotics on most inputs, and even when it is not, it is a major constant factor. Profiling information on various inputs verifies that this prediction holds in practice in the current implementation: well over 20% of the runtime – and 33% of the cache misses – are directly due to point location, even ignoring any ancillary costs (malloc/free; memory overhead; cache evictions causing slowdowns elsewhere; etc).

The PLC input is described using a standard *incidence poset*: segments know their endpoints and any internal vertices, polygons are described using a set of boundary segments (and possible internal segments or vertices), etc. For simplicity, we compute the transitive closure of the poset, linking facets to vertices, etc. The structure of this linking between meshes is shown in Figure 5.

To quickly handle point location queries, we need to keep track of intersections between elements of one mesh and elements of the mesh of a lower-dimensional feature (and vice-versa). The published SVR algorithm used the Voronoi cells of the mesh vertices as elements, to make the proofs most succinct. Here we report on the differences between several choices, all asymptotically equivalent but with significant constant-factor differences.

We formalize this notion: we maintain a bipartite map between the abstract types of an Upper container and a Lower cell. See Figure 6. At the moment, we only have two Lower types: circumspheres (for features of dimension 1 and higher) and points (for features of dimension 0). Points are of course merely special cases of sphere with radius 0, but they are sufficiently special to merit distinct consideration. This formalism involving Upper and Lower is likely to extend to more interesting inputs such as curves.

The point location data structure will need to perform the following three operations quickly:

FINDWARPPOINT: Given a Delaunay simplex s in a mesh M, determine whether there exists an point to be inserted in M (a child of M in the poset) that lies within the warp ball $B(c(s), k \cdot R(s))$.

FINDENCROACHED: Given a point p chosen by FINDWARPPOINT for s, determine the (possibly-empty) set of lower-dimensional protective balls b_i that p encroaches – that is, p lies in b_i.

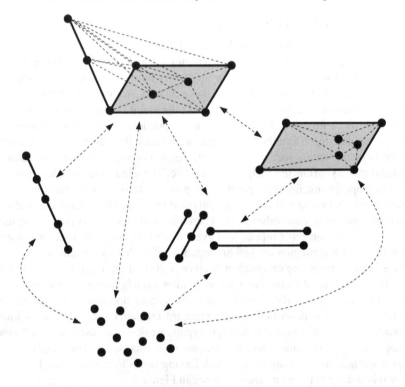

Fig. 5. The volume mesh at the top of the figure has features consisting of one facet, five segments, and several input points. The point location structure must link all of the meshes along the arrows shown. Note that these meshes may not all conform to each other; input points or feature refinements may not be resolved in the volume mesh. Linking two meshes involves tracking the intersections between every lower dimensional element and every higher dimensional element.

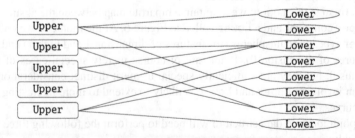

Fig. 6. We maintain a map, tracking intersections between Upper containers and Lower cells.

UPDATE: Given a point p chosen by FINDWARPPOINT in a mesh M, and given the set of simplices before (C, the *cavity*) and after (S, the *star*) inserting p, update the point-location structures.

We compare four different choices of Upper in the following subsections. We have experimented with all four for maintaining the point location structure for points. For maintaining protected balls, we so far have only used the first method; trying other techniques remains future but high-priority work.

8.1 Circumballs as Upper

The easiest way to implement FINDWARPPOINT and FINDENCROACHED is to use the circumball $B(c(s), R(s))$ of each simplex s as the Upper elements. Indeed, any point that we may warp to is in $B(c(s), k \cdot R(s))$, a strict subset of the circumball. Similarly, any lower-dimensional feature encroached by a point chosen by FINDWARPPOINT necessarily intersects the circumball.

UPDATE is only slightly more complicated: to compute the Lower s intersecting a new simplex s, we must find every lower-dimensional protected ball that intersects the circumball of s. It is a standard fact that the set of circumballs of the cavity C (the simplices before inserting p), plus the neighbours C_N of the cavity, covers the set of new circumballs. Thus, we can compute the Lower s intersecting $C \cup C_N$ and for each new simplex, filter this set to compute those intersecting the circumball of s.

8.2 Tetrahedra as Upper

When maintaining the uninserted points of a mesh, the approach in the prior section stores a point repeatedly since the circumballs of the mesh intersect. If instead we use the simplices directly, we avoid duplicates; this is the basis for in-simplex point location being the most traditional kind.

The lack of overlap between regions greatly speeds up the UPDATE cost on contained points: we can perform about half as many intersection queries on average since we can shortcut execution at the first simplex of the star that matches the vertex. Furthermore, the star and cavity cover the same area, so we need only relocate the Lower s intersecting C, ignoring the neighbours. Finally, computing the set of Lower vertices interesting C is easier since we need not eliminate duplicates.

The cost of this approach is that now in FINDWARPPOINT(s) we must find all the simplices that intersect the circumball of s. There are only a constant number of these, but finding them all is still expensive. Nevertheless, note that every simplex ever created is involved in an UPDATE call, whereas only a small minority (empirically about one in 18, in three dimensions) ever have FINDWARPPOINT called on them.

For maintaining lower-dimensional features, we do not gain the benefit of not having duplicates, which makes this technique unlikely to be useful. For uninserted points, however, this point location structure is substantially faster than in-sphere in our experiments.

8.3 Voronoi Cells as Upper

Traditionally, the Delaunay triangulation is the most common structure to use for point location, which made it the natural choice for the implementation. However, determining whether a point p is owned by a simplex s is expensive: either an in-sphere or an in-simplex test, both of which are essentially determinants of matrices of rank d, which becomes a major cost in UPDATE. Another choice is to use the dual

Voronoi diagram, as described in the original paper. A point p is in the Voronoi cell of a mesh vertex v if v is the nearest neighbour to p. If we know the old Voronoi cell in which p lies (call it $V(v)$), then updating after the insertion of one mesh vertex v' requires only two distance calculations: the distance $|p - v|$ and the distance $|p - v'|$. Furthermore, there are many fewer vertices than simplices, by a ratio of, empirically, 1:6 in three dimensions (both in the experiments we ran, and in prior reports [Bla05, *e.g.*]). Thus, the overhead of the lists is greatly reduced compared to using the simplices as the basic objects, which becomes important near the end of the algorithm.

Searching the Upper Voronoi cells for FINDWARPPOINT has a disadvantage: we must search every Voronoi cell that intersects the warp ball. We can do this by searching the set associated with every vertex on every simplex whose circumball intersects the warp ball. In terms of implementation, this is only a slight increase in code compared to using the interior of simplices; but this is a much larger area being searched, so we may visit many more vertices that are not in the warp ball. In practice, we found this technique to be equal in runtime to using simplices for point location, although the memory usage is slightly reduced.

Beware of one pitfall when using Voronoi cells: Unlike simplices, vertices have long lifetime. Therefore, if each vertex has its own memory pool for the list of uninserted points in its Voronoi cell – or uses the STL STD::VECTOR class, which never shrinks –, vertices inserted early in the run of the algorithm will reserve a large amount of memory even when they no longer need to track many vertices. This is easily avoidable by actually releasing memory when it is no longer needed, or by using a single memory pool shared among all vertices.

8.4 Shelled Voronoi Cells as Upper

When looking for a point to warp to, we take a Delaunay ball and shrink it by a factor k. Therefore, any input point that is "near" a vertex need not be examined at all. However, the previously-described approaches take no note of this. An easy way to implement this is to store uninserted points in the Voronoi cells, as described above, but within each Voronoi cell, bucket the points according to distance from the Voronoi site – that is, into concentric shells of geometrically increasing radius (see Figure 7). Upon a FINDWARPPOINT query, we ignore any bucket that lies entirely outside the query region; similarly, on UPDATE we do not try to reassign points in buckets that are closer to the old vertex than to the new. Asymptotically speaking, this converts $O(\lg L/s)$ distance calculations on an uninserted point chosen very late, into $O(1)$ distance calculations and $O(\lg L/s)$ divide-by-two operations. Indeed, in practice we saw the total runtime of the algorithm fall by half when we implemented shelling of Voronoi cells as compared to either Voronoi cells or in-simplex point location.

8.5 Further Refinements

As noted earlier, using the Voronoi diagram has the disadvantage that uninserted points rather far from the query ball can be accessed. Shelling does not reduce this tendency: shelling only means that uninserted points very close to a vertex will be

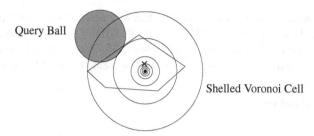

Fig. 7. Illustration of shelling and of a query on a shelled Voronoi cell. The shells are concentric around the Voronoi site, with radius halving at each step toward the center. While all the shells exist mathematically, in the implementation we only store those that include at least one uninserted point. During FindWarpPoint, the uninserted point marked by an X is not visited because the annulus that contains X does not intersect the query ball.

ignored. Another view of the problem is that we now understand well how to store points near the mesh vertices, but the best strategy for storing point far from vertices is not yet clear. One possibility is a hybrid, using shelling for points close to mesh vertices, and in-sphere or in-simplex for points far from the vertices. Another possibility is to locate in the Voronoi cells of both the mesh vertices and the element circumcenters.

We have identified three times when we can completely avoid calling FindWarp-Point. Clearly, if there are no uninserted vertices remaining anywhere, we can simply skip the call. This occurs in about 10% of the insertions on the bunny dataset, 30% on the pathological input, and merely requires keeping a global or per-mesh counter. If we are retrying an insertion that caused a small sliver, we know we will not warp since the previous iteration would have warped if needed. How often this occurs depends on the sliver parameters. Finally, when we insert a point due to a crowded Steiner point, we can blindly insert any uninserted vertex in the Steiner point's Voronoi cell. This happens 5% percent of the time on the bunny dataset, almost never on the pathological input. Together, then, we see that about 15-30% of split operations can avoid this call.

9 Experiments

We performed some experiments on our implementation to determine the runtime and output size of our algorithm as compared to a few other equally-available codes. The experiments are on point-cloud inputs: as of this writing, the point-cloud code of both SVR and Pyramid is more mature than the feature-set code. We ran the experiments on a desktop 3.2 GHz Pentium D with 2 GB RAM running Linux 2.6. We compiled all applications using the gcc compiler version 4.2.1 with compiler flags `-m32 -O2 -g -fomit-frame-pointer -mtune=native -DNDEBUG` for both C and C++, except for one file in TetGen which cannot be optimized and had to be compiled `-m32 -O0`. All three codes currently assume 32-bit pointers, though this should be easy to fix in all cases. We compare our implementation (SVR) to Pyramid 0.50 [She05a] and to TetGen 1.4.2 [Si07]. For all meshes here, we required an output radius/edge quality of 2.0; unless otherwise noted, SVR did not perform sliver removal in these experiments. SVR used a value of 0.9 for its k parameter (see Section 5); Pyramid and TetGen use default values. We measure time by using the UNIX 'time' utility and summed "user" (cpu) and "system" times. All reported times are averaged over five runs.

9.1 Point cloud results

Input	SVR	Pyramid	TetGen	SVR	Pyramid	TetGen
Stanford Bunny ($n = 34890$)	4.62	6.35	12.4	59702	59040	74269
Line & Circle ($n = 2000$)	0.80	4.79	6.5	12119	14003	14573
Line & Circle ($n = 20000$)	7.62	N/A	N/A	120933	N/A	N/A
$50 \times 50 \times 50$ Grid ($n = 125000$)	11.30	15.96	45.9	129839	129929	130140
$100 \times 100 \times 100$ Grid ($n = 10^6$)	97.71	179.04	400.3	1016262	1017799	1018684

Table 1. Comparison of the SVR, Pyramid, and TetGen codes on a few point-cloud inputs. Both Pyramid and TetGen ran out of memory on the $n = 20000$ Line & Circle example, and could not complete; otherwise, all examples fit in memory. **Left:** Execution times (seconds of CPU plus system time) versus inputs. Average of 5 runs. **Right:** Output size, in vertices. All three methods produce meshes of approximately the same size.

Table 1 shows a comparisons of timings and output sizes for TetGen, Pyramid, and SVR on three inputs: (1) the Stanford Bunny, (2) line-and-circle (a pathological quadratic example), and (3) points on a grid. All these inputs are then bounded by a $6 \times 6 \times 6$ bounding box to avoid all boundary effects. SVR is the fastest on all inputs, increasingly so as complexity increases. Furthermore, as predicted by theory, both Pyramid and TetGen crash on very modestly-sized pathological inputs. Even with inputs of just 20,000 points, they can be made to try to allocate more memory than can be addressed with 32-bit pointers.

TetGen has a performance bug for meshing point clouds: it requires the user first invoke TetGen to produce and output a Delaunay mesh; then invoke TetGen again to load that mesh and refine it. On non-pathological input, this is a low-order effect. However, on pathological input the reloading of the mesh runs in time cubic in the number of vertices. This is a front-end issue and is not fundamental, so we subtract out the time of the intermediate output and reload in the TetGen times we report.

The number of vertices and tetrahedra output by all three codes is similar. During the development of SVR, we found that relatively minor changes in the code can often change the output size by 20% in either direction (often in opposite directions for different inputs), but not reliably so. As noted by Har-Peled and Üngör [HPU05], one of the few consistently useful heuristics is to work on the smallest simplex first, as measured by its shortest edge; in our algorithm, this happens to also improve cache efficiency since the smallest simplex is usually the most recently-create one.

Finally, we investigated the issue of slivers. On the bunny, radius/edge refinement using either SVR or Pyramid created several thousand simplices with dihedral angles flatter than $175°$; in both, dihedral angles ranged from $0.01°$ to $179.98°$. TetGen has a separate mesh optimization post-process which produced dihedral angles from 4.64° to 168.56°. In our implementation of SVR with Li-Teng sliver removal, we find we can achieve angles ranging only between 12.09° to 154.21° on the bunny when asking for radius/radius quality of 9.0 (recall from elementary geometry that an equilateral tetrahedron has radius/radius ratio 3.0) . Unfortunately, Li-Teng only provably works with very weak guarantees: at the best quality we demanded, the technique is very brittle and often fails depending on small changes in the parameter settings. This suggests two avenues for improvement: tweaking Li and Teng's algorithm to produce more practical bounds so that we need never go beyond the provable envelope, and/or the addition of a mesh optimization post-process to the SVR code.

9.2 PLC results

The vast majority models available in mesh repositories – even those that use the Pyramid input format we adopt – are triangulated surface meshes. Unfortunately, this is automatically an illegal input: the edges of a triangle obviously do not all meet at non-acute angles. TetGen merges adjacent triangles that are (nearly) coplanar; time did not permit us to implement this type of input grooming. Instead, we ran SVR on several PLC examples provided to us by Jonathan Shewchuk. None are strictly legal according to our input restrictions (see Section 4); in particular, they all have facets meeting at acute angles. Nevertheless, we were able to produce quality meshes of about the same size as Pyramid's output. We also synthesized our own examples, such as the ten-barbells example shown in the introduction, which satisfy all the theoretical requirements. On these examples, we were also able to remove slivers with dihedrals worse than about 5° or 175°, albeit not entirely reliably. The major outstanding issues in the feature-set SVR code come down to implementing some of the tricks that TetGen has to work with input that is, strictly speaking, illegal; and to speed up the provably correct code using many of the techniques we have discussed and used in the point-cloud code.

9.3 SVR profiling

We made heavy use of profiling (particularly using the Valgrind toolset) while optimizing the underlying data structures for point-cloud refinement. Similar optimization of the feature-set refinement code remains future work. On various examples, we find the following trends. First, a large fraction (about 20%) of the time is spent writing the output to an ASCII file. Applications for which time is critical will output in a faster and smaller format, so let us ignore this cost. On the platforms we tested on (Intel, AMD, and PowerPC), we find the processor issues about 0.7 to 1.0 instructions per clock cycle, which is in line with what we would expect from an unstructured code. The cache performance was good: essentially no I-cache misses, while 1.5% of data reads reach to L2 and only 0.3% to main memory. This can be explained by the fact that SVR is fundamentally parallel [HMP07]. Therefore, merely by using a work stack instead of a work queue, we achieve good data locality. Another illustration of this is that SVR could maintain about 4% CPU utilization in tests where the mesh did not fit in memory. By contrast, Pyramid under the same conditions suffers a 0.6% miss rate to main memory, and achieves less than 1% CPU utilization when it hits swap.

On the Stanford bunny example, the code issues about 5.5 billion instructions. Three major components each take almost exactly one billion instructions each. (1) Calls to Shewchuk's numerical predicates library, mostly in-sphere tests. (2) Computing circumcenters, and performing distance calculations for point location. (3) Topologically updating the Delaunay triangulation. The remainder of the time is taken up traversing the mesh to perform point location queries, maintaining the work queues, allocating memory and maintaining reference counts, and reading the input. We note that it is critical that we use memory pools to allocate small objects (list nodes, simplices, and so on).

Major improvements in point-cloud meshing time will require three improvements. Most prosaically, we need to make our code run on 64-bit architectures to take advantage of large memories; as the mesh size grows, we expect our lead over previous codes to widen. Next, using a compressed mesh structure [Bla05] would improve cache performance and allow us to mesh much larger examples entirely in memory. Finally, we cache circumcenters because we would otherwise repeatedly recompute the circumcenter; but the same effect is true of in-sphere tests. It is highly likely we could substantially reduce the cost of category (1) above by merging it with category (2). Major improvements in feature-set meshing time are low hanging fruit at the moment. After some more work to eliminate any remaining factors of two in sequential runtime, we expect eventually to look to parallelize the code, as we have theoretically proved can be done.

10 Conclusions

We have shown that it is possible to implement the theoretically described SVR algorithm to achieve practical robust software. The strong asymptotic runtime guarantees underlying the SVR algorithm are clearly evident on simple examples. Performance is competitive with existing codes on relatively well shaped inputs. For pathological inputs, the superiority of SVR is unquestionable.

In the SVR implementation, we have combined strong theory with a great deal of software engineering. Devising and implementing appropriate data structures was crucial to implementing practical software. As well, attention to geometric predicates is necessary for any robust code.

10.1 Extensions

The most obvious extension of the SVR code is to handle a richer set of input descriptions. Modern geometries require a mesher that can handle curved input surfaces and sharp corners; SVR on the other hand requires input to come from a highly restricted class (as do the algorithms underlying Pyramid and TetGen).

One of the elegant features of our SVR implementation is that although it is tuned for performance in three dimensions, the data structures are versatile enough for use in four or more dimensions, which is of possible use in emerging space-time meshing applications. The only routines we need for a higher dimensional implementation are fast yet robust in-sphere predicates in higher dimensions. Automated techniques for generating these predicates are known [NBH01], but the implementation is not currently available. Using exact arithmetic kernels such as are available with CGAL is a possibility, for small problems where the runtime is not critical.

Finally, we close by noting that SVR is known to be parallelizable [HMP07]. At the moment we feel there are still large constant factors to attack in the sequential implementation of the code; but as soon as those have been vanquished, the next obvious step will be to implement a multi-core version of this code. The current implementation has been explicitly geared toward leaving open that possibility.

Acknowledgments

We wish to acknowledge the help of Jonathan Shewchuk in offering many useful tips for the code and geometric predicates, for making available to us pre-release version of Pyramid, and for sending us some sample PLC inputs. We also acknowledge Hang Si's help in understanding TetGen; mesh drawings and mesh quality statistics were generated by TetGen.

References

[BEG94]   Marshall Bern, David Eppstein, and John Gilbert. Provably good mesh generation. *J. Comput. Syst. Sci.*, 48(3):384–409, 1994.

[Bla05]   Daniel K. Blandford. *Compact Data Structures with Fast Queries*. PhD thesis, Computer Science Department, Carnegie Mellon University, Pittsburgh, Pennsylvania, October 2005. CMU CS Tech Report CMU-CS-05-196.

[CD03]   Siu-Wing Cheng and Tamal K. Dey. Quality meshing with weighted delaunay refinement. *SIAM J. Comput.*, 33(1):69–93, 2003.

[Che89]   L. Paul Chew. Guaranteed-quality triangular meshes. Technical Report TR-89-983, Department of Computer Science, Cornell University, 1989.

[CP03]   Siu-Wing Cheng and Sheung-Hung Poon. Graded Conforming Delaunay Tetrahedralization with Bounded Radius-Edge Ratio. In *Proceedings of the Fourteenth Annual Symposium on Discrete Algorithms*, pages 295–304, Baltimore, Maryland, January 2003. Society for Industrial and Applied Mathematics.

[ELM+00] Herbert Edelsbrunner, Xiang-Yang Li, Gary L. Miller, Andreas Stathopoulos, Dafna Talmor, Shang-Hua Teng, Alper Üngör, and Noel Walkington. Smoothing and cleaning up slivers. In *Proceedings of the 32th Annual ACM Symposium on Theory of Computing*, pages 273–277, Portland, Oregon, 2000.

[HMP06]   Benoît Hudson, Gary Miller, and Todd Phillips. Sparse Voronoi Refinement. In *Proceedings of the 15th International Meshing Roundtable*, pages 339–356, Birmingham, Alabama, 2006. Long version available as Carnegie Mellon University Technical Report CMU-CS-06-132.

[HMP07]   Benoît Hudson, Gary L. Miller, and Todd Phillips. Sparse Parallel Delaunay Refinement. In *19th Annual ACM Symposium on Parallelism in Algorithms and Architectures*, pages 339–347, San Diego, June 2007.

[HPU05]   Sariel Har-Peled and Alper Üngör. A Time-Optimal Delaunay Refinement Algorithm in Two Dimensions. In *Symposium on Computational Geometry*, 2005.

[Li03]   Xiang-Yang Li. Generating well-shaped d-dimensional Delaunay meshes. *Theor. Comput. Sci.*, 296(1):145–165, 2003.

[LT01]   Xiang-Yang Li and Shang-Hua Teng. Generating well-shaped Delaunay meshes in 3D. In *SODA '01: Proceedings of the twelfth annual ACM-SIAM symposium* on Discrete algorithms, pages 28–37, Philadelphia, PA, USA, 2001. Society for Industrial and Applied Mathematics.

[Mil04]   Gary L. Miller. A time efficient Delaunay refinement algorithm. In *SODA '04: Proceedings of the fifteenth annual ACM-SIAM symposium on Discrete algorithms*, pages 400–409, Philadelphia, PA, USA, 2004. Society for Industrial and Applied Mathematics.

[MPW02]   Gary L. Miller, Steven E. Pav, and Noel J. Walkington. Fully Incremental 3D Delaunay Refinement Mesh Generation. In *Eleventh International Meshing Roundtable*, pages 75–86, Ithaca, New York, September 2002. Sandia National Laboratories.

[MTTW99] Gary L. Miller, Dafna Talmor, Shang-Hua Teng, and Noel Walkington. On the radius–edge condition in the control volume method. *SIAM J. Numer. Anal.*, 36(6):1690–1708, 1999.

[MV00]   Scott A. Mitchell and Stephen A. Vavasis. Quality Mesh Generation in Higher Dimensions. *SIAM Journal on Computing*, 29(4):1334–1370, 2000.

[NBH01]   Aleksandar Nanevski, Guy E. Blelloch, and Robert Harper. Automatic Generation of Staged Geometric Predicates. In *International Conference on Functional Programming*, pages 217–228, Florence, Italy, September 2001.

[Rup95]   Jim Ruppert. A Delaunay refinement algorithm for quality 2-dimensional mesh generation. *J. Algorithms*, 18(3):548–585, 1995.

[She97]   Jonathan Richard Shewchuk. Adaptive Precision Floating-Point Arithmetic and Fast Robust Geometric Predicates. *Discrete & Computational Geometry*, 18(3):305–363, October 1997.

[She98]   Jonathan Richard Shewchuk. Tetrahedral Mesh Generation by Delaunay Refinement. In *Proceedings of the Fourteenth Annual Symposium on Computational Geometry*, pages 86–95, Minneapolis, Minnesota, June 1998. Association for Computing Machinery.

[She99]   Jonathan Richard Shewchuk. Lecture notes on geometric robustness, 1999.

[She05a]   Jonathan R. Shewchuk. Pyramid, 2005. Personal communication.

[She05b]   Jonathan R. Shewchuk. Triangle, 2005. http://www.cs.cmu.edu/~quake/triangle.html.

[Si06]   Hang Si. On refinement of constrained Delaunay tetrahedralizations. In *Proceedings of the 15th International Meshing Roundtable*, 2006.

[Si07]   Hang Si. TetGen, 2007. tetgen.berlios.de.

[STÜ07]   Daniel Spielman, Shang-Hua Teng, and Alper Üngör. Parallel Delaunay refinement: Algorithms and analyses. *IJCGA*, 17:1–30, 2007.

[Vav00]   Stephen A. Vavasis. QMG, 2000. http://www.cs.cornell.edu/home/vavasis/qmg-home.html.

Construction of Sparse Well-spaced Point Sets for Quality Tetrahedralizations

Ravi Jampani and Alper Üngör

Dept. of Computer & Information Science & Engineering
University of Florida
{rjampani,ungor}@cise.ufl.edu

Summary. We propose a new mesh refinement algorithm for computing quality guaranteed Delaunay triangulations in three dimensions. The refinement relies on new ideas for computing the goodness of the mesh, and a sampling strategy that employs numerically stable Steiner points. We show through experiments that the new algorithm results in sparse well-spaced point sets which in turn leads to tetrahedral meshes with fewer elements than the traditional refinement methods.

1 Introduction

We consider the following three dimensional geometric problem:

Problem 1. [QUALITY STEINER TRIANGULATION] *Compute a small size triangulation of a given three dimensional domain such that all the tetrahedra in the triangulation are of good quality.*

The quality constraint is motivated by the numerical methods used in many engineering applications. Among various criteria, the following two are widely used to describe the goodness of a tetrahedron:

1. *Radius-ratio.* A tetrahedron is said to be *good* if its radius ratio (circumradius over inradius) is bounded from above.
2. *Radius-edge-ratio.* A tetrahedron is said to be *good* if its radius-edge ratio (circumradius over shortest edge length) is bounded from above.

An upper bound on the former criterion implies an upper bound on the latter, but not vice versa. Moreover, a bound on the radius ratio is equivalent to a bound on the dihedral angles of a tetrahedron. In this paper we mainly, focus on the former quality criterion. Point sets whose triangulation has bounded radius-edge ratio are also known as the well-spaced point set [41]. We can easliy meet the latter criterion using a sliver-removal algorithm [5, 12] as a postprocessing step to our algorithm. Under the quality constraint, our objective is to make the triangulation size (the number of tetrahedra) as small as possible for its efficient use in applications.

The quality triangulation problem has been extensively studied both in two dimensions [1, 7, 11, 14, 28, 42] and three dimensions [3, 5, 12, 16, 19, 25, 30]. There are two main techniques that solve the mesh generation problem and provide theoretical guarantees: (i) Delaunay refinement and (ii) quadtree refinement. Among these Delaunay refinement method seems to be more popular due to its superior performance in practice. There are also a number of heuristic solutions addressing the mesh generation problem, such as the advancing front methods [20] and the sphere packing methods [18, 36]. While these algorithmic and heuristic methods are generally effective computing quality triangulations in two dimensions, it is hard to claim the same in three dimensions. There are two main reasons for the shortcomings in three dimensional mesh generation. First, algorithmic solutions for computing optimal triangulations remain as a major open research topic. There is no known algorithm for computing triangulations in three dimensions that maximize the minimum dihedral angle, minimize the maximum dihedral angle, or any other useful optimization criteria. Second, it is hard to come up with a point sampling strategy that leads to a good triangulation. Here, we address the second issue and give a Steiner point selection strategy effective in computing Delaunay meshes.

Delaunay refinement method involves first computing an initial Delaunay triangulation of the input domain, and then iteratively adding points called *Steiner points* to improve the quality of the triangulation. Traditionally, circumcenters of bad simplices are used as Steiner points [28, 30]. Recently, alternative techniques have been studied in two dimenions, with great benefits both in theory and in practice. We elaborate more on this in the next section.

In this paper, we present a new tetrahedral mesh refinement algorithm, which relies on the following main ideas:

1. We first present a new criterion to check whether a triangulation is good or not. We show that this new criterion is equivalent to the traditional way (a triangulation is good if all tetrehedra have good radius-edge ratio). The new criterion considers edges and the shape of their dual Voronoi facets.

2. Unlike the traditional algorithms, which insert one vertex at a time to fix a bad tetrahedron, we insert multiple Steiner vertices to fix all bad tetrahedra incident to a "loose" edge. Moreover, Steiner points used here are often different from the circumcenters (Voronoi vertices). The proposed strategy results in sparser well-spaced point sets and meshes with fewer tetrahedra.

2 Previous Work vs. Our Focus

In our review, we mainly focus on the Delaunay based mesh generation. The first generation of tetrahedral refinement algorithms [9, 30] came as extensions of the planar Delaunay refinement algorithms [7, 28]. These algorithms employed circumcenters of bad tetrahedra as Steiner points and provided upper bounds on the radius-edge ratio. Then, the tetrahedral meshing research progressed mainly in addressing two important challenges described below. Following those, we describe two other challenges, the latter of which is our focus in this paper.

Slivers

There are tetrahedra, called *slivers*, with bounded radius edge ratio but unbounded dihedral angles (and radius ratio). Slivers are both undesirable and ubiquitous in three dimensional (Delaunay) triangulations. Even when the point set is well-spaced slivers may result. Recently, several algorithms have been developed that are guaranteed to result in meshes with no slivers [3, 5, 12, 16, 19]. These can be grouped in three classes (i) those that employ weighted Delaunay triangulations [3, 5], (ii) those that rely on a perturbation on the Steiner points, and (iii) those that use a more structured Steiner point packing such as octree or lattice structures [16, 17, 25]. The theoretical guarantees provided by the first two classes are too small to be relevant in practice. The rigid structure of the Steiner point packing employed by the last class of algorithms leads to dense points and hence meshes with significantly large number of elements.

Input Constraints

Computing triangulations of three dimensional domains is a major challenge in itself even without enforcing any quality measure or output size requirement. This is mainly due to the input boundary constraints. For instance, there are polyhedra which cannot be triangulated unless Steiner points are allowed [2, 27, 29]. Delaunay triangulation of the vertices of an input domain may not conform the domain boundary (facets) in general. The conforming and constrained Delaunay triangulation algorithms [8, 26, 33, 35, 39] address this issue. These algorithms are integrated within the Delaunay refinement framework. Earlier Delaunay refinement algorithms had strong constraints on the input type which they can handle. For instance, the dihedral angles between the input facets are assumed to be reasonably large [30]. Later, algorithms that can handle smaller input angles and larger classes of inputs are studied. [4, 32]

Time efficiency

The original Delaunay refinement algorithm has quadratic time complexity [28]. This compares poorly to the time-optimal quadtree refinement algorithm

Well-spaced points in two dimensions

| (a) Packing circumcenters | (b) Packing locally optimal points |

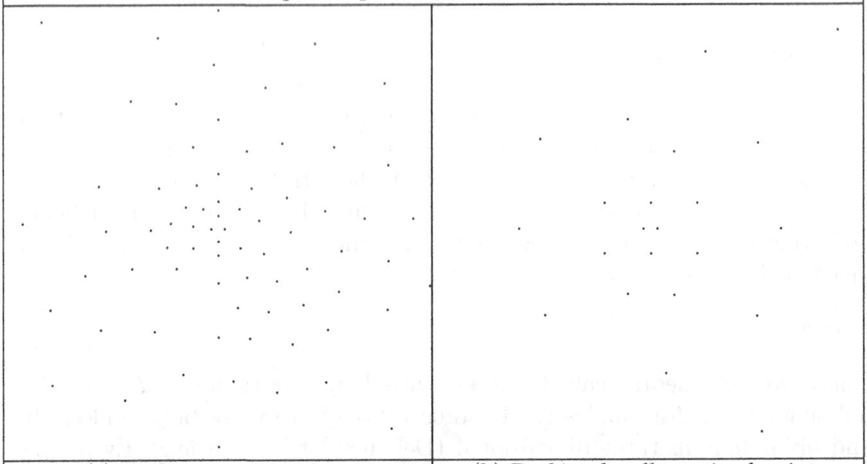

Fig. 1. Vertices of the output of the planar Delaunay refinement algorithms are shown for a constraint angle of 30°. Input in this case is a pair of points surrounded by a large empty space. The traditional circumcenter insertion packs many more Steiner points (a) than the locally optimal Steiner point placement strategy (b).

of Bern *et al.* [1] which runs in $O(n \log n + m)$ time, where m is the minimum size of a good quality mesh. The first improvement was given by Spielman *et al.* [37, 38] as a consequence of their parallelization of the Delaunay refinement algorithm. Their algorithm runs in $O(m \log m \log^2(L/h))$ time (on a single processor), where L is the diameter of the domain and h is the smallest feature in the input. Later, Miller [21] further improved this describing a new sequential Delaunay refinement algorithm with running time $O((n \log(L/h) + m) \log m)$. Then, Har-Peled and Üngör [14] presented the first time-optimal Delaunay refinement algorithm, which runs in $O(n \log n + m)$ time. This algorithm, originally presented in two dimensions has a natural extension to three dimensions for point sets. More recently, alternative time-efficient Delaunay refinement algorithms for higher dimensionsional meshing have been introduced [15]. The implications and benefits of the time-efficient algorithms in practice have yet to be explored. We should note that the Steiner point placement algorithm we propose here complements the time-optimal algorithm framework proposed in [14], and promises to be a good algorithm both in theory and practice.

Point Packing

Our focus in this study is on creating good packings of the Steiner points. We want the output point sets to be well-spaced [41], as the Delaunay triangulation of such point sets has bounded radius-edge ratio. On the other hand, having the packing sparse leads to meshes with fewer elements. Recent

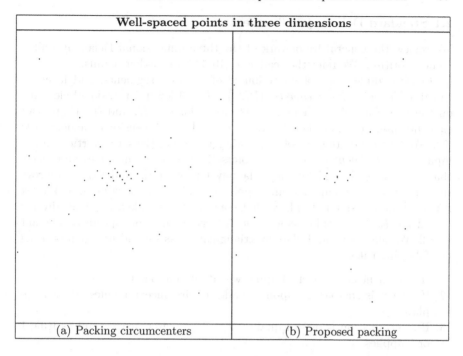

Well-spaced points in three dimensions	
(a) Packing circumcenters	(b) Proposed packing

Fig. 2. A careful packing results in sparse well-spaced point sets in three dimensions. Here input is a pair of points inside a cube. We zoomed in to illustrate the difference in our packing strategy (b) with respect to the traditional iterative circumcenter packing (a). See Figure 6 for the complete view of this data set.

research indicated that a more careful packing strategy than the traditional circumcenter packing has great benefits in two dimensional mesh generation [42, 13]. Thanks to the sparsity of the generated point sets, the meshes tend to be a factor of two or more smaller than the meshes generated by iterative circumcenter insertions. This implies substantial reduction not only in triangulation time, but also in the running time of the subsequent application algorithms. Moreover, the new Steiner point packing strategies help us design software that are effective for stronger constraint angles than the traditional methods. The original Delaunay refinement algorithm of Ruppert is proven to terminate with size-optimal quality triangulations for $\alpha \leq 20.7°$. In practice, it generally works for $\alpha \leq 34°$ and fails to terminate for larger constraint angles. The new variant of the Delaunay refinement algorithm generally terminates for constraint angles up to $42°$. Figures 1 and 2 illustrate the benefits of a locally optimal Steiner packing in two dimensions, and potential benefits of the proposed packing in three dimensions. At a recent workshop, the second author sketched ideas to illustrate the potential benefits in three dimensions [43].

2.1 Standard Delaunay refinement

We review the general framework of the three dimensional Delaunay refinement algorithm. We refer the reader to [10, 30] for further details.

In three dimensions, a collection Ω of vertices, segments, and facets is called a *piecewise linear complex* (PLC) if (i) all lower dimensional elements on the boundary of an element in Ω also belong to Ω, and (ii) if any two elements intersect, then their intersection is a lower dimensional element in Ω [24]. We first compute the Delaunay triangulation of the set of vertices of the input PLC Ω. Then, we add new points (i) to recover the edges and facets that are not conformed by the Delaunay triangulation and (ii) to improve the quality of the triangulation. A point is said to *encroach* upon a simplex if it is inside the smallest sphere that contains the simplex. A tetrahedron is considered *bad* if its radius-edge ratio is larger than a pre-specified constant $\beta \geq 2$. We maintain the Delaunay triangulation as we add new points using the following rules.

1. If a segment is encroached upon, we add its midpoint.
2. If a facet is encroached upon, we add its circumcenter unless Rule 1 applies.
3. If a tetrahedron is of bad quality, we add its circumcenter unless Rule 1 or 2 applies.

Shewchuk [30, 31] showed that this algorithm is guaranteed to compute tetrahedral meshes with bounded radius-edge ratio. That is, the output vertex set is well-spaced. He also showed that the point set has good grading with respect to input feature size distribution. The key component of the termination guarantee and the correctness of the Delaunay refinement algorithms is that a Steiner point insertion does not introduce features smaller than the existing ones. In this paper, we show that this property holds for the our Steiner point packing strategy. The termination property and the correctness of the algorithm follows, replacing the corresponding lemmas in the standard Delaunay refinement framework. For simplicity, we do not repeat the entire framework, and refer to [10, 30] for the structure of the complete proof of correctness for Delaunay refinement algorithms.

3 Preliminaries for our Algorithm

We assume the user is familiar with the definitions of Delaunay triangulations and Voronoi diagrams [10]

3.1 Edges and their Voronoi Facets

Consider a pair of points p and q in the mesh that are connected to each other in the Delaunay triangulation. (See Figure 3.) Let $Vor(pq)$ denote the Voronoi facet dual to the Delaunay edge pq.

Definition 1. *Let S be the set of spheres that have radius $\beta|pq|$ and go through the points p and q. The disk formed by the centers of the spheres in S is called the* **core disk** *of pq, denoted as $CoreDisk(pq)$.*

Note that $CoreDisk(pq)$ is coplanar with $Vor(pq)$.

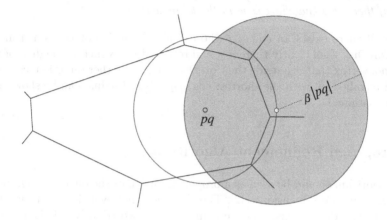

Fig. 3. Let pq be an edge (orthogonal to the view plane) in a 3D triangulation. The Voronoi face dual of the edge pq is shown as a polygonal region. The shaded region represents a sphere with radius $\beta|pq|$ that goes through points p and q. The centers of all such spheres form a circle, i.e. $CoreDisk(pq)$.

Definition 2. *A Delaunay edge pq is said to be* **short** *if it is the shortest edge of a tetrahedron incident to it.*

Definition 3. *A Delaunay edge pq that is not on the boundary is said to be* **bad** *if $Vor(pq) \nsubseteq CoreDisk(pq)$. A boundary Delaunay edge is* **bad** *if there exists a vertex of $Vor(pq)$ that is not in $CoreDisk(pq)$.*

We prove below that the Delaunay triangulation of a point set includes a bad tetrahedron if and only if it includes a bad edge. We propose a local sampling strategy which introduces several Steiner points at once to ensure $Vor(pq) \subseteq CoreDisk(pq)$ after the insertion for every Delaunay edge pq. Our sampling strategy should be locally optimal maximizing the smallest new pairwise point distance.

Let R_c denote the radius of the $CoreDisk(pq)$. Then, $R_c = \sqrt{\beta^2 - 1/4}|pq|$.

Theorem 1. *Delaunay triangulation of a point set has bad tetrahedra if and only if there is a bad edge in the triangulation.*

Indeed we can prove the following stronger statement, which considers only the short edges. This modification plays a role in the efficiency of the sampling algorithm for fixing the bad edges. That is we only need to address the short edges.

Theorem 2. *Delaunay triangulation of a point set has bad tetrahedra if and only if there is a bad short edge in the triangulation.*

Proof. If there exists an edge pq whose Voronoi cell is not contained inside the CoreDisk then clearly there is a bad tetrahedron (which is the dual of the Voronoi vertex that is outside the CoreDisk). For the other direction, consider a bad tetrahedron with the shortest edge pq. Then, its dual can be shown not to be contained.

4 Proposed Refinement Algorithm

Our algorithm should be seen as a proper extension of the offcenter algorithm presented in two dimensions [42]. The extension, however, is not straightforward. In two dimensions, an edge is incident to two triangles. This makes it easy to compute Steiner points such that an existing "short" edge is guaranteed to be neighbour to two good triangles. The main challenge in three dimensions is there could be a varying number of tetrahedra incident to an edge. That there are very many ways to pack points around a small feature so that it is surrounded by good tetrahedra. Our algorithm (presented as Algorithm 1) first detects a collection of consecutive bad tetrahedra incident to a short edge and then introduces potentially multiple Steiner points so that the edge is incident to a set of good tetrahedra (or it dissappears from the triangulation). We should note that we always maintain the Delaunay property of the triangulation.

Algorithm 1

Compute the Delaunay triangulation of the input
while ∃ a bad short edge pq
  Compute the connected components of $Vor(pq) - CoreDisk(pq)$
  Select the connected component with the largest arc boundary
  Sample Steiner points on this component
  Recompute the Delaunay triangulation of the modified point set

---

Traditional Delaunay refinement algorithms and implementations consider the triangulation as a list of tetrahedra, and maintain the list of bad tetrahedra in a priority queue. We can transform Algorithm 1 to Algorithm 2 to make our packing strategy immediately compatible with the existing Delaunay refinement framework. The two algorithms are equivalent as a consequence of Theorem 2.

Algorithm 2

Compute the Delaunay triangulation of the input
while ∃ a bad tetrahedron $pqrs$ with shortest edge pq
    Compute the connected components of $Vor(pq) - CoreDisk(pq)$
    Select the connected component with the largest arc boundary
    Sample Steiner points on this component
    Recompute the Delaunay triangulation of the modified point set

---

4.1 Connected Components

Let pq be an edge incident to a bad tetrahedron. Connected components of $Vor(pq) - CoreDisk(pq)$ can be computed through a linear scan on the polygonal chain boundary of $Vor(pq)$. (See Figure 4 (left).) Each connected component can be characterized by the intersection points of the boundary of $Vor(pq)$ and the boundary of $CoreDisk(pq)$. In our notation, we use ∂ to denote the boundary of a region. If $\partial Vor(pq) \cap \partial CoreDisk(pq) = \varnothing$, then there is one connected component. Unless $Vor(pq) \cap CoreDisk(pq) = \varnothing$, this component is a polygonal region with a hole, whose outer boundary is $\partial Vor(pq)$ and interior boundary is $\partial CoreDisk(pq)$. One can detect whether $Vor(pq) \cap CoreDisk(pq) = \varnothing$. In practice this case occurs rarely, thanks to the strategy we implement for prioritizing the handling of bad tetrahedra. If $\partial Vor(pq) \cap \partial CoreDisk(pq) \neq \varnothing$, then each connected component can be represented by two intersection points a and b, and a polygonal sub-chain $v_1, v_2, \ldots, v_k$ of $\partial Vor(pq)$. (See Figure 4.)

We call the angle of the arc that bounds a connected component the *span angle* of the connected component. The span angle is a real number in the interval $(0, 2\pi]$. The span angle is 2π, either when $\partial Vor(pq) \cap \partial CoreDisk(pq) = \varnothing$ or when $a = b$.

4.2 Sampling the Connected Components

Making sure that the sampling is sufficiently sparse is important for the termination guarantee of the Delaunay refinement algorithm. The distance between the sampling points should be at least $|pq|$. On the other hand, it is desirable to have a dense enough sampling so that the new Voronoi dual of pq is contained inside the core. Here we sketch a strategy that computes the sampling points for a given connected component specified by its spanning angle γ.

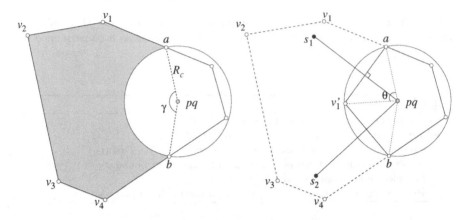

Fig. 4. An edge pq whose Voronoi CoreDisk difference consists of a single connected component is shown in (left). The span angle of the component is γ. A two sampling is shown in (right) for a sampling angle of $\gamma/2$. After the insertion of sample points s_1 and s_2, the Voronoi dual of pq, $Vor(pq)$, is modified to include new vertices a, b, and v_1'.

The sampling angle

We choose an angle called the *sampling angle*, denoted by θ. If $\gamma < \pi/2$ then $\theta = \gamma$ and we use only one sample point. Otherwise, we use $l = \lfloor 4\gamma/\pi \rfloor$ sample points. In this case we subdivide the arc ab of the connected component into equal pieces of angle $\theta = \gamma/l$. So, whenever $\gamma \geq \pi/4$, we have $\theta \in [\pi/4, \pi/2)$.

Location of the sampling points

The number of points and the sampling angle determines the exact location of where we expect the new Voronoi vertices of $Vor(pq)$. These will be $a, v_1', \ldots, v_{l-1}', b$, where l is a positive integer. See Figure 4 (right). For each expected Voronoi edge, there is a unique point in the plane of $Vor(pq)$ which leads to the formation of the edge. We take each such point as the sampling points. The sampling point is on a ray originating from the midpoint of pq and orthogonal to the projected Voronoi edge. It is of distance $(x + \cos(\theta/2)R_c)$ from the midpoint of pq, where $x = \sqrt{\cos^2(\theta/2)(\beta^2 - 1/4) + 1/4}\, |pq|$. This calculation is given in Lemma 1 in more detail.

Numerical errors and perturbation

Roundoff errors in the calculation of the sample point coordinates might lead to undesired result, e.g., the projected Voronoi vertices might fall slightly outside the $CoreDisk(pq)$. To avoid this we perturb the sample point coordinates toward the midpoint of pq.

Sampling points outside the connected components

It is also possible that sample points might land outside $Vor(pq)$. To prevent inserting points outside the $Vor(pq)$, we compute the intersection e of the ray ms_i and the $\partial Vor(pq)$. If e is far enough from the $CoreDisk(pq)$, we can still use this sample point together with the other sample points and guarantee that distance in between the sampling points is sufficiently large. Otherwise, we choose to use one sample point at a time.

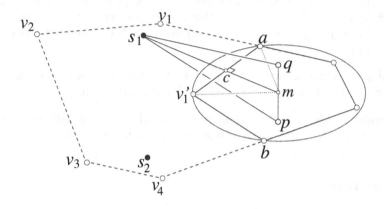

Fig. 5. The exact locations of the sampling points depend on the span angle sampling angle and the radius-edge ratio constraint β.

5 Analysis

Lemma 1. *Let $s_1, \ldots, s_l$ be the set of sampling points inserted at the same iteration. The distance between each of the sampling points s_i, $i = 1, \ldots, l$, and existing vertices is at least $\beta|pq|$.*

Proof. Each sampling point belongs $Vor(pq)$. The nearest neighbours of a point in this region are p and q. Moreover, the sampling points are in $Vor(pq) - CoreDisk(pq)$. Observe that for all $x \in (Vor(pq) - CoreDisk(pq))$, we have $xp > \beta|pq|$ Hence, $|ps_i| > \beta|pq|$ for every sampling point s_i.

We can indeed specify the distance $|ps_i|$ in terms of sampling angle θ, and β. Recall that $R_c = \sqrt{\beta^2 - 1/4}|pq|$. Let x be the distance of p to the new Voronoi edge dual to pqs_i. See Figure 5. By the Pythagoras Theorem,

$$x = \sqrt{(\cos^2(\theta/2)(\beta^2 - 1/4) + 1/4}\,|pq|.$$

Then, applying the same theorem once more, we get

$$ps_i = \sqrt{(x + \cos(\theta/2)R_c)^2 + |pq|^2/4}.$$

It is easy to verify that this value is greater than $|pq|$ for sufficiently large β and θ values.   □

Lemma 2. *Let $s_1, \ldots, s_l$ be the set of sampling points inserted at the same iteration where $l \geq 2$. The distance between any two sampling points s_i, and s_j, where $i, j \in \{1, \ldots, l\}$ and $i \neq j$, is at least $\beta|pq|$.*

Proof. The distance between any two sampling points is at least $s_i s_j = 2\sin(\theta/2)(x + \cos(\theta/2)R_c)$. (Note this is the case when either $j = i + 1$, or $i = 1$ and $j = l$.) For $\theta = [\pi/4, \pi/2)$ and $\beta > 2$, it is easy to verify that $s_i s_j > \beta|pq|$.

Integrating the above Lemmas with the Delaunay refinement analysis framework [10, 30] we conclude with the following theorem.

Theorem 3. *The proposed Delaunay refinement algorithm outlined in Algorithm 2 terminates with a guaranteed quality tetrahedral mesh.*

6 Experiments

Our implementation of the proposed algorithm relies on a beta version of the **Pyramid** software (provided by Jonathan Shewchuk). The current version is not very robust and was a limitation conducting comprehensive experimental study. We should emphasize that the results presented in this section are quite preliminary. Nevertheless, we believe the study is representative of the strength of the new packing strategy. We give a comparison with the circumcenter insertion algorithm as implemented in the **Pyramid** software. In our implementation, we handle bad tetrahedra starting with the ones with the shortest edges first.

Data Sets

We ran our experiments on various data sets including the following:

1. *Helix* data set consists of 1008 points forming a double helix. The two helices are very close to each other forming small features. See this model in Figure 7.
2. *Cube1* consists of ten points two of which are located unit distance from each other and at the center of a cube of side length 10^5 units. See this model in Figure 6.
3. *Cube8* is a cube that has eight small features, each one is close to a corner of the cube. See this model in Figure 6.
4. *Random Points* consists of 10,000 points spread uniformly at random inside a cube.

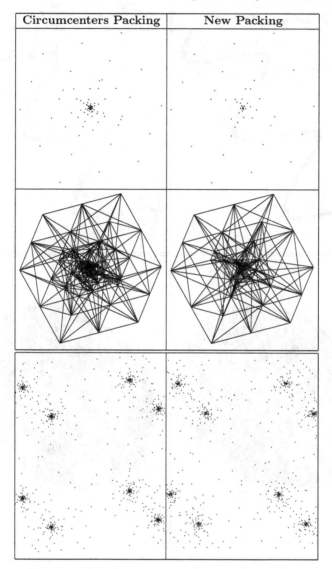

Circumcenters Packing	New Packing

Fig. 6. Output mesh vertex set of the previous and the new Delaunay refinement algorithm for the Cube1 (top row) and Cube8 (bottom row) data sets. Delaunay triangulation is also shown for the Cube1 data set (middle row). See Table 1 for the statistics.

5. *Ellipsoid* consists of 10,000 points spread uniformly at random on an ellipsoid which is inside a cube formed by 8 additional points. See this data set in Figure 7.

Circumcenters Packing	New Packing

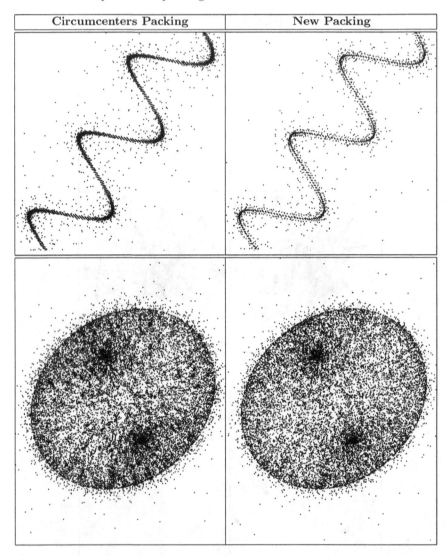

Fig. 7. Output of the previous and the new Delaunay refinement algorithm for the Helix (zoomed in) and the Ellipsoid data sets. The radius-edge ratio constraint in all triangulations is 4. See Table 1 for the statistics.

Table 1 presents a summary of our experiments on two data sets. The first block of four columns provide information about the input data sets. The fifth column shows the imposed radius-edge ratio constraint β. Columns six and seven list the output size of the circumcenter insertion algorithm in the number of vertices and in the number of tetrahedra, respectviely. Columns eight and nine list these quantities for the new algorithm. The last column

Data Set				Quality	CC algorithm		New algorithm		Imp.
name	v	e	f	R/l	v	t	v	t	%
Cube1	12	12	6	1.5	596	3,420	570	3,280	4.3
				2	265	1,440	220	1,196	16.9
				4	153	730	79	330	48.3
				8	133	607	51	195	61.6
Cube8	32	12	6	1.5	2,164	12,016	1,952	10,803	9.7
				2	1,042	5,531	881	4,537	15.4
				4	610	2,836	360	1,605	40.9
				8	483	2,074	233	924	51.7
Ellipsoid	10,008	12	6	1.5	69,075	431,908	60,861	379,718	11.8
				2	39,710	247,031	32,394	200,960	18.4
				4	21,396	131,553	17,256	103,380	19.3
				8	15,202	89,501	13,328	75,238	12.3
Helix	1,008	12	6	1.5	50,527	310,743	38,965	239,330	22.8
				2	26,259	157,163	16,596	99,521	36.7
				4	12,763	71,812	5,938	37,738	53.4
				8	8,727	49,998	3,923	37,738	55.0
Random	1,008	12	6	1.5	27,851	176,846	25,884	164,301	7.0
				2	17,743	113,743	16,572	106,420	6.5
				4	11,463	75,257	11,118	73,484	3.0
				8	10,377	68,595	10,326	68,202	0.4

Table 1. The number of vertices (v) and the number of tetrahedra (t) in the output is listed for the proposed algorithm in comparison to the circumcenter insertion (CC) algorithm. The last column reports the percentage improvement in the number of output vertices.

lists the percentage improvement in the number of output vertex size. In some cases, the percentage improvement listed seems insignificant. Note that in such cases, the input size is very large compared to the Steiner vertices inserted by each of the algorithm. If we report the improvements in the number of Steiner points, the percentages listed would be much larger. For instance, the last entry would be 13.7% instead of 0.4%. We get similar improvements for other data sets that are not listed here.

7 Discussions

We proposed a new Delaunay refinement algorithm for computing quality-guaranteed three dimensional tetrahedral meshes. It differs from previous algorithms on several fronts:

- Unlike the traditional algorithms, we handle potentially more than one bad tetrahedra at a time.
- We insert multiple Steiner points at each iteration. While this seems to be a good match within a parallel Delaunay refinement framework such

as [38], our sequential strategy should not be confused with a parallel algorithm.

- The point set of the output is not only well-speced but also significantly sparse compared to the output of the previous algorithms.

Further research is needed to fully utilize the benefits of the approach proposed here. We plan to complete a robust implementation of the three dimensional meshing software and run more comprehensive experiments. We would like to explore the potential benefits of our approach on computing sliver-free meshes.

Acknowledgements

Thanks to Jonathan Shewchuk for providing the Pyramid software.

References

1. M. Bern, D. Eppstein, and J. Gilbert. Provably good mesh generation. *J. Comp. System Sciences* 48:384–409, 1994.
2. B. Chazelle, Convex partitions of polyhedra: a lower bound and worst-case optimal algorithm, *SIAM Journal on Computing* 13: 488–507, 1984.
3. S.-W. Cheng and T. K. Dey. Quality meshing with weighted Delaunay refinement. *Proc. 13th ACM-SIAM Sympos. Discrete Algorithms*, 137–146, 2002.
4. S.-W. Cheng, T.K. Dey, E.A. Ramos, and T. Ray. Quality Meshing of Polyhedra with Small Angles. *Int. J. Computational Geometry & Applications* 15(4): 421–461, 2005.
5. S.-W. Cheng, T.K. Dey, H. Edelsbrunner,M.A. Facello, and S.-H. Teng. Sliver exudation. *Proc. 15th ACM Symp. Comp. Geometry*, 1–13, 1999.
6. S.-W. Cheng and Sheung-Hung Poon. Three-Dimensional Delaunay Mesh Generation. 419-456 *Discrete & Computational Geometry*, 36(3): 419–456, 2006.
7. L.P. Chew. Guaranteed-quality triangular meshes. TR-89-983, Cornell Univ., 1989.
8. D. Cohen-Steiner, E. C. de Verdiére, and M. Yvinec. Conforming Delaunay triangulations in 3d. *Proc. 18th ACM Symp. Comp. Geometry*, 199–208, 2002.
9. T.K. Dey, C. L. Bajaj, and K. Sugihara. On good triangulations in three dimensions. *Int. J. Computational Geometry & Applications* 2(1):75–95, 1992.
10. H. Edelsbrunner. *Geometry and Topology for Mesh Generation*. Cambridge Univ. Press, 2001.
11. H. Edelsbrunner and D. Guoy. Sink insertion for mesh improvement. *Proc. 17th ACM Symp. Comp. Geometry*, 115–123, 2001.
12. H. Edelsbrunner, X. Li, G.L. Miller, A. Stathopoulos, D. Talmor, S.-H. Teng, A. Üngör, and N. Walkington. Smoothing and cleaning up slivers. *Proc. 32nd ACM Symp. on Theory of Computing*, 273–277, 2000.
13. H. Erten and A. Üngör. Delaunay Refinement with Locally Optimal Steiner Points. *(to appear) In Proc. EUROGRAPHICS Symposium on Geometry Processing, Barcelona, Spain* July 2007.

14. S. Har-Peled and A. Üngör. A time-optimal Delaunay refinement algorithm in two dimensions. *Proc. ACM Symposium on Computational Geometry*, 228–236, 2005.

15. B. Hudson, G.L. Miller, and T. Phillips. Sparse Voronoi Refinement. *Proc. 15th Int. Meshing Roundtable*, 339–356, 2006.

16. F. Labelle. Sliver Removal by Lattice Refinement. *Proc. ACM Symposium on Computational Geometry*, 347–356, 2006.

17. F. Labelle and J.R. Shewchuk. Isosurface Stuffing: Fast Tetrahedral Meshes with Good Dihedral Angles *Proc. ACM SIGGRAPH 2007*.

18. X.-Y. Li, S.-H. Teng, and A. Üngör. Biting: Advancing front meets sphere packing. *Int. J. Numer. Meth. Eng.* 49:61–81, 2000.

19. X.-Y. Li and S.-H. Teng. Generating well-shaped Delaunay meshed in 3D. *Proc. ACM-SIAM Symp. on Discrete Algorithms*, 28–37, 2001.

20. R. Löhner. Progress in grid generation via the advancing front technique. *Engineering with Computers* 12:186–210, 1996.

21. G.L. Miller. A time efficient Delaunay refinement algorithm. *Proc. ACM-SIAM Symp. on Disc. Algorithms*, 400–409, 2004.

22. G.L. Miller, S. Pav, and N. Walkington. When and why Ruppert's algorithm works. *Proc. 12th Int. Meshing Roundtable*, 91–102, 2003.

23. G.L. Miller, S. Pav, and N. Walkington. Fully Incremental 3D Delaunay Refinement Mesh Generation. *Proc. 11th Int. Meshing Roundtable*, 75–86, 2002.

24. G.L. Miller, D. Talmor, S.-H. Teng, N. Walkington, and H. Wang. Control volume meshes using sphere packing: Generation, refinement, and coarsening. *Proc. 5th Int. Meshing Roundtable*, 47–61, 1996.

25. S. Mitchell and S. Vavasis. Quality mesh generation in three dimensions. *Proc. 8th ACM Symp. Comp. Geometry*, 212–221, 1992.

26. M. Murphy, D. M. Mount, and C. W. Gable. A point-placement strategy for conforming Delaunay tetrahedralization. *Proc. 11th ACM-SIAM Symp. on Discrete Algorithms*, 67–74, 2000.

27. M.S. Paterson and F.F. Yao, Binary partitions with applications to hidden-surface removal and solid modeling, *Proc. 5th ACM Symp. Computational Geometry*, 23–32, 1989.

28. J. Ruppert. A new and simple algorithm for quality 2-dimensional mesh generation. *Proc. 4th ACM-SIAM Symp. on Disc. Algorithms*, 83–92, 1993.

29. E. Schönhardt. Über die Zerlegung von Dreieckspolyedern in Tetraeder, *Math. Annalen*, 98: 309–312, 1928.

30. J.R. Shewchuk. *Delaunay Refinement Mesh Generation*. Ph.D. thesis, Carnegie Mellon University, 1997.

31. J.R. Shewchuk. Tetrahedral mesh generation by Delaunay refinement. *Proc. 14th Annual ACM Symposium on Computational Geometry*, 86–95, 1998.

32. J.R. Shewchuk. Mesh generation for domains with small angles. *Proc. 16th ACM Symposium on Computational Geometry*, 1–10, 2000.

33. J.R. Shewchuk. Sweep algorithms for constructing higher-dimensional constrained Delaunay triangulations. *Proc. 16th ACM Symposium on Computational Geometry*, 350–359, 2000.

34. J.R. Shewchuk. Constrained Delaunay Tetrahedralizations and Provably Good Boundary Recovery. *Proc. 11th Int. Meshing Roundtable*, 193–204, 2002.

35. J.R. Shewchuk. Delaunay refinement algorithms for triangular mesh generation. *Computational Geometry: Theory and Applications* 22(1–3):21–74, 2002.

36. K. Shimada. *Physically-based Mesh Generation: Automated Triangulation of Surfaces and Volumes via Bubble Packing*. Ph.D. thesis, MIT, 1993.
37. D.A. Spielman, S.-H. Teng, and A. Üngör. Parallel Delaunay refinement: Algorithms and analyses. *Proc. 11th Int. Meshing Roundtable*, 205–217, 2002.
38. D.A. Spielman, S.-H. Teng, and A. Üngör. Parallel Delaunay refinement: Algorithms and analyses. *Int. J. Comput. Geometry Appl.* 17(1): 1-30 (2007)
39. H. Si. On Refinement of Constrained Delaunay Tetrahedralizations. *Proc. 11th Int. Meshing Roundtable*, 2006.
40. D.A. Spielman, S.-H. Teng, and A. Üngör. Parallel Delaunay refinement with off-centers. *Proc. EUROPAR*, LNCS 3149, 812–819, 2004.
41. D. Talmor. *Well-Spaced Points for Numerical Methods*. Ph.D. thesis, Carnegie Mellon University, 1997.
42. A. Üngör. Off-centers: A new type of Steiner points for computing size-optimal quality-guaranteed Delaunay triangulations. *Proc. LATIN*, LNCS 2976, 152–161, 2004.
43. A. Üngör. Quality meshes made smaller. *Proc. 20th European Workshop on Computational Geometry*, 5–8, 2005.

Optimization and Adaptivity

1B.1

Interleaving Delaunay Refinement and Optimization for 2D Triangle Mesh Generation

Jane Tournois, Pierre Alliez, and Olivier Devillers

INRIA Sophia-Antipolis, France
firstname.lastname@sophia.inria.fr

Summary. We address the problem of generating 2D quality triangle meshes from a set of constraints provided as a planar straight line graph. The algorithm first computes a constrained Delaunay triangulation of the input set of constraints, then interleaves Delaunay refinement and optimization. The refinement stage inserts a subset of the Voronoi vertices and midpoints of constrained edges as Steiner points. The optimization stage optimizes the shape of the triangles through the Lloyd iteration applied to Steiner points both in 1D along constrained edges and in 2D after computing the bounded Voronoi diagram. Our experiments show that the proposed algorithm inserts fewer Steiner points than Delaunay refinement alone, and improves over the mesh quality.

Key words: 2D triangle mesh generation, Delaunay refinement, optimization, mesh smoothing, bounded Voronoi diagram.

1 Introduction

We consider the problem of generating 2D triangle meshes from a bounded domain and a set of geometric and sizing constraints provided respectively as a planar straight line graph (PSLG) and a sizing function. This problem is mainly motivated by several practical applications, such as numerical simulation of physical phenomena. The latter applications require *quality* meshes, ideally of smallest size. Quality herein refers to the shape and size of the elements: A triangle is said to be good if its angles are bounded from below, and if the length of its longest edge does not exceed the sizing function. The smallest size triangulation, conforming the sizing function, is sought after for efficiency reasons as the number of elements governs the computation time.

Quality triangulation is a well-studied problem, and many meshing strategies have been proposed and studied. Existing algorithms could be roughly classified as being *greedy*, *variational*, or *pliant*. Greedy algorithms commonly perform one local change at a time, such as vertex insertion, until the initial stated goal is satisfied. Variational techniques cast the initial problem into the one of minimizing an energy functional such that low levels of this energy correspond to good solutions for this problem (reaching a global optimum is in general elusive). A minimizer for this energy may perform global relaxation, i.e., vertex relocations and re-triangulation until convergence. Finally, an algorithm is pliant when it combines both refinement and decimation, possibly interleaved with a relaxation procedure [1, 2].

Although satisfactory bounds have been obtained for the triangle angles using Delaunay refinement algorithms [3], one recurrent question is to try lowering the number of points (so-called *Steiner points*) added by the meshing algorithm. Recall that the Steiner points are inserted either on the constrained edges or inside the domain, to satisfy sizing and quality constraints, as well as to preserve the input constrained edges. At the intuitive level, inserting "just enough" Steiner points requires generating triangles which are everywhere as big as possible, while preserving the sizing constraints. Luckily enough, the triangle which covers the biggest area for a given maximum edge length is the equilateral triangle. In other words, and by using a simple domain covering argument, trying to generate *large and well-shaped triangles* simultaneously serves the goal of lowering the number of Steiner points. Our approach builds upon this observation.

1.1 Contributions

This paper proposes to mesh a 2D domain into large and well-shaped triangles by alternating Delaunay refinement and optimization. We pursue the goal of generating large triangles by performing refinement in a multilevel manner, i.e., by inserting a subset of the Voronoi vertices batch-wise at each refinement stage. Each refinement stage is parameterized with a decreasing size, which is computed from the current mesh. Each mesh optimization stage is performed by applying the Lloyd iteration both in 1D along constrained edges, and in 2D inside the domain. To ensure that the constraints are preserved during mesh optimization, the bounded Voronoi diagram (Voronoi diagram with constraints) is computed using a novel robust and effective algorithm. For the sake of efficiency, the optimization stages are applied with increasing accuracy. Our experiments show that, for the required sizing, our algorithm generates in general meshes with fewer Steiner points and better quality than Delaunay refinement alone. We now briefly review Delaunay refinement and mesh optimization techniques.

1.2 Delaunay Refinement

A mesh refinement algorithm iteratively inserts well chosen points, so-called *Steiner points* in a given coarse mesh until all constraints are satisfied. One trend in refinement algorithms is to insert as few Steiner points as possible. One popular approach is to take as initial coarse mesh the constrained Delaunay triangulation of the input PSLG, and to refine it. Refinement algorithms of this kind are called *Delaunay Refinement algorithms*. They were pioneered by Chew [4], and later extended by many authors [5, 6]. Delaunay refinement algorithms proceed as follows: During refinement, an edge is said to be encroached if a point of the triangulation (not its endpoints) is on or inside its diametral circle. As long as the current mesh contains encroached edges, the algorithm inserts their midpoints. It then iteratively inserts the circumcenter of the "worst" triangle of the triangulation (according to size and shape criteria), unless it encroaches an edge. These steps are performed until all triangles are good, i.e., until each triangle has the size and shape criteria satisfied. Shewchuk [6] shows that this algorithm terminates with a finite number of Steiner points and with bounds on the triangles angles (provided the input PSLG does not contain any small angle). Several studies have been made on the insertion order of the Steiner points, and a satisfactory choice is to insert the circumcenter of the worst triangle first.

Another way to grasp the problem of minimizing the number of Steiner points is to define another type of Steiner points than circumcircle centers. The main idea is that if a triangle contains a large angle, Ruppert's algorithm inserts a Steiner point far away from it, in a place which may be irrelevant. Üngör [7] defines a so-called *off-center* as follows. The off-center off_c of a triangle $\triangle(p, q, r)$ of shortest edge pq is defined as the circumcenter of $\triangle(p, q, r)$ if the radius-edge ratio of $\triangle(p, q, r)$ is lower or equal to β, the angle quality bound. Otherwise, off_c is the point v on the bisector of pq, and inside its circumcircle, such that the radius-edge ratio of $\triangle(p, q, off_c)$ equals β. This technique is shown to insert fewer Steiner points than Ruppert's algorithm.

Chernikov and Chrisochoides generalize and improve over Üngör's off-centers solution [8]. They show that, more generally, any point in the *selection disk* of a bad triangle $\triangle$ can be chosen as a Steiner point. It is shown that, choosing any point inside the *selection disk* as Steiner point eliminates $\triangle$, and that the algorithm terminates. They propose an example of new Steiner point inside the *selection disk* and show that the corresponding refinement algorithm in general inserts fewer points than when using off-centers.

Some other methods combine both insertion and removal of mesh elements (including Steiner points) to obtain higher quality meshes. Such combination was used by Bossen and Heckbert [1] as well as by Coll et al. [2], to cite a few. Finally, Erten and Üngör [9] have recently introduced a new method which first tries relocating the vertices of bad triangles. If the relocations would not improve the mesh, they are canceled and a Steiner point is inserted inside $\triangle$.

1.3 Optimization

Many different triangulations covering a given domain and satisfying a set of constraints may exist, each of them of different quality. When high quality meshes are sought after, it is therefore desirable to resort to an optimization procedure so as to optimize a specific quality measure (see [10] for a comprehensive study of quality measures). Two questions now arise: Which criterion should we optimized? By exploiting which degrees of freedom? The optimized criterion can be directly related to the shape and size of the triangles [11], but other criteria have been proposed as well. We refer the reader to [12] for a comprehensive survey of mesh optimization techniques. As the number of degrees of freedom are both continuous and discrete (vertex positions and mesh connectivity), there is often a need for narrowing the space of possible triangulations. For example, Chen proposes to cast the isotropic meshing problem as an (isotropic) function approximation problem, optimizing within the space of Delaunay triangulations [13].

Eppstein [12] highlights the fact that, in 2D, evenly distributed points lead to well-shaped triangles, assuming an isotropic triangulation such as the Delaunay triangulation. Isotropic meshing can therefore be casted into the problem of isotropic point sampling, which amounts to distribute a set of points on the input domain in as even a manner as possible. One way to distribute a set of points isotropically and in accordance with a sizing function is to apply the Lloyd iteration procedure (described in Section 2.2) over an initial Voronoi diagram. Du et al. [14] have shown how the Lloyd iteration transforms an initial ordinary Voronoi diagram into a centroidal Voronoi diagram, where each generator happens to coincide with the center of mass of its Voronoi cell. Another interesting feature of the Lloyd iteration in our context is that it can be applied to any dimension, i.e., in 1D for the Steiner points inserted on the constrained edges, and in 2D for the Steiner points inserted inside the domain. One drawback of the Lloyd iteration is its slow convergence. Moreover, it only converges to a local minimum of a certain energy functional [15]. Convergence accelerations are possible either by using Newton-Lloyd iterations or by using specific types of multilevel refinements [16, 17].

2 Algorithm

We introduce an algorithm which interleaves Delaunay refinement and optimization, in order to generate a triangle mesh satisfying both shape and size properties for each triangle. We sketch our algorithm in pseudo-code as follows:

Algorithm 1 Meshing algorithm

Input: A PSLG domain $\Omega \subset \mathbb{R}^2$.
    Let T be the constrained Delaunay triangulation of Ω.
    repeat
        Batch refinement of the triangulation T (Section 2.1).
        repeat
            Optimization of T by the Lloyd algorithm (Section 2.2),
        until Stopping criterion S.
    until Refinement does not insert any new Steiner point in T.
    repeat
        Optimization of T by the Lloyd algorithm,
    until Stopping criterion S' stronger than S.
Output: The final triangle mesh.

The complete algorithm sequence is illustrated by Figure 1. Figure 2 compares the output of our algorithm with a mesh obtained by Delaunay refinement alone. Our method improves over the number of Steiner points as well as on the shape of the elements.

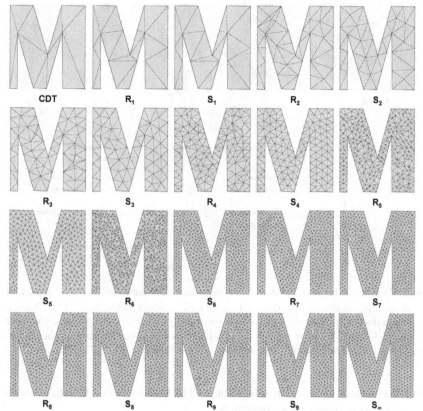

Fig. 1. Interleaved Delaunay refinement and optimization with a uniform sizing parameter. In the reading order: The constrained Delaunay triangulation (CDT) of the initial PSLG (13 vertices), then refinement (R_i) and smoothing (S_i) in alternation. The last step (S_∞) is the final optimization until the stronger stopping criterion. The final mesh contains 645 vertices.

2.1 Refinement

The refinement steps are specifically designed to insert "just enough" Steiner points to respect the sizing function, and to provide the optimization steps with good initial solutions. We achieve this goal by inserting batches of Steiner points chosen as Voronoi vertices. The roles are separated: Each refinement step acts on the size of the elements, while each optimization step acts on the shape of the elements (see Section 1.3).

Let us call $\mu(p)$ the desired mesh sizing specified at point p. We assume that this sizing function is provided by the user. During meshing we call *sizing ratio* the ratio between the current size of an element and the local desired sizing. An element can be an edge or a triangle, i.e., the sizing ratio is computed as $sr(f) = \frac{\text{longest edge}(f)}{\mu(centroid(f))}$ for a triangle and $sr(e) = \frac{\text{length}(e)}{\mu(centroid(e))}$ for an edge. After meshing, the sizing ratios of all elements must be lower or equal to 1 in order to satisfy the sizing constraints,

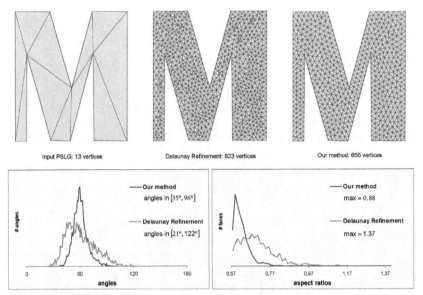

Fig. 2. Comparison between Delaunay refinement alone and our algorithm. The sizing function is uniform and equals 0.05. Our method inserts 20% fewer Steiner points than Delaunay refinement does. The distribution of angles is narrower at the end of our algorithm. The angles are within $[35°, 96°]$ with our method, vs $[21°, 122°]$ with Delaunay refinement. The practical upper bound on aspect ratios is lower with our method: 0.88 vs 1.37.

and a low number of Steiner points is obtained by keeping the sizing ratios as close to one as possible under these constraints.

Before each refinement step, the algorithm iterates over each triangle and constrained edge of the mesh in order to compute the maximum sizing ratio sr_{max}. From this value we compute the current *target sizing ratio* as $sr_{target} = \max(\frac{sr_{max}}{\sqrt{3}}, 1)$, which is then used as a slowly decreasing sizing criterion for refinement. The refinement factor $1/\sqrt{3}$ is justified by the fact that inserting all circumcenters in a triangulation made of equilateral triangles operates a $\sqrt{3} - section$ of the edges (compared to a bisection on the constrained edges). In 2D, the refinement corresponding to the target sizing ratio can therefore not be faster than this value. As we aim at refining the mesh with sizing ratios as uniform as possible, we parameterize the 1D refinement with the same target sizing ratio. The target sizing ratio is thresholded to 1 in order not to insert points that would introduce new triangles or edges with a sizing ratio lower than 1, i.e. smaller than the desired sizing. Hence, at each refinement iteration, the target sizing ratio slowly decreases, until it reaches the value of 1, which corresponds to satisfying the input sizing requirement, via the sizing function μ, for each simplex of the mesh.

Refinement is first performed in 1D along the constraint edges, then in 2D. Call $\mathcal{L}$ the list of Steiner points to be inserted. The algorithm iterates over all triangles and constrained edges and measure their sizing ratio. If it is larger than the current target sizing ratio sr_{target}, we add its midpoint or circumcenter to $\mathcal{L}$, and alter the sizing ratio of its (unprocessed) incident elements in order to avoid over-refinement.

More specifically, and relying on the fact that a local point insertion involves neighboring triangles, we inform the neighboring triangles of a split triangle that they should take upon themselves a part of this refinement. To do so, when a triangle f is processed and eligible for splitting, the quantity $1/3 * (sr_f/\sqrt{3} - sr_{target})$ is added to the sizing ratios of each of its unprocessed neighboring triangles. In this way, we divide between f's neighboring triangles the difference between what we expect to be the sizing ratio of f after splitting $(sr_f/\sqrt{3})$ and the target sizing ratio. Then, these triangle sizing ratios may be increased or decreased, and respectively get more or less likely to be split during the current refinement step. A similar method is used in 1D over the edges, with $1/2$ as refinement factor. When all elements have been visited, all points in $\mathcal{L}$ are inserted to the constrained Delaunay triangulation. The next step is mesh optimization.

2.2 Optimization

The mesh optimization step involves the Lloyd iteration applied to the cells of a bounded Voronoi diagram (the pseudo-dual of the constrained Delaunay triangulation).

Lloyd iteration.

The Lloyd iteration [18] is a minimizer for the following energy functional:

$$\mathcal{E}(\{z_i\}_{i=1}^N) = \sum_{i=1}^N \int_{y \in V_i} \rho(y) \|y - z_i\|^2 dy,$$

where $\{z_i\}_{i=1}^N$ are a set of generators and $\{V_i\}_{i=1}^N$ the corresponding Voronoi cells. The Lloyd iteration minimizes this energy by alternately moving the generators to the centroid of their Voronoi cells, and recomputing the Voronoi diagram. The centroid z^* of the cell V is defined as:

$$z^* = \frac{\int_V y\rho(y)dy}{\int_V \rho(y)dy},$$

where ρ is a density function defined on the domain Ω. After convergence, the space subdivision obtained is a *centroidal Voronoi tessellation* [14, 15]. As it corresponds to a critical point of the energy $\mathcal{E}$, it is a necessary condition for optimality, in the sense of minimizing $\mathcal{E}$.

We choose to stop the Lloyd iteration when all generators move less than a user-defined distance threshold. We first define the notion of move ratio of a generator z in its Voronoi cell as $m_r(z) = \frac{\|z-z^*\|}{size(cell(z))}$, where the size of the cell is defined as the longest distance between any pair of points of this cell. The Lloyd iteration is stopped when $\max_{z \in \{z_i\}_{i=1}^N} m_r(z) < p$ (p is typically set to 3% during refinement,

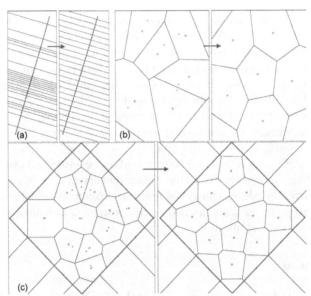

Fig. 3. Lloyd iteration with a uniform density. (a) Steiner points randomly inserted on a constrained edge, and after convergence of the Lloyd iteration. (b) Steiner points randomly inserted on a 2D domain, and after convergence of the Lloyd iteration. Sites are filled dots, and centroids are outlined dots (they coincide after convergence). (c) Lloyd iterations applied both in 1D and 2D.

and to 1% for the final relaxation step). In practice, the 1D Lloyd iteration, which allows moving the Steiner points to the centroids of their 1D cells (along the constraints), plays an important role at minimizing the number of Steiner points inserted. We could exhibit an extreme example where combining refinement with 1D Lloyd iteration inserts half of the number of 1D Steiner points inserted by the common recursive edge bisection of the constraints.

Our algorithm applies the Lloyd iteration both in 1D by moving Steiner points along the input constrained edges (Voronoi cells are line segments), and in 2D by moving Steiner points at the centroid of their bounded Voronoi cell. Figure 3 illustrates both 1D and 2D iterations.

Bounded Voronoi Diagram.

While the ordinary Voronoi diagram and Delaunay triangulation do not take constraints into account, we wish here to prevent the Voronoi regions to cross over the constraints. To this aim we use a *constrained Delaunay triangulation* [4,19,20] and a variant of its dual, the *bounded Voronoi diagram* (BVD), defined by Seidel [21]. The common duality between Delaunay triangulation and Voronoi diagram links each triangle to its circumcenter. In our context, each triangle may have its circumcenter on the other side of a constrained edge, therefore its corresponding incident Voronoi edge must be clipped by this constraint. The BVD is defined as follows: each cell V_i of a generator x_i is composed by the points of the domain Ω which are closer to x_i

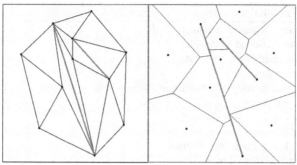

Fig. 4. Constrained Delaunay triangulation of a set of points (left) and its pseudo-dual bounded Voronoi diagram (right).

than to any other generator. As for the constrained Delaunay triangulation, the distance incorporates visibility constraints. The distance $d_S(x, y)$ between two points x and y of $\mathbb{R}^2$ is defined as:

$$d_S(x,y) = \begin{cases} ||x - y||_{\mathbb{R}^2} & \text{if } x \text{ "sees" } y, \\ +\infty & \text{otherwise.} \end{cases}$$

In this definition, x "sees" y when no constrained edge intersects the segment $[x, y]$. This visibility notion can be extended to triangles. We will see later how the notion of triangle sight, or symmetrically triangle "blindness", is important to construct the bounded Voronoi diagram. Figure 4 illustrates a constrained Delaunay triangulation and its pseudo-dual bounded Voronoi diagram. Notice that constructing the standard Voronoi diagram by joining the circumcenters of all pairs of incident triangles would not even form a partition. The notion of triangle blindness is pivotal for constructing the bounded Voronoi diagram.

Definition 2.1 (Blind triangle) *A triangle $\triangle$ is said to be* blind *if the triangle and its circumcenter c lie on the two different sides of a constrained edge E. Formally, $\triangle$ is blind if and only if there exists a constrained edge E such that one can find a point p in $\triangle$ (not an endpoint of E), such that the intersection $[p, c] \cap E$ is non empty.*

The BVD construction algorithm initially tags all triangles of the triangulation as being blind or not blind (Algorithm 2). It then constructs each cell of the diagram independently using these tags (Algorithm 3). Finally, all cells are assembled to build the complete bounded Voronoi diagram of a given set of points and constrained edges.

Algorithm 2 tags all triangles of the triangulation as being either *blind* or *nonblind*. In addition, each *blind* triangle stores which constrained edge in the triangulation acts as a visibility obstacle, i.e., which closest edge prevents it to *see* its circumcenter. Notice how the algorithm only needs to iterate over the constrained edges of the triangulation, as all sets of blinded triangles form connected components incident to constrained edges (the corresponding parts of the Voronoi diagram "hidden" by constrained edges are trees rooted from dual rays of these constraints).

Algorithm 2 Tag blind triangles

Input: Constrained Delaunay triangulation cdt.
  Tag all triangles non-blind by default.
  for each Constrained edge e of cdt **do**
    Create a stack: $faces$
    for each Adjacent triangle f_e to e tagged non-blind **do**
      Push f_e into $faces$
      while $triangles$ is non-empty **do**
        Pop f from stack $triangles$
        if f is blinded by e (use $\mathcal{P}$) **then**
          Tag f as blinded by e
          for each Adjacent triangle f' to f **do**
            if f' is finite and non-blind
            and the common edge between f and f' is unconstrained **then**
              Push f' into $triangles$.
            end if
          end for
        end if
      end while
    end for
  end for

We define a robust predicate, called $\mathcal{P}$ in the sequel, to test if a triangle is blinded by a constrained edge. More specifically, $\mathcal{P}$ takes as input a triangle and a segment, and returns a Boolean indicating whether or not the circumcenter of the triangle lies on the same side of the segment than the triangle. The circumcenter is never constructed explicitly in order to obtain a robust tagging of the blind triangles. Each cell of the bounded Voronoi diagram can be constructed by circulating around vertices of the triangulation, and by choosing as cell vertex either circumcenters or intersections of the standard Voronoi edges with the constrained edges. Note that we do not need to construct bounded Voronoi cells incident to input constrained vertices as the latter are constrained and therefore not relocated by the Lloyd iteration. Algorithm 3 describes this construction, and Figure 5 illustrates the construction of a single bounded Voronoi cell. Figure 6 illustrates a bounded Voronoi diagram.

Quadratures

The Lloyd iteration requires computing centroids of line segments in 1D and of (possibly bounded) Voronoi cells in 2D. Such computations require quadrature formulas when a variable density function is specified, i.e., when the input sizing function is not uniform. The density function ρ and the sizing function μ are linked by the following formula [22]: $\mu(x) = \frac{1}{\rho(x)^{d+2}}$, where d is the dimension of the domain. In 2D, we have

$$\mu(x) = \frac{1}{\rho(x)^4}.$$

Algorithm 3 Construct a BVD cell

Input: Unconstrained vertex z of the constrained Delaunay triangulation cdt.

Call P the polygon (cell) in construction,

Call f the current triangle and f_{next} the next triangle in the circulation,

Call $\mathcal{L}_{f,f_{next}}$ the line going through the circumcenters of f and f_{next}.

for each Incident triangle f to z in cdt **do**

    if f is tagged non-blind **then**

        Insert the circumcenter of f into P.

        if f_{next} is blind **then**

            Call $S_{f_{next}}$ the constrained edge blinding f_{next},

            Insert point $\mathcal{L}_{f,f_{next}} \cap S_{f_{next}}$ into P.

        end if

    else

        Call S_f the constrained edge blinding f.

        if f_{next} is tagged non-blind **then**

            Insert $\mathcal{L}_{f,f_{next}} \cap S_f$ into P.

        else

            Call $S_{f_{next}}$ the constrained edge blinding f_{next},

            if $S_f \neq S_{f_{next}}$ **then**

                Insert $\mathcal{L}_{f,f_{next}} \cap S_f$ and $\mathcal{L}_{f,f_{next}} \cap S_{f_{next}}$ into P.

            end if

        end if

    end if

end for

Output: Bounded Voronoi cell of z.

The key idea behind a quadrature is to decompose a simple domain (be it an edge or a triangle in our case) into smaller sub-domains (so-called quadrature primitives) where simple interpolation schemes are devised. The number N of quadrature primitives used for each element allows the user to tune the computation accuracy of the centroids. In practice this number is increased during refinement so that a low precision is used at coarse levels (typically $N = 10$), and a high precision is used at fine levels ($N = 100$ for the final level).

We use the trapezium rule in 1D, with N sub-segments, and the midpoint approximation rule in 2D, with a decomposition of each bounded Voronoi cell into N sub-triangles. More precisely, an initial step triangulates the cell by joining each of its vertices to its generator. The next step recursively bisects the longest edge of these triangles until the number of quadrature triangles reaches N. On each quadrature triangle, the midpoint approximation formula is applied:

$$\int_{\triangle} f(x)dx \approx \frac{|\triangle|}{3}(f(x_{12}) + f(x_{23}) + f(x_{13})),$$

where x_{12}, x_{23} and x_{13} are the midpoints of a quadrature triangle edges. Finally, we sum the integrals on each quadrature triangle in order to obtain an approximate centroid of the whole bounded Voronoi cell.

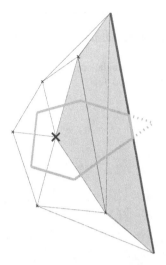
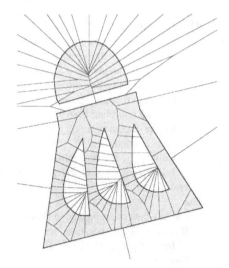

Fig. 5. Construction of a cell of the bounded Voronoi diagram. The standard Voronoi diagram is truncated on the constrained edge (colored triangles are blind).

Fig. 6. Bounded Voronoi diagram of a PSLG.

3 Results

Our algorithm is implemented in C++, using the *Computational Geometry Algorithms Library* CGAL [23]. The Delaunay refinement algorithm that we have used to compare our meshes to is the terminator algorithm defined by Shewchuk [24], implemented as a CGAL package [25].

Figures 7 and 8 show uniform meshes, and compare our algorithm with the standard Delaunay refinement algorithm. The number of Steiner points inserted are respectively 22% and 23% fewer with our method, and the angle distributions are better centered around 60 degrees, as expected. The intervals in which the angles lie are also tighter: $[33°, 99°]$ vs $[20°, 124°]$ for Figure 7, and $[29°, 103°]$ vs $[22°, 127°]$ for Figure 8. The distributions of aspect ratios (circumradii to shortest edge ratios) show that our algorithm produces many more triangles with aspect ratios close to the optimal value, which is $1/\sqrt{3} \approx 0.57$ for an equilateral triangle. This shows that, although our algorithm does not provide theoretical bounds on the triangle angles or aspect ratios, the practical bounds are improved.

Obviously, any mesh generation algorithm which incorporates smoothing or optimization is expected to obtain meshes of higher quality than a greedy algorithm such as Delaunay refinement, at the price of higher computation times (10 times slower for our algorithm in the uniform cases, and close to 100 times slower in the nonuniform cases which involve costly quadratures). In our experiments, interleaving refinement and optimization, similar in spirit to multilevel algorithms, consistently leads to higher quality meshes than refining, then optimizing at the finest level. In addition to the mesh quality obtained, the choice of the Lloyd iteration instead of the

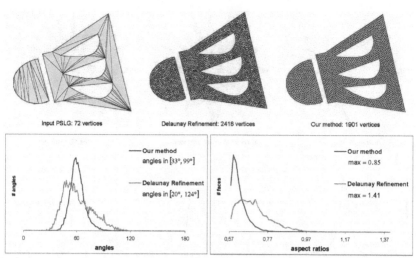

Fig. 7. The sizing function is uniform and equals 0.02. Our method inserts 22% fewer Steiner points than Delaunay refinement does. The angles are in $[33°, 99°]$ with our method, vs $[20°, 124°]$ with Delaunay refinement. The maximum aspect ratio is lower: 0.85 vs 1.41.

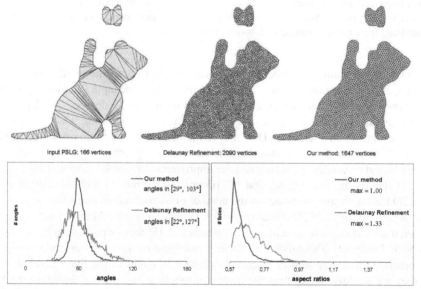

Fig. 8. The sizing function is uniform and equals 0.02. Our method inserts 23% fewer Steiner points than Delaunay refinement does. The angles are in $[29°, 103°]$ with our method, vs $[22°, 127°]$ with Delaunay refinement. The maximum aspect ratio is lower: 1.00 vs 1.33.

many other mesh optimization techniques is justified by the possibilities to apply it in 1D and 2D, and to take a sizing function as input with variable precision quadratures.

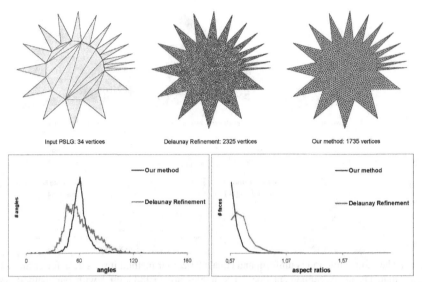

Fig. 9. The input PSLG constraints form acute angles from $5°$ to $37°$. The sizing function is uniform and equals 0.02. Our method inserts 26% fewer Steiner points than Delaunay refinement does. The angle distribution shows small angles due to the input. The angles range in $[5°, 128°]$ with our method, vs $[5°, 131°]$ with Delaunay refinement. As expected the upper bound on aspect ratios is similar with both methods (5.73).

One of the main problems in mesh generation is to limit the number of small angles, even when small angles are part of the input [10]. The input PSLG of Figure 9 contains 17 "spin", with inside angles ranging from 5 to 37 degrees. As depicted by the angles distributions, small angles do not hurt our algorithm, which does not produce more small angles than Delaunay refinement does. In addition, our algorithm inserts 26% fewer Steiner points than the terminator algorithm does. The practical upper bound on angles is lower, albeit the improvement is small ($128°$ vs $131°$).

As already discussed in Section 2.2, optimizing the mesh simultaneously in 1D and 2D has an important impact on the number of Steiner points inserted. For example, when the input PSLG contains two long constraints parallel and close to each other, the local improvement in terms of number of 1D Steiner points can be as high as 50%. Intuitively, Delaunay refinement is restricted to recursive constrained edge bisection, which may lead to bisect all constrained edges at the finest level, even if they are slightly longer than the local admissible sizing. In our algorithm, each refinement step is followed by an optimization step, which prevents the next refinement step to over-refine. This behavior is illustrated by Figure 10, where our method inserts 25% fewer Steiner points than Delaunay refinement alone.

For an input domain Ω and a sizing function μ, one can compute the minimum number of triangles needed to cover Ω. After computing the mean edge length over the domain: $\bar{\mu} = \frac{\int_\Omega \mu(x)dx}{area(\Omega)}$, the minimum number of triangles needed is given by $m_\triangle = \frac{area(\Omega)}{\sqrt{3}/4.\bar{\mu}^2}$, where the denominator is the mean area of an equilateral trian-

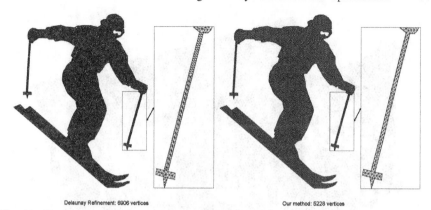

Delaunay Refinement: 6906 vertices   Our method: 5228 vertices

Fig. 10. The sizing function is uniform and equals 0.01. The input PSLG contained 248 vertices. Our method inserts 25% fewer Steiner points than Delaunay refinement does. The closeup depicts two constrained edges, parallel and close to each other. Combining 1D and 2D optimization allows inserting fewer Steiner points in this area.

gle inside the domain. In the uniform case, we observe that our method produces meshes with $1.60m_\triangle$ triangles, whereas Delaunay refinement produces meshes with $2.20m_\triangle$ triangles. These values are very consistent over all the examples we have tested. In the non-uniform case, $\bar{\mu}$ depends too much on the gradation of μ and on the domain geometry to be able to exhibit a representative average.

Figures 11, 12, 13 and 14 illustrate results from larger inputs and with non-uniform sizing functions, such as the one described by Alliez et al. [26] as: $\mu(x) = \inf_{s \in \partial\Omega}[kd(s, x) + lfs(s)]$, where $\partial\Omega$ is the domain boundary, d the Euclidian distance, lfs the local feature size, and k a constant. This sizing function is shown [26] to be the maximum k-Lipschitz function that is smaller or equal to lfs on $\partial\Omega$. In the non-uniform case, our algorithm inserts on average 28% fewer Steiner points than the standard Delaunay refinement.

4 Conclusion

We have presented a new 2D triangle mesh generation algorithm, interleaving Delaunay refinement and optimization using the Lloyd iteration. The meshes generated are bound to cover the input constraints provided as a PSLG, and to satisfy sizing constraints provided as a sizing function. Although our algorithm comes with no theoretical bounds on the triangle angles, we show experimental evidences that it produces meshes with fewer Steiner points (25% on average) than the standard Delaunay refinement algorithm, and of better quality. Another contribution of this paper is an efficient and robust algorithm for computing bounded Voronoi diagrams. We plan to integrate the latter into the CGAL library.

The main added value provided by the interleaved refinement and optimization steps in a multilevel manner is to provide the Lloyd optimization step with good

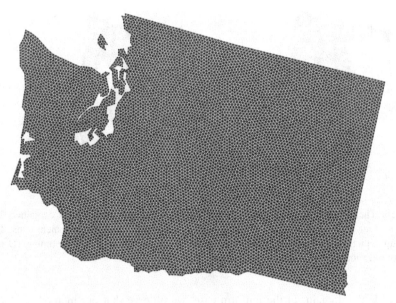

Fig. 11. The sizing function is uniform and equals 0.01. The initial PSLG contained 256 vertices. The final mesh contains 9761 vertices.

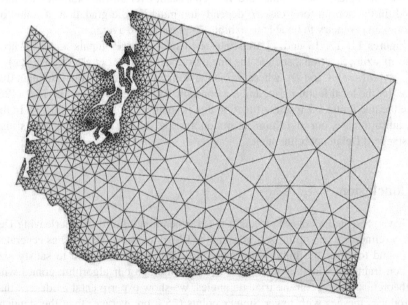

Fig. 12. Mesh generated with respect to the sizing function $\mu(x) = \inf_{s \in \partial\Omega}[0.45 * d(s, x) + lfs(s)]$. The initial PSLG contained 256 vertices. The final mesh contains 1440 vertices.

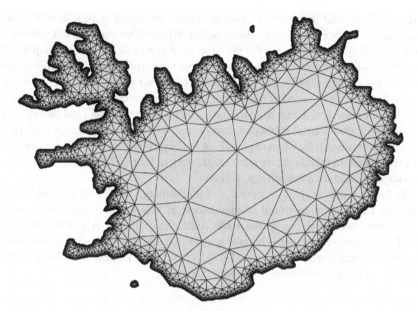

Fig. 13. Mesh generated with respect to the sizing function $\mu(x) = \inf_{s \in \partial \Omega}[d(s,x) + lfs(s)]$. The initial PSLG contained 3632 vertices. The final mesh contains 9691 vertices.

Fig. 14. Mesh generated with respect to the sizing function $\mu(x) = \inf_{s \in \partial \Omega}[0.6 * d(s,x) + lfs(s)]$. The initial PSLG contained 205 vertices. The final mesh contains 981 vertices.

initial solutions, and therefore to generate meshes of higher quality than the ones optimized once after refinement. More specifically, such interleaving not only contributes to obtain lower minima for the functional energy described in Section 2.2, but also prevents the Lloyd iteration to perform long range vertex relocations, and therefore to converge slowly. The final number of Steiner points is (non-trivially) related to the decreasing speed of the "target sizing" parameter used by the refinement step. The ultimate goal being to insert just enough Steiner points (and therefore to generate large well-shaped elements), it is desirable to slow down the decreasing of this parameter, especially when approaching the objective sizing function. Although our experiments in this direction were satisfactory in terms of Steiner points added, the computation times substantially increase.

As future work we plan to derive a "hill-climbing" version of our algorithm, where the Lloyd iteration would be modified so as to move each generator *toward* their centroid while staying within some fixed angle bounds. A challenging goal would be to provide theoretical bounds greater than the ones provided by Delaunay refinement. Finally, we plan to extend some of the ideas presented in this paper to isotropic tetrahedral mesh generation with constraints.

References

1. F.J.BOSSEN and P.S.HECKBERT. A pliant method for anisotropic mesh generation. In *Proceedings of 5th International Meshing Roundtable*, pages 63–74, oct 1996.
2. N.COLL, M.GUERRIERI, and J.A.SELLARÈS. Mesh modification under local domain changes. In *Proceedings of the 15th International Meshing Roundtable*, pages 39–56, 2006.
3. S.E.PAV G.L.MILLER and N.J.WALKINGTON. When and why Delaunay refinement algorithms work. *International Journal of Computational Geometry and Applications (IJCGA)*, 15(1):25–54, feb 2005.
4. P.CHEW. Constrained delaunay triangulations. *Algorithmica*, 4(1):97–108, 1989.
5. J.RUPPERT. A delaunay refinement algorithm for quality 2-dimensional mesh generation. *Journal of Algorithms*, 18(3):548–585, 1995.
6. J.R.SHEWCHUCK. Delaunay refinement algorithms for triangular mesh generation. *Computational Geometry*, 22:21–74, 2002.
7. A.ÜNGÖR. Off-centers: A new type of steiner points for computing size-optimal quality-guaranteed delaunay triangulations. In *LATIN: Latin American Symposium on Theoretical Informatics*, 2004.
8. A.N.CHERNIKOV and N.P.CHRISOCHOIDES. Generalized delaunay mesh refinement: From scalar to parallel. In *Proceedings of the 15th International Meshing Roundtable*, pages 563–580, 2006.
9. H.ERTEN and A.ÜNGÖR. Triangulations with locally optimal steiner points. *Eurographics Symposium on Geometry Processing*, 2007.
10. J.R.SHEWCHUCK. What is a good linear element? interpolation, conditioning, and quality measures. *11th International Meshing Roundtable*, pages 115–126, 2002.
11. M.BERN N.AMENTA and D.EPPSTEIN. Optimal point placement for mesh smoothing. *Journal of Algorithms*, 30(2):302–322, 1999.
12. D.EPPSTEIN. Global optimization of mesh quality. *Tutorial at the 10th International Meshing Roundtable*, 10, 2001.

13. L.CHEN. Mesh smoothing schemes based on optimal delaunay triangulations. In *Proceedings of 13th International Meshing Roundtable*, pages 109–120, 2004.
14. V.FABER Q.DU and M.GUNZBURGER. Centroidal voronoi tesselations: Applications and algorithms. *SIAM review*, 41(4):637–676, 1999.
15. M.EMELIANENKO Q.DU and L.JU. Convergence of the lloyd algorithm for computing centroidal voronoi tesselations. *SIAM Journal, Numerical Analysis*, 44(1):102–119, 2006.
16. Q.DU and M.EMELIANENKO. Acceleration schemes for computing centroidal Voronoi tessellations. *j-NUM-LIN-ALG-APPL*, 13(2–3):173–192, mar-apr 2006.
17. Q.DU and M.EMELIANENKO. Uniform convergence of a nonlinear energy-based multilevel quantization scheme via centroidal voronoi tessellations. *submitted to SIAM J. Numer. Anal.*, 2006.
18. S.LLOYD. Least square quantization in pcm. *IEEE Trans. Inform. Theory*, 28:129–137, 1982.
19. J.R.SHEWCHUK. Triangle: Engineering a 2d quality mesh generator and delaunay triangulator. *Applied Computational Geometry: Towards Geometric Engineering*, 1148:203–222, 1996.
20. M.GUNZBURGER Q.DU and L.JU. Constrained centroidal voronoi tessellations for surfaces. *SIJSSC: SIAM Journal on Scientific and Statistical Computing*, 24, 2003.
21. R.SEIDEL. Constrained Delaunay triangulations and Voronoi diagrams with obstacles. pages 178–191. 1988.
22. Q.DU and D.WANG. Recent progress in robust and quality delaunay mesh generation. *Journal of Computational and Applied Mathematics*, 2005.
23. A.FABRI, G.-J.GIEZEMAN, L.KETTNER, S.SCHIRRA, and S.SCHÖNHERR. On the design of CGAL, a computational geometry algorithms library. *Softw.–Pract. Exp.*, 30(11):1167–1202, 2000.
24. J.R.SHEWCHUK. Mesh generation for domains with small angles. *Proceedings of the sixteenth annual Symposium on Computational Geometry*, pages 1–10, 2000.
25. L.RINEAU. 2d conforming triangulations and meshes. In CGAL Editorial Board, editor, *CGAL User and Reference Manual*. 3.3 edition, 2007.
26. M.YVINEC P.ALLIEZ, D.COHEN-STEINER and M.DESBRUN. Variational tetrahedral meshing. *ACM Transactions on Graphics*, 24(3):617–625, july 2005.

A New *Meccano* Technique for Adaptive 3-D Triangulations

J.M. Cascón[1], R. Montenegro[2], J.M. Escobar[2], E. Rodríguez[2] and
G. Montero[2]

[1] Department of Mathematics, Faculty of Sciences, University of Salamanca,
Spain, `casbar@usal.es`
[2] Institute for Intelligent Systems and Numerical Applications in Engineering,
University of Las Palmas de Gran Canaria, Campus Universitario de Tafira,
Las Palmas de G.C., Spain, `rafa@dma.ulpgc.es`, `jescobar@dsc.ulpgc.es`,
`barrera@dma.ulpgc.es`, `gustavo@dma.ulpgc.es`

This paper introduces a new automatic strategy for adaptive tetrahedral mesh
generation. A local refinement/derefinement algorithm for nested triangulations and a simultaneous untangling and smoothing procedure are the main
involved techniques. The mesh generator is applied to 3-D complex domains
whose boundaries are projectable on external faces of a coarse object *meccano*
composed of cuboid pieces. The domain surfaces must be given by a mapping
between *meccano* surfaces and object boundary. This mapping can be defined
by analytical or discrete functions. At present we have fixed mappings with orthogonal, cylindrical and radial projections, but any other one-to-one projection may be considered. The mesh generator starts from a coarse tetrahedral
mesh which is automatically obtained by the subdivision of each hexahedra, of
a *meccano* hexahedral mesh, into six tetrahedra. The main idea is to construct
a sequence of nested meshes by refining only those tetrahedra which have a
face on the *meccano* boundary. The virtual projection of *meccano* external
faces defines a valid triangulation on the domain boundary. Then a 3-D local
refinement/derefinement is carried out such that the approximation of domain
surfaces verifies a given precision. Once this objective is reached, those nodes
placed on the *meccano* boundary are really projected on their corresponding
true boundary, and inner nodes are relocated using a suitable mapping. As the
mesh topology is kept during node movement, poor quality or even inverted
elements could appear in the resulting mesh. For this reason, we finally apply
a mesh optimization procedure. The efficiency of the proposed technique is
shown with several applications to complex objects.

1 Introduction

In finite element simulation in engineering problems, it is crucial to automatically adapt the three-dimensional discretization to geometry and to solution. Many authors have devoted great efforts in the past to solve this problem in different ways [3, 11, 13, 30], but automatic 3-D mesh generation is still an open problem. Generally, as the complexity of the problem increases (domain geometry and model), the methods for approximating the solution are more complicated. At present, it is well known that most mesh generator are based on Delaunay triangulation and advancing front technique.

In the last few years, we have developed a tetrahedral mesh generator that approximates the orography of complex terrains with a given precision [21, 22]. To do so, we only use digital terrain information. The generated mesh have been applied for numerical simulation of environmental phenomena, such as wind field adjustment [26], fire propagation or atmospheric pollution [25]. The following procedures were mainly involved in this former automatic mesh generator: a Delaunay triangulation method [5, 12], a 2-D refinement/derefinement algorithm [8], based on the 4-T Rivara's algorithm [27], and a simultaneous untangling and smoothing algorithm [6]. Many ideas were introduced in this mesh generator and they have been taken into account for developing the new mesh generator proposed in this paper.

On the other hand, local adaptive refinement strategies are employed to adapt the mesh to singularities of numerical solution. These adaptive methods usually involve remeshing or nested refinement [14, 17, 18, 28]. Another interesting idea is to adapt simultaneously the model and the discretization in different regions of the domain. A perspective about adaptive modeling and meshing is studied in [4]. The main objective of all these adaptive techniques is to achieve a good approximation of the *real* solution with a minimal user intervention and a low computational cost. For this purpose, the mesh element quality is also an essential aspect for the efficiency and numerical behaviour of finite element method. The element quality measure should be understood depending on the isotropic or anisotropic character of the numerical solution.

In this paper we present new ideas and applications of an innovative tetrahedral mesh generator which was introduced in [24]. This automatic mesh generation strategy uses no Delaunay triangulation, nor advancing front technique, and it simplifies the geometrical discretization problem in particular cases. The main idea of the new mesh generator is to combine a local refinement/derefinement algorithm for 3-D nested triangulations [17] and a simultaneous untangling and smoothing procedure [6]. 3-D complex domains, which surfaces can be mapped from a *meccano* face to object boundary, are discretized by the mesh generator. Resulting adaptive meshes have an appropriate quality for finite element applications.

At present, this idea has been implemented in ALBERTA code [1, 29]. This software can be used for solving several types of 1-D, 2-D or 3-D problems with adaptive finite elements. The local refinement and derefinement can be done by evaluating an error indicator for each element of the mesh and it is based on element bisection. To be more specific, the newest vertex bisection method is implemented for 2-D triangulations [20]. Actually, ALBERTA has implemented an efficient data structure and adaption for 3-D domains which can be decomposed into hexahedral elements as regular as possible. Each hexahedron is subdivided into six tetrahedra by constructing a main diagonal and its projections on its faces, see Figure 1 (a). The local bisection of the resulting tetrahedra is recursively carried out by using ideas of the longest edge [27] and the newest vertex bisection methods. Details about the local refinement technique implemented in ALBERTA for two and three dimensions can be analyzed in [17, 20]. This strategy works very efficiently for initial meshes with a particular topology and high-quality elements (obtained by subdivision of regular quadrilateral or hexahedral elements). In these cases, the degeneration of the resulting 2-D or 3-D triangulations after successive refinements is avoided. The restriction on the initial element shapes and mesh connectivities makes necessary to develop a particular mesh generator for ALBERTA. In this paper we summarize the main ideas introduced for this purpose. Obviously, all these techniques could be applied for generating meshes with other types of codes. Besides, these ideas could be combined with other type of local refinement/derefinement algorithms for tetrahedral meshes [14, 18, 28].

In the following section we present a description of the main stages of the new mesh generation procedure. In section 3 we show test problems and practical applications which illustrate the efficiency of this strategy. Finally, conclusions and future research are presented in section 4.

2 Description of the Mesh Generator

In this section, we present the main ideas which have been introduced in the mesh generation procedure. In section 2.1 and 2.2, we start with the definition of the domain and its subdivision in an initial 3-D triangulation that verifies the restrictions imposed in ALBERTA. In section 2.3, we continue with the presentation of different strategies to obtain an adapted mesh which can approximate the boundaries of the domain within a given precision. We construct a mesh of the domain by projecting the boundary nodes from a *meccano* plane face to the true boundary surface and by relocating the inner nodes. These two steps are summarized in section 2.4 and 2.5, respectively. Finally, in section 2.6 we present a procedure to optimize the resulting mesh.

2.1 Object *Meccano*

In order to understand the idea of the proposed mesh generator, it is convenient to first consider a domain whose boundary can be projected on the faces of a cube. A second simple case is to consider a cuboid instead of a cube. We can generalize the previous cases with a *meccano* constructed by

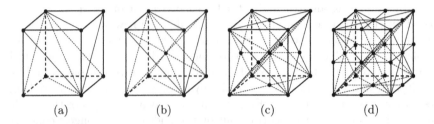

    (a)   (b)   (c)   (d)

Fig. 1. Refinement of a cube by using Kossaczky's algorithm: (a) cube subdivision into six tetrahedra, (b) bisection of all tetrahedra by inserting a new node in the cube main diagonal, (c) new nodes in diagonals of cube faces and (d) global refinement with new nodes in cube edges

cuboid pieces. In this last case, we suppose that an automatic decomposition of the *meccano* into cubes (or hexahedra) can be carried out. At present, we have implemented these cases in ALBERTA and we have defined the *meccano* input data as connected cuboids, such that the boundary of the object is obtained by a one-to-one projection (or mapping) from the boundary faces of the *meccano*.

2.2 Coarse Tetrahedral Mesh of the *Meccano*

Once the *meccano* decomposition into cubes is done, we build an initial coarse tetrahedral mesh by splitting all cubes into six tetrahedra [17]. For this purpose, it is necessary to define a main diagonal on each cube and the projections on its faces, see Figure 1(a). In order to get a conforming tetrahedral mesh, all cubes are subdivided in the same way maintaining compatibility between the diagonal of their faces. The resulting initial mesh τ_1 can be introduced in ALBERTA since it verifies the imposed restrictions about topology and structure. The user can introduce in the code the necessary number of recursive global bisections [17] for fixing a uniform element size in the whole initial mesh. In Figures 1 (b), (c) and (d) are presented three consecutive global bisections in a cube. The resulting mesh of Figure 1(d) contains 8 similar cubes to the one represented in Figure 1(a).

If we consider a *meccano* composed of hexahedra pieces instead of cuboids, a similar technique can be applied. In this case, the recursive local refinement [17] may produce poor quality elements depending on the initial mesh quality. The minimum quality of refined meshes is function of the initial mesh quality. A study about this aspect can be seen in [19, 31]. In this paper, as a first approach, we have used a decomposition of the object *meccano* into cuboids.

2.3 Local Refined Mesh of the *Meccano*

The next step of the mesh generator includes a recursive adaptive local refinement strategy of those tetrahedra with a face placed on a boundary face of the initial coarse mesh. The refinement process is done in such a way that the true surfaces are approximated with a linear piece-wise interpolation within a given precision. That is, we look for an adaptive triangulation on the *meccano* boundary faces, such that the resulting triangulation after node projection (or mapping) on the object true boundary is a good approximation of this boundary. The user has to introduce as input data a parameter ε that defines the maximum separation allowed between the linear piece-wise interpolation and the true surface. We remark that the true surface may be given by an analytical or a discrete function, such that each point of *meccano* boundary faces corresponds only to one point on the true surface. We propose two different strategies for reaching our objective.

The first one consists on a simple method. We construct a sequence of tetrahedral nested meshes by recursive bisection of all tetrahedra which contain a face located on the *meccano* boundary faces; see Figure 1. The number of bisections is determined by the user as a function of the desired resolution of the true surface. So, we have a uniform distribution of nodes on these *meccano* faces. Once all these nodes are *virtually* projected on the true surface, a generalization of the derefinement criterion developed in [8], with a given derefinement parameter ε, defines different adaptive triangulations for each *meccano* face. We remark that the generalized derefinement criterion fixes which nodes, placed on *meccano* faces, can not be eliminated in the derefinement process in order to obtain a good approach of the true surface.

To illustrate the new derefinement criterion, we have represented the projection of a *meccano* external node P on its true position P' placed on the object surface Σ, see Figure 2(a). We consider that node P is located in the middle point of its *surrounding edge ac*. The insertion of node P produces the bisection of the *father* triangle abc and, consequently, the bisection of the *father* tetrahedron (T) which contains the face abc. This tetrahedron is subdivided in two *sons* $(T_1$ and $T_2)$ which are defined by the partition of face abc into the two faces abP and bcP, respectively. If we project nodes a, b and c on surface Σ, we get a planar approximation given by the triangle $a'b'c'$, but if we consider the insertion of node P, we obtain an improved approximation given by the triangles $a'b'P'$ and $b'c'P'$. Therefore, we have introduced the following generalized derefinement condition:

Node P could be eliminated if the volume of *virtual* tetrahedron $a'b'c'P'$ is less than ε (although perhaps node P can not be finally removed due to conformity reasons). On the other hand, nodes belonging to the coarse initial mesh are not removed from the sequence of nested meshes.

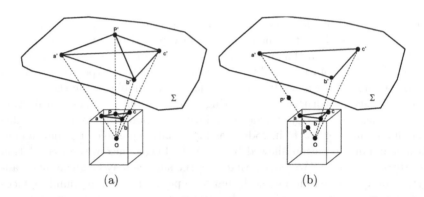

(a)   (b)

Fig. 2. Node mapping from a cube to real domain: (a) transformation of an external node P and (b) of an inner node P

As the derefinement criterion in ALBERTA is associated to elements, we mark for derefining all tetrahedra containing a node which can be eliminated (i.e., if node P can be eliminated, we mark tetrahedra T_1 and T_2 for derefining). In particular, we make a loop on tetrahedra during the derefinement process from the penultimate level of the sequence to the coarse initial mesh. We only analyze tetrahedra with two sons, such that if the node introduced in their *father*'s bisection verifies the derefinement condition, then we mark their two *sons* for derefining.

The second strategy only works with the local refinement algorithm. In this case, the refinement criterion for tetrahedron T should be:

Node P must be inserted if the volume of *virtual* tetrahedron $a'b'c'Q'$, being Q' the projection on Σ of any point Q belonging to face abc, is greater or equal than ε. Then, tetrahedron T should be subdivided in T_1 and T_2.

The first strategy is simpler, but it could lead to problems with memory requirements if the number of tetrahedra is very high before applying the derefinement algorithm. Suppose for example that we have the information of a surface defined on a meccano face by a discrete function on a very fine grid. In order to capture the surface geometry without losing information, we need to construct a global refined mesh as fine as the grid associated to the discrete function. Nevertheless, the user could control the number of recursive bisections and the maximum lost volume for the resulting object discretization.

On the other hand, the problem of the second strategy is to determine if a face placed on *meccano* boundary must be subdivided attending to the approximation of the true surface. This analysis must be done every time that a face is subdivided into its two *son* faces. Suppose for example that the true surface is given by a discrete function. Then, the subdivision criterion should stop for a particular face when all the surface discretization points, defined on this face, have been analyzed and all of them verify the approximation criterion. So, this second strategy has the inconvenient that each surface discretization point should be studied many times and, therefore, it generally implies a higher computational cost than the first strategy.

2.4 External Node Mapping on Object Boundary

Although ALBERTA has already implemented a node projection on a given boundary surface during the bisection process, it has two important restrictions: nodes belonging to the initial mesh are not projected and inverted elements could appear in case of projecting new nodes on complex surfaces (i.e. non-convex object). In the latter case, the code does not work properly, since it is only prepared to manage *valid* meshes.

For this reason, a new strategy must be developed in the mesh generator. The projection (or mapping) is really done once we have defined the local refined mesh by using one of the two methods proposed in the previous section. Then, the nodes placed on the *meccano* faces are projected (or mapped) on their corresponding true surfaces, maintaining the position of the inner nodes of the *meccano* triangulation. We have remarked that any one-to-one projection can be defined: orthogonal, spherical, cylindrical, etc. For example, spherical projection from point O has been used in Figure 2.

After this process, we obtain a valid triangulation of the domain boundary, but it could appear a tangled tetrahedral mesh. Inner nodes of the *meccano* could be located now even outside the domain. Thus, an optimization of the mesh is necessary. Although the final optimized mesh does not depend on the initial position of the inner nodes, it is better for the optimization algorithm to start from a mesh with a quality as good as possible. Therefore, we propose to relocate in a reasonable position the inner nodes of the *meccano* before the mesh optimization.

2.5 Relocation of Inner Nodes

There would be several strategies for defining an appropriate position for each inner node of the domain. An acceptable procedure is to modify their relative position as a function of the distance between boundary surfaces before and after their projections. This relocation is done attending to proportional criteria along the corresponding projection line. For example, relocation of inner node P in its new position P', such that $OP' = OP \times Oa' / Oa$, is represented in Figure 2(b). Although this node movement does not solve the tangle mesh problem, it normally makes it decrease. That is, the number of resulting inverted elements is less and the mean quality of valid elements is greater.

2.6 Object Mesh Optimization: Untangling and Smoothing

An efficient procedure is necessary to optimize the current mesh. This process must be able to smooth and untangle the mesh and it is crucial in the proposed mesh generator.

The most usual techniques to improve the quality of a *valid* mesh, that is, one that does not have inverted elements, are based upon local smoothing. In short, these techniques consist of finding the new positions that the mesh nodes must hold, in such a way that they optimize an objective function. Such a function is based on a certain measurement of the quality of the *local sub-mesh*, $N(v)$, formed by the set of tetrahedra connected to the *free node v*. As it is a local optimization process, we can not guarantee that the final mesh is globally optimum. Nevertheless, after repeating this process several times for all the nodes of the current mesh, quite satisfactory results can be achieved. Usually, objective functions are appropriate to improve the quality of a valid mesh, but they do not work properly when there are inverted elements. This is because they present singularities (barriers) when any tetrahedron of $N(v)$ changes the sign of its Jacobian determinant. To avoid this problem we can proceed as Freitag et al in [9, 10], where an optimization method consisting of two stages is proposed. In the first one, the possible inverted elements are untangled by an algorithm that maximises their negative Jacobian determinants [9]; in the second, the resulting mesh from the first stage is smoothed using another objective function based on a quality metric of the tetrahedra of $N(v)$ [10]. After the untangling procedure, the mesh has a very poor quality because the technique has no motivation to create good-quality elements. As remarked in [10], it is not possible to apply a gradient-based algorithm to optimize the objective function because it is not continuous all over $\mathbb{R}^3$, making it necessary to use other non-standard approaches.

We have proposed an alternative to this procedure [6], so the untangling and smoothing are carried out in the same stage. For this purpose, we use a suitable modification of the objective function such that it is regular all over $\mathbb{R}^3$. When a feasible region (subset of $\mathbb{R}^3$ where v could be placed, being $N(v)$ a valid submesh) exists, the minima of the original and modified objective functions are very close and, when this region does not exist, the minimum of the modified objective function is located in such a way that it tends to untangle $N(v)$. The latter occurs, for example, when the fixed boundary of $N(v)$ is tangled. With this approach, we can use any standard and efficient unconstrained optimization method to find the minimum of the modified objective function, see for example [2].

In this work we have applied, for simultaneous smoothing and untangling of the mesh by moving their inner nodes, the proposed modification [6] to one objective function derived from an *algebraic mesh quality metric* studied in [16], but it would also be possible to apply it to other objective functions which have barriers like those presented in [15].

Besides, a smoothing of the boundary surface triangulation could be applied before the movement of inner nodes of the domain by using the new procedure presented in [7] and [23]. This surface triangulation smoothing technique is also based on a vertex repositioning defined by the minimization of a suitable objective function. The original problem on the surface is transformed into a two-dimensional one on the *parametric space*. In our case, the parametric space is a plane, chosen in terms of the local mesh, in such a way that this mesh can be optimally projected performing a *valid* mesh, that is, without *inverted* elements.

3 Test Examples

The performance of our new mesh generator is shown in the following applications. The first example corresponding to the discretizacion of a *16th IMR conmemoration* glass and the second one is an Earth-Atmosphere system.

3.1 16th IMR conmemoration glass

We consider the discretization of a glass filled with a liquid, we include a legend and a face profile on the external face of the glass (see Figure 3(d)). The input data consists on a *meccano* of five cuboids for the glass and one cuboid for the liquid, see Figure 3(a). This *meccano* is included in a parallelepiped which dimensions are $5 \times 5 \times 8$. The mesh generation strategy automatically defines the boundary between the two materials and gets a good mesh adaption to the geometrical domain characteristics. A one-to-one projection between the *meccano* and the object is defined by a cylindrical projection for the vertical faces and an orthogonal one for the horizontal faces. We modify the original projection in order to include the legend *16th IMR* and the 3-D face profile of Igea on the external wall of the glass. The profile of Igea has been obtained from its 3D surface triangulation provided at *http://www.cyberware.com/*.

The *meccano* is splitted into a 3-D triangulation of 1146 tetrahedra and 320 nodes. We apply 6 recursive bisections on all tetrahedra which have a face placed on the *meccano* boundary or on the material interface. Additionally, we bisect 12 times all tetrahedra which have a face on the side of *meccano* where the legend and the face profile are included. This mesh contains 1189989 nodes and 5426256 tetrahedra. The derefinement parameter is fixed to $\varepsilon = 5 \; 10^{-6}$. The resulting adapted mesh contains 24258 nodes and 112388 tetrahedra and it is shown in Figure 3(b). Note that the refinement/derefinement algorithm captures the letters and the face profile. The projection of this *meccano* surface triangulation on the true surface produces a 3-D tangled mesh with 48775 inverted elements, see Figure 4. Relocation of inner nodes reduces the number of inverted elements to 2648. After ten iterations of the optimization procedure, the mesh generator converts the tangled mesh into the one presented in

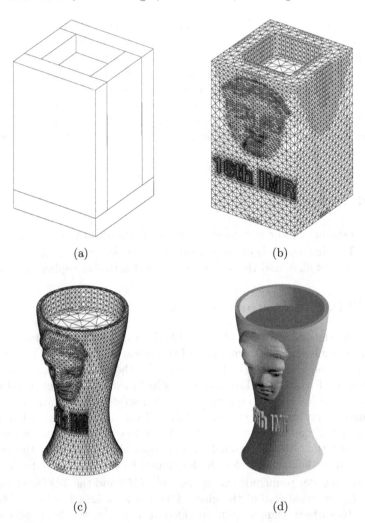

(a)   (b)

(c)   (d)

Fig. 3. Main stages of the mesh generator for the 16th IMR glass: (a) *meccano*, (b) mesh adaption after applying the refinement/derefinement procedure, (c) resulting optimized mesh, (d) texture map of the resulting mesh

Figure 3(c). The mesh quality is improved to a minimum value of 0.02 and an average $\overline{q}_\kappa = 0.55$. The quality curves for the initial tangled mesh and the final optimized triangulation are shown in Figure 5. There are a few elements (less than one hundred) with a poor quality (less than 0.1). This is due to rough profile of the letters, and could be avoided with a smoother definition.

The CPU time for constructing the initial mesh is approximately 1 minute and for its optimization is less than 3 minutes on a laptop with Intel proces-

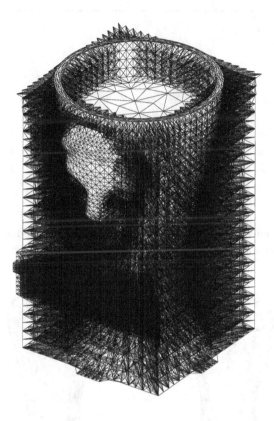

Fig. 4. Tangled mesh with 48775 inverted elements after the projection on glass surface and before relocation of inner nodes. The optimization process is able to convert this mesh in a valid one

sor 1.6 GHz, 1 Gb RAM memory on Linux Fedore Core 4 system. In order to show the efficiency of our method running on a workstation, we report in Table 1 data (number of nodes, tetrahedra, partial CPU times) of several meshes corresponding to the same example with different values of derefinement parameter. Note that the mesh optimization dominates the procedure.

The crucial step of the mesh generation process is the optimization algorithm. As a measurement of its capability notice that the procedure is able to convert the tangled mesh of Figure 4, without any relocation, into the mesh represented in Figure 3(c). However, it requires some more iterations of the optimization algorithm, and the corresponding increasing of CPU time. In Figure 6, we can appreciate the effect of the optimization process into the glass.

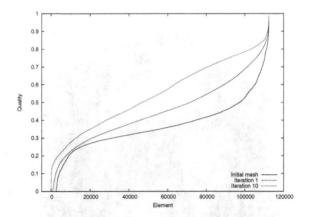

Fig. 5. Quality curves for the initial tangled mesh, an intermidiate one, and the optimized triangulation after ten iterations for the 16th IMR glass

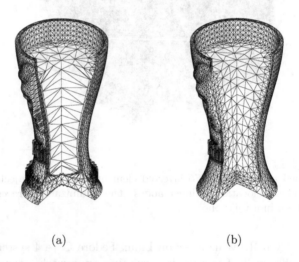

(a)   (b)

Fig. 6. Cross sections of 16th IMR glass: (a) After relocation of inner nodes, (b) after application of the optimization process

Finally, we remark that the mesh is made up of two different material: glass and liquid. The mesh generator automatically conserves the material interface and generates a conforming mesh between both materials.

3.2 Earth-Atmosphere

We have applied our technique to construct a 3-D triangulation of the Earth and its atmosphere. To define the topography we have used GTOPO30

ϵ	Nodes	Tetrahedra	CPU time (seconds)			
			Meccano Mesh	Projection/Relocation	Untangling	Smoothing
$5\ 10^{-6}$	24.258	112.388	54.3	0.3	35.4	39.5
10^{-6}	51.756	235.360	57.8	0.7	119.8	64.2
10^{-7}	94.488	419.632	61.3	1.3	393.3	84.3
10^{-8}	166.964	732.104	64.1	2.2	592.9	201.8

Table 1. Data corresponding to meshes of 16th-IMR glass example for several values of derefinement parameter. These experiments have been done on a Dell Precison 960, with two Intel Xeon doble kernel processor, 3.2 GHz, 64 bits and 8 Gb RAM, on a Red Hat Enterprise Linux WS v.4 system and using the compiler gcc v.3.4.6

(*http://edc.usgs.gov/products/elevation/gtopo30/gtopo30.html*). GTOPO30 is a global digital elevation model (DEM) with a horizontal grid spacing of 30 arc seconds (approximately 1 kilometer).

In this case, the *meccano* is made up of six cuboids for the atmosphere and one cuboid for the Earth. The union of all of them is a cube of dimension $8000 \times 8000 \times 8000$ kilometers. We use a radial projection to generate the topography of the surface of the Earth from GTOPO30, and to project the external faces of the *meccano* on a sphere of ratio 8000 kilometers.

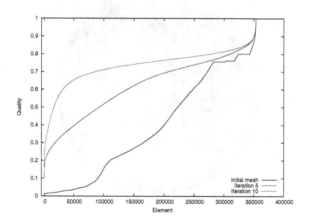

Fig. 7. Quality curves for the initial mesh, intermediate one and the optimized triangulation after ten iterations for the Earth-Atmosphere

The *meccano* is splitted into a 3-D triangulation of 3072 tetrahedra and 729 nodes. We apply 15 recursive bisections on all tetrahedra which have a face placed on the *meccano* boundary or on the interface between the cuboids representing the Earth and the atmosphere. This mesh contains 1119939 nodes and 5053392 tetrahedra. In order to obtain a final mesh with a good repre-

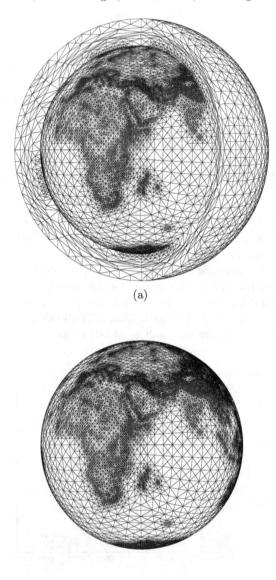

(a)

(b)

Fig. 8. Two views of the final mesh of the Earth and atmosphere

sentation of land areas but with an acceptable number of nodes, we used in this example different values of the derefinement parameter ε. The value of ε on land areas is fixed to $2.5\ 10^2$ and on ocean areas to $5\ 10^5$. The resulting mesh has 64901 nodes and 353142 tetrahedra; see Figure 8.

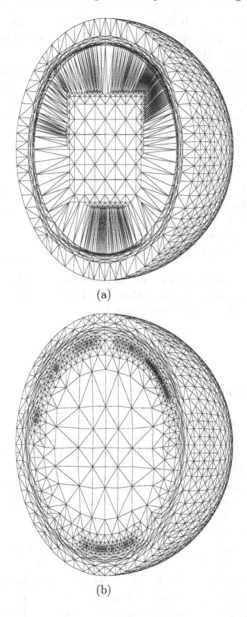

(a)

(b)

Fig. 9. Cross section of the Earth and atmosphere before (a) and after (b) the application of the mesh optimization process. The refined region in the interface corresponds to land areas

The relocation of inner nodes is enough to obtain a untangled mesh, however its quality is very poor. After ten iterations, the optimization procedure improves the value of the minimum quality from 0.01 to 0.17 and its average from $\overline{q}_\kappa = 0.38$ to $\overline{q}_\kappa = 0.73$; see Figure 7. The efficiency of this mesh optimization can be appreciated in Figure 9 where we represent two cross sections before (a) and after (b) its application.

The CPU time for constructing the mesh is approximately 10 minutes on a laptop, with Intel processor 1.6 GHz, 1 Gb RAM memory on Linux Fedore Core 4 system. In Table 2 we show data (number of nodes, tetrahedra, partial CPU times) of several meshes corresponding to different values of derefinement parameter.

ϵ				CPU time (seconds)			
Land	Sea	Nodes	Tetrahedra	Meccano Mesh	Projection/Relocation	Untangling	Smoothing
250	$5\,10^4$	64.901	353.142	57.3	1.3	0.0	281.0
50	10^5	105.013	563.092	63.1	2.1	0.0	427.7
5	10^4	165.704	863.818	63.6	3.1	0.0	637.0
0.5	10^3	319.851	1.601.800	67.7	5.8	0.0	1196.3

Table 2. Data corresponding to meshes of Earth-Atmosphere example for several values of derefinement parameter. These experiments have been done at the same workstation specified in caption of Table 1

4 Conclusions and Future Research

The proposed mesh generator is an efficient method for creating tetrahedral meshes on domains with boundary faces projectable on a *meccano* boundary. At present, the user has to define the meccano associated to the object and projections between meccano and object surfaces. Once these aspects are fixed, the mesh generation procedure is fully automatic and has a low computational cost. In future works, we will develop a special CAD package for minimizing the user intervention. Specifically, object surface patches should be defined by using meccano surfaces as parametric spaces.

The main ideas presented in this paper for automatic mesh generation, which have been implemented in ALBERTA, could be used for different codes which work with other tetrahedral or hexahedral local refinement/derefinement algorithms. Taking into account these ideas, complex domains could be meshed by decomposing its *outline* into a set of connected cubes or hexahedra. In future works, new types of pieces and connections could be considered for constructing the *meccano*.

Although this procedure is at present limited in applicability for high complex geometries, it results in a very efficient approach for the problems that fall within the mentioned class. The mesh generation technique is based on sub-processes (subdivision, projection, optimization) which are not in themselves new, but the overall integration using a simple shape as starting point is an original contribution of this paper and it has some obvious performance advantages. Besides, we have introduced a generalized derefinement condition for a simple approximation of surfaces. On the other hand, the new mesh generation strategy automatically fixes the boundary between materials and gets a good mesh adaption to the geometrical domain characteristics.

Acknowledgments

This work has been supported by the Spanish Government and FEDER, grant contracts: CGL2004-06171-C03-03/CLI and CGL2004-06171-C03-02/CLI. Besides, we thanks to the authors of ALBERTA [29] for the code availability in internet [1] and for their suggestions.

References

1. ALBERTA - An Adaptive Hierarchical Finite Element Toolbox, http://www.alberta-fem.de/
2. Bazaraa MS, Sherali HD, Shetty CM (1993) Nonlinear programing: theory and algorithms. John Wiley and Sons Inc, New York
3. Carey GF (1997) Computational grids: generation, adaptation, and solution strategies. Taylor & Francis, Washington
4. Carey GF (2006) A perspective on adaptive modeling and meshing (AM&M). Comput Meth Appl Mech Eng 195:214–235
5. Escobar JM, Montenegro R (1996) Several aspects of three-dimensional Delaunay triangulation. Adv Eng Soft 27:27–39
6. Escobar JM, Rodríguez E, Montenegro R, Montero G, González-Yuste JM (2003) Simultaneous untangling and smoothing of tetrahedral meshes. Comput Meth Appl Mech Eng 192:2775–2787
7. Escobar JM, Montero G, Montenegro R, E. Rodríguez E (2006) An algebraic method for smoothing surface triangulations on a local parametric space. Int J Num Meth Eng 66:740–760
8. Ferragut F, Montenegro R, Plaza A (1994) Efficient refinement/derefinement algorithm of nested meshes to solve evolution problems. Comm Num Meth Eng 10:403–412
9. Freitag LA, Plassmann P (2000) Local optimization-based simplicial mesh untangling and improvement. Int J Num Meth Eng 49:109–125
10. Freitag LA, Knupp PM (2002) Tetrahedral mesh improvement via optimization of the element condition number. Int J Num Meth Eng 53:1377–1391
11. Frey PJ, George PL (2000) Mesh generation. Hermes Sci Publishing, Oxford
12. George PL, Hecht F, Saltel E (1991) Automatic mesh generation with specified boundary. Comput Meth Appl Mech Eng 92:269–288
13. George PL, Borouchaki H (1998) Delaunay triangulation and meshing: application to finite elements. Editions Hermes, Paris
14. González-Yuste JM, Montenegro R, Escobar JM, Montero G, Rodríguez E (2004) Local refinement of 3-D triangulations using object-oriented methods. Adv Eng Soft 35:693–702

15. Knupp PM (2000) Achieving finite element mesh quality via optimization of the jacobian matrix norm and associated quantities. Part II-A frame work for volume mesh optimization and the condition number of the jacobian matrix. Int J Num Meth Eng 48:1165–1185

16. Knupp PM (2001) Algebraic mesh quality metrics. SIAM J Sci Comput 23:193–218

17. Kossaczky I (1994) A recursive approach to local mesh refinement in two and three dimensions. J Comput Appl Math 55:275–288

18. Löhner R, Baum JD (1992) Adaptive h-refinement on 3-D unstructured grids for transient problems. Int J Num Meth Fluids 14:1407–1419

19. Maubach J (1995) Local bisection refinement for n-simplicial grids generated by reflection. SIAM J Sci Comput 16:210–227

20. Mitchell WF (1989) A comparison of adaptive refinement techniques for elliptic problems. ACM Trans Math Soft 15:326–347

21. Montenegro R, Montero G, Escobar JM, Rodríguez E (2002) Efficient strategies for adaptive 3-D mesh generation over complex orography. Neural, Parallel & Scientific Computation 10:57–76

22. Montenegro R, Montero G, Escobar JM, Rodríguez E, González-Yuste JM (2002) Tetrahedral mesh generation for environmental problems over complex terrains. Lecture Notes in Computer Science 2329:335-344

23. Montenegro R, Escobar JM, Montero G, Rodríguez E (2005) Quality improvement of surface triangulations. Proc 14th Int Meshing Roundtable 469–484, Springer, Berlin

24. Montenegro R, Cascón JM, Escobar JM, Rodríguez E, Montero G (2006) Implementation in ALBERTA of an automatic tetrahedral mesh generator. Proc 15th Int Meshing Roundtable 325–338, Springer, Berlin

25. Montero G, Montenegro R, Escobar JM, Rodríguez E, González-Yuste JM (2004) Velocity field modelling for pollutant plume using 3-D adaptive finite element method. Lecture Notes in Computer Science 3037:642–645

26. Montero G, Rodríguez E, Montenegro R, Escobar JM, González-Yuste JM (2005) Genetic algorithms for an improved parameter estimation with local refinement of tetrahedral meshes in a wind model. Adv Eng Soft 36:3–10

27. Rivara MC (1987) A grid generator based on 4-triangles conforming. Mesh-refinement algorithms. Int J Num Meth Eng 24:1343–1354

28. Rivara MC, Levin C (1992) A 3-D refinement algorithm suitable for adaptive multigrid techniques. J Comm Appl Numer Meth 8:281–290

29. Schmidt A, Siebert KG (2005) Design of adaptive finite element software: the finite element toolbox ALBERTA. Lecture Notes in Computer Science and Engineering 42, Springer, Berlin

30. Thompson JF, Soni B, Weatherill N (1999) Handbook of grid generation, CRC Press, London

31. Traxler CT (1997) An algorithm for adaptive mesh refinement in n dimensions. Computing 59:115–137

1B.3

Well-centered Planar Triangulation – An Iterative Approach

Evan VanderZee[1], Anil N. Hirani[2], Damrong Guoy[3], and Edgar Ramos[4]

[1] vanderze@uiuc.edu, Department of Mathematics
[2] hirani@cs.uiuc.edu, Department of Computer Science
[3] guoy@uiuc.edu, Center for Simulation of Advanced Rockets
[4] eramosn@cs.uiuc.edu, Department of Computer Science

University of Illinois at Urbana-Champaign

Summary. We present an iterative algorithm to transform a given planar triangle mesh into a well-centered one by moving the interior vertices while keeping the connectivity fixed. A well-centered planar triangulation is one in which all angles are acute. Our approach is based on minimizing a certain energy that we propose. Well-centered meshes have the advantage of having nice orthogonal dual meshes (the dual Voronoi diagram). This may be useful in scientific computing, for example, in discrete exterior calculus, in covolume method, and in space-time meshing. For some connectivities with no well-centered configurations, we present preprocessing steps that increase the possibility of finding a well-centered configuration. We show the results of applying our energy minimization approach to small and large meshes, with and without holes and gradations. Results are generally good, but in certain cases the method might result in inverted elements.

1 Introduction

A *well-centered* mesh is a simplicial mesh in which each simplex contains its circumcenter. A 3D example is a tetrahedral mesh in which the circumcenter of each tetrahedron lies inside it, and circumcenter of each triangle face lies inside it. In this paper we address the case of planar well-centered triangulations, i.e., triangle meshes in which each triangle is acute angled. Typical meshing algorithms do not guarantee this property. For example, a Delaunay triangulation is not necessarily well-centered. We present an iterative energy minimization approach in which a given mesh, after possible preprocessing, is made well-centered by moving the internal vertices while keeping the boundary vertices and mesh connectivity fixed. The preprocessing step may add vertices or change the mesh connectivity, as this is sometimes necessary to permit a well-centered mesh.

A well-centered (primal) mesh has a corresponding dual mesh assembled from a circumcentric subdivision [14]. For an n-dimensional primal mesh, a k-simplex in the primal corresponds to an $(n - k)$-cell in the dual. In a well-centered planar triangle mesh, the dual of a primal interior vertex is a convex polygon with boundary edges that are orthogonal and dual to primal edges. This orthogonality makes it possible to discretize the Hodge star operator of exterior calculus [1] as a diagonal matrix which simplifies certain computational methods for solving partial differential equations. Some numerical methods that mention well-centered meshes in this context are the covolume method [17] and discrete exterior calculus [8, 14]. Well-centered meshes are not strictly required for these or other related methods. However, some computations may be easier if such meshes were available. Another example from scientific computing is space-time meshing. When tent-pitching methods for space-time meshing were first introduced, the initial spatial mesh was required to be well-centered [19]. More recently, this requirement has been avoided, although at the expense of some optimality in the construction [12].

2 Previous Results

We are mostly concerned with planar triangulations where the domain is specified by a polygonal boundary or, more general, a *straight line graph*. In addition to the triangulations being acute, we are also mostly interested in *quality triangulations* in which a lower bound on the triangle angles is achieved. Relevant work can be divided into constructive and iterative approaches.

Constructive approaches start with the specified input boundary/constraints and generate additional points and a corresponding triangulation. Normally a point is committed to a position and never moved afterwards. An algorithm for non obtuse triangulations based on circle packings is described in [3], and more recent works describe improved constructions while also describing how to derive an acute triangulation from a non obtuse one [15, 20]. In fact, these algorithms aim to achieve a triangulation of size linear in the input size, and so the smallest angle can be arbitrarily close to zero. It is not straightforward to extend this class of algorithms to acute quality triangulations. There are also algorithms that achieve acute quality triangulations for limited domains for point sets [4], or non obtuse quality triangulations [16]. Also relevant is an algorithm that, given a constraint set of both points and segments, finds a triangulation that minimizes the maximum angle [11], without adding points. If an acute triangulation exists for the input constraints, the algorithm will find it, otherwise it fails. Thus, there appears to be a no complete constructive solution for the generation of acute quality triangulations.

On the other hand, there are *iterative* or *optimization* approaches which allow an initial triangulation (possibly the canonical Delaunay) and then move the points while possibly changing the connectivity. This is the class of algorithms in which we are primarily interested: there are well-known algorithms to generate quality triangulations [10, 18] for which reliable implementations exist and so they are good candidates as starting points for iterative approaches that seek to achieve acute angles. In this class there are optimization approaches like centroidal Voronoi diagrams [9] and variational triangulations [2]. Each approach has a global energy function that it attempts to minimize through an iterative procedure that alternates between updating the location of the mesh vertices and the triangulation of those vertices. The energy functions optimized in these approaches are designed to optimize certain qualities of the mesh elements, but they do not explicitly seek well-centered simplices. Though these methods appear to produce nice experimental results, there is no guarantee that they construct quality triangulations, much less acute ones. In fact, in Sect. 5 we show examples in which the converging triangulation is not acute. Also, only limited convergence results are known (there are indeed local minima that can be reached). In addition to the optimization approaches that work directly with a mesh, there are several algorithms that generate circle packings or circle patterns by optimizing the radii of the circles. In particular the algorithms for creating circle patterns that were proposed in [7] and [5] can be adapted to create triangulations. These algorithms produce circle patterns that have specified combinatorics but they do not permit a complete specification of the domain boundary. Thus they are not appropriate to our purpose.

3 Iterative Energy Minimization

Given a simplicial mesh of a two-dimensional planar domain, we iteratively modify the mesh guided by minimizing a cost function defined over the mesh. We'll refer to the cost function as *energy*. Our method is somewhat similar to the methods of [9] and [2] in that it uses an iterative procedure to minimize an energy defined on the mesh, but it differs in that the mesh connectivity and boundary vertices remain fixed as the energy is minimized. We also minimize a different energy, one designed to achieve well-centeredness. We next describe this energy, which is the main component of our method. In Sect. 4 we describe preprocessing steps that may add vertices and change the connectivity of the mesh before energy minimization is begun.

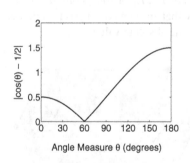

Fig. 1. Graph of $|\cos(\theta) - 1/2|$

Given a simplicial mesh $\mathcal{M}$ of a planar region, the mesh is well-centered if and only if every angle θ in every triangle in the mesh has angle measure strictly less than 90°. This suggests minimizing the maximum angle in the mesh. The cosine of an angle, which can be computed with a dot product and division by vector norms, is easier to compute from a mesh than the actual angle, so a good alternative to directly minimizing the maximum angle is to minimize the maximum $|\cos(\theta) - 1/2|$ over all angles θ. As the graph of $|\cos(\theta) - 1/2|$ in Fig. 1 shows, if any angle of the mesh is obtuse, then the largest angle of the mesh will dominate the expression, but when the mesh is well-centered and all angles are acute, the expression penalizes small angles as well as large ones. This is an added benefit, since small angles may be considered poor quality in some scientific computing applications. We also note the nice feature that a mesh composed of entirely equilateral triangles achieves the minimum energy, taking a value of 0. The key feature, though, is that the energy maintains the property that any acute triangulation is of lower energy than any nonacute triangulation.

As stated, this energy faces several difficulties, in particular some problems with differentiability. In many cases the gradient is not well-defined in areas of the mesh away from the current maximum angle. This problem could probably be addressed by using techniques of nonsmooth analysis such as the taking the minimum norm member of a generalized gradient [6], but a simpler alternative is to note that

$$\lim_{p\to\infty} \left(\sum_i |\cos(\theta_i) - 1/2|^p \right)^{1/p} = \max_i |\cos(\theta_i) - 1/2|,$$

and make a reasonable modification of the energy to use, for some finite p, the energy

$$E_p(\mathcal{M}) = E_p(\mathcal{V}, \mathcal{T}) = \sum_{\theta \in \mathcal{M}} |\cos(\theta) - 1/2|^p. \tag{1}$$

The mesh $\mathcal{M}$ is defined by the collection of vertex locations $\mathcal{V}$ and collection of triangles $\mathcal{T}$. The sum here is taken over every angle θ of every triangle $T \in \mathcal{T}$.

Unfortunately, in taking a fixed p we lose the property that every well-centered mesh is of lower energy than any mesh with an obtuse angle. On the other hand, using E_p we gain some sense of the quality of a triangle in regions of the mesh away from the maximum angle. Moreover, we know that for any particular domain and mesh connectivity that admit a well-centered configuration of the vertices, there is a finite p such that a given well-centered configuration is of lower energy than any non-well-centered arrangement of the vertices. It is computationally infeasible to compute E_p as p approaches infinity, but globally minimizing E_4, E_6, E_8 or some combination of them has sufficed in our experiments to date. Note that an even power p is preferable, since that makes it unnecessary to explicitly take the absolute value.

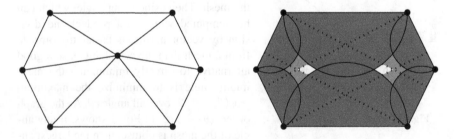

Fig. 2. Necessity of a Nonconvex Energy

The energy we have defined has the undesirable feature of being nonconvex. For energy minimization, one might hope to develop an energy that is convex or at least has a unique minimum. It is not possible, though, to define an energy that accurately reflects the goals of well-centered meshing and also has a unique minimum. Consider the mesh shown on the left in Fig. 2 where only the interior vertex may move. We want to relocate the interior vertex and obtain a well-centered mesh. The right side of Fig. 2 shows the constraints on where the vertex can be placed to produce a well-centered mesh. The lighter gray regions are forbidden because placing the interior vertex there would make some boundary angle nonacute. (The dotted lines show the four boundary angles that are most important in defining this region.) The darker gray regions, which overlay the lighter gray regions, are forbidden because placing the interior vertex there would make some angle at the interior vertex nonacute.

If the interior vertex is placed in either of the two small white regions that remain, the mesh will be well-centered. We see that the points permitted for well-centeredness form a disconnected set in $\mathbb{R}^2$. Moreover, the mesh is radially symmetric, so there is no way to create an energy that prefers one white region over the other unless we violate the desired property that the energy be insensitive to a rotation of the entire mesh. Any symmetric energy that has minima in only the white regions must have at least two distinct global minima and is not convex.

In most planar meshes there is an interior vertex v that has exactly six neighbors, all of which are interior vertices. If all interior vertices are free to move, as we assume in the method we propose, then the six neighbors of v can be rearranged to match a scaled version of the boundary of the mesh in Fig. 2. Moving v around when its neighbors have such a configuration should exhibit nonconvexity in whatever energy for well-centeredness we might define. One certainly can define convex energies on a mesh, but for the mesh in Fig. 2, no symmetric strictly convex energy will have a minimum at a well-centered mesh.

4 Neighborhood Improvement

In some cases the mesh connectivity or the fixed boundary vertices are specified in a way that no well-centered mesh exists. A simple example of this is an interior vertex with fewer than five neighbors. Any such vertex has some adjacent angle of at least $90°$. Similarly, boundary vertices need to have enough interior neighbors to divide the boundary angle into pieces strictly smaller than $90°$. We will refer to a vertex that does not have enough neighbors as a *lonely* vertex.

To address the problem of lonely vertices we propose a preprocessing step involving three simple operations for increasing the valence of vertices in the mesh. By applying these three operations – edge flip, edge split, and triangle subdivision – one can guarantee that no mesh vertex is lonely. The authors plan to include a formal proof of that in a sequel paper, though a thoughtful consideration of the double triangle subdivision (Sect. 4.3) may make it clear to the reader. The condition that each vertex have enough neighbors is not sufficient to guarantee that a well-centered arrangement of vertices exists. In our experience this situation – in which there is no lonely vertex and no well-centered configuration possible – appears to occur rarely in meshes we encountered in practice. This issue, too, will be addressed in a future paper. We next describe the three preprocessing steps and an algorithm for applying these steps in a way that limits the number of new vertices introduced.

4.1 Edge flip

Figure 3 is a graphical illustration of the well-known edge flip operation. The initial mesh is shown on the left, with the edge that will be deleted drawn as a dotted line rather than a solid one. The initial mesh has exactly one lonely vertex. The final mesh is displayed at right, with the edge that replaced the deleted one shown as a dotted line. The final mesh has no lonely vertices, and is, in fact, a well-centered mesh.

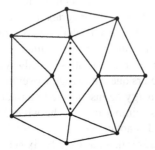 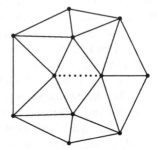

Fig. 3. Edge Flip

In the edge flip operation there are two vertices whose valence decreases and two vertices whose valence increases. We permit an edge flip for increasing valence of a lonely vertex only if no new lonely vertices are introduced. For example, if the vertices whose valence decreases are both interior vertices, they must initially have valence at least six. We also require that the edge-flip not introduce an inverted triangle. Triangle inversion is rarely a problem, but it is possible.

4.2 Edge split

The edge split operation shown in Fig. 4 is a more versatile operation than the edge flip. In the initial mesh shown at left, the edge that will be deleted through the edge split operation is drawn as a dotted line. There is exactly one lonely vertex in the initial mesh. In the final mesh shown on the right, there are three new edges (the dotted lines), and one new vertex (the empty circle). The final mesh is not well-centered, but it has no lonely vertices, and it is quite easy to find a well-centered configuration of the mesh by relocating the new vertex. Note that we do not permit an edge flip operation in this case since all the candidate edges for flip have an endpoint with valence five.

For definiteness of location, the new vertex is introduced at the midpoint of the edge. The edge being split and the one being deleted both have to be interior edges (we do not want to split a boundary edge this way, because the new vertex would be lonely, having only one interior neighbor). Thus the new vertex is always an interior one, and it always gets exactly five neighbors, so it is not a lonely vertex. Of the vertices that appeared initially there are two whose valence increases and one whose valence decreases. We permit the edge split only if it will not introduce a new lonely vertex and will not introduce an inverted triangle. Again, triangle inversion is rare but possible.

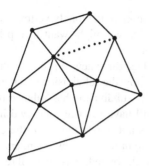
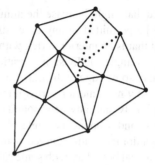

Fig. 4. Edge Split

4.3 Triangle subdivision

The edge split operation is quite versatile, since in the interior of the mesh there is usually a choice of both which edge to split and which endpoint of the edge will decrease in valence. There are some infrequent cases, though, when something more than the edge split may be needed. Except for the small example meshes we created to illustrate these cases, we have not needed anything else in our experiments, but for completeness we propose a third operation, triangle subdivision, that will work in all cases.

Figure 5 shows how a triangle subdivision can be used to increase the valence of a vertex in the mesh. The initial mesh (left) has exactly one lonely vertex. The final mesh (right) has six new edges, shown as dotted lines, and three new vertices, shown as empty circles. The final mesh has some angles that are much larger than the largest angle of the initial mesh, but it has no lonely vertices, and one can obtain a well-centered configuration of the mesh by relocating all of the interior vertices appropriately.

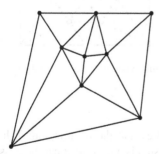
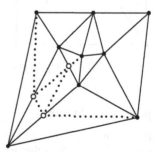

Fig. 5. Triangle Subdivision

If the mesh shown in Fig. 5 were a submesh of a larger mesh, it might be possible to do an edge split to increase the valence of the lonely vertex, but if this submesh occurs with the two edges at the top along the boundary, then no edge split will be permitted that might increase the number of neighbors of the lonely vertex. No edge flip will be permitted in any case, since every edge we might want to flip has an endpoint that is an interior vertex with valence five.

In the triangle subdivision operation we introduce three new vertices, each at the midpoint of some edge of the initial mesh. We do allow the subdivision of a boundary triangle. Any new interior vertex introduced by the subdivision has valence five, so it is not lonely, and any new boundary vertex will have valence four (two interior neighbors) and a boundary angle of 180°, so it is not lonely either. In the case of meshing a domain with a curved boundary, one might move a new boundary vertex onto the actual boundary with some rule, but unless the boundary is not well resolved, the boundary angle will still measure much less than 270°, so a new boundary vertex will not be lonely.

There is a case when a single triangle subdivision is not sufficient to increase the valence of a lonely vertex. In that case, one can perform a pair of triangle subdivisions to increase the number of neighbors of the lonely vertex. This double triangle subdivision can be performed within a single triangle, so we see that we can increase the valence of any lonely vertex. The double subdivision is illustrated in Fig. 6, where the initial mesh on the left is taken to be the entire mesh, i.e., a mesh of the square with four triangles. Again we see that the final mesh has some very large angles. Increasing the valence of lonely vertices is only a preprocessing step, though, and we note that there is a well-centered mesh with the same connectivity and boundary vertices as the final mesh in Fig. 6.

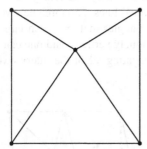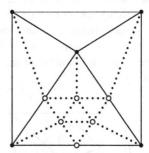

Fig. 6. Double Triangle Subdivision

4.4 Preprocessing algorithm

The preprocessing algorithm in Fig. 7 guarantees that the resulting mesh has no lonely vertices. The result of the preprocessing step is a mesh that can be passed to the energy minimization method. The repeat-until loop of the algorithm is necessary because a mesh may have an interior vertex with only three neighbors. For such a vertex we would need to increase the valence twice to keep the vertex from being lonely.

Using an operation increases the valence of vertices other than the targeted lonely vertex. This can make it possible to perform an operation at another vertex where

```
for each lonely vertex v
   repeat
      perform first permitted operation among:
           edge flip,
           edge split,
           triangle subdivision,
           double triangle subdivision;
   until v is not lonely
end
```

Fig. 7. Preprocessing Algorithm

previously that operation was not permitted. For example, sometimes flipping one edge lets us flip another edge that we were not initially allowed to flip. For this reason it may be advantageous to implement the algorithm in stages. Each stage would look like the algorithm in Fig. 7, but the earlier stages would have a shorter list of operations. For example, one might allow only edge flips in the first stage, then allow edge flips or edge splits in the second stage, and allow any operation in a third stage. In the earlier stages, the until condition would need to be modified to detect the possibility that vertex v was unchanged, i.e., that no operation was permitted among the list of allowed operations.

Repeating stages in the algorithm can also be advantageous. Our current implementation is the three-stage example just mentioned, with the first stage repeated twice before moving on to the second stage. We implement the algorithm this way in an attempt to limit the number of new vertices. To limit the number of new vertices even more, one could implement the preprocessing step to continue flipping edges until no more edge flips were allowed, then split an edge and return to flipping edges, etc., using the more versatile operations (and adding vertices) only when the simpler operations cannot be applied anywhere in the mesh. On the other hand, if addition of vertices is not a concern, one can perform a triangle subdivision for each lonely vertex and be done with the preprocessing.

5 Experimental Results

In this section we give some experimental results of applying our energy minimization to a variety of meshes. The algorithm for the preprocessing step has been already described in Section 4 and it was implemented in MATLAB.

We also used MATLAB for the energy minimization, implementing the conjugate gradient method with the Polak-Ribiere formula for modifying the search direction [13]. If $\mathbf{g}_k$ is the gradient at step k and $\mathbf{s}_k$ is the search direction at step k in the conjugate gradient method, then $\mathbf{s}_{k+1} = -\mathbf{g}_{k+1} + \beta_{k+1}\mathbf{s}_k$ for some scalar β_{k+1}. In the Polak-Ribiere formula, we have $\beta_{k+1} = ((\mathbf{g}_{k+1} - \mathbf{g}_k) \cdot \mathbf{g}_{k+1})/(\mathbf{g}_k \cdot \mathbf{g}_k)$ instead of the more common Fletcher-Reeves formula, $\beta_{k+1} = (\mathbf{g}_{k+1} \cdot \mathbf{g}_{k+1})/(\mathbf{g}_k \cdot \mathbf{g}_k)$. For a problem with n free variables, we reset the search direction to the negative gradient after every n iterations. We discovered that MATLAB's fminbnd function was not

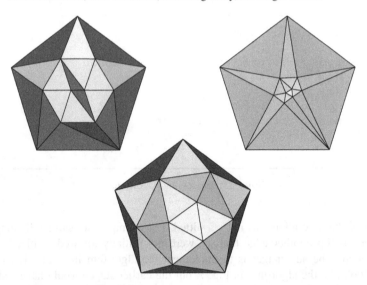

Fig. 8. Regular pentagon on top left is not well-centered. On top right is the well-centered mesh obtained by our method by applying 14 steps of conjugate gradient minimization of E_4. This particular well-centered mesh is not ideal for the finite element method (which is not our target application anyway), but some well-centered meshes produced by our method are appropriate for that application. See for example, the mesh in Fig. 10. We have not yet studied the effects of our algorithm on element aspect ratios. The second row shows the result of applying variational triangle meshing to the initial mesh. We used a boundary fixing variant of the 2D version of the variational tetrahedral meshing algorithm [2]. The resulting mesh shown is not well-centered.

performing a good line search, so we implemented our own (rather expensive) line search. Our line search takes samples of the function on a log scale to determine where the function decreases and where it increases. Then near the minimum we take evenly spaced samples of the function to get better resolution of where precisely the minimum lies. The line search should eventually be replaced by something more efficient. In what follows, *number of iterations* refers to the number of iterations of the conjugate gradient method.

In all the experiments, our MATLAB code can successfully decrease maximum angle below 90° with little or no degradation of minimum angle of the mesh. Some experiments show impressive improvement in both maximum and minimum angle.

Shading scheme: For all the meshes shown hereafter the shading indicates triangle quality with regard to well-centeredness. The shade of a triangle is based on the cosine of the largest angle of the triangle. Darker shade indicates greater largest angle and there is a noticeable jump at 90° so that well-centered triangles can be distinguished from those that are not. For example, the ten triangles that are not well-centered in the initial mesh on the left in Fig. 8 should be easily identifiable.

5.1 Small Meshes

The top row of Fig. 8 shows a test involving a small mesh of a regular pentagon and the well-centered mesh we obtained. Fourteen iterations using the energy E_4, results in the well-centered mesh shown. The final mesh on the right in top row of Fig. 8 has some long, thin isosceles triangles and a rather abrupt change from small triangles in the center to large triangles along the boundary. These features may be unusual compared to an intuitive idea of a nice mesh, but they are permitted in a well-centered planar triangulation, and, in this case, essential to getting a well-centered triangulation with the given boundary vertices and connectivity. We acknowledge that this particular well-centered mesh would not be of high quality for the finite element method, but in some cases, such as the mesh in Fig. 10, making a mesh well-centered improves the aspect ratio of the majority of the mesh elements. We have not yet systematically studied the effects of our algorithm on element aspect ratios, since the finite element method is not the primary motivation of the work on well-centered triangulations.

In Sect. 2 we mentioned variational triangulations. These are based on an iterative energy minimization algorithm introduced in [2] for tetrahedral meshing. We adapted it for comparison with our method. Our adaptation keeps boundary vertices fixed, but is otherwise analogous to the algorithm given in [2]. The bottom row of Fig. 8 shows the result of applying variational triangle meshing to the initial mesh on top left. The result is shown after 10 iterations, which is quite near convergence. The vertices are spread out, and the triangles in the middle of the mesh are nice, but the boundary triangles are all obtuse. The energy used for variational triangle meshing does not detect a benefit of clustering the interior vertices near the center of the pentagon. We tried variational triangle meshing for several of our meshes, sometimes obtaining good results and sometimes not. In most of the failures the method converged to a mesh containing at least one lonely vertex.

We also tried Centroidal Voronoi Tessellation (CVT) [9] for several of our meshes. For our implementation of CVT, we kept boundary vertices fixed, along with any other vertices that had unbounded Voronoi cells. Vertices in the interior of the mesh with bounded Voronoi cells were moved to the centroid of their full Voronoi cells (not clipped by the domain). The preliminary CVT experiments were inconclusive regarding an explanation for why CVT produces a well-centered mesh in some cases and not in others and more analysis is required in this direction. The mesh shown in top left of Fig. 9 is one of the cases for which centroidal Voronoi tessellation did not work. The actual initial mesh is shown on the left in middle row of Fig. 9, but CVT depends on only the vertex positions and uses the Delaunay triangulation of the point set, so we show the Delaunay triangulation and the bounded Voronoi cells in Fig. 9. We see that this mesh is, in fact, a fixed point of our implementation of CVT while being far from well-centered. Note also that the pattern of the mesh can be extended, and it will remain a fixed point of CVT.

There are other extensible patterns that are not well-centered but are fixed points of the CVT variant that clips Voronoi cells to the domain and allows boundary vertices to move within the boundary. Our method yields a well-centered mesh with 30 iterations of energy minimization using E_4 (Fig. 9). That figure also shows how the energy and the maximum angle evolve. The graphs show that the method is nearing convergence at 30 iterations and that decreases in the energy E_4 do not always correspond to decreases in the maximum angle of the mesh.

5.2 Larger Meshes

These first two examples shown in Figures 8 and 9 are for small toy meshes. We have also done experiments with larger meshes. For the largest of these meshes it is difficult to see detail in a small figure, hence for this paper we include only midsize meshes such as the meshes in Fig. 10 and Fig. 11. The largest mesh we have used in our experiments has 2982 vertices and 5652 triangles (about four times as many as the mesh in Fig. 11). That mesh and several others of similar size that we tried all became well-centered without complications.

The initial mesh in the top row of Fig. 10 has some lonely vertices, and the result of preprocessing the mesh with the operations described in Sect. 4 is shown in middle row. The well-centered mesh obtained by 30 iterations minimizing E_4 appears in the bottom row of Fig. 10. Histograms of the minimum and maximum angles of each triangle are included beside each mesh. We see that the preprocessing step introduces several new large angles, but the energy minimization takes care of these, finding a mesh with maximum angle approximately $82.38°$, and having most triangles with maximum angle in the range $[62, 76]$.

5.3 Meshes Requiring Retriangulation

Next, we show a mesh for which our energy could not find a well-centered configuration. However, when we applied our method after a retriangulation of the same set of vertices, we did obtain a well-centered mesh. The initial mesh is show on the top left in Fig. 11. The mesh has no lonely vertices, so we apply the energy minimization directly. After 500 iterations using the energy E_4, we obtain the mesh shown at top center in Fig. 11. The shading shows that the general quality is much improved, but there is a problem. In the top right corner of the result mesh there are some inverted triangles. A zoom on that portion of the mesh is displayed at top right in Fig. 11. Inversion of triangles is rare, since it requires some angle of the mesh to reach $180°$, but for the same reason, when inversion does occur, the inverted triangles tend to stay inverted.

The standard method does not work, but there are other ways to get a well-centered mesh. One way is to try a completely different connectivity for the same vertex set. The middle row of Fig. 11 shows the Delaunay triangulation of the two holes mesh after preprocessing has been applied. Along with more than twenty edge flips, the preprocessing step included eight edge splits, which produced the eight groups of bad triangles along the inner boundary. Five hundred iterations with energy E_4 produce a fairly good result with several slightly obtuse triangles, so we follow that with 500 iterations using the energy E_8. The E_8 energy focuses more on the largest angles of the mesh and less on the general quality of the triangles, producing a well-centered result, which appears at middle right in Fig. 11.

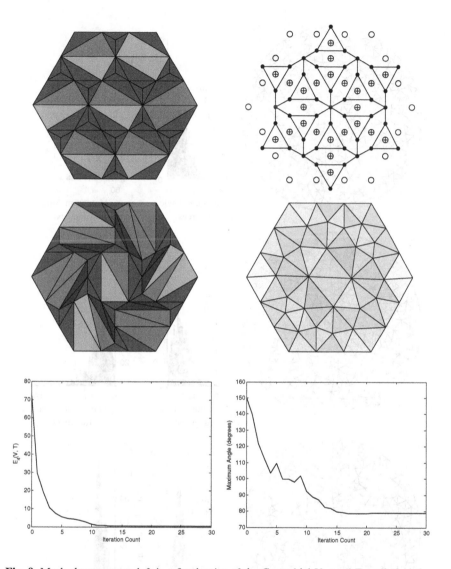

Fig. 9. Mesh shown on top left is a fixed point of the Centroidal Voronoi Tessellation algorithm [9] but it is far from being well-centered. It is the Delaunay triangulation corresponding to the initial mesh which is shown in middle left (CVT uses Delaunay triangulation). Bounded Voronoi cells are shown in top right figure, with vertices denoted by empty circles and centroids of Voronoi cells by plus symbols. Starting with initial mesh in middle left, 30 iterations of our energy minimization using E_4 yields the well-centered mesh in middle right. The evolution of energy and maximum angle observed during energy minimization is shown in the bottom row.

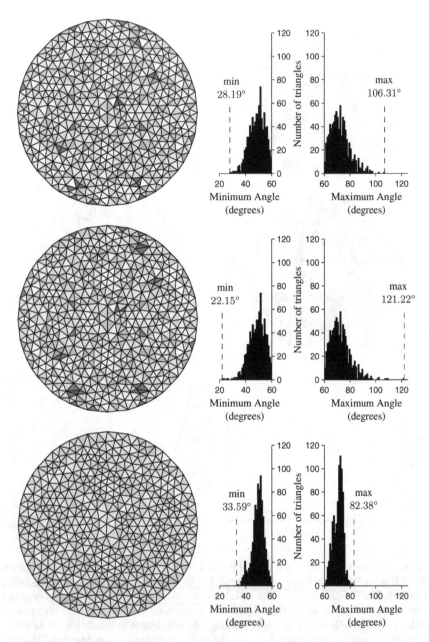

Fig. 10. The initial mesh in the top row has some lonely vertices which are removed by the preprocessing algorithm described in Section 4. The result of preprocessing is shown in middle row. The well-centered mesh resulting from 30 iterations of E_4 minimization is shown in the bottom row. Histograms of the minimum and maximum angles are shown next to each mesh.

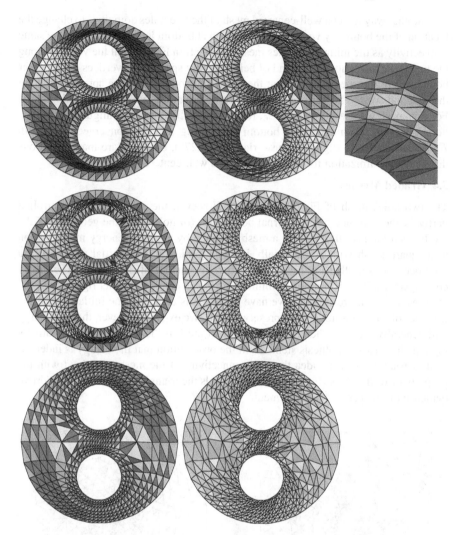

Fig. 11. Energy minimization applied to the two holes mesh on top left does not yield a well-centered mesh. Result after 500 iterations of E_4 minimization is shown at top center. This resulting mesh has some inverted triangles which are shown in the close-up at top right. With a different connectivity for the same vertex set, our minimization does produce a well-centered mesh. This is shown in the middle row. Middle left shows a Delaunay triangulation of the original vertex set after preprocessing. Using this mesh as the initial mesh and applying 500 iterations of E_4 followed by 500 iterations of E_8 minimization yields the well-centered mesh shown in middle right. The bottom row shows the two holes mesh with a different boundary. The mesh has the same mesh connectivity as the initial mesh at top left, but the vertices along the boundary have been moved. The well-centered mesh at bottom right was obtained by applying 100 iterations of E_6 minimization to the mesh at bottom left.

Another way to get a well-centered mesh of the two holes domain is to change the location of the boundary vertices. The mesh on at bottom left in Fig. 11 has the same connectivity as the initial two holes mesh at top left in Fig. 11, but the vertices along the boundary have moved. Instead of being equally spaced, the vertices on the outer boundary are more concentrated at the north and south and more spread out along the east and west. The vertices along the inner boundary have also moved slightly. For this mesh we use the energy E_6, reaching a well-centered configuration by 100 iterations. This mesh appears at bottom right in Fig. 11. The converged result with E_6 actually has one slightly obtuse triangle (90.27°), but there are many iterations during the minimization for which the mesh is well-centered.

5.4 Graded Meshes

The two holes mesh of Fig. 11 is graded. However, the gradation was controlled partly by the presence of the internal boundaries (of holes) and the geometry of the mesh. As a final result we show a mesh obtained by applying energy minimization to a square mesh with an artificially induced gradation. The initial mesh (after a preprocessing step that used only edge flips) appears at left in Fig. 12. The nearly converged result of 50 iterations minimizing E_4 is displayed to its right. From the other experimental results that we have shown, it is clear that the initial size of the triangles of a mesh is not always preserved well. We expect, however, that the energy will generally preserve the grading of an input mesh if the initial mesh is relatively high quality. This hypothesis stems from the observation that the energy is independent of triangle size, the idea that the connectivity of the mesh combined with the property of well-centeredness somehow controls the triangle size, and the supportive evidence of this particular experiment.

 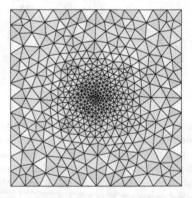

Fig. 12. A graded mesh of the square, in which the gradation is not due to internal boundaries. The initial mesh on left becomes the well-centered mesh on right after 50 iterations of E_4 minimization.

6 Conclusions and Future Work

This paper introduces a new energy function that measures the well-centeredness of planar triangulations. The authors are preparing a paper that describes an extension of this energy to three and higher dimensions. We also intend to do theoretical analysis and proofs of when and how quickly the energy minimization converges to a well-centered configuration. As for optimization, in our experiments shown here we have used relatively simple optimization techniques. It would be good to investigate other options for optimization. In addition, implementation in some language faster than MATLAB may significantly change the execution time of the algorithm, so collecting meaningful data about efficiency remains future work as well.

The current paper also discussed lonely vertices and the general problem that some meshes have no well-centered configuration that preserves both the mesh connectivity and the positions of the boundary vertices. The paper proposes a preprocessing algorithm that eliminates lonely vertices from the mesh. A complete characterization of which meshes permit a well-centered configuration is still lacking, however. Moreover, there may be a more efficient or effective way to perform preprocessing. In particular, the preprocessing algorithm proposed here employs only refinement operators, but it is possible that coarsening operators, such as the edge-collapse operator, might be helpful. An analysis of the number of points added during preprocessing would be interesting as well.

Our experiments shown here demonstrate that the proposed energy function can be effective in finding a well-centered configuration of a mesh. Unfortunately, inverted elements may be introduced and avoiding or handling inversions is a topic for future research. Additional experiments with larger, more complex meshes are also planned. A systematic study of the change in aspect ratio would be worthwhile as well. Although the finite element method is not a primary motivation for the work on well-centered meshes, it is possible that the method could be used effectively to improve finite element meshes in some cases.

Acknowledgment

The work of Evan VanderZee was supported by a fellowship from the Computational Science and Engineering Program, and the Applied Mathematics Program of the University of Illinois at Urbana-Champaign. The work of Anil N. Hirani was supported in part by NSF grant DMS-0645604. The authors thank the reviewers for their comments and suggestions.

References

[1] ABRAHAM, R., MARSDEN, J. E., AND RATIU, T. *Manifolds, Tensor Analysis, and Applications*, second ed. Springer–Verlag, New York, 1988.

[2] ALLIEZ, P., COHEN-STEINER, D., YVINEC, M., AND DESBRUN, M. Variational tetrahedral meshing. *ACM Transactions on Graphics 24*, 3 (2005), 617–625.

[3] BERN, M., MITCHELL, S., AND RUPPERT, J. Linear-size nonobtuse triangulation of polygons. In *Proceedings of the tenth annual ACM Symposium on Computational Geometry* (New York, 1994), ACM Press, pp. 221–230.

[4] BERN, M. W., EPPSTEIN, D., AND GILBERT, J. Provably good mesh generation. *J. Computer and Systems Sciences 48*, 3 (June 1994), 384–409. Special issue for 31st FOCS.

[5] BOBENKO, A. I., AND SPRINGBORN, B. A. Variational principles for circle patterns and Koebe's theorem. *Trans. Amer. Math. Soc. 356* (2004), 659–689.

[6] CLARKE, F. H. *Optimization and Nonsmooth Analysis*. Society for Industrial and Applied Mathematics (SIAM), 1990.

[7] COLLINS, C. R., AND STEPHENSON, K. A circle packing algorithm. *Computational Geometry: Theory and Applications 25*, 3 (2003), 233–256.

[8] DESBRUN, M., HIRANI, A. N., LEOK, M., AND MARSDEN, J. E. Discrete exterior calculus. *arXiv:math.DG/0508341* (August 2005). Available from: http://arxiv.org/abs/math.DG/0508341.

[9] DU, Q., FABER, V., AND GUNZBURGER, M. Centroidal Voronoi tessellations: applications and algorithms. *SIAM Review 41*, 4 (1999), 637–676.

[10] EDELSBRUNNER, H. *Geometry and Topology for Mesh Generation*. Cambridge University Press, 2001.

[11] EDELSBRUNNER, H., TAN, T. S., AND WAUPOTITSCH, R. An $O(n^2 \log n)$ time algorithm for the minmax angle triangulation. In *SCG90: Proceedings of the sixth annual ACM Symposium on Computational Geometry* (New York, 1990), ACM Press, pp. 44–52.

[12] ERICKSON, J., GUOY, D., SULLIVAN, J., AND UNGOR, A. Building space-time meshes over arbitrary spatial domains. In *Proceedings of the 11th International Meshing Roundtable* (2002), pp. 391–402.

[13] HEATH, M. T. *Scientific Computing: An Introductory Survey*, second ed. McGraw-Hill, 2002.

[14] HIRANI, A. N. *Discrete Exterior Calculus*. PhD thesis, California Institute of Technology, May 2003. Available from: http://resolver.caltech.edu/CaltechETD:etd-05202003-095403.

[15] MAEHARA, H. Acute triangulations of polygons. *European Journal of Combinatorics 23*, 1 (2002), 45–55.

[16] MELISSARATOS, E. A., AND SOUVAINE, D. L. Coping with inconsistencies: A new approach to produce quality triangulations of polygonal domains with holes. In *SCG '92: Proceedings of the Eighth Annual Symposium on Computational Geometry* (New York, NY, USA, 1992), ACM Press, pp. 202–211.

[17] NICOLAIDES, R. A. Direct discretization of planar div-curl problems. *SIAM J. Numer. Anal. 29*, 1 (1992), 32–56.

[18] RUPPERT, J. A Delaunay refinement algorithm for quality 2-dimensional mesh generation. *J. Algorithms 18*, 3 (1995), 548–585.

[19] ÜNGÖR, A., AND SHEFFER, A. Pitching tents in space-time: Mesh generation for discontinuous Galerkin method. *International Journal of Foundations of Computer Science 13*, 2 (2002), 201–221.

[20] YUAN, L. Acute triangulations of polygons. *Discrete and Computational Geometry 34*, 4 (2005), 697–706.

Generation of Quasi-optimal Meshes Based on a Posteriori Error Estimates

Abdellatif Agouzal[3], Konstantin Lipnikov[1], and Yuri Vassilevski[2]

[1] Los Alamos National Laboratory, lipnikov@lanl.gov
[2] Institute of Numerical Mathematics, vasilevs@dodo.inm.ras.ru
[3] Universite de Lyon 1, Laboratoire d'Analyse Numerique,
  Abdellatif.Agouzal@univ-lyon1.fr

Summary. We study methods for recovering tensor metrics from given error estimates and properties of meshes which are quasi-uniform in these metrics. We derive optimal upper and lower bounds for the global error on these meshes. We demonstrate with numerical experiments that the edge-based error estimates are preferable for recovering anisotropic metrics than the conventional element-based errors estimates.

1 Introduction

Challenging problems in predictive numerical simulation of complex systems require solution of many coupled PDEs. This objective is difficult to achieve even with modern parallel computers with thousand processors when the underlying mesh is quasi-uniform. One of the possible solutions is to optimize the computation mesh. The physically motivated optimization methods require error estimates and a space metric related to these estimates. In this article, we show that the quasi-optimal meshes or equivalently the optimal metric can be recovered from the edge-based error estimates.

The metric-based mesh adaptation is the most powerful tool for optimal mesh generation. It has been studied in numerous articles and books (see, e.g. [2, 7, 1, 6] and references therein). However, theoretical study of optimal meshes is relatively new area of research [1, 13, 15, 4, 3, 8, 11]. An optimal metric may be derived from *a posteriori* error estimates and/or solution features. Also, metric modification is a simple way to control mesh properties. Eigenvalues and eigenvectors of the tensor metric allow to control the shape and orientation of simplexes. The impact of metric modification on error estimates has been studied in [15].

In this paper, we study piecewise constant metrics recovered from edge-based error estimates and properties of meshes which are quasi-uniform in these metrics. We prove that these meshes are quasi-optimal, in a sense that the global error is bounded from above and below by $|\Omega|^2_M N_T^{-1}$, where N_T is the number of simplexes and $|\Omega|_M$ is the volume of computational domain in metric M.

We study numerically three methods for recovering a continuous piece-wise linear metric from given error estimates. In the first method, we take edge-based error estimates and combine ideas from [14] and [1] to build an anisotropic metrics. In the second method, we use again edge-based error estimates and the least square approach to build another anisotropic metric. Recently, we found that the least square approach has been already employed in [5] where a particular edge-based error estimates are derived. In our method, supported by Theorem 1, we may use any edge-based error estimates which makes it more general. Note that the metric generated with the least-square approach results in less sharper estimates.

Our experience shows that simple methods like the least square approach, applied to element-based error estimates, result frequently in isotropic metrics even for anisotropic solutions. Therefore, in the third method, we use element-based error estimates and the ZZ-interpolation method to generate a metric. This metric in inherently isotropic and used only for comparison purposes. We also show that there are cases when the element-based error estimates provide *no* information about an anisotropic solution, while the edge-based estimates allow to generate optimal anisotropic metrics.

The article outline is as follows. In Section 2, we recall basic concepts of metric-based mesh adaption. In Section 3, we present the main result for piecewise constant metrics. In Section 4, we consider methods for recovering continuous metrics. In Section 5, we illustrate our findings with numerical experiments.

2 Metric-based mesh generation

Let Ω be a polygonal domain in $\Re^2$, and $M(x)$ be a symmetric positive definite 2×2 matrix (tensor metric) defined for every $x = (x_1, x_2)$ in Ω. In metric M, the volume of a domain $D \subset \Omega$ and the length of a curve ℓ are defined as follows:

$$|D|_M = \int_D \sqrt{\det M(x)}\, dx, \qquad |\ell|_M = \int_0^1 \sqrt{(M(\gamma(t))\gamma(t), \gamma(t))}\, dt, \qquad (1)$$

where $\gamma(t) : [0, 1] \to \Re^2$ is a parametrization of ℓ.

Let Ω_h be a conformal triangular partition (triangulation) of Ω. There are a number of ways to generate a triangulation which is quasi-uniform in metric $M(x)$ (see, for example, [7, 1]). One of the robust metric-based mesh generation methods uses a sequence of *local* mesh modifications [2, 1]. The list of mesh modifications includes alternation of topology with node deletion/insertion, edge swapping, and node movement. The topological operation is accepted if it increases a mesh quality which is a measure of mesh M-quasi-uniformity. Different mesh qualities may be used [2, 1, 12, 9].

It has been shown in [1, 12, 13] that for a particular choice of metric M, the M-quasi-uniform meshes provide the same asymptotic reduction of the piecewise linear interpolation error as the optimal mesh. The objective of this paper is to discuss how such a metric may be generated based on robust and reliable a posteriori error estimates.

3 A posteriori error estimates and mesh quasi-optimality

In this section we assume that a robust and reliable *a posteriori* error estimate η_e to a true error ξ_e is assigned to every mesh edge e:

$$C_1 \xi_e \leq \eta_e \leq C_2 \xi_e \tag{2}$$

and that constants C_1, C_2 do not depend on Ω_h. The estimates of these type are not popular in scientific computations since they measure error in unusual way (on mesh edges). However, we show below that these estimates are more preferable for building anisotropic meshes.

For every triangle Δ with edges e_1, e_2, e_3, we define a constant (tensor) metric M_Δ which satisfies the following equations:

$$(M_\Delta \, e_i, e_i) = \eta_{e_i}, \quad i = 1, 2, 3. \tag{3}$$

Hereafter, we use e both for a mesh edge and for a vector from one edge-end point to the other. Formula (3) implies that entries of matrix

$$M_\Delta = \begin{pmatrix} \mathrm{m}_{11} & \mathrm{m}_{12} \\ \mathrm{m}_{12} & \mathrm{m}_{22} \end{pmatrix}$$

are the unknowns of the linear system:

$$(\mathrm{m}_{11} e_{i,x} + \mathrm{m}_{12} e_{i,y}) \, e_{i,x} + (\mathrm{m}_{12} e_{i,x} + \mathrm{m}_{22} e_{i,y}) \, e_{i,y} = \eta_{e_i}, \quad i = 1, 2, 3,$$

where $e_i = (e_{i,x}, e_{i,y})^T$. This linear system is non-singular since its determinant is nonzero and equals to $16|\Delta|^3$ [14], where $|\Delta|$ denotes the area of Δ. Let $|M_\Delta|$ be the spectral module of M_Δ. The metric M is defined as the piecewise constant metric on Ω_h with values $|M_\Delta|$.

For simplicity, we assume that $\det(M_\Delta) \neq 0$. Otherwise, we must introduce a perturbation of zero eigenvalues of M_Δ [1] and take it into account in the analysis.

For every element Δ, we introduce the secondary error estimator:

$$\chi_\Delta^2 = \sum_{e \subset \partial \Delta} \eta_e^2$$

and the global error estimator

$$\chi^2 = \frac{1}{3} \sum_{\Delta \subset \Omega_h} \chi_\Delta^2.$$

Using the definition of edge length in the constant metric $|M_\Delta|$,

$$|e|_{|M_\Delta|}^2 = (|M_\Delta|e, e),$$

and the inequality

$$\max_{e \subset \partial \Delta} (M_\Delta e, e)^2 \geq C |\Delta|_{|M_\Delta|}^2,$$

proved in [4], we can estimate χ_Δ^2 from both sides:

$$\chi_\Delta^2 = \sum_{e \subset \partial \Delta} (M_\Delta e, e)^2 \leq \sum_{e \subset \partial \Delta} (|M_\Delta|e, e)^2 \leq |\partial \Delta|_{|M_\Delta|}^4$$

and

$$\chi_\Delta^2 = \sum_{e \subset \partial \Delta} (M_\Delta e, e)^2 \geq \max_{e \subset \partial \Delta} (M_\Delta e, e)^2 \geq C |\Delta|_{|M_\Delta|}^2.$$

Here $|\partial \Delta|_{|M_\Delta|}$ is the perimeter of Δ measured in metric $|M_\Delta|$.

Assume that triangulation Ω_h with N_T triangles is M-quasi-uniform, i.e., for any Δ in Ω_h we have

$$|\partial \Delta|_M^2 \simeq |\Delta|_M \qquad \text{and} \qquad |\Delta|_M \simeq N_T^{-1} |\Omega|_M.$$

Then

$$\chi^2 \leq \frac{1}{3} \sum_{\Delta \subset \Omega_h} |\partial \Delta|_{|M_\Delta|}^4 \leq C \sum_{\Delta \subset \Omega_h} |\Delta|_{|M_\Delta|}^2 \leq C N_T^{-1} |\Omega|_M^2$$

and

$$\chi^2 \geq C \sum_{\Delta \subset \Omega_h} |\Delta|_{|M_\Delta|}^2 \geq C N_T \min_{\Delta \subset \Omega_h} |\Delta|_{|M_\Delta|}^2 \geq C N_T^{-1} |\Omega|_M^2.$$

We proved the following result.

Theorem 1. *For any edge e, let η_e be a given a posteriori error estimator satisfying (2). Let the piecewise constant metric M be defined by (3) and $|M_\Delta|$. If Ω_h is a M-quasi-uniform triangulation with N_T triangles, then it is quasi-optimal:*

$$c N_T^{-1} |\Omega|_M^2 \leq \sum_{\Delta \subset \Omega_h} \sum_{e \subset \partial \Delta} \xi_e^2 \leq C N_T^{-1} |\Omega|_M^2$$

with constants c, C independent of N_T and Ω_h.

The theorem holds for any definition of error ξ_e. In reality, the sum of all ξ_e^2 represents some norm of error, $\|u - u_h\|_{*,\Omega}^2$, where u is the exact solution, and u_h is its approximation. For instance, this is true if ξ_e^2 is proportional to $|\sigma_e|$ where σ_e is the union of triangles sharing e.

The global error on the M-quasi-uniform mesh is also the sum of approximately equal element-based errors $\chi_\Delta^2 \approx |\Delta|^2_{|M_\Delta|}$. However, it is not clear how to use only these element-based error estimates (without using η_e^2) to recover an anisotropic metric (see also discussion at the end of Section 4). Finally, we emphasize that the N_T^{-1}-asymptotic error reduction holds on anisotropic meshes as long as $|\Omega|_M$ is independent of N_T.

To produce a quasi-optimal mesh, we suggest the following adaptive algorithm.

Initialization Step. Generate an initial triangulation Ω_h. Set $\chi = +\infty$. Choose the final number N_T of mesh elements.

Iterative Step.
1. Compute the approximate solution u_h.
2. Compute the estimators η_e and χ. Stop, if χ is not reduced.
3. Otherwise, compute metric M from η_e.
4. Generate a M-quasi-uniform mesh $\tilde{\Omega}_h$ with N_T elements.
5. Set $\Omega_h := \tilde{\Omega}_h$ and go to step 1.

This iterative method requires an initial mesh which may be arbitrary and very coarse.

4 Recovery of a continuous metric

Our experience shows that continuous metrics are more beneficial in the adaptive metric-based generation. To this end, we suggest a simple technique to generate a continuous piecewise linear metric $\widehat{M}$ first by recovering it at mesh nodes and then by interpolating it linearly inside mesh elements.

We consider three methods to recover $\widehat{M}$ at mesh nodes. First, for a mesh node a_i we define the nodal *tensor* metric $\widehat{M}_i$ by taking $|M_\Delta|$ from one of the surrounding elements:

$$\widehat{M}_i = \arg \max_{|M_\Delta|, a_i \in \Delta} \det\left(|M_\Delta|\right). \tag{4}$$

Second, for a mesh node a_i, the nodal *tensor* metric $\widehat{M}_i$ is defined using η_e on mesh edges e incident to a_i. Let κ_i be the number of these edges. As in (3), we would like to have

$$(\widehat{M}_i e_j, e_j) = \eta_{e_j}, \quad j = 1, \ldots, \kappa_i. \tag{5}$$

In algebraic terms (5) is a linear system

$$Am = b$$

where $m \in \Re^3$ is the vector of unknown entries of matrix $\widehat{M}_i$, and $A \in \Re^{\kappa_i \times 3}$. For $\kappa_i > 3$ this system may be overdetermined. In this case, we use the least squares solution:

$$m = \arg\min_m \|Am - b\|^2. \qquad (6)$$

We have learned recently that the least square techniques has been used in [5]. The authors use the discrete solution and its gradient to generate vector b. In our method, supported by Theorem 1, we may use any edge-based error estimates which makes it more general.

Third, we recover a *scalar* (isotropic) metric at each triangle based on a given element-based error estimate η_Δ:

$$M_\Delta = |\Delta|^{-1} \eta_\Delta I.$$

Let $\bar{M}$ be a piecewise constant scalar metric composed of M_Δ. Note that arguments of Theorem 1 may be applied to $\bar{M}$-quasi-uniform triangulations with N_T elements if

$$C_1 \xi_\Delta \le \eta_\Delta \le C_2 \xi_\Delta,$$

where ξ_Δ is the true element-based error. The definition (1) implies that $|\Delta|^2_{M_\Delta} = |\Delta|^2 \det(M) = \eta_\Delta^2$. Thus,

$$\sum_{\Delta \subset \Omega_h} \xi_\Delta^2 \simeq \sum_{\Delta \subset \Omega_h} \eta_\Delta^2 = \sum_{\Delta \subset \Omega_h} |\Delta|^2_{M_\Delta} \simeq N_T^{-1} |\Omega|^2_{\bar{M}}.$$

However, these shape-regular meshes will be quasi-optimal if and only if $|\Omega|_{\bar{M}} \simeq |\Omega|_M$, that is, if the optimal mesh is shape-regular. For anisotropic solutions, $|\Omega|_{\bar{M}}$ will depend on the number of simplexes.

In order to define a nodal *scalar* (isotropic) metric from the element-based metric $\bar{M}$, we consider triangles with a common mesh node a_i and construct the best (in terms of the least squares) linear function which approximates given errors at triangle centers. The value of the linear function at node a_i defines the scalar metric $\widehat{M_i}$. This type of interpolation is known as the ZZ-interpolation method [16]. The use of scalar metrics represents a group of methods which generate adaptive meshes using a size function.

Let us show that there are cases when the element-based error estimates provide no information about solution anisotropy. Let $u(x_1, x_2) = x_1^2$ and

$$\eta_\Delta = \int_\Delta (u - u_I)^2 \, dx$$

where u_I is the continuous piecewise linear interpolant of u on a structured triangular mesh. The triangular mesh is obtained from a square mesh by dividing each square into two triangles. The direct calculations show that η_Δ is the same for all elements. On the other hand, $\eta_e = 0$ on vertical edges which indicates the direction of solution anisotropy. More complicated examples are consider in the next section.

5 Numerical experiments

In this section we consider a model problem of piecewise linear interpolation of an anisotropic function and three nodal metrics. For the first two metrics, we assume that the edge-based error estimates η_e are given. The anisotropic metric $\widehat{M}^{(1)}$ is recovered by (3) and (4). The anisotropic metric $\widehat{M}^{(2)}$ is recovered by (5) and (6). The isotropic metric $\widehat{M}^{(3)}$ is defined with the ZZ-interpolation method using given element-based error estimates η_Δ.

To generate a $\widehat{M}$-quasi-uniform mesh, we use the algorithms described in articles [1, 10] and implemented in the software package ani2d (http://source-forge.net/projects/ani2d). In all experiments, we start with a quasi-uniform unstructured mesh with 2792 triangles. We request that the final adaptive mesh must have 1000 triangles. Depending on the mesh quality, the final number of elements in the quasi-optimal mesh may deviate from the requested number [10]. In the experiments, the target mesh quality was 0.7 on a scale from 0 to 1, where quality 1 corresponds to an ideal mesh.

Let us consider the function

$$u(x_1, x_2) = \exp(2(x_1^\alpha + x_2^\alpha)), \qquad \alpha > 0,$$

in the unit square. The function is constant along curves $x_1^\alpha + x_2^\alpha = const$ and becomes one-dimensional when $\alpha \to 1$. We expect that the optimal mesh be shape-regular for $\alpha = 2$ and strongly stretched for $\alpha = 1.01$. For simplicity, we choose the edge-based, η_e, and element-based, η_Δ, estimators as the properly scaled exact interpolation errors:

$$\eta_e^2 = \max_{(x_1, x_2) \in e} |u - u_I|^2 |\sigma_e|, \quad \eta_\Delta^2 = \max_{(x_1, x_2) \in \Delta} |u - u_I|^2 |\Delta|.$$

We consider two cases, $\alpha = 1.01$ (see Fig. 1) and $\alpha = 2$ (see Fig. 2). For both values of α, the adaptive iterative method converged in 2 iterations for all metrics. According to Table 1, the tensor metrics results in sharper estimates for the interpolation error. The mean interpolation error on the anisotropic meshes in 10 times smaller then on the isotropic mesh. Fig. 2 shows that the shape-regular quasi-optimal mesh corresponding to $\widehat{M}^{(1)}$ is aligned better with the function u.

Table 1. Mean interpolation errors for different metrics.

α	metric $\widehat{M}^{(1)}$	metric $\widehat{M}^{(2)}$	metric $\widehat{M}^{(3)}$
2.00	1.48e-2	1.54e-2	3.38e-2
1.01	1.99e-3	2.08e-3	2.25e-2

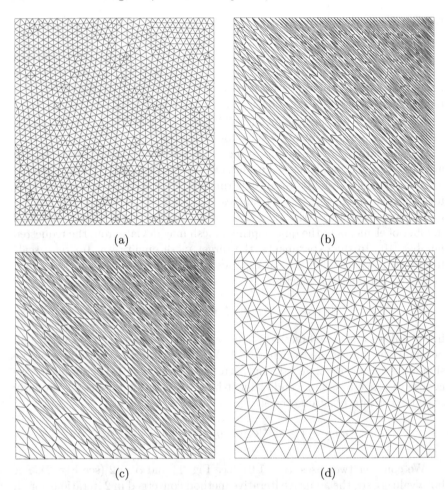

(a)

(b)

(c)

(d)

Fig. 1. Adaptive meshes for anisotropic function with $\alpha = 1.01$: (a) Initial mesh with 2792 triangles. (b) $\widehat{M}^{(1)}$-quasi-uniform mesh with 1042 triangles. (c) $\widehat{M}^{(2)}$-quasi-uniform mesh with 1018 triangles. (d) $\widehat{M}^{(3)}$-quasi-uniform mesh with 1006 triangles.

Conclusion

We showed that the robust and reliable edge-based a *posteriori* error estimates may be used for generation of quasi-optimal (possibly anisotropic) meshes. We considered three methods for recovering a continuous piecewise linear tensor

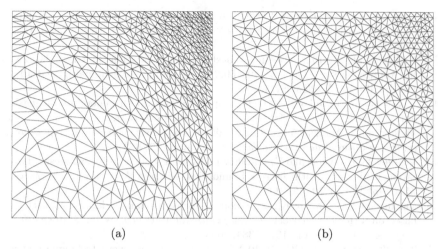

(a)   (b)

Fig. 2. Adaptive meshes for isotropic function with $\alpha = 2$: (a) $\widehat{M}^{(2)}$-quasi-uniform mesh with 1073 triangles. (d) $\widehat{M}^{(3)}$-quasi-uniform mesh with 941 triangles.

metric and demonstrated with numerical experiments efficiency of two of them for generating meshes which minimize the interpolation error.

Acknowledgments. Research of the third author has been supported partially by the RFBR project 04-07-90336 and the academic program "Optimal methods for problems of mathematical physics".

References

1. Agouzal A, Lipnikov K, Vassilevski Y (1999) Adaptive generation of quasi-optimal tetrahedral meshes, *East-West J. Numer. Math.* **7**, 223–244.
2. Buscaglia GC, Dari EA (1997) Anisotropic mesh optimization and its application in adaptivity, *Inter. J. Numer. Meth. Engrg.* **40**, 4119–4136.
3. Chen L, Sun P, Xu J (2007) Optimal anisotropic meshes for minimizing interpolation errors in L^p-norm. *Mathematics of Computation* **76**, 179-204.
4. D'Azevedo E (1991) Optimal triangular mesh generation by coordinate transformation, *SIAM J. Sci. Stat. Comput.* **12**, 755–786.
5. Frey PJ, Alauzet F (2005) Anisotropic mesh adaptation for CFD computations. *Comput. Methods Appl. Mech. Engrg.* **194**, 5068–5082.
6. George PL (1991) *Automatic mesh generation.* Chichester: John Wiley & Sons Ltd., 344p.
7. Hecht F (2005) *A few snags in mesh adaptation loops,* In proceedings of the 14th International Meshing Roundtable, Springer-Verlag Berlin Heidelberg.
8. Huang W (2005) Metric tensors for anisotropic mesh generation. *J. Comp. Physics.* **204**, 633–665.
9. Knupp P (2001) Algebraic mesh quality metrics. *SIAM Journal on Scientific Computing* **23**, 193–218.

10. Lipnikov K, Vassilevski Y (2003) Parallel adaptive solution of 3D boundary value problems by Hessian recovery. *Comput. Methods Appl. Mech. Engrg.* **192**, 1495–1513.
11. Nadler E (1986) Piecewise linear best L_2 approximation on triangulations. In: *Approximation Theory* **V**, 499–502, (C. K.Chui, L.L. Schumaker, and J.D. Ward, editors), Academic Press.
12. Vassilevski Y, Lipnikov K (1999) Adaptive algorithm for generation of quasi-optimal meshes, *Comp. Math. Math. Phys.* **39**, 1532–1551.
13. Vassilevski Y, Agouzal A (2005) An unified asymptotic analysis of interpolation errors for optimal meshes, *Doklady Mathematics* **72**, 879–882.
14. Vassilevski Y, Dyadechko V, Lipnikov K. Hessian based anisotropic mesh adaptation in domains with discrete boundaries, *Russian J. Numer. Anal. Math. Modelling* (2005) **20**, 391–402.
15. Vassilevski Y, Lipnikov K. Error bounds for controllable adaptive algorithms based on a Hessian recovery. *J. Computational Mathematics and Mathematical Physics*, V.45, No.8, pp. 1374-1384, 2005.
16. Zhu JZ, Zienkiewicz OC (1990) Superconvergence recovery technique and a posteriori error estimators, *Inter. J. Numer. Meth. Engrg.* **30**, 1321–1339.

Geometry

An Efficient Geometrical Model for Meshing Applications in Heterogeneous Environments

Andrey Mezentsev

CeBeNetwork Engineering & IT, Bremen, D-28199, Germany

AMezentsev@cebenetwork.com

ABSTRACT

This paper introduces a new neutral hybrid discrete (in the limit continuous) solid CAD model for meshing applications within the Integrated Computational Environments, based on subdivision surfaces. The model uses the Boundary Representation for the CAD model topology and the Butterfly Interpolating subdivision scheme for definition of underlying curves and surfaces. It is automatically derived from the original solid model, based on parametric surfaces, using a fast loop-traversal approach for identification of geometrical discontinuities. A curvature-based sizing function is introduced for generation of an optimal control mesh for subdivision surfaces. A new hybrid CAD model has significantly fewer faces, uses robust discrete structure, which simplifies computational meshing and geometrical model transfer within the heterogeneous components of computational environments.

Keywords: solid modeling, boundary representation, subdivision surfaces, surface mesh generation, surface data interpolation.

1. INTRODUCTION

Recent advances in numerical solution of differential equations and significant increase of power of affordable computers have largely extended application of the Finite Element (FEM) and the Finite Volume Method (FVM) in Computational Fluid Dynamics (CFD) and Computational Structural Mechanics (CSM) to simulation of a new physical phenomena on significantly more complex geometry. Nowadays it is common to deal in the FEM/FVM simulations with full configurations of aircrafts, cars, etc. defined by composition of thousands of free form faces [1,2]. Application of the FEM and the FVM requires meshing to be performed on the geometry and CAD models provide an effective input for the process [1,2]. However, the problem of efficient definition of the geometrical input for downstream engineering applications (i.e. mesh generation) is not solved for the Integrated Computational Environments (ICEs), frequently found in aerospace and automotive domains. Indeed, the ICEs are composed from numerous heterogeneous in-house and commercial software components and therefore require efficient exchange of geometrical and computational data [2] in the production cycle. This paper addresses the problem of automatic creation and seamless integration of an exchangeable solid CAD model to the ICEs with elements of virtual reality. The new model is designed for shape definition for volumetric mesh generation and virtual reality applications.

2. OVERVIEW OF THE EXISTING GEOMETRICAL MODELS

For a modern computational process unambiguous volumetric shape of an object in heterogeneous environments is defined in digital form, using a number of different approaches such as solid modeling, discrete modeling, etc. Most of the modern CAD engines uses the Boundary Representation (BREP) [3], when a solid is defined by the object boundary. The BREP definitions mostly rely on parametric surfaces [3-6]; however in many computational applications discrete models without parameterization and hierarchical topology play an important role due to robustness and simplicity of processing [7-10]. While parametric CAD models with topology are suitable for further geometrical modeling, discrete mesh-based representations are mainly targeted on efficient geometrical data transfer [6].

2.1 Parametric CAD models

In the BREP format [3] the boundary is typically composed in a hierarchical way from conformal parametric (as a rule topologically rectangular) free form faces (see Fig. 1 for an example of a topological tree, Fig. 2 for an example of a typical BREP geometry). Non-Uniform Rational B-Spline (NURBS) is a standard choice for definition of curves and faces [4], providing an accurate and robust framework for geometry representation. Unfortunately the original CAD in the BREP format often contains errors such as gaps, overlaps of faces, incorrect faces topology, etc. and requires special pre-processing to enforce conformal boundary definition, known as CAD repair [5,6]. For the mentioned complex CAD models the number of geometrical errors build up in the non-liner manner with the increase of the number of faces in the model, forming a major bottleneck in the FEM/FVM analyses workflow [5,6].

The ICEs require multiple export/import of geometrical models via exchange formats, such as IGES and STEP, thus introducing extra CAD errors due to tolerance problems and different CAD representation in different modules of the ICE. Therefore ICEs are obliged to use internal CAD repair tools after each geometrical model transfer to maintain consistent watertight CAD models [5,6]. Avoiding multiple CAD repair operations is a key requirement for the ICE efficiency. The repaired model approximates the initial shape within the given tolerance. Typically it removes or re-defines some features of the original CAD that affects quality of computational mesh – i.e. small and high aspect ratio faces, etc. The repaired meshing ready CAD model here is called a Neutral CAD Geometrical Model (NCGM). Indeed, the NCGM slightly changes shape and topology of the initial CAD within the given tolerance, so *a priory* the NCGM is not the same CAD, imported to the environment. We can formulate the following requirements for an NCGM [5]:

1. The NCGM should accurately represent complex 3D shapes with a minimal number of faces.

2. The NCGM should be suitable for computational meshing.

3. Interchange of the NCGM should be simple and efficient.

4. The size of the NCGM should not be prohibitively large.

5. The NCGM should be efficient for visualization.

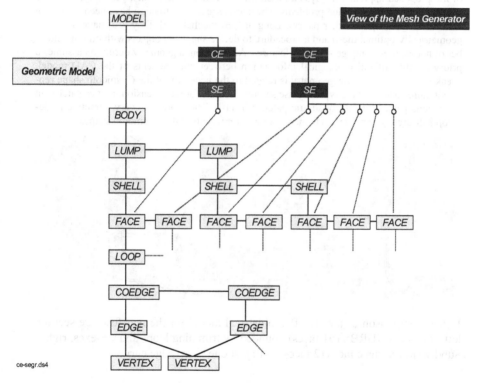

Fig. 1 The topological tree of a classical BREP geometrical model: CAD and mesh generation representations. Meshing requires automatic unification of faces to Super and Constructive Elements (SE and CE) for better mesh quality.

2.2 Discrete CAD models

Considering complexity of the solid NCGM exchange within engineering applications, many authors advocate polyhedral discrete mesh-based geometry without topological tree. It is the most stable format for exchange of geometrical data between inhomogeneous components of the ICEs [2,7,8]. The idea of a discrete geometry definition is certainly not new. There are a large number of publications on the subject to mention few: Lohner [7] has effectively used polyhedral representation for geometry definition in the context of advancing front mesh generation scheme. Recently Owen and White [8] have developed an efficient polyhedral geometry definition and provided an algorithm to extract model topology from the volumetric mesh-based definition. Also the motivation of [8] is in support of the legacy FEM geometrical data, when traditional CAD definition is no longer available. Strong point of the proposed approach in [8] is a possibility to deal with deformed geometry, resulting from the FEM/FVM coupled problems. The surface feature extraction have been studied in References [19,20] and other papers. Lang and Borouchaki [9], Frey [10] have proposed geometrically optimal mesh and a procedure to define shapes of objects with minimal number of nodes, preserving geometrical features. All mentioned geometry definition assumes a priory absence of full topological information between the elements of the CAD model. Weak point of the discrete geometry is related to the necessity of the C_I smooth shape representation, so the authors typically use smooth faces as format extension: Bezier-patches in [8], Coons patches in [9], quadratic patches in [10]. Therefore it is very attractive to develop a discrete geometrical scheme capable of representing the C_I smooth shapes.

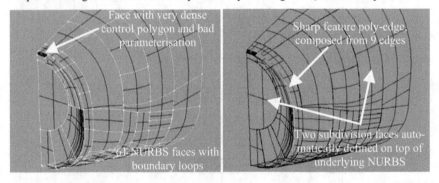

Fig. 2 Comparison of the BREP geometrical model for the rear fuselage section: left - classical NURBS (61 faces) bounded by trimming loops and vertexes, right - Subdivision Surface faces (2 faces) with just one sharp edge curve.

2.3 Closing the gap between parametric and discrete geometry

On the other hand, there is a large class of problems, when the original topological CAD model information could be used for an effective definition of the NCGM. One of the most important cases is found in the context of the ICEs, where elements of the CAD repair are used to maintain consistency of the CAD model during transfer. However, we want to avoid application of the CAD repair after each export-import operation in the ICE workflow. This paper assumes that we have a watertight BREP geometry definition, resulting for example from the basic CAD repair [5], developed by the authors. The challenge is to define an effective hybrid geometrical model that could combine robustness of the discrete geometry representation with the strength of the full topological information of the BREP model tree (Fig. 1).

Researches have proposed a number of alternative schemes for the definition of smooth faces. One of such approaches uses Subdivision Surfaces [11,12,13]. Originally, subdivision surfaces were developed in computer graphics for visualization of complex free form objects [15,17]. In the context of mesh generation Kobbelt et al. [11] pioneered the usage of interpolating subdivision surfaces as the basis for geometry definition, Rypl and Bittnar [12] have used subdivision geometry for advancing front meshing in physical space. Later Lee have developed an effective parametric mesh generation approach, based on the Butterfly subdivision geometrical model [13,14]. However, all mentioned work did not use initial topological information of the solid model. Mezentsev et al. [16] have proposed to combine subdivision surfaces with elements of topological information in the classical BREP model (the so-called S-BREP definition) however limited to scanned objects and to sub-set of the BREP tree. This paper develops an idea of a hybrid geometrical model further on, introducing the methodology for generic geometry definition, subsequent surface meshing and computational data interpolation. It focuses on application of the S-BREP geometry for direct mesh generation, using local subdivision rules formulated in [17]. Initial control mesh for the S-BREP faces is generated on the NURBS BREP geometry.

The paper is outlined as follows: Section three discusses specifics of the NURBS BREP model and provides fundamentals of the Butterfly subdivision scheme. It gives the background for a proposed hybrid subdivision surface model with the boundary representation. Section four discusses automatic generation of the S-BREP model from the solid BREP model. In Section five examples of the S-BREP models are given. Section six gives some implementation details and Section seven provides conclusions.

3. Geometrical BREP model based on subdivision surfaces

Developed geometrical model is tailored for an efficient application within the ICEs as discussed in Section 1 and is targeted for the mesh generation and the FEM computational data applications.

3.1 The Boundary Representation - BREP

As it has been discussed in Section 2, the BREP model is an efficient way of geometry definition for the ICEs. Downstream engineering applications, i.e. meshing or computational data related, frequently operate on the level of faces, therefore most of the NCGMs [5,8] apply a middle range subset of the BREP topological tree – just faces with boundary loops, as shown in Fig. 1. Frequently the BREP models contain faces that are too small or badly parameterized for quality mesh generation. In the process of the CAD repair such faces are typically logically grouped to form bigger entities (in terms of Fig. 1 the so-called Super Elements (SE) and Constructive Elements (CE)). In our approach we retain the complete topological tree of the initial BREP model, performing geometric continuity analyses and generating bigger subdivision geometry faces over the smooth regions of the model. The S-BREP model tree will be similar to the tree, shown in Fig. 1, however, subdivision faces will be larger spanning over a number of underlying NURBS faces.

3.2 Subdivision Surfaces

Interpolating subdivision represents a smooth curve or a surface as a limit of successive subdivisions of the initial mesh. By starting with the coarse (so-called control) mesh new positions of the inserted points are calculated according to pre-defined rules. In most cases local rules how to insert points [15,17] (i.e. weighted sum of surrounding nodes coordinates) and how to split the elements of the previous mesh are used. The resulting subdivision mesh will be "smoothed" out so the angles between adjacent elements will be nearly flattened (see Section 4). Eventually, after an infinite number of refinements, a smooth curve or surface in differential geometrical sense can be obtained.

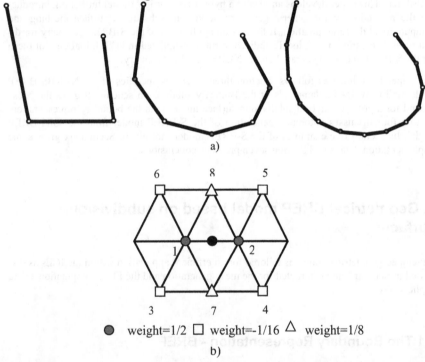

Fig. 3 Principles of subdivision: a) Successive subdivisions of interpolating subdivision curve b) Butterfly subdivision scheme

For example, Fig. 3 a) shows a number of successive subdivisions for a curve. Initial coarse mesh (left, nodes are represented by hollow points) is refined by insertion of new nodes (shown as filled points) to obtain smooth curve representation (right). The advantages of subdivision algorithms are that the schemes are local and surface representation will be good enough for most applications after a small number of subdivision steps. On first steps the initial position of control mesh nodes could affect surface quality, therefore it is advisable to provide subdivision-optimal control mesh as discussed in Section 4.3. Moreover, surface at any point can be improved arbitrarily by applying more local refinements (see Reference [16] for details). As a geometrical basis for the BREP-type geometry the so-called interpolating Butterfly scheme is used, initially proposed by Dyn et al. [15] and later modified by Zorin et al. [17]. The scheme could be applied for an arbitrary connectivity pattern of initial triangular mesh and uses eight points of the coarse level (Fig. 3 b), hollow points, triangles and quads) to compute position of the node on the new level of refinement (filled point). Note that the position of nodes on the previous subdivision level is retained.

The following formula is used for computation of the regular node position:

$$x_p = \frac{1}{2}\left(x_1 + x_2\right) - \frac{1}{16}\left(x_3 + x_4 + x_5 + x_6\right) + \frac{1}{8}\left(x_7 + x_8\right) \qquad (1)$$

Where x_p is the coordinate (or nodal variable, see Section 6 for example) of the interpolated subdivision point and x_1 - x_8 are coordinates (nodal variables) of the points in the vicinity of the interpolated point. For nodes with valence different from six (extraordinary internal and external boundary nodes) different subdivision rules with different weights are applied. A complete set of rules for the modified Butterfly scheme could be found in [17].

Interpolating subdivision surface is a generalization of spline surfaces for control net (polygon) of arbitrary topology [15,17]. Modified Butterfly scheme gives in the limit a C_1- continuous surface and tangent vectors could be computed at any point of the surface. With reference to the triangular surface meshes considered in this study, it is also possible to apply the Loop [18] scheme. However, the modified Butterfly scheme provides better results on sharp corners without dedicated insertion of boundary curves with different subdivision pattern, producing only minor smoothing (Fig. 4, right). Our approach models C_0 features of geometry (sharp corners) using the NURBS BREP to S-BREP curve mapping process, as described in Section 4. This is the main difference between the models developed by Rypl and Bitnar [12] and by Lee [13], who proposed application of subdivision geometry for meshing without dedicated discussion of the CAD model creation.

4. Automatic generation of the S-BREP models

Subdivision surfaces provide an effective framework for free form faces representation in the BREP model. However up to now formalisms for automatic definition of the composite CAD models, based on subdivision surfaces were not developed. Open problem is mostly related to automatic detection and representation of the BREP model discontinuities requiring definition of the adequate subdivision surfaces boundary curves and corner points. The problem could be illustrated by the following example of a typical rear part of the fuselage geometry (Fig. 4). Should the whole rear fuselage be represented just by one subdivision surface, certain smoothing of sharp features will occur on further levels of subdivision. To overcome this difficulty, an automatic procedure for analyses of a classical NURBS BREP model is developed. The general idea is rather simple: detection of a classical BREP model geometrical discontinuities and later definition of the respective boundary edges for subdivision faces. A similar procedures for discrete mesh based geometry have been developed in Reference [8]. Frey [10] and Lee [13] have proposed respectfully geometrical mesh simplification and tagging processes, not using the initial parametric CAD model. In terms of the methodology, most of the proposed algorithms for feature edges extraction are working on the 3D discrete polyhedral models (see, for example Owen and White [8] for volumetric, Baker [19] or Yamakawa and Shimada [20] for surface features). In our approach we are interested in the extraction of feature edges from the parametric BREP models, keeping in mind that cylindrical or closed surfaces could be effectively represented by the subdivision surfaces. In the area of the BREP analyses for sharp features, the work of Lu, Gadh and Tautges [21] is rather close to our BREP traversing, however current paper focuses on a sharp features extraction for subdivision geometry definition and on generation of an optimal control mesh, therefore it differs from the approach in Reference [21].

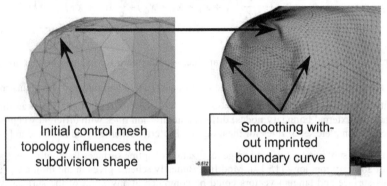

Fig. 4 The rear fuselage, represented by one subdivision surface face on two levels of subdivision without imprinting of subdivision curve at rear tip of the geometry. Smoothing of the C_1 discontinuity is clearly visible on the cylindrical part.

Proposed method works on a classical solid BREP (for example CATIA or ACIS with a full topological tree, similar to shown in Fig. 1) model and contains the following main stages:

1. CAD repair for enforcing conformal properties of the NURBS BREP model.

2. The NURBS BREP model geometrical analyses with automatic flagging of geometrical discontinuities: cusp and corner features (see [13] for definitions).

3. Automatic geometrical surface mesh generation on the NURBS BREP model, using a special curvature-based mesh sizing function.

4. Identification of the flagged discontinuity curves and assignment of subdivision faces. Typically the S-BREP faces are composed of 10-100 NURBS faces.

5. Application of the generated mesh as the subdivision surface control mesh.

6. Storage of the S-BREP model within the ICE.

Due to the assignment of the S-BREP faces to larger geometrically smooth regions, the number of faces in the S-BREP model is significantly reduced. The coarse nature of the S-BREP control mesh generated on stage 3 of the proposed approach, guarantees low storage requirements of the NCGM model. Further subdivision refinement of the initial control mesh provides in the limit smooth faces as in the Butterfly subdivision scheme [17]. As compared to the NURBS BREP model, each face is defined by control points and weights, while in the S-BREP only the control mesh (polygon) is stored and weights are constant and are not variable in the geometry, therefore reducing storage requirements.

4.1 CAD repair

The automated CAD repair process is carried out only once in the proposed workflow. Later on numerous exchanges of the watertight S-BREP geometrical model are performed in the discrete form and no geometrical data exchange errors are introduced. Specifics of the initial CAD repair concept developed and implemented by the authors are described in Reference [5] and are shortly repeated here for consistency. The basic CAD repair has received extra features, ensuring efficient geometrical model preparation for further analyses and conversion to the S-BREP format. The CAD repair is also tailored for the efficient usage in the ICEs as a pre-processing stage of the NCGM creation. It works in line with the concept of the abstract meshing interface [5] and can be summarized as following:

- CAD repair contains two separate interacting modules for independent (insuring global conformity of the model boundary) and dependent repair (targeting on a specific requirements of the application, in our case the S-BREP mesh generation), with the possibility of intermediate storage of the repaired model

- CAD repair is as a cyclic process involving both modules as presented in Fig. 5

- CAD repair provides an automated repair functionality, with minimal user interaction

- All operations of the CAD repair are performed on the NURBS BREP model and geometrical elements are considered equal within the given tolerance

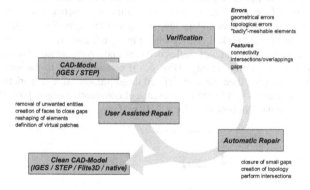

Fig. 5 CAD repair loop with interaction of independent and dependent modules

- Empirical realization of specific CAD repair features are ensured, mainly oriented on the complementary hierarchy of the parts and units of the CAD model, based on the grouping of the CAD model elements in engineering sense (see Fig. 1)

- The CAD model is accessible at any moment to the mesh generation process through the CAD/Mesh abstract interface.

The processing of the CAD-model into a form which is suitable for downstream applications is performed in three major steps (see Fig. 5):

- Verification

- Repair of geometric and topological errors - independent repair

- Modification and removal of the of problematic configurations, posing difficulty for further model continuity analyses– dependent repair.

We use the cyclic workflow of the classical CAD repair in the process of model preparation for the automated S-BREP generation. As a result of CAD repair process the conformal but rather constrained in terms of possible variants of geometrical configurations NURBS BREP model is received. Typically certain geometric curve configurations are eliminated, like T-junctions of edges, thus reducing ambiguity and variations of cases for the loop edges analyses, described in section 4.2.

It is also important, that most of modern CAD engines (i.e CATIA, ACIS, Parasolid and others) are capable of creation of fully watertight CAD models. As the S-BREP model is targeted to be a generic internal geometrical format for the ICEs it is possible to upload the conformal NURBS BREP definition to the CADR module and perform only dependent part of the CAD repair. Later on the S-BREP model can be created in a straight forward manner, with further possibilities of the model export to heterogeneous components of the ICEs, such as volume mesh generation tools, surface data interpolation tools, etc.

4.2 BREP model analysis

Geometrical analysis identifies C_0 discontinuous features of a solid model. Analyses is started on the level of faces (see Fig. 1), it is based on the angle between surface normal on two sides of a co-edge, forming boundary loop of a given face (see Fig. 6, a)). The loop is traversed, loop co-edges are picked one by one (Fig. 6, b)) and an angle F between the faces is defined as a maximal value of angles between underlying surface normals at n discrete parameter positions along the edge (Fig. 6, c)) corresponding to the given co-edge. Providing an angle is higher then the user-defined threshold, the edge is marked as discontinuous C_0 feature and later on is tested for presence of consecutive edges, possibly forming a composite subdivision curve boundary for a S-BREP face.

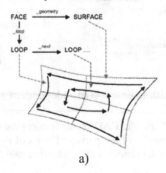

a)

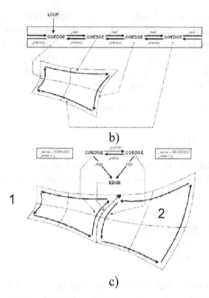

b)

c)

Fig. 6 The sequence of the BREP model analyses steps: a) – loop extraction on face, b) – loop transversal with co-edge extraction, c) – computation of the mean angle between normals to faces 1 and 2 along the given edge, referenced by co-edges.

For example, on Fig. 2 (right) the extracted sharp edge feature is composed from 9 segments, presented in the framework of the original NURBS model on the left. Corner nodes of the geometry are defined during analyses of the loop co-edges (Fig. 6, b)). Traversal of the co-edges is performed in physical space using automatically defined arc length parameter, which is dependent on the bounding box of the curve. During this process the NURBS curve C_0 discontinuities, related to inserted knots are also flagged as corner nodes of a subdivision curve. The corner position for a standard junction of edges is picked via normal position variation, similar to the detection of discontinuous edges as described above.

Extracting feature edges by the BREP loops traversal requires assessment of approximately 10-1000 less geometric elements, then in mesh-based volumetric element traversal, described in [8] and in surface polygon crawling algorithm in [20]. Presence of the topological tree of the BREP model permits to apply fast tree searching algorithms, further improving speed of geometry analyses. In our approach we compute the fundamental tree of the BREP topological graph and use it for the initial navigation in the search process for the geometry with multiple connected domains. Due to space limitations it is not possible to present further details of the algorithm.

4.3 Control mesh generation

The control subdivision mesh generation is performed on a watertight NURBS BREP model, obtained during the CAD repair stage of the process (Section 4.1). The MezGen advancing front surface mesh generation code [22] is used for this purpose. As obtained mesh is targeted for efficient definition of underlying subdivision surfaces, general requirements for the control mesh generation are different from the mesh generation for a generic computational application. To some extent, the control mesh of an interpolating subdivision surface defines how close the shape resembles a NURBS geometry on the initial levels of subdivision [13] (see also Fig. 4) and requires a specially tailored sizing function to control triangular element size of the control mesh.

Let us consider the following parametric surface:

$$r(u,v) = [x(u,v), y(u,v), z(u,v)] \qquad (2)$$

The sizing function is related to the principal curvature of the underlying NURBS surface and can be described as follows. First we define in the standard differential geometry notation [22] the first and the second fundamental forms of (2):

$$f_1 = Edu^2 + 2Fdudv + Gdv^2$$
$$f_2 = Ldu^2 + 2Mdudv + Ndv^2 \qquad (3)$$

where E, F and G are the first fundamental form coefficients, and L, M and N are the second fundamental form coefficients. The Gaussian (K) and the mean (H) curvatures are given by:

$$K = \frac{LN - M^2}{EG - F^2}$$
$$H = \frac{1}{2}\left(\frac{2FM - EN - GL}{EG - F^2}\right) \qquad (4)$$

The sizing function $S(u,v)$ for a given point in the parametric space of a NURBS face for control mesh generation is taken as follows:

$$S(u,v) = \frac{1}{2}\left[K + C\ H^{\frac{1}{2}}\right] \qquad (5)$$

where: C is a constant, defined from sizing geometrical considerations, i.e. dimensions of the bounding box of the smallest face in the repaired and processed NURBS BREP model. The value of S is dependent on the curvature of the underlying NURBS model, ensuring smaller control mesh cells in the curved areas. Strictly speaking, this is not absolutely required for adequate definition of a subdivision face, as after a certain number of subdivision steps its shape will converge to the underlying smooth surface. However, our experience shows that using our sizing function the mesh is closer to the underlying shape just after 2 subdivision steps and this coincides with information from [13] on the influence of the initial control mesh on the shape quality after limited number of subdivisions.

An important problem in application of subdivision surfaces for mesh generation is related to the absence of global maps for surface parameterization, therefore early applications are mostly using meshing in the physical space (see [11,12]). However, in Reference [14] Lee has developed an efficient parameterization scheme, based on an idea of the associated parametric mesh, which is basically a result of the application of the known flattening algorithms and interested readers may refer to References [24,25] for more details. Based on such a parameterization Lee has efficiently used an advancing front mesh generation scheme on subdivision surfaces geometry. Interestingly parameterization of the subdivision surface faces could be dynamic, providing more flexibility for surface mesh generation [14]. Further on, as subdivision surfaces are considered as generalization of splines to the parametric spaces of arbitrary shape [17,18] the problem of more generic parameterization appears to be possible. However, this problem is outside the scope of this paper and for the concept of the ICEs we expect direct application of the subdivision surface meshes for further generation of volumetric computational meshes and surface data interpolation.

4.4 Flagged discontinuities identification

During mesh generation, nodes of a control mesh are positioned on the edges of the underlying NURBS model. Providing an edge is flagged as discontinuity, respective nodes of geometrical mesh are flagged as interior cusp node, boundary cusp node, corner node or corner cusp node similar to the tagging process, described in [13]. The main difference in our approach is in the direct analyses the underlying NURBS BREP model.

5. Examples of the S-BREP geometry

Due to space limitations, current section provides only two example of the S-BREP geometry, automatically generated from the NURBS BREP model using the algorithm outlined in section 4.

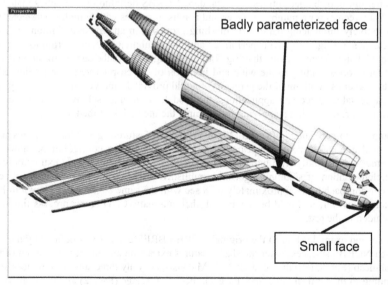

a)

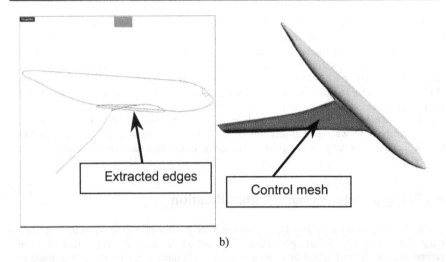

b)

Fig. 7 a) A blow-off of CAD faces in the original NURBS BREP model (41 faces) b) automatically extracted curves (left) and the S-BREP model, containing 5 subdivision faces, corresponding to the wing, fuselage, belly, window and rear fuselage tip.

In the first example (Fig. 7) a simplified wing-body aerospace configuration has been analysed for extraction of sharp features. The configuration contains 41 NURBS faces with different parameterisation and size (Fig. 7 a) gives a blow-up of the faces, a number of small and badly parameterised faces could be observed). Minimisation of the number of faces requires definition of bigger entities; in a standard NURBS definition that implies matching of the knot vectors for a number of faces and results in splines with complex parameterisation negatively influencing parametric meshing process. An automatic definition of sharp feature edges in (Fig. 7 b) left) is performed and just 5 subdivision faces permit to simplify CAD model significantly. Note, that Fig. 7 b) on the right shows the control mesh on the S-BREP face corresponding to the wing and insertion of the cusp curve at the trailing edge permits to avoid smoothing of the geometry. Should provided subdivision mesh quality will not be acceptable for certain applications, the model on different subdivision levels could be used for surface meshing, applying for example the method described in [14].

As compared to published in [8] and [20] algorithms of feature edges extraction, presented topological tree based method appears to be rather effective. For configuration, shown on Fig. 6, meshed internally with tetrahedral mesh (112321 elements), feature extraction with our implementation of skinning algorithm, published in [8] takes approximately 18 sec., while our approach takes approximately 0.78 sec with the same edges result, shown in Fig. 6 b), left. However, it should be mentioned, that automatic CAD repair process timing is not included in the test.

In our second example (Fig. 8) the original NURBS BREP geometry is defined by the 427 trimmed NURBS surfaces. After the sharp features extraction and further generation of the control mesh (Fig. 8, left) the obtained NCGM consists of only three subdivision faces, corresponding to the vertical tail plane (Face 1), the rare fuselage (Face 2) and the front vertical tail plane (Face 3) sections of the geometry. Flexibility of the automatic S-BREP creation approach is also demonstrated by the operator-driven definition of the Face 3 (Fig. 8), when a certain group of faces could be assigned to a separate SE group, i.e. having different boundary conditions or physical properties for the downstream FEM solver. Apparently, there is no sharp feature on the boundary of the Faces 1 and 3 and a separation is enforced due to the requirements of the simulations. The colour scheme in Fig. 8 gives the value of pressure provided by CFD computation on the surface of the geometry. The concept of the Butterfly interpolation of scalar/vector variables permits to use geometrical model directly for the CFD or other relevant aerodynamic data storage. The colour scheme of Face 3 is taken different for visualisation purposes.

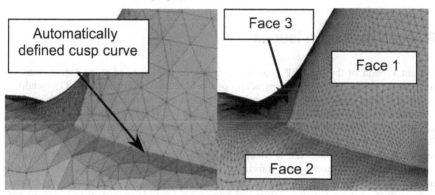

Fig. 8 The rear fuselage and vertical tail plane sections of the NURBS CAD model (427 faces). The S-BREP geometry is represented by just 3 faces on initial and second subdivision levels. C_0 internal cusp curve is automatically defined on step 2 of the NURBS CAD analyses. Face 3 is operator-defined entity, reflecting requirements of the FEM model.

The interpolation idea is to apply equation (1) for computation of scalars/vectors on the S-BREP geometry on new levels of subdivision. For example, pressure could be efficiently represented on the NCGM model using firstly control mesh of the subdivision surface and respectively interpolated values on different levels of subdivision directly using interpolation rules of the Butterfly subdivision scheme. Though certain smoothing effect could be observed on the data, subdivision interpolation could be directly used for storing the data on the geometry without fitting or extra processing. Original nodal values of the CFD aerodynamic data are obtained by projection of subdivision surface control mesh on the underlying CFD surface grid. Detailed discussion of the data storage using subdivision interpolation is outside the scope of the current paper.

The main motivation of the new model development has been the robust exchange of geometrical data and it has been tested for input to undisclosed Reynolds Averaged Navier-Stokes solver, aero elasticity and virtual reality modules. Using traditional CAD exchange format, input of the CAD model to individual component of the ICE required automated CAD repair process, which contained a number of user-driven operations. Each CAD repair cycle takes 0.3 – 2.7 person hours and overall production cycle within the ICE has not satisfied technical requirements for time. After introduction of the new model repeated CAD repair process has been excluded from the cycle due to stability of geometry exchange. Surface mesh, used in the S-BREP model has been directly used for volumetric CFD computational mesh generation, while discrete geometry has provided robust input for structural mesh generation and rendering in virtual reality applications.

In general, the proposed S-BREP model is extremely efficient for visualisation in the virtual reality environments, providing *a priori* tessellated surfaces for virtual reality engine. For example, the initial loading and rendering of the model on the 4th level of subdivision takes only 0.27 seconds for 57960 triangles with the Visualisation Toolkit (VTK) engine. Initial loading and rendering of the same geometry in the NURBS BREP with approximately 50000 surface triangles on the same workstation requires approximately 1.89 seconds.

6. Implementation issues

Current version of the S-BREP geometrical model is implemented in an object-oriented framework, similar to the coding paradigm of the MezGen mesh generator [22]. Subdivision BREP object is derived from the base virtual class mtkMesh [22] and in its topological structure coincides with the ACIS 8.0 BREP model. The S-BREP model is directly integrated to the CAD repair module [5], developed in collaboration with Dr. Thomas Woehler, Fraunhofer Institute for Production Systems and Design Technology – IPK, and works in the workflow of the ICE. The S-BREP here is an extension of the traditional NURBS model, combining full topological information with robust transfer and direct surface meshing capabilities. In our opinion there is no need for alternative surface meshing algorithms to be applied on the S-BREP as initial and subsequent subdivision surface tessellations are adequate for most downstream computational applications. Insuring optimal geometrical and computational mesh quality, the MezGen mesh generator applies a number of especially designed cosmetics operation on the generated control mesh, i.e. edge swapping and removal of triangles, as described in [22].

7. Conclusions

We have proposed a generic highly automatic method for complex geometry definition using Subdivision Surfaces based on the BREP data structure of the underlying NURBS model. Due to robustness of the mesh-like geometry transfer, proposed model combines fidelity of CAD, based on parametric surfaces with stability of discrete mesh-based geometry exchange. A set of extensions is proposed for the existing CAD repair process to enforce subdivision-compliant structure of the repaired model. Developed method of geometry definition directly uses the repaired NURBS BREP model for fully automatic extraction and flagging of the discontinuous geometrical features, thus providing complex C_1 in the limit free form CAD models with C_0 discontinuous features utilizing a new concept of the S-BREP faces. A heuristically defined curvature-based sizing function is proposed for an optimal definition of the control meshes for the S-BREP faces of the new model. A set of cosmetics operations, related to the optimal connectivity pattern of the generated control mesh is introduced in the MezGen surface mesh generation code.

The developed S-BREP NCGM model has been effectively tested on a set of complex CATIA and ACIS BREP input geometries with thousands of trimmed faces and has proved to be stable, efficient for direct surface data interpolation and capable of variable global/local resolution of the surface meshes.

Provided generalization of the S-BREP generation forms the main contribution of the paper together with the idea of subdivision interpolation of variables directly on the subdivision mesh using subdivision rules.

Acknowledgements

The author would like to acknowledge his support by CeBeNetwork Engineering & IT, Bremen, Germany during paper preparation and Dr. Thomas Woehler, Fraunhofer Institute for Production Systems and Design Technology - IPK, Berlin, for the CAD repair related discussions and help in preparation of some images on the BREP structure.

REFERENCES

[1] O. Zienkiewicz and R. Taylor. *The Finite Element Method, Volume 1, The Basis*. Fifth Edition, Butterworth-Heinemann: Oxford, (2000).

[2] A.A. Di Sessa. "Principle Design for an Integrated Computational Environment", *Human-Computer Interaction*, 1(1):1-47, (1985).

[3] A. Requicha, J. Rossignac. "Solid modelling and beyond". *IEEE Computer graphics and beyond*, 12(5): 31–44, (1992).

[4] L. Piegel, W. Tiller. "Curve and surface constructions using rational B-splines". *Computer-Aided design*, 19:485-498, (1987).

[5] Mezentsev, T. Woehler. "Methods and algorithms of automated CAD repair for incremental surface meshing". *Proceedings of the 8th International Meshing Roundtable*; South Lake Tahoe, USA, (1999).

[6] M. Beall, J. Walsh, M. Shepard. "Accessing CAD geometry for mesh generation." *Proceedings of the 12th International Meshing Roundtable*; Santa Fe, USA, (2003).

[7] R. Lohner. "Surface gridding from discrete data". *Proceedings of the 4th International Meshing Roundtable*; Albuquerque, USA, (1995).

[8] S. Owen, D. White. "Mesh-based geometry". *International Journal for Numerical Methods in Engineering*; 58:375-395, (2003).

[9] P. Lang, H. Bourouchaki. "Interpolating and meshing 3D surface grids". *International Journal for Numerical Methods in Engineering*; 58:209-225, (2003).

[10] P. Frey. "About surface remeshing". *Proceedings of the 9th International Meshing Roundtable*, New Orleans, USA, (2000).

[11] L. Kobbelt, T. Hesse, H. Prautzsch, K. Schweizerhof. "Iterative mesh generation for FE-computations on free form surfaces". *Engineering Computations*, 14(7):806-820, (1997).

[12] D. Rypl, Z. Bittnar. "Discretization of 3D surfaces reconstructed by interpolating subdivision". *Proceedings of the 7th International Conference on Nu-

merical Grid Generation in Computational Field Simulations. Whistler, Canada, pages 679-688, (2000).

[13]C.K. Lee. "Automatic metric 3D surface mesh generation using subdivision surface geometrical model. Part 1: Construction of underlying geometrical model". *International Journal for Numerical Methods in Engineering*; 56: 1593-1614, (2003).

[14]C.K. Lee. "Automatic metric 3D surface mesh generation using subdivision surface geometrical model. Part 2: Mesh generation algorithm and examples." *International Journal for Numerical Methods in Engineering*; 56: 1615-1646, (2003).

[15]N. Dyn, D. Levin, J. Gregory. "A butterfly subdivision scheme for surface interpolation with tension control". *Proceedings of SIGGRAPH 90,* Annual Conference series, ACM, *SIGGRAPH 90*, pages 160-169, 1990.

[16]Mezentsev, A. Munjiza, J.-P. Latham. "Unstructured Computational Meshes For Subdivision Geometry Of Scanned Geological Objects", *Proceedings of the 14th International Meshing Roundtable,* San Diego, California, USA, Springer, pages 73 – 89, (2005).

[17]D. Zorin, P. Schroder, W. Sweldens. "Interpolating subdivision with arbitrary topology". In *Proceedings of SIGGRAPH 96,* New Orleans, Annual Conference series, ACM SIGGRAPH, pages 189-192, (1996).

[18]C. Loop. "Smooth subdivision surface, based on triangles". Master Thesis. University of Utah, Department of Mathematics, (1987).

[19]T.J. Baker. "Identification and preservation of surface features". *Proceedings of the 13th International Meshing Roundtable,* Williamsburg, USA, pages 299-309, (2004).

[20]S. Yamakawa, K. Shimada. "Polygon crawling: Feature-edge extraction from a general polygonal surface for mesh generation". *Proceedings of the 14th International Meshing Roundtable,* San Diego, California, USA, Springer, pages 257 – 274, (2005).

[21]Y. Lu, R. Gadh, T. Tautges. "Volume decomposition and feature recognition for hexahedral mesh generation". *Proceeding of the 8th International Meshing Roundtable*; South Lake Tahoe, USA, pages 269-280, (1999).

[22]"MezGen – An unstructured hybrid mesh generator with CAD interface". http://cadmesh.homeunix.com, (2007).

[23]D. Struik. *Lectures on classical Differential geometry.* Second edition, Dover Publications: Dover, UK, (1988).

[24]Sheffer, E. de Sturler. "Surface parameterization for meshing by triangulation flattening", *Proceedings of the 9th International Meshing Roundtable,* New Orleans, USA, (2000).

[25]R.J. Cass, S.E. Benzley, R.J. Meyers, T.D. Blacker. "Generalized 3-D paving: An automated quadrilateral surface mesh generation algorithm", *International Journal for Numerical Methods in Engineering*; 56: 1615-1646, (2003).

2.2

A Hole-filling Algorithm Using Non-uniform Rational B-splines

Amitesh Kumar, Alan Shih, Yasushi Ito, Douglas Ross and Bharat Soni

Department of Mechanical Engineering
University of Alabama at Birmingham, Birmingham, AL, U.S.A.
{amitesh, ashih, yito, dhross, bsoni}@uab.edu

Abstract

A three-dimensional (3D) geometric model obtained from a 3D device or other approaches is not necessarily watertight due to the presence of geometric deficiencies. These inadequacies must be repaired to create a valid surface mesh on the model as a pre-process of computational engineering analyses. This procedure has been a tedious and labor-intensive step, as there are many kinds of deficiencies that can make the geometry to be non-watertight, such as gaps and holes. It is still challenging to repair discrete surface models based on available geometric information. The focus of this paper is to develop a new automated method for patching holes on the surface models in order to achieve watertightness. It describes a numerical algorithm utilizing Non-Uniform Rational B-Splines (NURBS) surfaces to generate smooth triangulated surface patches for topologically simple holes on discrete surface models. The Delaunay criterion for point insertion and edge swapping is used in this algorithm to improve the outcome. Surface patches are generated based on existing points surrounding the holes without altering them. The watertight geometry produced can be used in a wide range of engineering applications in the field of computational engineering simulation studies.

Key Words: Hole Filling, NURBS, Geometry Repair, Watertight, Mesh Generation, CFD, CSM, Delaunay Triangulation, Edge Swapping

1. Introduction

Mesh-based computational technologies, such as Computational Fluid Dynamics (CFD) and Computational Structural Mechanics (CSM), require a high quality mesh to achieve numerical stability and accuracy. Most of mesh generators need a properly prepared underlying geometry so that a valid surface mesh respecting the geometry definition can be generated. The geometry needs to be watertight. Overlapping or gaps between geometry entities, for example, must be treated properly before the mesh generation process. A geometry or surface mesh is considered to be not watertight in two situations:

1. It has edges shared by only one polygon, *i.e.*, they lie on boundaries. The occurrences of this kind of connected edges create holes in the surface.
2. It has edges shared by more than two polygons. The occurrences of these kinds of edges create non-manifold by virtue of hanging triangles. This kind of non-manifold mostly occurs in Computer-Aided Design (CAD) applications due to improper stitching of surface patches to generate a desired geometric model.

The surface mesh not only needs to maintain its mesh quality by achieving the required geometric quality measures, but it also needs to represent the geometry with high fidelity. It is still challenging to repair defective geometric surface models automatically and robustly while maintaining high geometric fidelity. This is especially true for the holes and gaps found on discrete surface models, as we have limited information upon which to estimate the missing geometric information.

Several algorithms have been proposed for filling polygonal holes [1-11]. Most of these algorithms cater to very specific sets of problems, *i.e.*, patching holes from only a particular source of geometry [1-6]. These algorithms broadly fall in two main categories: volume-based repair methods [5-7] and mesh-based repair methods [1-4, 8-11].

The key to all volume based methods lies in converting a surface model into a volume representation and generating a sign at each voxel representing whether the particular voxel lies inside, outside or on the surface of the geometry. The signs are generated with the help of distance map of each point on the geometry using line-of-sight information, which is usually obtained from range-finding devices. This crucial piece of information may not be available for a purely computational geometric model. The uncertain voxels are assigned signs based on volumetric diffusion [5, 6]. Once all the voxels are assigned signs, the volume-based methods simply extract the contour to find a closed surface. Curless and Levoy [6] propose a hole filling algorithm based on volumetric diffusion optimized for patching holes in models reconstructed by range-finding devices. These devices generally create small holes with complex topologies. However, these holes can be large compared to of the size of the polygons in the mesh. The volumetric methods are not suitable for this research because of the lack of line-of-sight information. Ju [7] presents a method for generating signs of voxels for repairing a polygonal mesh using an octree. However, this method is suitable only for small sized holes.

A number of algorithms have been suggested for filling holes in a triangular mesh using a surface triangulation-based approach. Leipa [8] describes a method for filling holes by a weight-based hole triangulation, mesh refinement based on the Delaunay criterion and mesh fairing based on energy minimization as used in [9]. The algorithm was confined to holes in an oriented connected mesh in relatively smooth region with the assumption that the holes are relatively small compared to the entire model. Barequet and Sharir [1] use a dynamic programming method to find minimum area of triangulation of a three-dimensional (3D) polygon in order to fill mesh holes. Barequet and Kumar [2] describe an interactive system that closes small cracks by stitching corresponding edges and fills big holes by triangulating the hole boundary similar in approach to Barequet and Sharir [1]. Unfortunately, this method cannot provide satisfactory results on relatively large-sized holes with complex geometric shapes. Jun [4] describes an algorithm based on stitching planar projection of complex holes and projecting back the stitched patch. Bruno [10] attempts to fill a hole and blend surface based on global parameterization for complete geometry approximation and then energy minimization for surface blending based on the assumption that global parameterization of the complete model is available or possible. Branch et al. [11] suggest a method for filling holes in triangular meshes using a local radial basis function. The method works quite well with skinny holes, but fails miserably when the holes are fatter in shape. All these algorithms are limited by their assumption that holes are relatively small-sized.

In this paper, a new automated method is proposed for filling holes on a triangulated surface model to make it watertight [12]. The existing points around the holes are used to obtain a set of Non-Uniform Rational B-Splines (NURBS) surfaces approximating the missing smooth surface patches. A Delaunay triangulation method is used to generate internal points which are then projected on to the set of NURBS surfaces to obtain the desired patch. The patches generated by this method are achieved without altering the geometric information of the surrounding geometry. This algorithm is currently applicable to topologically simple holes in the discrete geometry as a triangular mesh, while holes with more sophisticated topology will be considered in the future study. Such topologically simple holes are common in the geometry obtained from 3D scanners or geometry extraction from the medical image datasets using the marching cubes algorithm [14].

2. Hole Filling Processes

In order to properly fill the holes presented in the geometry, we need to first identify the boundaries of these holes, obtain its neighboring geometric information and attempt to estimate the missing surface patches. For the description purpose, a generic surface with a circular hole as shown in Figure 1 is used as a test case.

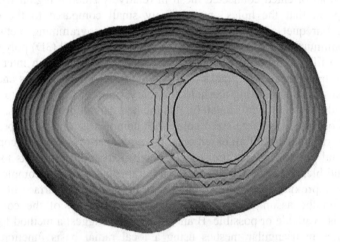

Figure 1. Discrete ordered points in the form of rings

2.1. Identification of a Hole

A hole is identified by checking the edge list to locate each boundary edge that is shared by only one triangle. A set of the boundary edges creates a hole on the surface. Once a closed edge list is identified, these edges are used to perform initial triangulation using the vtkTriangleFilter algorithm in the Visualization ToolKit [15]. Point insertion and edge swapping based on the Delaunay criterion are repeated on the initially triangulated mesh to produce a fine mesh that approximately (but not accurately) represents the missing hole surface. The fine mesh will be used later in Section 2.4 to project onto a more accurately approximated surface.

2.2. Identification of Neighboring Geometry into Ring Curves

The points on the original discrete surface are selected to form rings around the hole identified in Section 2.1. The hole is considered as 0th ring and only the 2nd, 4th and 6th rings from the hole are used to avoid the NURBS representation of the rings crossing each other. The rings are represented by NURBS curves to form a set of NURBS surfaces approximating the missing surface in the hole region (see Section 2.3). Figure 1 shows part of a surface with a hole in black and the rings in blue, red and green around it. The approach is robust even when two holes lie in the vicinity of each other as the rings would simply go around the boundary of the other hole.

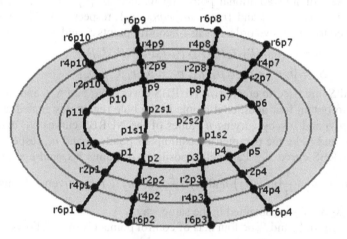

Figure 2. A simplistic NURBS representation of a hole, rings surrounding it and various points lying on them

2.3. Non-Uniform Rational B-Spline (NURBS) Surfaces

Once the rings around the hole are obtained as described in Section 2.2, they are used to construct a set of NURBS surfaces approximating the missing region. This is first done by calculating spline curves of the rings to ensure that they all possess the same number of points, hence producing smoother curves. Each of these closed curves is then divided into four segments at u values of 0.0, 0.25, 0.5 and 0.75 to form a four-sided surface inside, as illustrated in Figure 2, where u is the parametric value used in the NURBS curve formulation [13].

Let the NURBS curve representation of the hole and the 2nd, 4th and 6th rings be represented by R_0, R_2, R_4 and R_6, respectively. Let P be the set of 12 points found on R_0 by using a set of scalar values u_i ($i = 1, 2,..., 12$) in the parametric NURBS curve. A parameter, t, increases with each incremental rotation of the rings around the hole.

$$P = \{p_1, p_2,..., p_{12}\}$$

$$u_i = \begin{cases} u_i' & (0 \le u_i' \le 1) \\ u_i' - 1 & (u_i' > 1) \end{cases}, \quad u_i' = t + \frac{i-1}{12} \quad (0 \le t \le 1) \tag{1}$$

Each of the points p_2, p_3, p_8 and p_9 is projected on the NURBS curve approximation of rings R_2, R_4 and R_6 using nearest point projection to obtain three sets of four additional points designated as $\{r_2p_2, r_2p_3, r_2p_8, r_2p_9\}$, $\{r_4p_2, r_4p_3, r_4p_8, r_4p_9\}$ and $\{r_6p_2, r_6p_3, r_6p_8, r_6p_9\}$, respectively. These points are used to obtain two ordered sets of control points S_1 and S_2:

$$S_1 = \{r_6p_2, r_4p_2, r_2p_2, p_2, p_9, r_2p_9, r_4p_9, r_6p_9\}$$
$$S_2 = \{r_6p_3, r_4p_3, r_2p_3, p_3, p_8, r_2p_8, r_4p_8, r_6p_8\} \tag{2}$$

S_1 and S_2 are used to create two NURBS curves, N_{S1} and N_{S2}, and they are used to obtain four additional points in the hole region designated as p_1s_1, p_2s_1, p_1s_2 and p_2s_2 by varying u values of the NURBS curves N_{S1} and N_{S2}. Let u_2 and u_9 be the parametric positions of points p_2 and p_9, respectively, on NURBS curve N_{S1}. Parametric positions of points p_1s_1 and p_2s_1 are obtained as $u_2 + \frac{1}{3}(u_9 - u_2)$ and $u_2 + \frac{2}{3}(u_9 - u_2)$, respectively. Points p_1s_2 and p_2s_2 on N_{S2} are obtained in the same way.

Let S_3, S_4, S_5 and S_6 be four sets of control points given as follows:

$$S_3 = \{p_1, p_{12}, p_{11}, p_{10}\}$$
$$S_4 = \{p_2, p_1s_1, p_2s_1, p_9\}$$
$$S_5 = \{p_3, p_1s_2, p_2s_2, p_8\} \tag{3}$$
$$S_6 = \{p_4, p_5, p_6, p_7\}$$

These four sets of control points are used to obtain four interpolating NURBS curves, which are collectively used to create a lofted NURBS surface as shown in Figure 3.

A number of lofted NURBS surfaces are obtained by varying t. In this paper, 18 control surfaces were created for each hole in an attempt to obtain better estimation for the missing region. When each of the ring curves is divided into four segments for constructing a NURBS surface, the starting points are obtained by varying t corresponding to $10°$ of increment ($t = 0, 1/18, 2/18,..., 17/18$). It can be further studied to improve the algorithm efficiency by reducing the number of NURBS surfaces needed while maintaining the quality of the final surface generated.

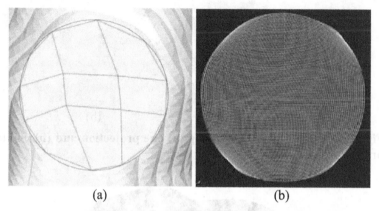

(a)   (b)

Figure 3. NURBS surface and control net: (a) Control net in the hole region; (b) Lofted NURBS surface and underlying control surface

2.4. Projection of Interior Points

All interior points generated in Section 2.1 are projected on to a set of 18 lofted NURBS surfaces to find a set of 18 coordinates. A simple average of these 18 projections is used to find the final coordinates and form a good approximation for the hole region. Figure 4 shows patches generated using only one projection and eighteen projections respectively. It clearly shows a distinct difference in the patches where the one generated with only a single NURBS surface tends to be very sensitive to the manner in which the ring curves are split, while the averaged one reflects the geometric information from the neighboring geometry more accurately. This is clear when the patch is placed on the original surface, as shown in Figure 5: the ridge in the middle of Figure 4b reflects the ridge on the neighboring geometry. The connectivity of each point on the mesh remains the same as the original point after the projection.

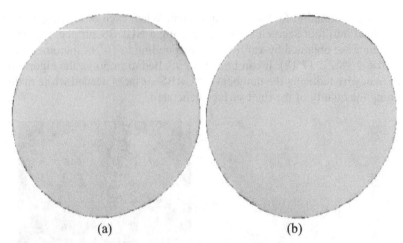

(a)   (b)

Figure 4. Patches generated using (a) one projection and (b) eighteen projections

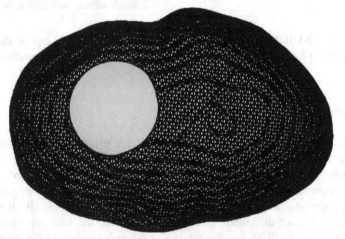

Figure 5. Surface patch generated for the hole. It reflects the ridge line feature that is present in the neighboring geometry.

3. Benchmark

In order to evaluate the errors of this algorithm, ellipsoids are chosen as benchmark cases to calculate the average errors in terms of radius and the standard deviation of errors. An ellipsoid of semi-axes a, b and c is defined as:

$$\frac{x^2}{a^2} + \frac{y^2}{b^2} + \frac{z^2}{c^2} = 1 \qquad (4)$$

Error in the location of points on the ellipsoids can be quantified as

$$E_i = \frac{x_i^2}{a^2} + \frac{y_i^2}{b^2} + \frac{z_i^2}{c^2} - 1 \qquad (5)$$

The average error can be measured as:

$$\overline{E} = \frac{\sum\limits_i^N E_i}{N} \qquad (6)$$

The standard deviation is the measure of the spread in a set of values. The standard deviation in the error of coordinate position on the patches can be obtained as follows:

$$\sigma = \sqrt{\frac{\sum\limits_i^N (\overline{E} - E_i)^2}{N}} \qquad (7)$$

Figure 6. Patches on a sphere with $a = b = c = 1$

Figure 7. Patches on an ellipsoid with $a = 1$, $b = 0.5$ and $c = 1$

Figure 8. Patches on an ellipsoid with $a = 1$, $b = 0.2$ and $c = 1$

A sphere is a special case of an ellipsoid when $a = b = c$. The sphere with a radius of 1 is shown in Figure 6. Figures 7 and 8 show two ellipsoids obtained by changing b values. Each of these three ellipsoids has 26,826 nodes and four holes that have been filled by patches (green) using the proposed algorithm. Table 1 shows average errors and standard deviations in the position of points on the patches generated for the three cases. It shows that the average errors lie within 0.35%, while the standard deviations are within 0.5%. We can also notice that the standard deviation increases with the increasing curvature of the surfaces surrounding the hole. This can be seen by comparing Figure 6, which has the lowest curvature, with Figure 8, which has the highest curvature.

Table 1. Average errors and standard deviations in the position of points on the patches

Cases	# of new nodes	$\overline{E}$	σ
$a = b = c = 1$ (sphere)	8,500	-3.16299 x 10^{-3}	2.71724 x 10^{-3}
$a = 1, b = 0.5, c = 1$	8,607	-2.91879 x 10^{-3}	3.52282 x 10^{-3}
$a = 1, b = 0.2, c = 1$	11,838	-2.97327 x 10^{-3}	4.84794 x 10^{-3}

4. Applications

This section aims to demonstrate the proposed algorithm with several complex geometric models. Section 4.1 shows a model of white matter of a human brain with 13 holes of various sizes punched on its surface. Section 4.2 shows a human pelvis mesh with 10 holes. These models were chosen to demonstrate the robustness of the algorithm in patching holes in areas of high curvature as well as larger-sized holes in the manner that the patches reflect the neighboring surface characteristics.

4.1. Brain White Matter Model

Figure 9 shows a simplistic model of white matter of a human brain with holes on its surface [16]. The model was obtained after segmentation of a series of Magnetic Resonance Imaging (MRI) data [17]. Once the segmentation was done, the marching cubes algorithm was used to obtain the triangulated surface. Thirteen holes were punched on the white matter model—some of them in regions of very high curvature—to test the robustness of the algorithm. The proposed algorithm successfully generated smooth patches to fill all the holes. The resulting surface is shown in

Figure 10. This test case emphasized the capability of this algorithm to generate smooth patches for holes with challenging geometry and curvature without smoothing out the local features of the geometry.

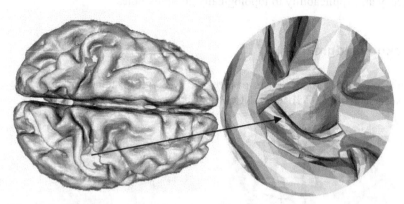

Figure 9. White matter model with 13 holes on its surface

4.2. Pelvis

Figure 11 shows a surface rendering of a patched unstructured surface mesh of a human pelvis [16]. The pelvis model is made of 1,090,700 triangles and 546,093 nodes with ten holes. The patches for the ten holes are made of 31,099 triangles and 16,331 nodes in total. Upon looking closely, it can be noticed that the three patches in dark green, red and yellow provide patches conforming to the surrounding geometry. Although they must be flat surfaces, the proposed method does not have a capability to identify such surfaces. A graphical user interface (GUI) will be added to exclude these holes beforehand.

5. Conclusion and Future Work

A new fully-automated hole-filling algorithm was proposed for triangulated surface models and was applied to several cases successfully. It utilizes the neighboring geometric information by identifying rings of curves around the holes to create a set of NURBS surfaces, which are subsequently used for projection of the triangulated surface patches. The 18 sets of projected coordinates are then averaged to obtain the final point coordinates for the hole region. Several benchmark cases are studied to assess the accuracy of the algorithms with satisfactory results. Finally, complex bio-medical geometries are used to test the robustness of the algorithms. Further investigation will improve the algorithm in terms of computation efficiency and applicability to topologically complex holes.

Acknowledgments

This research is supported in part by the NASA Constellation University Institutes Project (CUIP) No. NCC3-994.

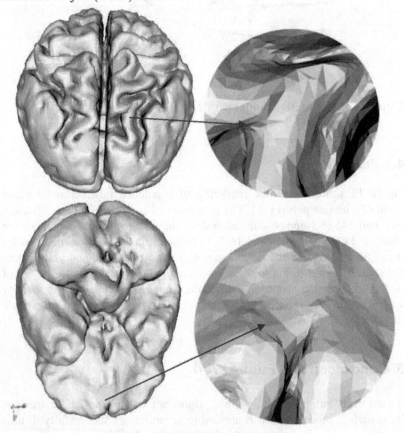

Figure 10. Patches for 13 holes on the white matter model

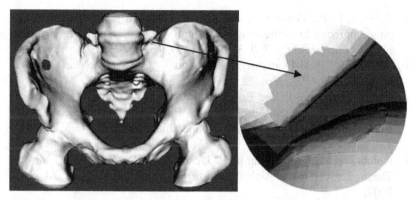

Figure 11. Human pelvis model with 10 holes patched on its surface

References

1. Barequet, G. and Sharir M., "Filling Gaps in the Boundary of a Polyhe-dron," *Computer Aided Geometric Design*, Vol. 12, 1995, pp. 207-229.
2. Barequet, G. and Kumar, S., "Repairing CAD Models," Proceedings of IEEE Visualization 1997, 1997, pp. 363-370.
3. Lee, Y. K., Lim, C. K., Ghazialam, H., Vardhan, H. and Eklund, E., "Surface Mesh Generation for Dirty Geometries by Shrink Wrapping using Cartesian Grid Approach," Proceedings of the 15th International Meshing Roundtable, Springer, 2006, pp. 393-410.
4. Jun, Y., "A Piecewise Hole Filling Algorithm in Reverse Engineering," *Computer-Aided Design*, Vol. 37, 2005, pp. 263-270.
5. Davis, J., Marschner, S. R., Garr, M. and Levoy, M., "Filling Holes in Complex Surfaces using Volumetric Diffusion," Proceedings of First International Symposium on 3D Data Processing, Visualization, Transmission, 2002, pp. 428-861.
6. Curless, B. and Levoy M., "A Volumetric Method for Building Complex Models from Range Images," *Computer Graphics*, Vol. 30, 1996, pp. 303-312.
7. Ju, T., "Robust Repair of Polygonal Models," Proceedings of ACM SIGGRAPH, 2004 ACM Transactions on Graphics, Vol. 23, pp. 888-895.
8. Liepa, P., "Filling Holes in Meshes," Proceedings of the 2003 Euro-graphics/ACM SIGGRAPH Symposium on Geometry processing, Eurographics Association, 2003, pp. 200-205.

9. Kobbelt, L. P., Vorsatz, J., Ulf, L. and Seidel, H.-P., "A Shrink Wrapping Approach to Remeshing Polygonal Surfaces," *Computer Graphics Forum (Eurographics '99)*, Vol. 18, 1999, pp. 119-130.

10. Bruno, L., "Dual Domain Extrapolation," *ACM Transactions on Graphics (SIGGRAPH)*, Vol. 22, 2003, pp. 364-369.

11. Branch, J., Prieto, F. and Boulanger, P., "A Hole-Filling Algorithm for Triangular Meshes using Local Radial Basis Function," Proceedings of the 15th International Meshing Roundtable, Springer, 2006, pp. 411-431.

12. Kumar, A., "Hole Patching in 3D Unstructured Surface Mesh," Masters Thesis, University of Alabama at Birmingham, Birmingham, AL, May 2007.

13. Piegel, L. A. and Tiller, W., *The NURBS Book*, 2nd ed., Springer, Berlin, 1996.

14. Lorenson, W. E. and Cline, H. E., "Marching Cubes: A High Resolution 3D Surface Construction Algorithm," *Computer Graphics*, Vol. 21, 1987, pp. 163-169.

15. Visualization ToolKit (VTK), http://www.vtk.org/.

16. Ito, Y., Shum, P. C., Shih, A. M., Soni, B. K. and Nakahashi, K., "Robust Generation of High-Quality Unstructured Meshes on Realistic Biomedical Geometry," *International Journal for Numerical Methods in Engineering*, Vol. 65, Issue 6, 2006, pp. 943-973.

17. Yoo, T. S., *Insight into Images. Principles and Practice for Segmentation, Registration, and Image Analysis*, A K Peters, Wellesley, MA, 2004.

Removing Small Features with Real CAD Operations

Brett W. Clark

Sandia National Laboratories
P.O. Box 5800 MS 0376
Albuquerque, New Mexico
87185-0376
bwclark@sandia.gov

Abstract

Preparing Computer Aided Design models for successful mesh generation continues to be a crucial part of the design to analysis process. A common problem in CAD models is features that are very small compared to the desired mesh size. Small features exist for a variety of reasons and can require an excessive amount of elements or inhibit mesh generation all together. Many of the tools for removing small features modify only the topology of the model (often in a secondary topological representation of the model) leaving the underlying geometry as is. The availability of tools that actually modify the topology and underlying geometry in the boundary representation (B-rep) model is much more limited regardless of the inherent advantages of this approach. This paper presents a process for removing small featrues from a B-rep model using almost solely functionality provided by the underlying solid modeling kernel. The process cuts out the old topology and reconstructs new topology and geometry to close the volume. The process is quite general and can be applied to complex configurations of unwanted topology.

1 Introduction

Preparing CAD models for mesh generation involves various processes which may include CAD format translation, geometry and topology generation and repair, defeaturing, and decomposition into meshable pieces. Defeaturing is usually required because the meshing algorithms rely heavily on the topology and underlying geometry of the boundary representation (B-rep) model and in many cases the B-rep model resulting from design does not meet the stringent requirements of the meshing algorithm. Of particular interest in this work is the removal of features that are very small compared to the desired mesh size.

These features require unnecessary resolution in the mesh or inhibit mesh generation completely.

Because B-rep models used for meshing can originate from a number of different sources, small features also have many different origins. Small features can be actual features that were intended in the design but which have little meaning in the analysis. They can also be features that were not intended in the design but which were a result of a careless CAD designer. During design, sliver surfaces and curves are often introduced by the solid modeling kernel to ensure a water-tight volume. Furthermore, small features can result from translating a model from a CAD format with loose tolerances to a CAD format with tight tolerances. As the analyst usually does not have control over these issues it is crucial to have tools for eliminating small features.

One place to eliminate small features is in the native CAD system where design took place. There, unwanted CAD features can simply be suppressed or modified directly. However, CAD systems do not generally provide many tools for fixing small topology that is not one of the provided CAD features. Therefore, the analyst is forced to use whatever geometry cleanup functionality is available in downstream applications. As a result, any modifications to the model will generally be done to the topology at the B-rep level and not at the feature level. Most defeaturing capabilities in applications downstream of the CAD design system fall into 1 of 3 categories: 1) modify the topology and geometry of the B-rep model directly using solid modeling or "real" operations if available, 2) modify a virtual representation of the topology in the B-rep model using "virtual" topology operations [6], and 3) modify the generated mesh to remove adverse affects of unwanted features in the B-rep model.

This work presents a small feature removal process that removes unwanted topology and associated geometry from a B-rep model. The proposed process works directly on the B-rep model reconstructing the geometry and topology in the vicinity of the feature(s) being removed using a combination of real operations provided by the underlying solid modeling kernel and virtual operations provided by the meshing application. The implementation was done in the CUBIT mesh generation package [1] developed at Sandia National Labs which supports the ACIS solid modeling kernel [2]. The solid modeling kernel operations employed and referred to during the description of the algorithm are operations provided by ACIS. However, the process could be implemented with any solid modeling kernel.

2 Related Work

As stated in the introduction, most defeaturing capabilities in applications downstream of the CAD design system fall into 1 of 3 categories: 1) modify the topology and geometry of the B-rep model directly using solid modeling or "real" operations if available, 2) modify a virtual representation of the topology in the B-rep model using "virtual" topology operations (composites/merges and partitions/splits), and 3) modify the generated mesh to remove adverse affects of unwanted features in the B-rep model.

The first approach, modifying the B-rep model directly with real operations, has the advantage that modifications made can be saved in the native solid modeling kernel format for later use in other applications. These types of modifications also have the advantage that in most cases not only is the topology modified but the underlying geometric definitions are modified maintaining a consistent relationship between the topology and geometry. Despite these advantages, however, the breadth of tools that modify the B-rep topology and geometry directly is limited. Research in this area includes work by Eccles and Steinbrenner [3]. They describe tools for generating B-rep topology information for models that don't provide it (some IGES files for example). The tools include algorithms for merging curves within a given tolerance and reconstructing single surfaces from networks of surfaces and curves. The main benefit of these tools appears to be in generating topology information but could probably also be extended to do general defeaturing tasks. These capabilities all appear to be for surface meshing applications. Butlin and Stops [4] describe tools for fixing common problems resulting from CAD format translation. They also mention defeaturing capabilities such as "joining patchworks of faces and chains of edges" and "collapsing of edges and faces" but do not go into much detail about these capabilities. Similarly, Jones et al. [5] describe tools for general gap cleanup and defeaturing including tools to make a single NURBS curve out of a chain of curves. Analogous capabilities for surfaces are discussed but not yet provided. Lee et al. [6] describe a system for representing feature modification. The system is independent of whether the implementation is done with real B-rep operations or virtual operations but they give some examples using the former. This work is focused more on the theory of a system that can provide reversible operations and not on the introduction of new operators. Venkataraman and Sohoni [7] present a delete face operator that is similar to the delete face operators provided by commercial solid modeling kernels but which handles a wider variety of cases. This operator appears to be quite useful for cases where the faces neighboring the deleted faces can intersect to close the volume.

The next area of research is very similar to the first except the topology operations are applied to a secondary representation of the B-rep topology. This secondary representation is sometimes referred to as virtual topology and allows for topological modifications without modifying the underlying geometry. Various incarnations of this approach have been implemented [6, 8, 9, 10, 11, 12]. The topological operations provided are generally based on two main operators: composite or merge and partition or split. With these two building blocks various topological modifications can be accomplished. As powerful as these tools have become there are still some limitations. First, the topology modifications can usually only be used in the application that defines them. In most cases they can not be saved out with the B-rep model to be used in other applications. Second, just modifying the topology and not the underlying geometry can lead to difficulties in geometric evaluations (surface normals, for example) in downstream algorithms such as mesh generation. Third, virtual topology operators are generally not interoperable with real solid modeling operations. This means that real solid modeling operations usually cannot be applied after virtual operations have been applied. The reason for this is that the virtual operations are typically applied to a secondary layer of topology that sits on top of the original B-rep model. As a result, the solid modeling kernel knows nothing about the virtual topology and real operations often invalidate the virtual topology. At the time of this writing, however, the author is aware of yet unpublished advancements in interoperability between real and virtual operators made in the CUBIT mesh generation toolkit developed at Sandia National Labs.

The third area of research is somewhat different from the first two in that modifications are made to the mesh after mesh generation and not to the original B-rep model. Eccles and Steinbrenner [3] use this approach to help generate model topology from meshed surfaces. The meshed surfaces are "stitched" together after modifying the mesh at the junction. Mobley et al. [12] first perform curve meshing to help guide them through the defeaturing process. The surface meshing then follows using the modified curve meshes. Dey et al. [13] first generate a mesh and then use mesh quality to determine where to modify the mesh to remove the adverse effects from small features in the geometry. In a follow-on work [14] Shephard et al. took this approach a step further and modified the topology (with virtual operations) based on modifications to the mesh.

3 Small Feature Removal Process

The small feature removal process presented here has two main parts: 1) small feature recognition and preprocessing via a surface splitting algorithm and 2) small feature removal via the "remove topology" operator. Both of these will now be described.

3.1 Small Feature Recognition and Preprocessing

Prior to removing unwanted small features in the B-rep model it is often necessary to examine the neighboring topology and make modifications that will facilitate the removal process. This is especially true when removing complex configurations of small features. This section describes a surface splitting algorithm that was developed for this purpose.

Small features can show up in a number of different ways. The simplest types to detect are small curves and small surfaces where the curve length and surface area can be calculated and compared to a reference value. However, small features can also exist when topological entities come in close proximity to one another. For example, the surface in Fig. 1 necks down to a very narrow region even though there is not a small curve length or small surface area to detect. The surface splitting algorithm developed as a part of this research attempts to find narrow regions of surfaces that can be split off into individual surfaces. These splits are "real" splits done using the solid modeling kernel. Fig. 1 shows how the algorithm would split the current surface into three

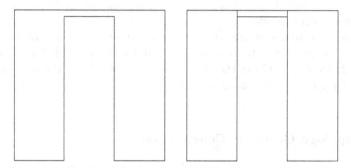

Fig. 1. Surface with narrow region

different surfaces. The algorithm checks the distance from endpoints of curves to other curves to find close proximities. When endpoints are found to be close to other curves, additional points along the curves are compared to locate the extents of the close regions. There are cases that this approach will not catch (like curves coming into close proximity to one another at their mid sections) but it has proven sufficient for the current needs. Fig. 2 shows an example of a model that has been modified by the surface splitting algorithm. The effect is to break up surfaces with narrow regions into the simplest set of surfaces based on surrounding topology. This usually results in reducing the number of curves in a surface to 3 or 4.

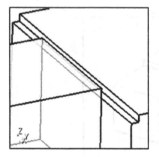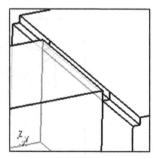

Fig. 2. Before and after performing automatic splitting of surfaces

3.2 Small Feature Removal

After the surface splitting algorithm has been applied the "remove topology" operator is used to remove the unwanted small features from the B-rep model. This is done using a combination of functionality provided by the underlying solid modeling kernel and virtual composites. As all of the critical modifications to topology and geometry are done directly to the B-rep model using the solid modeling kernel, the above-mentioned advantages of "real" operations are realized with this operator.

The remove topology operation has four main stages: 1) construct new topology and geometry to replace the "old topology" (topology being removed), 2) locally "cut" out the old topology, 3) "stitch" in the new topology, and 4) clean up extraneous curves. Each of these stages will be discussed below.

New Topology/Geometry Construction

The remove topology operator is similar to the remove surface operator provided in many solid modeling kernels. A remove surface operator generally removes a surface from the model and then reintersects adjacent surfaces to create a water-tight topology. However, as shown in Fig. 3, this approach will not always work. In the model shown in Fig. 3 there is a small step represented with a surface. The narrow surface may be removed from the model, but the adjacent surfaces will not intersect to close the model because they are parallel.

The remove topology operator is different from the remove surface operator in that it does not rely on reintersecting the adjacent surfaces to close the topology. The first part of the remove topology operation is to determine what "new topology" to use to replace the "old topology" (topology being removed). In the example in Fig. 3 the remove topology operator will generate new topology to replace the narrow surface representing the step. The remove

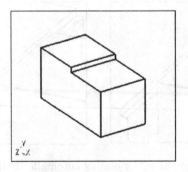

Fig. 3. Model with a small step surface

topology operator tries to reduce the dimension of the old topology wherever possible. In this example it will try to reduce the narrow step surface to a single curve as depicted by dotted lines in Fig. 4. The algorithm for doing

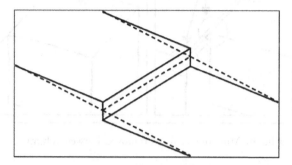

Fig. 4. Dimension reduction of narrow surface

this first looks at all of the topology specified for removal and finds vertices that are closer than some characteristic small curve length. These vertices are "clumped" together. Each clump of vertices will result in a single new vertex in the new topology. The clumps for the example above are shown in Fig. 5. The second step is to decide what new topology will exist between the

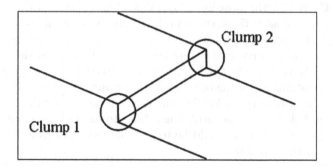

Fig. 5. Clumps of vertices in original topology

clumps. The old topology is examined to see how vertices were connected with curves. Curves that were between vertices in the same clump will be removed from the model as they are smaller than the characteristic small curve size. As a rule, in the new topology, clumps will be connected by a single curve. Therefore, surfaces whose vertices are completely contained in two clumps, as in the example in Fig. 5, will be removed and replaced by the new curve connecting the two clumps. Multiple surfaces can be removed simultaneously in this manner as shown in Fig. 6. Because the algorithm will handle networks of small features at the same time it is often advantageous to use the automatic

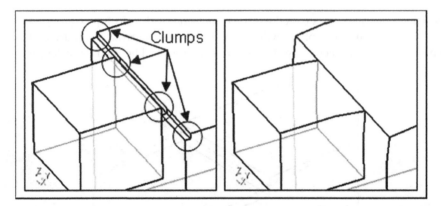

Fig. 6. Multiple surfaces removed between clumps

surface splitting algorithm described above prior to doing the remove topology operation. For example, the model on the left in Fig. 6 had some of its surfaces split before applying the remove topology operator. Fig. 2 shows this model before and after the splits. This splitting process produces vertices and curves that contribute to and simplify the remove topology operation.

Part of the difficulty in generating the new topology is deciding where to put the vertex representing a clump and how to define the geometry of the curves that connect clumps. There are various factors that could go into these decisions. Currently, the position of the new clump vertex is chosen as the average position of all of the vertices in the clump. As there are often multiple curves connecting clumps as in the example in Fig. 6, there can be multiple choices for the new curve connecting the clumps. Similar to the averaging approach for the vertices the new curve between two clumps is a spline defined by average positions along the old curves connecting the clumps. This choice for defining the geometry underlying the new topology is arbitrary and may not be the best choice in some situations. One alternative that would improve the flexibility of the process would be to give the user options for controlling the creation of the geometry.

"Cutting" out the Old Topology

The second part of the operation is to locally cut out the old topology. "Locally" is defined by a user-defined distance representing the distance that will be backed off from the old topology before making the "cuts." Each curve connected to the old topology is split at the specified distance and then the surfaces attached to these curves are split so that the entire network of old topology can be removed from the boundary representation of the volume. These split operations are done using common curve and surface splitting functionality provided by the solid modeling kernel. Fig. 7 shows an exploded view of the old topology being cut out of the boundary representation of the model. The cuts are made in such a way as to facilitate the use of general four-sided surfaces when stitching the new topology into the model. As a result the algorithm must examine the topology and geometry around the old topology and determine an optimal number of four-sided surfaces to use. The cuts are then made based on this analysis.

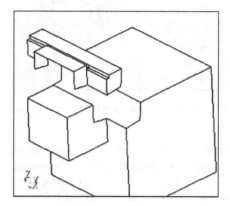

Fig. 7. Exploded view of old topology being cut out of boundary representation

"Stitching" in the New Topology

The last part of the operation is to "stitch" the new topology into the model. In the first part of the operation the new topology and corresponding geometry were defined. However, curves and surfaces to interface this new topology with the rest of the model were not defined. Four-sided (and on rare occasions three-sided) surfaces are used to connect the new topology to the rest of the model. In some cases the four curves bounding the surface will lead to a planar surface, but for all other cases four-sided splines are used. Most solid modeling kernels provide functionality for generating a spline surface from three or four bounding curves. Four-sided surfaces were chosen for their robustness when doing the final stitching to produce a water-tight volume. However, there is no limitation in the algorithm that requires four-sided surfaces. In fact any combination of surfaces could be used to connect the new topology to the rest of the model as long as the result is a water-tight volume. Fig. 8 shows the new topology stitched into the volume. The orange curves (new topology) were generated during the "new topology generation" stage of the operation and the red faces are the four-sided splines used to interface the new topology with the rest of the model. The solid modeling kernel's stitch operator was used to stitch everything together into a water-tight volume. A conscious decision was made to make the modifications to the model as local to the topology being removed as possible. One reason for this is to minimize the amount

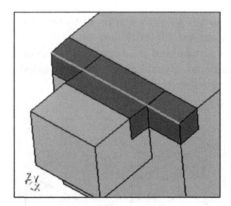

Fig. 8. New topology stitched to volume

of the model approximated by the four-sided surface patches. The interior of these surfaces is completely defined by their bounding curves. Therefore, these surfaces are not good for replacing other surfaces with lots of curvature unless the curvature is represented in the bounding curves.

Cleaning up Extraneous Curves

At this point in the operation the volume is water tight and the only thing left to do is get rid of the curves that were introduced by cutting out the old topology and stitching in the new topology. This step is optional and may not be critical depending on the downstream application. For example, if the downstream application is tetrahedral meshing the curves may not need to be removed depending on the mesh size requirements. The user has some control over how large a region is cut out when removing topology so that if the cutout region size is close to the mesh size there probably won't be a need to remove the curves. If the downstream application is hexahedral meshing it may be necessary to remove the extra curves to facilitate the volume meshing scheme.

If the geometric definition of the surfaces on either side of a given curve is the same the curve can usually be "regularized" away using a solid modeling kernel operation. However, when this is not possible, virtual composite operations are used to composite the surfaces on either side of the curve together into a single surface. Fig. 9 shows the final result after removing the extraneous curves.

4 Advantages of the Small Feature Removal Process

One advantage of this small feature removal process is that most of the operation is done using the solid modeling kernel. As a result, the modifications

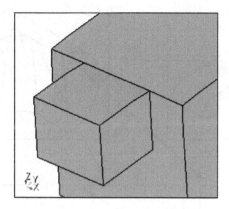

Fig. 9. Final result after removing extraneous curves

can be saved in the native B-rep format and reused in other applications. The only part of the operation that uses virtual topology is the clean-up of the extraneous curves at the end using composites. Obviously, these composites cannot be saved and used in other applications. However, the significant part of the topology/feature removal is persistent and the composites can be reapplied if necessary when needed.

Another advantage of this small feature removal process is that it can remove complex networks of "small" features and results in a fairly smooth transition from the new topology to the rest of the model. If the narrow surfaces in the above example were all eliminated using just composite surfaces, the composite surfaces would contain extreme C1 discontinuities where the surfaces meet at 90 degree angles. These types of discontinuities can greatly inhibit attempts to mesh the composite surface with advancing front surface meshers that rely heavily on surface normals. As the meshers advance across sharp corners like this they often break down.

Another advantage of the remove topology operator is that it could be incorporated into a CAD design system as it uses mainly real solid modeling kernel operations. Even though CAD systems provide powerful features for design they don't provide a large set of tools for removing unwanted topology not related to features.

5 Example

This example is of a part that contains many very small features on the order of 1e-5 but where the desired mesh size is about 0.2. Fig. 10 and Fig. 11 show zoomed in pictures of some of the small features. Fig. 10 (a) shows a cutout where a single curve is expected. The cutout propagates all the way along the top of the model. Fig. 10 (b) shows a cylindrical surface coming in tangent to another surface. The top of the cylindrical surface comes short of meeting

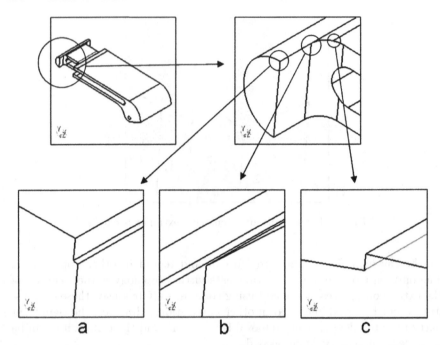

Fig. 10. Close-ups of problems in example model: a) unexpected cutout instead of single curve, b) features not connected as expected, c) unexpected small step

up with the cutout from Fig. 10 (a) creating a very narrow region. Fig. 10 (c) shows a very small step where a single curve is expected. Fig. 11 (a) shows a very narrow region that exists because two features don't line up exactly. Similarly, Fig. 11 (b) shows a step caused by the same misalignment. Fig. 11 (c) shows a small step with another small surface at the base of the step.

A tetrahedral mesher was used to generate a mesh on the unmodified geometry. However, the mesher could not succeed with the tiny features present. The automatic surface splitting algorithm was used to prepare the model for the remove topology operator which was then applied to the small features. Fig. 12 shows one of the small features as it goes through the split surface and remove topology operations. Higher zoom magnifications are shown in the progressing rows and the split surface and remove topology operation results in the progressing columns. Column 2 of Fig. 12 shows the model after applying the surface splitting algorithm. In row 2 (first zoom level) the surface with the curves meeting tangentially is split to remove the portion of the surface that is considered "narrow". The algorithm stepped back from the tangency until the distance between the two curves is the characteristic small curve length. It then split the surface at that location. As can be seen by this example, the process can be used to remove small angles in the model. In row 3 (second zoom level) the narrow region on the surface where the cylinder

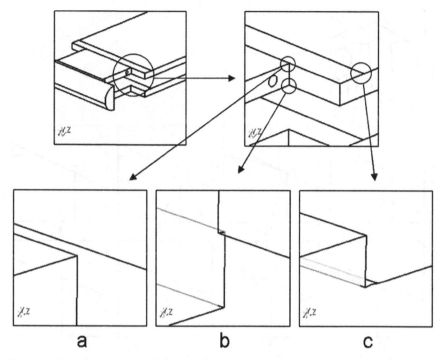

Fig. 11. More problems: a), b) features don't align as expected c) unexpected step and extra small surface

comes in tangent was split out and then the new vertex introduced by this split was propagated to the other adjacent narrow surfaces. All of these new vertices will form a single clump in the new topology.

Column 3 of Fig. 12 shows the model after the remove topology operator was applied. As can be seen, the topology was greatly simplified removing all of the little steps and misalignments that existed previously. The dotted lines indicate extraneous curves that were composited out and show where the original model was cut to remove the old topology Column 4 shows the final topology with the composited curves not drawn. This is the topology that the meshing algorithms interact with.

After applying the small feature removal process to all of the small features in the model the tetrahedral mesher was successful in generating the mesh with size 0.2 as shown in Fig. 13. Note that there is no unnecessary refinement which would have been necessary if the small features were present (the refinement in the far right frame is due to a hole that is part of the design).

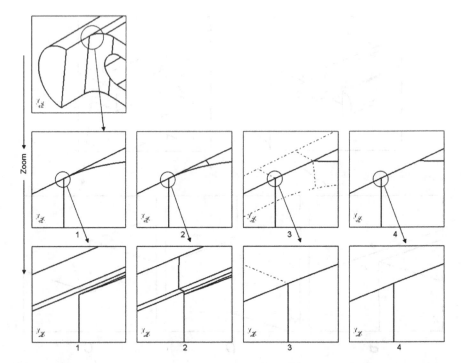

Fig. 12. Zoom in from top to bottom. Split surface and remove topology operations from left to right: 1) unmodified model 2) after split surface algorithm 3) after remove topology (showing extraneous curves as dotted lines) 4) after compositing out extraneous curves

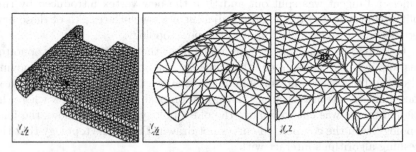

Fig. 13. Meshed example after removing small topology

6 Conclusion and Future Work

This paper introduced a new process for removing small features from a solid model using real operations provided by the solid modeling kernel. The process cuts out the specified topology and reconstructs the topology and geometry required to close the volume and maintain a water-tight topology. The process is built upon functionality that is commonly provided by solid modeling kernels. The process provides a general way to remove unwanted features and provides powerful capabilities for general geometry cleanup. Examples were given that demonstrated the process's ability to remove small curves, small/narrow surfaces, small regions in surfaces, and complex combinations of these. It was also shown to be able to remove small angles created by surface tangencies in the B-rep model.

One area for future research is to automate the application of this small feature removal process to a model in a global fashion. The algorithm relies on the user to provide the topology to be removed, a small curve length to decide what is "small", and the distance to backoff from the topology being removed when cutting it out of the model. These values will vary at different locations in the model and for different features to be removed. Having an algorithm that automatically generated and specified the input to the process based on the analysis of the B-rep model would be valuable.

A second area for future research is to develop a new method or incorporate an existing method for reconstructing the surfaces adjacent to the topology being removed so that they maintain their original shape as much as possible and also connect to the new topology. This capability could be used in 2 different ways: 1) the current algorithm could use this capability to reconstruct a single surface representing surfaces that are currently being composited together using virtual topology or 2) the current algorithm could be simplified to not "cut out" the old topology but rather just reconstruct surfaces adjacent to the old topology so that they connect to the new topology. This sort of reconstruction functionality would have an effect similar to a virtual composite operation but would be provided using real solid modeling operations.

A third area for future research is to extend the remove topology operator to work on assemblies. Adjacent volumes in an assembly often contain topology (vertices, curves, and surfaces) that occupies the same space within some tolerance. This topology is often "merged" together into a single piece of topology representing the two. The surrounding topology of both volumes is modified to incorporate the merged topology. This merging is often done in order to generate a contiguous mesh between the two volumes. If the topology being removed during a remove topology operation is topology that could be merged with topology on adjacent volumes, the remove operation should be performed on all such volumes at the same time so that the resulting topology can also be merged once the operation is complete. As well as maintaining mergeable topology this could be used as a tool to fix misalignments between volumes that prevent correct merging.

7 Acknowledgements

Sandia is a multiprogram laboratory operated by Sandia Corporation, a Lockheed Martin Company, for the United States Department of Energy's National Nuclear Security Administration under contract DE-AC04-94AL85000.

References

1. T.D. Blacker. *CUBIT Mesh Generation Environment, V. 1: User's Manual.* Sandia National Labs, 1994. SAND94-1100.
2. Spatial Corporation. Acis 3d toolkit. http://www.spatial.com/products/acis.html.
3. N.C. Eccles and J.P. Steinbrenner. Solid modeling and fault tolerant meshing-two complimentary strategies. In *17th AIAA Computational Fluid Dynamics Conference*, June 6-9, 2005.
4. G. Butlin and C. Stops. Cad data repair. In *Proceedings of 5th International Meshing Roundtable*, 1996.
5. M.R. Jones, M.A. Price, and G. Butlin. Geometry management support for auto-meshing. In *Proceedings of 4th International Meshing Roundtable*, pages 153–164. Sandia National Laboratories, 1995.
6. K.Y. Lee, C.G. Armstrong, M.A. Price, and J.H. Lamont. A small feature suppression/unsuppression system for preparing b-rep models for analysis. In *Proceedings 2005 ACM Symposium on Solid and Physical Modeling*, pages 113–124, June 13-15, 2005.
7. S. Venkataraman and M. Sohoni. Reconstruction of feature volumes and feature suppression. In *Proceedings 7th ACM Symposium on Solid Modeling and Applications*, pages 60–71, 2002.
8. A. Sheffer, T.D. Blacker, J. Clements, and M. Bercovier. Virtual topology construction and applications. In *Proceedings Theory and Practice of Geometric Modeling*, 1996.
9. T.J. Tautges. Automatic detail reduction for mesh generation applications. In *Proceedings 10th International Meshing Roundtable*, pages 407–418, 2001.
10. A. Sheffer. Model simplification for meshing using face clustering. *Computer Aided Design*, 33:925–934, 2001.
11. C.G. Armstrong, S.J. Bridgett, R.J. Donaghy, R.W. McCune, R.M. McKeag, and D.J. Robinson. Techniques for interactive and automatic idealisation of cad models. In *Proceedings 6th International Conference on Numerical Grid Generation in Computational Field Simulations*, pages 643–662, 2001.
12. A.V. Mobley, M.P. Carroll, and S.A. Canann. An object oriented approach to geometry defeaturing for finite element meshing. In *Proceedings 7th International Meshing Roundtable*. Sandia National Labs, 1998.
13. S. Dey, M.S. Shephard, and M.K. Georges. Elimination of the adverse effects of small model features by the local modification of automatically generated meshes. *Engineering with Computers*, 13:134–152, 1997.
14. M.S. Shephard, M.W. Beall, and R.M. O'Bara. Revisiting the elimination of the adverse effects of small model features in automatically generated meshes. In *Proceedings 7th International Meshing Roundtable*, pages 119–131. Sandia National Labs, 1998.

Automatic Extraction of Quadrilateral Patches from Triangulated Surfaces Using Morse Theory

John William Branch[1], Flavio Prieto[2], and Pierre Boulanger[3]

[1] Escuela de Sistemas, Universidad Nacional de Colombia - Sede Medellín,
   Colombia jwbranch@unalmed.edu.co
[2] Departamento de Eléctrica, Electrónica y Computación, Universidad Nacional
   de Colombia - Sede Manizales, olombia faprietoo@unal.edu.co
[3] Department of Computing Science, University of Alberta, Canada
   pierreb@cs.ualberta.ca

Summary. A method for decompose the triangulated surface into quadrilateral patches using Morse theory and Spectral mesh analysis is proposed. The quadrilateral regions extracted are then regularized by means of geodesic curves and fitted using a B-splines creating a new grid on which NURBS surfaces can be fitted.

1 Introduction

One of the most important phases in the reconstruction of 3–D model from triangular meshes is the the process of fitting high-level surface primitives such as NURBS (Non-Uniform Rational B-Splines) from them. Triangular meshes often exhibit several deficiencies. They are often too large or too high-resolution and typically contain elements with inadequate shapes. On the other hand, NURBS have become the standard in modern CAD/CAM systems, because of their ease of use and and their ability to deal at high-level with local surface modifications. Although NURBS have the capacity to represent arbitrarily curved surfaces, they still present problems when one wants to model fine details. This is due to how the NURBS surface are defined where parameters such as: knots, weights, and control points must be controlled by the CAD user to achieve a certain shape. In addition, NURBS surfaces also require to be placed on a networks of curves that usually have a quadrilateral topology.

The majority of the work reported in the literature on re-meshing methods, is focused on the problem of producing well formed triangular meshes (ideally Delaunay). However, the ability to produce quadrilateral meshes is of great importance as it is a key requirement to fit NURBS surface on a large 3-D mesh. Quadrilateral topology is the preferred primitives for modelling many objects and in many application domains. Many formulations of surface subdivision such as SPLINES and NURBS, require complex quadrilateral bases. Recently, methods to automatically quadrilateralize complex triangulated mesh have been developed such as the one proposed by Dong et al. [5]. These methods are quite complex, hard to implement, and have many heuristic components.

A method for decompose the triangulated surface into quadrilateral patches using Morse theory and Spectral mesh analysis is proposed. The quadrilateral regions obtained from this analysis is then regularized by means of computing the geodesic curves between each corner of the quadrilateral regions and then a B-splines curves are fitted to the geodesic curves on which NURBS surfaces are fitted. Such NURBS surfaces are optimized by means of evolutive strategies to guaranty the best fit as well as C^1 continuity between the patches.

This paper is organized as following: In Section 2, an introduction to Morse theory is presented. In Section 4, the proposed method for the adjustment of surfaces by means of optimized NURBS patches is presented. In Section 7.1 is presented a comparison between Branch's method and Eck and Hoppe's method. In Section 7, the results with the proposed model are discussed, and finally in Section 8, conclusions are presented.

2 Morse Theory

In a general way, spectral mesh analysis tries to infer topological features of the object through mathematical functions. This produces a spectrum which becomes a set of eigenvectors and eigenvalues of a matrix which has been inferred from the triangular mesh. The spectral analysis is supported by the Morse Theory. This theory chooses some representative points from the vertexes of the triangular mesh, critical Morse points.

Given a real function on a surface, Morse theory connects the differential geometry of a surface with its algebraical topology. This theory describes the connectivity of the surface from the configuration of the points where the gradient function decays. Such points are called critical points (these are: minimum, maximum and saddle points). The Morse theory has been used by the graphics and computer visualization community to analyze different real functions. For example, in terrain data analysis, Morse theory is used to identify topological features, while controlling the plane's simplification and organizing the features in a multi-resolution hierarchy [1], [4].

Let S be a smooth, compact 2-manifold without a boundary and let $h : S \rightarrow \mathbb{R}$ be a smooth map. The differential of h at the point a is a linear map $dh_a : TS_a \rightarrow TS_{h(a)}$, mapping the tangent space of S at a to that of $\mathbb{R}$ at $h(a)$ (The tangent space of $\mathbb{R}$ at a point is simply $\mathbb{R}$ again, with the origin shifted to that point). In a formal way: let $a \in S \subset \mathbb{R}^n$ a point in a continuous neighborhood parametrized by (u, v) [5]: a point $a \in S$ is called critical $h(a)$, if $h_u(a) = h_v(a) = 0$ (its calculated two partial derivatives, and are called critical when both are zero), otherwise, it is a regular point. The critical point a is degenerate if $h_{uu}(a)h_{vv}(a) - 2h_{uv}(a) = 0$, otherwise, it is a Morse point. If al critical points satisfies the Morse conditions, then the function h is a Morse function.

At a critical point a, we compute the local coordinates of the Hessian of h:

$$H(a) = \begin{bmatrix} \frac{\partial^2 h}{\partial x^2}(a) & \frac{\partial^2 h}{\partial y \partial x}(a) \\ \frac{\partial^2 h}{\partial x \partial y}(a) & \frac{\partial^2 h}{\partial y^2}(a) \end{bmatrix}. \tag{1}$$

The Hessian is a symmetric bilinear form on the tangent space TS_a of S at a. The matrix above expresses this function in terms of the basis $(\frac{\partial}{\partial x}(a), \frac{\partial}{\partial y}(a))$ for TS_a. A critical point a is called *non-degenerate* if the Hessian is non-singular at a; i.e., $det H(a) \neq 0$, a property that is independent from the coordinate system. The Morse Lemma [7], states that near a non-degenerate critical point a, it is possible to choose local coordinates so that h takes the form:

$$h(x,y) = h(a) \pm x^2 \pm y^2. \tag{2}$$

The number of minuses is called *index $i(a)$* of h at a, and equals the number of negative eigenvalues of $H(a)$ or, equivalently, the index of the functional $H(a)$. The existence of these local coordinates implies that non-degenerate critical points are isolated.

Let $\lambda_1 \leq \lambda_2$ be the two eigenvalues of the Hessian of h, with corresponding eigenvectors. The index of a critical Morse point is the number of negative eigenvalues of its Hessian. Therefore, this can be classified as: minimum, (index 0, h increases in every direction), saddle point (index 1, h changes in decrements and increments four times around a point), and maximum (index 2, h decreases in every direction). The function h is called Morse function if its critical points are not degenerated.

3 Morse Theory for Triangular Meshes

The application of the Morse theory for triangular meshes implies to discretize Morse analysis. The Laplacian equation is used to find a Morse function which describes the topology represented on the triangular mesh. In this sense, additional points of the feature of the surface might exist, which produce a basis domain which adequately represents the geometry of the topology itself and the original mesh. The mesh can also be grouped into improved patches. In this work, Morse theory is applied by representing the saddle points and its borders by a Morse function which can then be used to determined a number of critical points.

This approximation function is based on a discrete version of the Laplacian, to find the harmonic functions. In many ways, Morse theory relates the topology of a surface S with its differential structure specified by the critical points of a Morse function $h : S \rightarrow \mathbb{R}$ [11] and is related to the mesh spectral analysis.

The spectral analysis of the mesh is performed by initially calculating the Laplacian. The discrete Laplacian operator on piecewise linear functions over triangulated manifolds is given by:

$$\Delta f_i = \sum_{j \in N_i} W_{ij}(f_j - f_i) \tag{3}$$

where N_i is the set of vertices adjacent to vertex i and W_{ij} is a scalar weight assigned to the directed edge (i, j).

For graphs free of any geometry embedding, it is customary to use the combinatorial weights $W_{ij} = 1/\deg(i)$ in defining the operator. However, for 2-manifold mapped in $\Re^3$, the appropriate choice is a discrete sets of harmonic weights, suggested by Dong [5] and is the one used in this paper (see Equation 4):

$$W_{ij} = \frac{1}{2}(\cot \alpha_{ij} + \cot \beta_{ij}). \tag{4}$$

Here α_{ij} and β_{ij} are the opposite angles to the edge (i, j).

Representing the function f, by the column vector of its values at all vertices $f = [f_1, f_2, \ldots, f_n]^T$, one can reformulate the Laplacian as a matrix $\Delta f = -Lf$ where the Laplacian matrix L each elements are defined by:

$$L_{ij} = \begin{cases} \sum_k W_{ik} & \text{if } i = j, \\ -W_{ij} & \text{if } (i, j) \text{ is an edge of } S, \\ 0 & \text{in other case.} \end{cases} \tag{5}$$

where k is the number of neighbors of the vertex i. The Eigenvalues $\lambda_1 = 0 \le \lambda_2 \le \ldots \le \lambda_n$ of the matrix L forms the *spectrum* of mesh S. Besides describing the square of the frequency and the corresponding eigenvectors $e_1, e_2, \ldots, e_n$ of L, one can define piecewise linear functions over S using progressively higher frequencies [13].

4 Literature Review

There are many research results that deals with fitting surface model on triangular meshes. We will review some of them.

Loop [10] generates B-spline surfaces on irregular meshes. These meshes do not require a known object topology, and therefore, they can be configured arbitrarily without carrying a sequence of the 3D coordinates of the points set. The advantage of this method is that it uses different spline types for the surface approximation. The algorithm was tested using synthetic data with low curvature.

Eck and Hoppe [6] present the first solution to the fitting problem of B-spline surfaces on arbitrary topology surfaces from disperse and unordered points. The method builds an initial parametrization, which in turn is re-parametrized to build a triangular base, which is then used to create a quadri-lateral domain. In the quadrilateral domain, the B-spline patches adjust with a continuity degree of C^1. This method, although effective, is quite complex due to the quantity of steps and process required to build the net of B-spline patches on the adjustment surface.

Krishnamurthy and Levoy [9] presented a novel approach to adjust NURBS surface patches on cloud of points. The method consists of building a polyg-onal mesh on the points set first. Then on this mesh, a re-sampling is per-formed to generate a regular mesh, on which NURBS surfaces patches can be adjusted. The method has poor performance when dealing with complex sur-faces. Other limitations are the impossibility to apply the method to surfaces having holes, and the underlying difficulty to keep continuity on the NURBS surface patches.

Park [12] proposed a two-phase algorithm. In the first phase, a grouping of the points is performed by means of the k-means algorithm to create a polyhedral mesh approximation of the points, which is later reduced to a triangular mesh, on which a quadrilateral mesh is built. In the second phase, the initial model is used to build a net of NURBS patches with continuity C^1. Park's proposal assumes that the cloud-of-points is closed in such a way that the NURBS patches network is fully connected. This implies that the proposed method is not applicable to open surfaces. The use of NURBS patches implies an additional process keeping continuity at the boundary, making the method computationally expensive even when the irregularity of the surface does not require it.

Boulanger et al. [3] describe linear approximation of continuous pieces by means of trimmed NURBS surfaces. This method generates triangular meshes which are adaptive to local surface curvature. First, the surface is approxi-mated with hierarchical quadrilaterals without considering the jagged curves. Later, jagged curves are inserted and hierarchical quadrilaterals are trian-gulated. The result is a triangulation which satisfies a given tolerance. The insertion of jagged curves is improved by organizing the quadrilaterals' hi-erarchy into a *quad-tree* structure. The quality of triangles is also improved by means of a Delaunay triangulation. Although this method produces good results, it is restricted to surfaces which are continuous and it does not accu-rately model fine details, limiting its application for objects with an arbitrary topology.

Gregorski [8] proposes an algorithm which decomposes a given points-set into a data structure *strip tree*. The *strip tree* is used to adjust a set of minimal squares quadratic surfaces to the points cloud. An elevation to bi-cubic surfaces is performed on the quadratic surfaces, and they are merged to form a set of B-spline surfaces which approximates the given points-set. This proposal can not be applied to closed surfaces or surfaces which curve themselves. The proposal is highly complex because it has to perform a degree elevation and a union of patches on B-spline patches at the same time that a continuity degree C^1 is performed among adjacent patches.

Bertram [2] proposes a method to approximate in an adaptive way to disperse points by using triangular hierarchical B-splines. A non-uniform distribution of sampling on the surface is assumed, in such a way that zones with a high curvature present a denser sampling that zones with a low curvature. This proposal uses patches for data adjustment which add quality to the solution.

A different approach is presented by Yvart et al. [14], which uses triangular NURBS for dispersed points adjustment. Triangular NURBS do not require that the points-set has a rectangular topology, although it is more complex that NURBS. Similar to previous works, it requires intermediate steps where triangular meshes are reconstructed, re-parametrization processes are performed, and continuity patches G^1 are adjusted to obtain a surface model.

5 Quadrilateralization of Triangular Meshes Using Morse Theory

Representation by means of NURBS patches requires building a regular base on which it is possible to estimate the set of parameters which permit the calculation of surfaces segments which correctly represent every region of the objects. Representation of an object by means of small surface segments, improves the fitting quality in comparison with representation by means of a single continuous surface. Because each segment can be better fitted to local features, which permits modelling small details without excessive loss of surface smoothness.

Construction of a regular base consists of converting a triangular representation of the quadrilaterals set, which permits a complete description of the object's geometry. In general, NURBS surfaces requires a regular base. This does not indicate that the surface is rectangular. However, building a rectangular base permits to easily and directly regularize equidistantly each one of its sides. Because of complex and diverse forms which free form objects can take, obtaining a quadrilateral description of the whole surface is not a trivial problem (See Algorithm 1).

Localizing Critical Points

The procedure proposed at this paper estimates an initial quadrilaterization of the mesh, using spectral analysis by means of the Morse theory. Initially, the quadrilateral's vertexes are obtained as a critical points-set of a Morse

---

Algorithm 1: Quadrilateralization method of a triangular mesh.

```
Quadrilateralization();
begin
    1. Critical points computation;
    2. Critical points interconnection;
end
```

---

function. Morse's discrete theory guarantees that, without caring about topological complexity of the surface represented by triangular mesh, a complete quadrilateral description is obtained. That is to say, it is possible to completely divide objects' surfaces by means of rectangles. In this algorithm, an equation system for the Laplacian matrix is solved by calculating a set of eigen-values and eigen-vectors for each matrix (Equation 5).

Morse-Smale Complex is obtained from the connection of a critical points-set which belongs to a field of the Laplacian matrix. The definition of a field of the matrix is obtained by selecting the set of vectors associated to a solution value of the equation. As Morse function represents a function in the mesh, each eigen-value describes the frequency square of each function. Thus, selecting each eigen-value directly indicates the quantity of critical points which the function has. For higher frequency values, a higher number of critical points will be obtained. This permits representing each object with a variable number of surface patches. The eigen value computations assigns function values to every vertex of the mesh, which permits determining whether a vertex of the mesh is at critical points of the Morse function. In addition, according to a value set obtained as the neighborhood of the first ring of every vertex, it is possible to classify the critical points as maximum, minimum or "saddle points." Identification and classification of every critical point permits building the Morse-Smale complex.

Critical points Interconnection

Once critical points are obtained and classified, then they should be connected to form the quadrilateral base of the mesh. The connection of critical points is started by selecting a "saddle point" and by building two inclined ascending lines and two declined descending lines. Inclined lines are formed as a vertex set ending at a maximum critical point. Reversely, a descending line is formed by a vertex path which ends at a minimum critical point. It is allowed to join two paths if both are ascending or descending.

After calculating every paths, the triangulation of K surface is divided into quadrilateral regions which forms Morse-Smale complex cells. Specifically, every quadrilateral of a triangle falls into a "saddle point" without ever crossing a path. The complete procedure is described in Algorithm 2:

Algorithm 2: Bulding method of MS cells.

```
Critical points interconnection();
begin
    Let T={F,E,V} M triangulation;
    Initialize Morse-Smale complex, M=0;
    Initialize the set of cells and paths, P=C=0;
    S=SaddlePointFinding(T);
    S=MultipleSaddlePointsDivission(T);
    SortByInclination(S);
    for every s ∈ S in ascending order do
        CalculeteAscedingPath(P);
    end
    while exists intact f ∈ F do
        GrowingRegion(f, p0, p1, p2, p3);
        CreateMorseCells(C, p0, p1, p2, p3);
    end
    M = MorseCellsConnection(C);
end
```

---

6 Regularization of the Quadrilateral Mesh

Because the surface needs to be fitted using NURBS patches, it is necessary to regularize the quadrilaterals obtained from the mesh. In Algorithm 3, the proposed method to regularize these quadrilaterals is presented.

Algorithm 3: Quadrilateral mesh regularization method..

```
Regularization();
begin
    1. Quadrilateral selection;
    2. Selection of a border of the selected quadrilateral and its opposite;
    3. Regularization using B-splines with lambda density;
    4. Regularized points match by means of geodetics FMM;
        4.1 Smoothing of geodetic with B-splines;
    5. Points generating for every B-spline line with lambda density;
end
```

One of the quadrilaterals is selected from the mesh, and later a border is selected from each quadrilateral and its opposite. The initially selected border is random. The opposite order is searched as one which does not contain the vertexes of the first one. If the first selected border has vertexes A and B, it is required that the opposite border does not contain vertexes A and B, but the remaining, B and C.

Later, B-splines are fitted on selected borders with a λ density, to guarantee the same points for both borders are chosen, regardless of the distance between them. In general, a B-spline does not interpolate every control point; therefore,

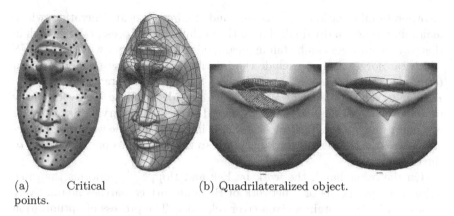

(a)   Critical   (b) Quadrilateralized object.
points.

Fig. 1. Quadrilaterization of a complex curve surface.

they approximate curves which permit a local manipulation of the curve, and
they require fewer calculations for coefficient determination.

Having these points at selected borders, it is required to match them. This
is done with FMM (*Fast Marching Method*). This algorithm is used to define
a distance function from an origin point to the remainder or surface in a
magnitude order of $O(nlogn)$. This method integrates a differential equation
to obtain the geodetic shortest path by traversing the triangles vertexes.

At the end of the regularization process, B-splines are fitted on geodetic
curves and density λ points are generated at every curve which unite the
border points of quadrilateral borders, to finally obtain the grid which is used
to fit the NURBS surface.

7 Results Analysis

The results obtained at each one of the intermediate stages of the proposed
algorithms in this paper are shown in Figure 1. This object called Mask is
composed of 84068 points. The reconstruction of the object took an average
time of 32 minutes on a dual Opteron PC .

7.1 Comparison Between Branch's Method and Eck and Hoppe's Method

The metric for adjustment error measurement previously described, was used
in this thesis to measure the adjustment error for each patch before and after
the optimization by means of an evolutive strategy. The obtained results show
the effectiveness of the proposed method. During the tests were used $(\mu + \lambda) -$
ES and $(\mu, \lambda) - ES$ evolutionary strategies, but the perform of the $(\mu + \lambda) - ES$
was superior in terms of error reduction. This behavior is because $(\mu + \lambda) - ES$

incorporate information from parents and children to next generation, which maintain the best individuals during the evolutionary process, even though in this way the process can be fall in local minimal. In contrast with $(\mu, \lambda) - ES$ which can forget and so exclude the information of the best individuals. In terms of execution time, $(\mu, \lambda) - ES$ had a negligible advantage over $(\mu + \lambda) - ES$. Only results of the $(\mu + \lambda) - ES$ are provided. During the evolutionary process, the weighting factors are restricted to the $[0, 1]$ interval. If as a results of a mutation, a recombination or a simple initialization, the weighting factors are outer to this interval, its value is set to zero, or set to one, according to the case.

On the other hand, the work by Eck and Hoppe [6] performs the same adjustment by means of a network of B-spline surface patches adaptively refined until they obtain a given error tolerance. The process of optimization performed by Eck and Hoppe reduces the error by generating new patches, which considerably augments the number of patches which represent the surface. The increment of the number of patches reduces the error because the regions to be adjusted are smaller and more geometrically homogeneous. In the method proposed in this thesis, the optimization process is focused on improving the adjustment for every patch by modifying only its parameterization (control points weight). Because of that, the number of patches does not augment after optimization process. The final number of patches which represent every object is determined by the number of critical points obtained in an eigenvector associated with the eigenvalue (λ) selected from the solution system of the Laplacian matrix, and it does not change at any stage of the process.

Figure 2 contains a couple of objects (foot and skidoo) reported by Eck and Hoppe. Every object is shown triangulated starting with the points cloud. The triangulation is then adjusted with a patch cloud without optimizing and the result obtained after optimization. The adjustment with the method proposed in this thesis, represents each object, with 27 and 25 patches, while Eck and Hoppe use 156 and 94 patches. This represents a reduction of 82% and 73% fewer patches respectively, in our work.

With respect to the reduction of the obtained error in the optimization process in each case, with the proposed method in this thesis, the error reduces an average of 77% (see Table 1), a value obtained in an experimental test with 30 range images (see Table 1). Among these appear the images included in Figure 2. The error reported in Eck and Hoppe for the same images of Figure 2 allow a error reduction of 70%. In spite of this difference which is given between our method with respect to Eck and Hoppe's method, we should emphasize that error metrics are not the same, Eck and Hoppe's method is a measurement of RMS, ours method corresponds to an average of distances of projections of points on the surface.

Another aspect to be considered in the method comparison is the number of patches required to represent the object's surfaces. In Eck's work, the number of patches used to represent the object's increase is an average of 485% in relation to the initial quadrilaterization, while in the method proposed in this thesis, the number of patches to represent the surface without optimization, and the optimized one, is constant.

Image	Initial Error	Optimization Error	Error Reduction(%)
1	19,43	4,41	77,30
2	12,15	2,21	81,81
3	14,25	3,12	78,11
4	13,47	2,74	79,66
5	11,45	2,12	81,48
6	15,26	3,21	78,96
7	14,84	3,14	78,84
8	18,45	3,54	80,81
9	12,12	2,04	83,17
10	14,32	3,31	76,89
11	15,31	3,91	74,47
12	16,70	4,09	75,51
13	20,05	4,48	77,66
14	18,23	4,27	76,58
15	13,24	3,12	76,44
16	19,32	4,45	76,97
17	17,32	4,01	76,85
18	15,24	3,45	77,36
19	16,24	3,69	77,28
20	11,25	2,65	76,44
21	17,32	3,56	79,45
22	14,25	3,25	77,19
23	11,22	2,35	79,06
24	13,26	3,21	75,79
25	14,15	4,25	69,96
26	16,25	4,21	74,09
27	14,25	3,69	74,11
28	18,23	4,56	74,99
29	12,20	2,98	75,57
30	19,43	4,41	77,30
Reduction Average			77,34

Table 1. Percentage of reduction of the error by means of optimization using our method.

8 Conclusion and Future Work

A novel method of quadrilateralization by means of spectral analysis of meshes and Morse theory has been proposed, starting from a triangular mesh. This method is topologically robust and guarantees that the complex base be always quadrilateral, thus avoiding ambiguities between quadrilaterals.

As future work, determination of the quantity of critical points in the following way:

- Quadrangulation refinement to take surface geometrical singularities (to be defined) into account, and optimization in terms of angles in quadrangles.

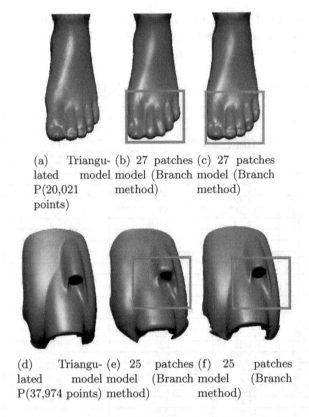

(a) Triangulated model P(20,021 points)   (b) 27 patches model (Branch method)   (c) 27 patches model (Branch method)

(d) Triangulated model P(37,974 points)   (e) 25 patches model (Branch method)   (f) 25 patches model (Branch method)

Fig. 2. Comparison Between Branch's Method and Eck and Hoppe's Method.

- To solve the problem of the determination of the control points, explore the use of nonlinear optimization methods, that can be applied efficiently by means of a parallel implementation.

References

1. C. Bajaj and D. Schikore. Topology preserving data simplification with error bounds. *Computers and Graphics*, 22(1):3–12, 1998.
2. M. Bertram, X. Tricoche, and H. Hagen. Adaptive smooth scattered-data approximation for large-scale terrain visualization. *EUROGRAPHICS IEEE TCVG Symposium on Visualization*, 2003.
3. P. Boulanger. Triangulating trimmed nurbs surfaces. *Curve and Surface Design*, 2000.
4. P. Bremer, H. Edelsbrunner, B. Hamann, and V. Pascucci. A topological hierarchy for functions on triangulated surfaces. TVCG 10, 4, 385396, 2004.

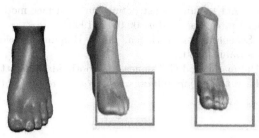

(g)   Triangu- (h) 29   patches (i) 156   patches
lated   model model (Eck and model (Eck and
P(20,021   Hopper's Method) Hopper's Method)
points)

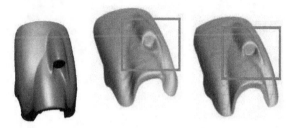

(j)   Triangu- (k) 15   patches (l) 94   patches
lated   model model (Eck and model (Eck and
P(37,974   Hopper's Method) Hopper's Method)
points)

Fig. 2. Comparison Between Branch's Method and Eck and Hoppe's Method.

5. S. Dong, P. Bremer, M. Garland, V. Pascucci, and J. Hart. Quadrangulating a mesh using laplacian eigenvectors. *Technical Report UIUCDCS-R-2005-2583*, 2005.

6. M. Eck and H. Hoppe. Automatic reconstruction of b-spline surface of arbitrary topological type. *ACM-0-89791-747-4*, 8, 1996.

7. H. Edelsbrunner, J. Harer, and A. Zomorodians. Hierarchical morse-smale complexes for piecewise linear 2-manifolds. *Discrete Comput. Geom*, 30:87–107, 2003.

8. B. Gregorski, B. Hamann, and D. Joy. Reconstruction of b-spline surfaces from scattered data points. *Proceedings of Computer Graphics International*, 2000.

9. V. Krishnamurthy and M. Levoy. Fitting smooth surfaces to dense polygon meshes. *In SIGGRAPH 96 Conference Proceedings*, ACM SIGGRAPH, Addison Wesley, 1996.

10. C. Loop. Smooth spline surfaces over irregular meshes. *Apple Computer Inc.*, 1994.

11. X. Ni, M. Garland, and J. Hart. Simplification and repair of polygonal models using volumetric techniques. *Proc. SIGGRAPH*, TOG 23,3,613-622, 2004.

12. I. Park, S. Lee, and I. Yun. Constructing nurbs surface model from scattered and unorganized range data. *3dim*, 00:0312, 1999.
13. G. Weber, G. Scheuermann, H. Hagen, and B. Hamann. Exploring scalar fields using critical isovalues. 2002.
14. A. Yvart, S. Hahmann, and G. Bonneau. Smooth adaptive fitting of 3d models using hierarchical triangular splines. *Shape Modelling International*, 2005.

Hexahedral Meshing

3A.1

An Extension of the Reliable Whisker Weaving Algorithm

Franck Ledoux and Jean-Christophe Weill

CEA/DAM Ile de France, DIF/DSSI, BP 12, 91680 Bruyères-Le-Châtel, France
{franck.ledoux@cea.fr|jean-christophe.weill}@cea.fr

Summary. In many FEM applications, hexahedral meshes are necessary to get the best results. However the automatic generation of an hexahedral mesh from an arbitrary 3D geometry still remains a challenge. In this paper, we propose an extension of the reliable Whisker Weaving algorithm [6]. The Whisker Weaving algorithm starts from a pre-meshed quadrilateral surface, it generates hexahedral meshes for a large spectrum of 3D geometries successfully, but it often creates poor-shaped hex elements. We show that considering geometric information in the shrinking loop selection process can provide better results. We introduce criteria to handle non-convex geometries, parallel loops and sheets bounded by several loops.

Key words: hex mesh generation, topology, whisker weaving, parallel loops, geometric properties

1 Introduction

For many years, the finite element method (FEM) has been successfully applied in simulation. Such a method needs a finite element mesh model obtained from a geometric CAD model and suitable for a numerical analysis. It is known that, in many FEM applications, hexahedral meshes give better results than tetrahedral or hex-dominant ones. Last decade showed many attempts to automatically generate hexahedral meshes from arbitrary 3D geometries [14, 2, 18, 16, 6, 5, 10, 15, 13, 12]. However, the automatic generation of hexahedral meshes is more complex than for tetrahedral meshes and still remains a challenge.

For our simulation problems, high mesh quality is required more near the boundary of the solid than deep inside. This property led us to study advancing-front algorithms that we classify in two main categories: algorithms which are essentially based on the geometry and algorithms which focus on the mesh topology. The former have the advantage of capturing the boundary very well but they are sensitive to numerical approximation and they need for expensive intersection calculations [14]. The latter do not have the drawbacks of the first ones but they often do not provide a satisfying result [6]. Actually, as authors of [6, 9, 13, 12], we think it is essential that any algorithm which attempts to automatically generate hexahedral meshes must take both the model topology and the model geometry into account.

In this paper, we introduce an extension of the reliable Whisker Weaving algorithm [6]. This algorithm has the advantages of being mathematically well-founded and successfully generating hexahedral topology for a wide spectrum of solid geometries. Its principle consists in driving the hexahedral element creation by introducing some complete layers of hexes in one go. The global mesh and model topology are well considered in the reliable Whisker Weaving algorithm but the global geometry may not be. Indeed, geometric information is not considered to select which layer of hexes to create. We suggest to extend the reliable Whisker Weaving by paying attention to the geometric information in the selection process. Moreover, we add new rules in order to manage parallel loops and non-convex regions. Even though we are using geometric information, the algorithm remains mainly topological. Note also that we do not attempt to extend any quadrilateral surface meshes to hexahedral meshes. Following [6, 9, 10], our approach is restricted to quadrilateral surface meshes without any self-intersecting loops in the dual of the surface mesh. Self-intersecting loops seem to be a difficulty to get a robust hexahedral meshing algorithm without degenerated hexahedral elements.

The remainder of this paper is organized as follows. Section 2 reminds of the notion of Spatial Twist Continuum and gives the rough lines of the reliable Whisker Weaving algorithm. Section 3 explains why we need to extend the reliable Whisker Weaving algorithm and thus introduces the concept of *high-level* and *intermediate* rules. We detail these rules respectively in Sects. 4 and 5 while Sect. 6 shows how basic and intermediate rules propagate the geometric information during the overall hex meshing process. A brief summary of this process is given in Sect. 7. Finally, we show some examples in Sect. 8 and we discuss future directions of this work in Sect. 9.

2 Whisker Weaving and Spatial Twist Continuum

The Whisker Weaving algorithm is based on the Spatial Twist Continuum [18], or STC, which structures the dual of a hexahedral mesh into surfaces or *sheets*. This algorithm is an advancing-front algorithm which attempts to address the hex meshing from a purely topological approach. Geometric characteristics are considered secondary to the overall mesh topology. This algorithm first generates the complete dual mesh, from which the primal mesh is obtainable. The spatial location of internal nodes is not computed until the mesh topology is totally defined. The first version of the Whisker Weaving algorithm [18, 16] only considered local element connectivities to create a single hex. This way, the global mesh topology was not taken into account.

The reliable Whisker Weaving version [6] corrects this drawback by introducing a global scheme to create some complete layers of hexes. The principle of the reliable Whisker Weaving algorithm consists in driving the hex creation by selecting a loop of quadrilateral elements on the surface front mesh. A loop of quadrilateral boundary elements corresponds to a layer of internal hexahedral elements. The rules to select the shrinking loop are mainly topological. Unlike the first version of the Whisker Weaving algorithm, the reliable Whisker Weaving algorithm starts from a quadrilateral boundary mesh without self-intersecting loop. This restriction makes it possible to get a robust algorithm that generates less wedges [3].

In the next Section, we begin by reminding the vocabulary relative to the STC before describing the reliable Whisker Weaving algorithm.

2.1 STC: the Fundamental Structure of any Hexahedral Meshes

The STC structures the dual of a hex mesh into *sheets*. Every sheet is dual to a layer of hexes. For example, consider Fig. 1 which shows an hex mesh (on the left) and its corresponding dual mesh (on the right).

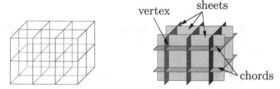

Fig. 1. An hexahedral mesh and the corresponding STC structure

The two green sheets in the STC are dual to the six lower hexes and the six upper hexes in the primal mesh. The intersection of two sheets is a *chord*. A chord is dual to the intersection of two layers of hexes, in other words, a line of hexes. A sheet which intersects oneself is a *self-intersecting sheet*. The intersection of three sheets is a *vertex* which is dual to a hex. In the primal mesh, the intersection of a layer of hexes with the quadrangular boundary mesh is formed by zero, one or more closed lines of quadrilateral elements. Such lines are defined by recursively passing from an edge of a quad to the edge opposite it. The dual of a boundary quad line is a *loop* which bounds a sheet. A sheet with no loop is a *blind sheet* and it contains at least one *blind chord* which is a chord which never meets the boundary.

2.2 A Brief Overview of the Reliable Whisker Weaving [6]

The reliable Whisker Weaving algorithm is based on a loop contraction algorithm which creates a complete layer of hexes in one go. We introduce this algorithm in this Section, but before we recall the basic weaving operations which are used to perform it.

Basic Weaving Operations

Three basic operations are useful for the reliable Whisker Weaving algorithm[1]: the *cross*, the *seam* (or *join*) and the *blind chord formation* (see Fig. 2). The cross operation creates an hex from three front faces which pairwise share edges. The seam operation completes a chord, i.e. a layer of hexes. In the primal mesh, it consists in joining two faces, which share exactly two adjacent edges, to make one face. The blind chord formation operation creates an hex from only two meshing front faces which share an edge. It introduces a blind chord which traverses the new hex by the two faces which are not opposite to an initial meshing front face in the hex. Note that every operation modifies the meshing front, i.e. the topological loop arrangement. This property is important and will be used in the loop contraction process.

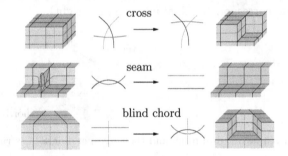

Fig. 2. The three basic rules used in reliable Whisker Weaving

Even though the Whisker Weaving algorithm is a topological advancing-front algorithm, some geometric pre-conditions are defined for every rule. For instance, the creation of an hex from three boundary faces is possible only if the angles between the faces are not too flat.

Loop Shrinking

The reliable Whisker Weaving algorithm is based on the observation that contracting a curve to a point on a sphere sweeps out a surface behind it inside the ball [7]. For this algorithm, the curves are the loops on the dual surface and the swept out surfaces are the sheets. In the primal mesh, shrinking a loop creates a complete layer of hexes in an advancing-front fashion. The

---

[1]Note that other rules are defined in the initial Whisker Weaving algorithm [16].

algorithm chooses which loop to shrink and decides to shrink it left or right. This operation is repeated until all the loops are shrunk.

Once a direction is selected, shrinking a loop consists in crossing it with all the other loops it came across from until it no longer intersects other loops. At this time the loop shrinks to a point and disappears. This process can also be interpreted as collapsing the faces which are adjacent to the loop in the shrinking direction. Collapsing a face consists in performing some basic weaving operations to reduce its number of vertices until it totally vanishes. For instance, consider the 5-sided face on Fig. 3. The red curve is the curve to shrink. The first operation to apply is a blind chord formation. It reduces a k-sided face to a $k-1$-sided face. In the example, the 5-sided face becomes 4-sided. A second blind chord formation transforms the gray face in a 3-sided face. Applying a cross operation completes the cell collapsing.

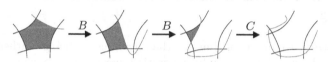

Fig. 3. Collapsing a cell

Now consider the overall shrinking process with Fig. 4 where we want to shrink the black loop inward. For every face adjacent to the shrinking loop, a figure indicates its number of sides. The idea is to begin by traversing once about the loop to find a face with the fewest number of sides and collapse it. Then we look again for the face with the fewest number of sides and we collapse it. This process is repeated until the loop shrinks to a point.

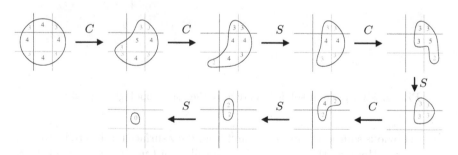

Fig. 4. An example of shrinking process

The authors of [6] claim that the best order to select the shrinking loop is by increasing *weight*. Each loop has two weights, corresponding to shrinking it left or right. Let n_{rp} be the number of points on the right of the curve, n_{lp} the number of points on the left of the curve and n_{pc} the number of points on the curve. The left weight is defined as $w_l = n_{lp} + n_{pc}$ and the right weight as $w_r = n_{lp} + n_{pc}$. The best choice is to select the loop with the smallest weight in a direction (left or right). This pure topological approach is balanced by introducing some geometric constraints. Actually, local rules associated with basic weaving operations are blended into the decision structure of the loop shrinking process. These rules are either geometric, which only apply to hexes near the model boundary, or topological, which apply universally. It seems that local rules are not used to select the shrinking loop but just to help drive the loop shrinking process. Note that, according to the authors of [6], it is usually necessary to turn the geometric rules off to complete the weave, but rarely necessary to turn the topological rules off.

3 Our Motivations to Extend the Reliable Whisker Weaving

The Whisker Weaving algorithm has the advantage of being mathematically well-founded and of successfully generating hexahedral topology in many cases. Still, generated hexes are often poorly shaped or even inverted. Two reasons could explain this drawback. First, the result of the Whisker Weaving algorithm strongly depends on the initial boundary mesh, this drawback is common to all meshing algorithms which start from a fixed boundary mesh. Second, the choice of the layer of hexes to introduce is only driven by topological information while geometric information may be useful. For instance, consider Fig. 5 which shows five topologically equivalent surface meshes. All these meshes have 14 faces, 28 edges and 16 vertices.

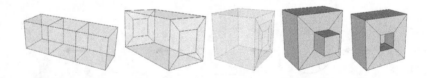

Fig. 5. Five different solids bounded by the same topological quad mesh

An obvious solution to mesh the first boundary surface mesh is to introduce three hexes. However this solution is not suitable for the other cases while they are topologically equivalent. A careful loop selection process has to take the surface geometry into account. Possible results obtained with our approach are shown on Fig. 6. To get such results, we consider geometric properties like convexity and particular topological configurations like parallel loops. These examples are quite typical of our concerns considering that we use this algorithm to create blocks in an hex meshing block-structured approach.

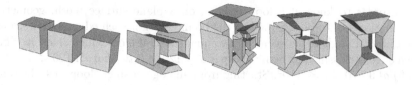

Fig. 6. Hex meshes which correspond to the surface meshes shown on Fig. 5

Our choices have led us to provide an algorithm built on three rule sets. First, we define *high level* rules to select which loop to shrink. These rules pay attention to geometric and topological information. Then, *basic rules* which are equivalent to the basic weaving operations are used to shrink a loop. Finally, we introduce *intermediate level rules* between the high level rules and the basic rules. They locally modify the arrangement of loops by introducing hexes or applying particulars seams. These local changes are driven by global geometric and topological considerations.

4 High-Level Rules to Drive the Shrinking Loop Selection

We are currently using five high-level rules which are useful for selecting the loop to shrink. They depend on some properties relative to each loop. As a loop can be shrunk left or right, each property is actually associated with a loop and a direction. Keep in mind that a difference between our approach and the Whisker Weaving algorithm is that we work on the primal mesh and not on the dual mesh. The convex rules (see below) and the properties necessary to these rules are the same ones as those in [10].

4.1 The Depth Rule

In many numerical applications, high mesh quality is required closer to the boundary than inside the mesh. In order to enforce that important characteristic, we defined the *depth rule*. It requires a value is assigned to each face, this depth value being smaller when the face is closer to the boundary.

A left depth index d_l and a right depth index d_r are associated with each loop. For a direction i, we consider all the faces on the i side of the loop and we define d_i as being equal to the least strictly positive depth index assigned to one of these faces. The depth rule consists in selecting the least depth loops.

4.2 The First Convex Rule

The *first convex rule* consists in selecting the loops that are not lying on a plane surface mesh. This rule attempts to keep the algorithm from building a sheet from one loop when it should be bounded by two loops at least.

In order to know if a loop is lying on a plane surface mesh, geometric information has to be assigned to the boundary. For each edge, we determine the dihedral angle α between the two quadrilateral faces which share it. Thus each edge is classified as being *sharp* if $\alpha \leq 120$, *a little sharp* if $120 < \alpha < 170$ and *flat* if $170 \leq \alpha \leq 180^2$. Starting from an edge, a single loop l can be built by recursively passing from an edge of a q quad to the edge opposite it in the q quad until the initial edge is reached. For instance, on the left of Fig. 7, starting from the e edge, the red loop is created. The edges of the quads crossed by the loop are separated in three sets: the set se_c^l contains the edges crossed by the loop l, the set se_l^l contains the edges on the left (the green edges on Fig. 7 left) and the set se_r^l contains the edges on the right (the blue edges on Fig. 7 left). In order to know if a loop is lying on a plane surface mesh, we count the number of sharp and a little sharp edges in the set se_c^l. The loop can be shrunk only when this number is positive.

Fig. 7. Left: The three sets of edges associated with a loop. Middle and right: Two non-convex solid geometries and some unacceptable loops shrinkings

4.3 The Second Convex Rule

The *second convex rule* first selects the loops which are along a sharp region of the solid. As [10], we assign a *side elimination weight* to each loop for both directions. For the right side (respectively the left side), this weight is the number of sharp edges in the set se_r^l (respectively se_l^l) divided by the loop length. The second convex rule consists in selecting the highest weight loops.

4.4 The Non-Convex Rule

The goal of the *non-convex* rule is to keep the algorithm from shrinking a loop towards a non-convex part of the solid. For instance, on Fig. 7 (middle and right), red and blue loops can not be shrunk in the indicated directions.

In order to do that, we consider the fact that a non self-intersecting loop separates the edges of the boundary mesh in two sets, se_{gr}^l and se_{gl}^l, which respectively contain all the meshing front edges on the right side and on the left side of the l loop. A loop can be shrunk to direction d only when

---

^{2}Note that the limit values can be modified.

the set sel^l_{gd} contains no non-convex edges. In order to manage non-convex property, we extend the edge classification introduced in Sec. 4.2. Consider that boundary face normals are directed to the outside of the solid, we classify each boundary edge e as a function of the dihedral angle α between the two faces which share it (see Fig. 8): if $\alpha \in]0, 120]$, the edge e is *sharp convex*; if $\alpha \in]120, 170]$, the edge e is *a little sharp convex*; if $\alpha \in]170, 190]$, the edge e is *flat*; if $\alpha \in]190, 240]$, the edge e is *a little sharp non-convex*; if $\alpha \in]240, 360[$, the edge e is *sharp non-convex*. The non-convex rule indicates that it is impossible to shrink a loop right (respectively left) if the number of non-convex edges in se^l_{gr} (respectively se^l_{gl}) is positive.

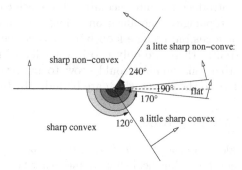

Fig. 8. Edge classification as a function of the dihedral angle

4.5 The Parallel Crossing Rule

Parallel loops have to be considered with attention in an hex meshing algorithm which starts from a fixed quad boundary mesh. This fifth rule has been added to manage the parallel loops separated by a sequence of sharp convex edges. Such configurations appear in some cylindrical mesh parts. For example, consider Fig. 9 which was already introduced on Figs. 5 and 6.

Fig. 9. A configuration where it is necessary to apply the parallel crossing rule

The meshing process must be driven so that it avoids to create a large hex which would completely fill in the geometry. Thus we first consider the blue and red loops which are topologically parallel and separated by a sequence of sharp convex edges. The idea is to introduce a layer of hexes in such a way that the first created chord is a torus of hexes adjacent to both blue and red loops. This result is obtained by shrinking one of the parallel loops in the direction of the other loop. We currently select the shrinking loop as being the loop which collapses the smallest number of faces when it is shrunk in the direction of the other loop.

5 Intermediate Rules to Locally Modify the Mesh

In this Section, we introduce the intermediate rules which clean up the mesh by applying a local topological modification before selecting a new loop to shrink. They do not focus on a single loop but consider the global topology and geometry. Currently there are only two intermediate rules. But we can legitimately think that this number could grow to improve the algorithm. Both rules are relative to the sheets which are bounded by two loops at least.

5.1 The Parallel Merging Rule

This rule is equivalent to the *double loop elimination* process introduced in [10]. It is efficient for cylindrical-shaped meshes when some sheets are bounded by two loops. More generally, this rule makes it possible to join certain configurations of two sheets into one sheet. Suppose that three loops are parallel and next to each other. Let l be the central loop, l_r be its right parallel loop and l_l be its left parallel loop. If, in addition, all the edges separating l and l_r on the one hand, and l and l_l on the other hand, are classified as sharp edges, the solution consists in eliminating loops l_r and l_l simultaneously. In [10], this elimination is interpreted as removing a chord which forms a torus of hexes, we choose rather to add the *parallel merging rule* shown on Fig. 10 which merges the two parallel loops l_r and l_l in one single loop. Afterward, we shrink this new loop to create a torus of hexes.

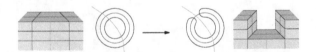

Fig. 10. Merging of two parallel loops

5.2 The Non-Convexity Seam Rule

The basic seam is performed when two faces share exactly two adjacent edges. It joins these faces to make one new face. As a consequence, it finishes creating a chord of hexes. In this Section, we introduce a new seam rule, called the *non-convexity seam rule*, which joins two faces sharing only one edge when particular geometric conditions are satisfied. This rule completes some chords in non-convex geometries. More precisely, when a chord arrives near a boundary face in a non-convex region, it may be better to directly join the chord and the boundary face. Without this rule, basic weaving rules introduce too many hexes in the final mesh and sometimes fill in non-convex areas.

Fig. 11. Example of non-convexity seam

Consider Fig. 11 where F_1 and F_2 are two faces which share[3] the single edge E. There are three possibilities: both faces are boundary faces, both faces are internal free faces or one face is a boundary face and the other one is an internal free face. If F_1 and F_2 are both boundary faces, they can not be sewed considering that their geometric positions are fixed. The case of both internal faces is out of the scope of the non-convexity rule. Indeed the goal of this rule is to complete chords which arrive near a non-convex boundary region. Actually, the non-convexity seam rule is only performed in the third case when a face is on the boundary and the other one is free. Let F_2 be the boundary face and F_1 be the internal free face. Some conditions have to be verified. The edge E has to be a convex[4] edge and an edge of F_2 adjacent to E has to be non-convex. Moreover the two vertices of F_2 which not belong to E must be free internal nodes. Under these conditions, the non-convexity seam rule can be used as illustrated on Fig. 11. Afterward, some faces adjacent to F_1 and F_2 share two edges. Such a degeneracy is removed with the basic seam.

6 Propagation of Geometric Information

The high-level and intermediate rules are based on some geometric properties of the initial surface mesh. These properties could be disposed of when the meshing front is no longer near the boundary, but we prefer to propagate them inside the mesh. This choice allows us to keep internal layer of hexes as parallel to the surface mesh as possible. Consider Fig. 12 where the sharp convex edges and the sharp non-convex edges are respectively colored in red

---

[3]If F_1 and F_2 share two edges, the basic seam is performed.

[4]If the edge E is convex by propagation, some tests are performed depending on its initial state and the number of hexes sharing it.

and green. We start from the left surface mesh. Creating a first layer of hexes modifies the meshing front. We propagate the geometric properties from the previous meshing front to the new one. In particular some front edges are now defined as being sharp and convex while they were not beforehand. Creating the second layer of hexes shows another characteristic of the propagation process: a convex property and a non-convex property cancel each other out. As a result, when a sharp convex property is propagated onto a sharp non-convex edge, this edge becomes flat.

Fig. 12. Propagation of geometric constraints during the meshing process

Since the basic and intermediate rules modify the surface meshing front, they have to propagate the geometric properties. The first property to propagate is the depth. At the beginning, the depth of every boundary face is equal to zero. When a rule is performed, either an hex is introduced or two faces are sewed. In the first case, each new face belongs to the new hex. Let F_1 be such a face and F_2 its opposite face in the hex. If F_2 is a former front face whose depth is equal to i, the depth of F_1 is equal to $i + 1$. If F_2 is a new front face too, the depths of F_1 and F_2 are equal to -1. A seam introduces no face, so we do not have to propagate the depth parameter.

Consider Fig. 13 where the red line represents the shrinking loop. The cross rule introduces a hex and then replaces three front faces with three new ones. As explained above, for an initial i-depth face, the corresponding new face depth (opposite in the created hex) is $i + 1$. The seam only removes faces. The blind chord formation rule introduces an hex and replaces two faces with four faces. Two new faces correspond to the initial ones. The two other faces are not topologically parallel to a front face. Their depth values are arbitrary fixed to -1. Note that when an initial front face is -1-depth, the parallel face is -1-depth too. The merging rule is similar to the blind chord formation rule and the non-convexity seam rule does not introduce any new faces.

The second property to propagate is the angle and the convex type[5] associated with every edge. As the basic rules are used to shrink a loop, they have to consider the shrinking direction to propagate the angle. Indeed, shrinking a loop creates a layer of hexes and the meshing front is then translated from one side of this layer to the other side. As a result, the basic rules just propagate the angle in a direction as illustrated on Fig. 14. The intermediate rules which are not associated with a shrinking loop, act quite differently. The par-

---

[5] convex or non-convex.

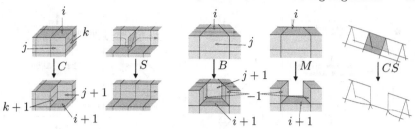

Fig. 13. Rules and depth

allel merging rule propagates the properties associated with the edges which separate the central loop and the parallel loops. The non-convexity seam rule propagates the properties associated with the edge which is shared by the two faces to sew.

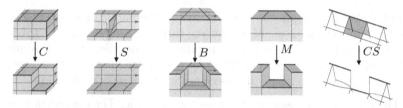

Fig. 14. Rules and convexity

7 A Brief Summary of the Algorithm

The algorithm is described in this Section and illustrated on Fig. 15. It essentially defines in which order the high level rules are applied.

The first step consists in initializing geometric properties. They are computed at the first time and then they are propagated inward without any other geometric computations.

The second step consists in applying the non-convexity rule when it is necessary. This rule is the only one which does not help to choose which loop to shrink. Note that the parallel merging loop rule which is an intermediate rule too implies to shrink the merged loop. This way, it contributes to choose which loop to shrink.

Selecting the shrinking loop needs to compute every loop properties. Actually we compute the properties of every couple (l, d) with l a loop and d a direction. Let S be the set of these couples. For a loop l, S contains the two couples (l, l) and (l, r) which correspond to shrinking l left or right. Then the parallel merging rule and the high-level rules allows us to select which loop

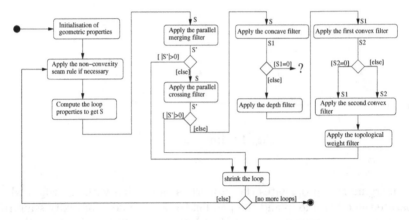

Fig. 15. Operation Sequence to shrink loops

to shrink. We first apply the rules relative to the parallel loops. These rules directly define the loop to shrink and the shrinking process can start. Otherwise, the high-level rules, except the parallel merging rule, are used. Applying such a rule r is equivalent to filter the set S in order to only keep the couples (l, d) which verify the r rule. If S becomes void, we relax some rules.

- The first filter is based on the non-convexity rule. The case when none of the couples (l, d) verifies this rule is currently not supported. We have not experienced yet this very particular case.
- The second filter is based on the depth rule. It keeps the couples (l, d) with the smallest depth. Then, among these couples, we select the couples (l, d) which verify the convex rules.
- The third and fourth filters are based on the first and second rules respectively. The first convex filter may be relaxed.
- The last filter is topological. It keeps the loops which collapse the smallest number of faces.

Note that only two filters can void S out. Once a loop is selected, it is shrunk by performing the same shrinking process as in the Whisker Weaving algorithm. Note that our basic rules have the same kind of pre-conditions as in the reliable Whisker Weaving algorithm. These pre-conditions can be turned off too.

8 Examples

The following examples illustrate the qualities and drawbacks of our approach. The first example on Fig. 16 shows the algorithm ability to work with non-convex geometries and to create sheets bounded by several loops.

The surface mesh of the examples on Figs. 17 and 18 are obtained by applying a paving-like algorithm [4], followed by a sheet removing algorithm [6]

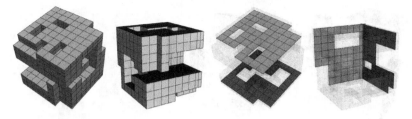

Fig. 16. View of a non convex mesh and several sheets bounded by more than one loop in its dual

coupled with pillowing [8]. This process guarantees to get a surface mesh without any self-intersections. Our algorithm is then used to generate hexahedral meshes. Taking care about geometric information does not prevent our algorithm from generating hexes with negative jacobians at nodes, making them unusable for most FEMs. In fact, far from boundary, topological rules have priority and even if we propagate geometric information inward our algorithm acts like the reliable Whisker Weaving one. So it has the same drawbacks. Note that we have not used any topological smoothing [6] yet to get a "better" mesh. Most of the meshes we have generated have an accumulation point which contains the last introduced layer of hexes by our algorithm. In fact, these meshes have too many sheets. Such a result tends to prove that geometric seams are necessary to reconnect some sheets and so decrease the number of sheets.

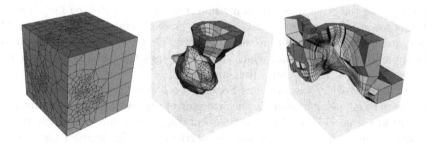

Fig. 17. View of a cube with a complicated surface mesh. It has 14022 elements and 681 with negative jacobians at nodes. The left picture shows the front mesh faces. The middle and right pictures show two layers of hexes. The former surrounds an accumulation point in the mesh while the latter goes through it

---

[6]In [6], degeneracies like *through-cells*, *non-simplicial meets* and *non-distinct sup-cells* are removed to get a well-defined hex mesh.

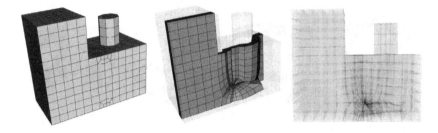

Fig. 18. Another mesh example. It has 2030 elements and 21 with negative jacobians at nodes. The left picture shows the front mesh faces. The middle picture shows a layer of hexes. The right picture shows the internal mesh structure where an accumulation point is visible

9 Future Work

The work presented in this paper is still in progress. Several main directions are identified to complete it. First we have to improve the algorithm robustness by testing it on a wider spectrum of geometries. Especially, we will test it on sparsely meshed boundary surfaces in order to build blocks in a block-structured meshing approach. We will also have to work on the case when the non-convexity filter indicates that none of the loops can be shrunk. A solution should be to perform an incomplete shrinking process which shrinks the loop until it reaches non-convex edges. The non-convexity seam or a new intermediate rule could then clean up the mesh. Such a solution would be useful for managing the parallel loops too. Indeed, in our algorithm, these loops have priority over the other ones and sometimes the selected shrinking loop can be shrunk towards a non-convex area. To get a more general algorithm, we could also attempt to manage self-intersecting loops. That would probably mean to deeply modify our algorithm which intensively uses the fact that a loop cuts the meshing front in two parts (left and right).

As illustrated by the previous Section, both topology and geometry are necessary to get a valid mesh. So the next step of this work will be to adapt our algorithm to get a classical geometric advancing-front algorithm which would be driven by both topological and geometrical rules. In such an algorithm, topological rules must have priority. As our algorithm works on the primal mesh directly, it can be easily extended in such a way.

The efficiency, the robustness and the result of an advancing front method starting from a pre-meshed boundary are strongly connected to the boundary mesh quality. However the surface mesh is generally created without considering that it will be the starting point of a hex mesh generation algorithm. In order to mesh the boundary of a 3D geometric solid, the traditional way consists in meshing its geometric boundary curves then meshing independently its geometric boundary surfaces with a robust quad meshing algorithm like the Paving algorithm [4] or the Q-Morph algorithm [11]. To get an acceptable boundary mesh for hexahedral meshing, we think that the boundary surfaces have to be meshed together considering the global topology and the geometry of the 3D object. It would be equivalent to create the loops on the boundary geometry in an element by element fashion. The loop could then be shrunk in their creation order.

It is also important to work on the optimization step. If the Whisker Weaving algorithm has the advantage of successfully generating hexahedral topology for a wide spectrum of solid geometries, it has the drawback of often creating nodes which are connected to a great number of hexes. Our algorithm has the same drawback. To improve the quality of hexes, local topological changes have to be performed. Few papers [17, 1] talk about this research field which is yet important. The main difficulty is that local re-meshing in hexahedral mesh can have severe global repercussions.

10 Conclusion

In this paper, we present an extension of the reliable Whisker Weaving algorithm. The new algorithm pays more attention to the geometric information to select the shrinking loop. It also attempts to manage non-convexity areas and parallel loops. The implementation of this algorithm has provided first results but it deserves to be enriched by introducing an incomplete loop shrinking process which seems to be a solution to manage both non-convex areas and parallel loops. Note that the algorithm structure built on different levels and rules with specific filters eases the addition of new rules.

Acknowledgments

The authors wish to thank people who shared their opinions and advices about the preliminaries of this work presented at the IMR'2006 Poster session.

References

1. M. Bern and D. Eppstein. Flipping Cubical Meshes. *ACM Computer Science Archive*, 2002. http://www.arXiv.org/cs.CG/108020.
2. T.D. Blacker and R.J. Meyers. Seams and Wedges in Plastering: A 3D Hexahedral Mesh Generation Algorithm. *Engineering with Computers*, 9:83–93, 1993.
3. T.D. Blacker, S.A. Mitchell, T.J. Tautges, P. Murdock, and S.E. Benzley. Forming and Resolving Wedges in the Spatial Twist Continuum. *Engineering With Computers*, 13:35–47, 1997.
4. T.D. Blacker and M.B. Stephenson. Paving: A New Approach to Automated Quadrilateral Mesh Generation. *International Journal for Numerical Methods in Engineering*, 32:811–847, 1991.

5. N.A. Calvo and S.R. Idelsohn. All-hexahedral element meshing: Generation of the dual mesh by recurrent subdivision. *Computer Methods in Applied Mechanics and Engineering*, pages 371–378, 2000.
6. N.T. Folwell and S.A. Mitchell. Reliable Whisker Weaving via Curve Contraction. *7th International Meshing Roundtable*, pages 365–378, 1998.
7. S.A. Mitchell. A Characterization of the Quadrilateral Meshes of a Surface Which Admit a Compatible Hexahedral Mesh of the Enclosed Volume. In Springer, editor, *13th Annual Symposium on Theoretical Aspects of Computer Science*, pages 465–476, 1996.
8. S.A. Mitchell and T.J. Tautges. Pillowing Doublets: Refining a Mesh to Ensure that Faces Share at Most One Edge. *4th International Meshing Roundtable*, pages 231–240, 1995.
9. M. Muller-Hannemann. Shelling Hexahedral Complexes for Mesh Generation. *Journal of Graph Algorithms and Applications*, 5(5):59–91, 2001.
10. M. Muller-Hannemann. Quadrilateral surface meshes without self-intersecting dual cycles for hexahedral mesh generation. *Computational Geometry*, 22:75–97, 2002.
11. S.J. Owen, M.L. Staten, S.A. Canann, and S. Saigal. Q-Morph: An Indirect Approach to Advancing Front Quad Meshing. *International Journal for Numerical Methods in Engineering*, 44:1317–1340, 1999.
12. M.L. Staten, R.A. Kerr, S.J. Owen, and T.D. Blacker. Unconstrained Paving and Plastering: Progress Update. *14th International Meshing Roundtable*, pages 469–486, 2006.
13. M.L. Staten, S.J. Owen, and T.D. Blacker. Unconstrained Paving and Plastering: A New Idea For All Hexahedral Mesh Generation. *14th International Meshing Roundtable*, pages 399–416, 2005.
14. M.B. Stephenson, S.A. Canann, and T.D. Blacker. Plastering: a new approach to automated 3D hexahedral mesh generation. *Progress Report*, 1992. SAND89-2192.
15. T. Suzuki, S. Takahashi, and J. Shepherd. An Interior Surface Mesh Generation Method For All-Hexahedral Meshing. *14th International Meshing Roundtable*, pages 377–398, 2005.
16. T.J. Tautges, T.D. Blacker, and S.A. Mitchell. The Whisker Weaving algorithm: A Connectivity-based Method for Constructing All-hexahedral Finite Element Meshes. *International Journal for Numerical Methods in Engineering*, 39:3327–3349, 1996.
17. T.J. Tautges and S.E. Knoop. Topology Modification of Hexahderal Meshes using Atomic Dual-Based operations. *12th International Meshing Roundtable*, pages 415–423, 2003.
18. T.J. Tautges and S.A. Mitchell. Whisker Weaving: Invalid Connectivity resolutions and Primal Construction Algorithm. *4th International Meshing Roundtable*, pages 115–127, 1995.

3A.2

Methods and Applications of Generalized Sheet Insertion for Hexahedral Meshing

Karl Merkley[1], Corey Ernst[2], Jason F. Shepherd[3], and Michael J. Borden[4]

[1] Computational Simulation Software, LLC., American Fork, UT
    karl@csimsoft.com
[2] Elemental Technologies, Inc., American Fork, UT corey@elemtech.com
[3] Sandia National Laboratories, Albuquerque, NM jfsheph@sandia.gov
[4] Sandia National Laboratories, Albuquerque, NM mborden@sandia.gov

Summary. This paper presents methods and applications of sheet insertion in a hexahedral mesh. A hexahedral sheet is dual to a layer of hexahedra in a hexahedral mesh. Because of symmetries within a hexahedral element, every hexahedral mesh can be viewed as a collection of these sheets. It is possible to insert new sheets into an existing mesh, and these new sheets can be used to define new mesh boundaries, refine the mesh, or in some cases can be used to improve quality in an existing mesh. Sheet insertion has a broad range of possible applications including mesh generation, boundary refinement, R-adaptivity and joining existing meshes. Examples of each of these applications are demonstrated.

Key words: hexahedra, meshing, dual, boundary layers, refinement

1 Introduction

Sheet insertion is a technique for modifying the topology of a hexahedral mesh and introducing new elements which geometrically correlate with the shape of the sheet. These topological changes and the new elements introduced by the insertion provide methods for defining, refining, and improving the quality of a mesh. This paper will review several applications of sheet insertion including pillowing, dicing, refinement, grafting, and mesh cutting and show how these methods are related by the sheet insertion technique.

Every hexahedral mesh is defined as a collection of hex sheets. A hexahedral sheet is dual to a layer of hexahedra and can be geometrically correlated to the shape of the hexahedral sheet. A sheet of hexahedra can be easily visualized by examining an equivalent structure in two dimensions. Given a quadrilateral mesh, a layer of quadrilaterals is found by starting at the center of a single quadrilateral and traversing through the opposite edge pairs of connected quadrilaterals in both directions. The layer will always end by returning to the original quadrilateral or reaching the boundary of the mesh, as shown in figure 1. A continuous line drawn through this layer of quadrilaterals can be used visually to represent all of the quadrilaterals in the layer. This line segment is known as a chord, and is dual to the layer of quadrilateral elements.

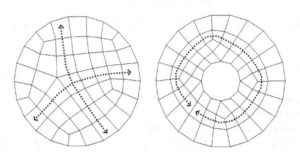

Fig. 1. Rows of quadrilaterals.

This description can be extended to hexahedra in three-dimensions. A layer of hexahedra can be visualized as a single manifold surface. This surface is known as a 'sheet', where a sheet is dual to a layer of hexahedra. Each hexahedron can be defined as a parametric object with 3 sets of 4 edges, where all edges in a set are normal to the same parametric coordinate direction, either i, j, or k. Starting from any hexahedron, we can create a sheet of hexahedra by first obtaining a set of edges in one of the three parametric directions. Next, we collect all neighboring hexahedra sharing these edges. For each neighboring hexahedron obtain the set of edges in the same parametric direction as the initial hexahedron. Using these edges, continue to propagate outwards, collecting neighboring hexahedra. When all adjacent hexahedra are collected in this manner, the result will be a layer of hexahedra that is manifold within the boundary of the mesh (that is the layer will either terminate at the boundary of the mesh, or will form a closed boundary within the mesh). This layer of hexahedra can be visualized as a single manifold surface, known as a sheet, where the sheet is dual to a layer of hexahedra (see figure 2).

Introduction of new sheets in a hexahedral mesh modifies the topology of the existing mesh and introduces new layers of elements within the mesh. As long as the sheets intersect according to a set of topological constraints for a hexeahedral mesh [1, 2], the resulting mesh with the new sheet will still be a hexahedral mesh. Using this principle, a hexahedral mesh may be modified by inserting new hex sheets in specific manners depending upon the application to realize new meshes, define new mesh boundaries, refine an existing mesh, or improve the quality of a hexahedral mesh.

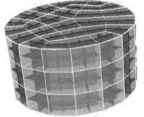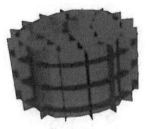

Fig. 2. Sheets of hexahedra.

2 Motivation

Hexahedral meshing has been an active research topic for many years [3]. There have been a number of attempts at generalized meshing algorithms such as plastering [4], whisker-weaving [5, 6, 7] and unconstrained plastering [8]. While significant progress as been made and some of these algorithms show promise for specific geometric types, hex meshing is still considered an open problem [9].

The most successful algorithms for hexahedral mesh generation are based on geometric primitives. For example, a mapping algorithm is based on a parametric cube [10]. Blocking algorithms wrap a geometry in a known hexahedral primitive [11, 12, 13, 14]. Submapping algorithms virtually decompose a geometry into cube-shaped primitives [15]. Sweeping algorithms are based on extrusions of 2D geometric shapes [16, 17, 18].

Sheet insertion offers techniques that can extend the capabilities of primitive based algorithms. Because sheet insertion provides methods for modifying hexahedral meshes, it can be considered as somewhat analogous to tetrahedral mesh modifications such as edge swapping. Tetrahedral mesh modification operations tend to be local and only affect the elements in the immediate area of the change. However, due to the topological structure of hexahedral meshes, modification of a hexahedral mesh affects the mesh in a non-local manner. Sheet insertion quantifies the scope of the non-local changes and provides effective tools for hexahedral mesh modification.

The remainder of this paper provides an overview of several algorithms utilizing sheet insertion. The scope of each algorithm is different, depending on the application for which the algorithm was initially intended. The common thread running through these algorithms is that they introduce new sheets into an existing hexahedral mesh. The new sheets are then utilized to affect changes to the original mesh including geometric boundary modification, refinement, quality improvement, and mesh adaptivity. We provide a survey of these sheet insertion algorithms, and then offer several examples of how sheet insertion can be utilized to generate hexahedral meshes in complex geometries, adaptively conform a mesh topology to specific supposed analytic

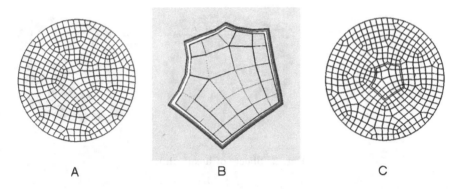

Fig. 3. A basic pillowing operation starts with an initial mesh (A) from which a subset of elements is defined to create a shrink set. The shrink set is separated from the original mesh and 'shrunk' (B), and a new layer of elements (i.e., a dual sheet) is inserted (C) to fill the void left by the shrinking process.

properties within a simulation, improve the mesh quality within an existing mesh, or control mesh sizing.

3 Methods

This section provides an overview of hexahedral mesh modification methods that rely on sheet insertion, including pillowing, dicing, refinement, mesh cutting and grafting.

3.1 Pillowing

Pillowing is a method for inserting a single sheet into an existing hexahedral mesh. Initially proposed by Mitchell et al. [19], a pillowing operation provides a means for eliminating problematic mesh topology.

The basic pillowing algorithm is as follows:

1. *Define a shrink set* - Divide the existing mesh into two sets of elements: one set for each of the half-spaces defined by the sheet to be inserted. Choose one set as the shrink set. The set with the fewest number of elements is typically chosen as the shrink set.
2. *Shrink the shrink set* - Create a gap region between the two element sets (see figure 3).
3. *Connect with a layer of elements* - Create a fully-conformal mesh with the new sheet inserted between the original two element sets.

It is often desirable to smooth the resulting mesh after the sheet insertion to obtain better nodal placement and higher quality elements. The speed of the pillowing algorithm is largely dependent on the time needed to find the shrink set. The number of new hexahedra created will be equal to the number of quadrilaterals on the boundary of the shrink set.

Pillowing is an operation for inserting a single sheet into an existing mesh. It is a useful multipurpose, foundational tool for operations on hexahedral meshes and is the basis for other mesh modification tools, including refinement, grafting, and mesh-cutting.

3.2 Dicing

The dicing algorithm [20] was created to efficiently generate large, refined meshes from existing coarse meshes by duplicating sheets within the coarse mesh. Each sheet is copied and placed in a parallel configuration to the sheet being copied. The basic method for dicing is as follows:

1. *Define the sheet to be diced* - See sheet definition in introduction.
2. *Dice the edges* - Split or dice the list of edges found in the previous step the specified number of times.
3. *Form the new sheets* - Redefine the hexahedra based on the split edges.

Utilizing the dicing method, the number of elements increases as the cube of the dicing value. For instance, if an existing mesh is diced four times (i.e., each of the sheets in the existing mesh is split four times), the resulting mesh would have 64X as many elements as the original mesh. Because all search operations can be performed directly, the dicing algorithm can produce large meshes at very efficient speeds (see figure 4).

3.3 Refinement

The size of a hexahedral mesh is relative to the sheet density in the mesh. Increasing the element count within a region of the mesh can be accomplished

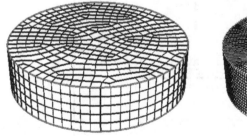

Fig. 4. The original mesh (left) contains 1,805 hex elements before dicing. Each sheet in the original mesh is copied three times resulting in a mesh that is 3^3 larger, with 48,735 hex elements.

by increasing the density of the sheets in the region. Refinement algorithms typically utilize predefined templates to increase the element density. These templates and the algorithms that insert them ensure manifold sheet structures within the mesh to maintain mesh validity [21, 22]. A basic refinement algorithm is constructed as follows:

1. *Define the region or elements to be refined* - Replace the hexahedra in the refinement region with a predefined template of smaller elements.
2. *Cap the boundary of the refinement region to ensure conformity to the rest of the mesh.* - Create a second set of templated elements which transition from the refinement template back to the original mesh size. Figure 5 demonstrates this process.

3.4 Mesh cutting

The mesh-cutting [23] method is an effective approach for capturing geometric surfaces within an existing mesh topology. The mesh-cutting method utilizes the pillowing and dicing methods mentioned previously to insert two sheets which are geometrically similar to the surface to be captured. By utilizing two sheets, a layer of quadrilaterals results that approximates the desired surface where each of the new quadrilateral is shared between the hexes in the two new sheets. The mesh-cutting method entails the following steps:

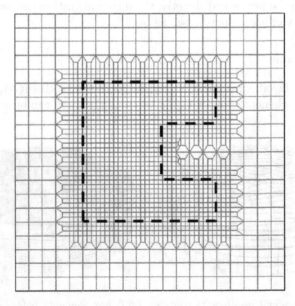

Fig. 5. Refinement of a hex mesh using refinement templates. Refinement region is shown as a dashed line.

1. *Define the pillowing shrink set* - Divide the existing mesh into two sets of elements. One of these element sets will be the shrink set, and a sheet (pillow) is inserted between the two sets of elements.
2. *Dice the new sheet* - Split the newly inserted sheet into two sheets utilizing an approach similar to dicing.
3. *Move the shared quadrilaterals to the surface* - Find all of the quadrilaterals that are shared by the hexes between the two sheets. These quadrilaterals become the mesh on the surface being cut (see figure 6).

Because the quadrilaterals between the two new sheets approximate the surface, the mesh topology must be fine enough to capture the detail of the surface being inserted. Because the resulting quadrilaterals only approximate the inserted surface, if the resulting quadrilateral mesh is too coarse, the surface may not be approximated adequately.

3.5 Grafting

The term 'grafting' is derived from the process of grafting a branch from one tree into the stem, or trunk, of another tree. In meshing, the grafting method was developed to allow a branch mesh to be inserted into the linking surface of a previously hexahedrally swept volume [24]. The grafting method offers a reasonably generalized approach to multiaxis sweeping. The method for creating a graft (i.e., capturing the geometric curve) can be outlined as follows (see also figure 7):

1. *Create a pillowing shrink set* - The shrink set is defined as the set of hexes which have one quadrilateral on the surface and which are interior to the bounding curves of the graft.

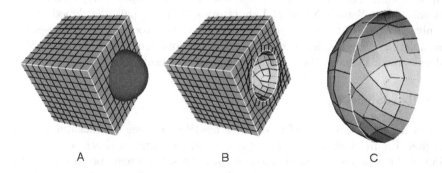

A   B   C

Fig. 6. Mesh cutting utilizes an existing mesh and inserts new sheets to capture a geometric surface (the existing mesh is shown in (A) where the spherical surface is the surface to be captured.) The resulting mesh after mesh cutting is shown in (B), with a close-up of the quadrilaterals on the captured surface being shown in (C).

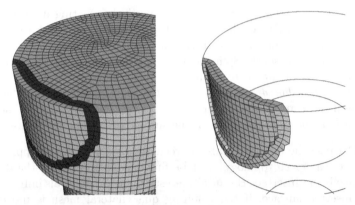

Fig. 7. A view of the hex sheets that were inserted into one of the trunk volumes by the grafting algorithm. The figure on the left highlights the sheets where they intersect the surface of the volume. The figure on the right shows the orientation of the sheets in the interior of the mesh.

2. *Insert the pillow (sheet)* -Inserting the second sheet to satisfy the hexahedral constraint for capturing the geometric curves used in the graft.

At this point, there are often database adjustments needed to ensure that the new mesh entities are associated with the correct geometric entities.

4 Applications

Because hexahedral sheet insertion operations are extremely flexible, there are numerous possible applications. This section demonstrates a few of these applications. The meshes in this section were developed using Cubit [25], developed at Sandia National Laboratories, SCIRun [26], developed at the University of Utah, or a combination of the two tools. The processes for completing these models are not yet fully automated, but there is ongoing development to make these processes robust and usable.

4.1 R-Adaptivity

Sheet insertion can be used to divide a model and refine the mesh in a given region. Figure 8 shows an R-adaptive mesh capturing a shock a wave. An initial sheet is placed that is aligned with the shock wave and mesh-cutting techniques are used to capture the shock wave. Additional sheets are then added to provide the required resolution around the shock wave.

Adaptive smoothing operations can be utilized to control orthogonality and mesh spacing close to the adaptive region. The mesh after several operations of layer smoothing is shown in figure 9.

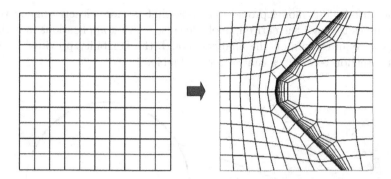

Fig. 8. Sheets can be placed within a mesh to capture analytic features, such as shock waves. The original mesh is shown on the left. After several sheet insertion operations to place layers of elements roughly conforming to a supposed shock, the newly modified mesh is shown on the right. Mesh generated using Cubit.

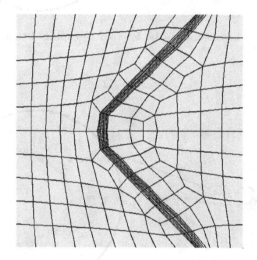

Fig. 9. Smoothing of R-adaptive region using Cubit

4.2 Boundary Layers

Sheet insertion techniques can be used in a similar manner to capture boundary layers as shown in airfoil mesh in figure 10. Inserted sheets can provide additional resolution at the boundary. The number of layers and the distribution of the sheets could be specified according to the requirements of the application.

Additionally, new geometric features can be added by the sheet insertion process. For example, the geometry in figure 11 was split through the center of the airfoil. This model provides a mesh that is more closely aligned with streamlines through the entire model. Because varying applications may pose different requirements on the analysis model, the flexibility provided by the sheet insertion techniques can be very valuable.

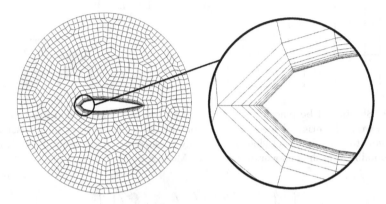

Fig. 10. Sheets of hexahedra can be inserted to capture boundary layers. Mesh generated in Cubit.

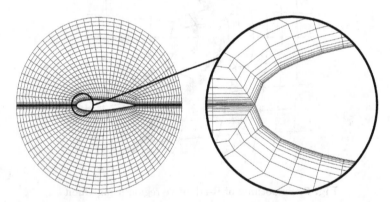

Fig. 11. Sheets of hexahedra added to better resolve streamlines of model. Mesh generated in Cubit.

4.3 Mesh Generation

Hexahedral sheet insertion also provides techniques that can be used to generate complex meshes. The typical mesh-cutting technique involves embedding the desired geometry into an existing hexahedral mesh that envelopes the geometry. One or more sheets are then inserted into the mesh which capture the missing, but desired, geometric features. Additional sheets may be inserted to improve the element quality at the boundary.

Figure 12 shows an all hexahedral mesh of a human hand while and figure 13 shows the internal detail of one finger of this model. This mesh was generated in SCIRun using a sheet insertion operation, similar to mesh cutting. Once the initial mesh was captured as desired, four additional layers of hexahedra were added at the boundary (for a total of five boundary layers) that are oriented orthogonally to the surface. This was accomplished by dicing the original sheet used to capture the surface of the hand. The resulting mesh contains 379,429 hexahedra elements, and the mesh contains all positive Jacobian and well shaped hexahedra. Figure 14 shows the spread of scaled Jacobian element quality.

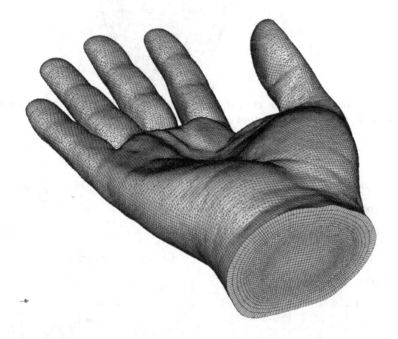

Fig. 12. All hexahedral mesh of a human hand created by mesh cutting. The original triangle mesh of the hand that was utilized in controlling the placement of the hexahedral boundary layer was provided courtesy of INRIA by the AIM@SHAPE Shapes Repository http://shapes.aim-at-shape.net/index.php). Mesh generated in SCIRun.

Biomedical applications are a source of very difficult modeling problems. Figure 15 shows a model of an all hexahedral mesh of a human skull and cranial cavity. The meshes were formed by first creating a rectilinear mesh that was sized such that the bounding box was slightly larger than the skull. Triangular meshes of the skull and the cranial cavity were then inserted into

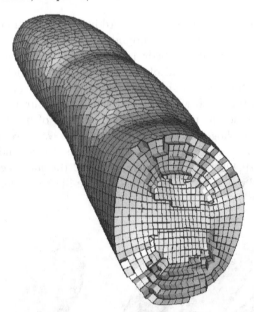

Fig. 13. Cut-away view of one of the fingers for the all-hexahedral mesh of a human hand showing the internal structure of the mesh.

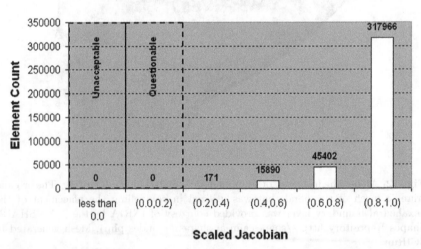

Fig. 14. Distibution of element quality by the scaled Jacobian metric for the hand model.

the rectilinear mesh to capture the surface data. The level of detail captured in this technique is proportional to the level of detail of the triangular surface mesh. Additional sheets where inserted at the surface to improve the quality of the mesh at the surface. Figure 16 shows a cut-away view of the mesh (accomplished again with a mesh-cutting operation) to give a view of the internal structure of the mesh. The mesh of the skull contains 77,775 elements consisting of well-shaped hexahedra with all positive Jacobian measures. The quality distribution for the elements of this mesh is shown in figure 17.

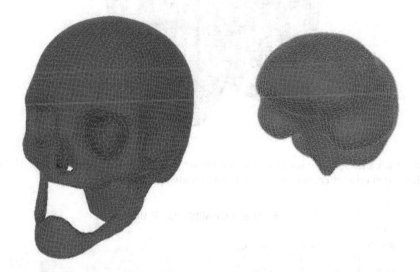

Fig. 15. All hexahedral mesh of a human skull and cranial cavity. The original triangle mesh of the skull that was utilized in controlling the placement of the hexahedral boundary layer was provided courtesy of INRIA by the AIM@SHAPE Shapes Repository http://shapes.aim-at-shape.net/index.php). Mesh generated in SCIRun.

4.4 Grafting

The grafting algorithm [27] was developed to allow a volume to be imprinted and merged with another volume after the initial volume was meshed. The grafting algorithm locally modifies the initial mesh to conform to the geometry of the second, unmeshed volume. One application of this algorithm has been generating swept meshes for models with two independent sweep directions.

For example, suppose an engineer is interested in analyzing the fluid flow through the valve housing shown in figure 18. The fluid region with the valve partially closed is shown in figure 19. Notice that there is an independent sweep direction corresponding to each pipe opening from the valve. To generate a

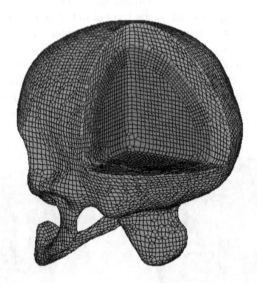

Fig. 16. Cutaway view showing the interior structure of the all hexahedral mesh of a human skull and cranial cavity. Cutaway created in Cubit.

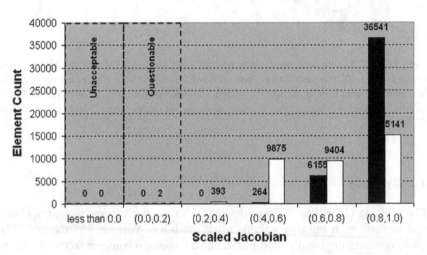

Fig. 17. Distribution of element quality by the scaled Jacobian metric for the skull model. The black bars represent the quality of the interior cranial mesh, while the white bars are representative of the skull bone.

mesh of this region with grafting, one would first mesh the two volumes that are split by the valve plate as shown in figure 19. The remaining unmeshed volume can then be grafted onto the meshed volumes as shown in figure 20. The grafting operation introduces a minimal, local change to the mesh as shown in figure 7.

Fig. 18. A valve housing

Fig. 19. The fluid region of the valve housing in figure 18 with the valve partially closed (left) and the initial, pre-grafting swept mesh (right)

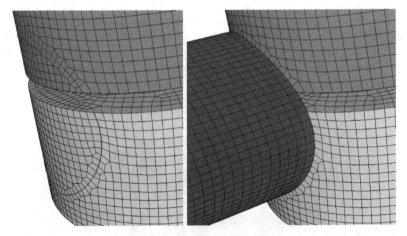

Fig. 20. The area of the mesh that has been modified by grafting. The figure on the left shows the imprint of the branch volume onto the trunk volumes. The figure on the right shows the final mesh on the branch. Mesh created in Cubit.

5 Conclusion

Generalized hexahedral sheet insertion is an effective technique for modifying hexahedral meshes, including defining new mesh boundaries, refining the mesh, and improving the quality of an existing mesh. The technique is very flexible and can be applied in a variety of methods to modify a mesh, including R-adaptivity, refinement, boundary layer insertion, mesh generation, and mesh improvement.

The examples presented here have shown that sheet insertion can greatly decrease the effort required to construct a high quality mesh on difficult models. In particular, sheet insertion as a mesh generation tool for biomedical models shows great promise. This method could easily be extended into other promising areas. Geotechnical analysis requires models similar to those found in biomedical applications—they are often irregularly shaped and possess little or no topology that can be used to define the structure needed by traditional hexahedral meshing schemes. Additionally, sheet insertion methods, like mesh cutting and grafting, can also be utilized to dramatically reduce the level of geometric decomposition that is traditionally required to generate a hexahedral mesh on complex geometries.

Acknowledgment

Sandia is a multiprogram laboratory operated by Sandia Corporation, a Lockheed Martin Company, for the United States Department of Energy's National Nuclear Security Administration under contract DE-AC04-94AL85000.

References

1. Scott A. Mitchell. A characterization of the quadrilateral meshes of a surface which admit a compatible hexahedral mesh of the enclosed volumes. In *13th Annual Symposium on Theoretical Aspects of Computer Science*, volume Lecture Notes in Computer Science: 1046, pages 465–476, 1996.

2. Jason F. Shepherd. *Topologic and Geometric Constraint-Based Hexahedral Mesh Generation*. Published Doctoral Dissertation, University of Utah, May 2007.

3. Steven J. Owen. A survey of unstructured mesh generation technology. http://www.andrew.cmu.edu/user/sowen/survey/index.html, September 1998.

4. Ted D. Blacker and Ray J. Meyers. Seams and wedges in plastering: A 3D hexahedral mesh generation algorithm. *Engineering With Computers*, 2(9):83–93, 1993.

5. Timothy J. Tautges, Ted D. Blacker, and Scott A. Mitchell. Whisker weaving: A connectivity-based method for constructing all-hexahedral finite element meshes. *International Journal for Numerical Methods in Engineering*, 39:3327–3349, 1996.

6. Matthias Muller-Hannemann. Hexahedral mesh generation by successive dual cycle elimination. In *Proceedings, 7th International Meshing Roundtable*, pages 365–378. Sandia National Laboratories, October 1998.

7. Nestor A. Calvo and Sergio R. Idelsohn. All-hexahedral element meshing: Generation of the dual mesh by recurrent subdivision. *Computer Methods in Applied Mechanics and Engineering*, 182:371–378, 2000.

8. Matthew L. Staten, Steven J. Owen, and Ted D. Blacker. Unconstrained paving and plastering: A new idea for all hexahedral mesh generation. In *Proceedings, 14th International Meshing Roundtable*, pages 399–416. Sandia National Laboratories, September 2005.

9. Erik D. Demain, Joseph S.B. Mitchell, and Joseph O'Rourke. http://maven.smith.edu/~orourke/TOPP/P27.html#Problem.27, 2006. The open problems project - problem 27: Hexahedral meshing.

10. W. A. Cook and W. R. Oakes. Mapping methods for generating three-dimensional meshes. *Computers In Mechanical Engineering*, CIME Research Supplement:67–72, August 1982.

11. John F. Dannenhoffer(III). A block-structuring technique for general geometries. In *29th Aerospace Sciences Meeting and Exhibit*, volume AIAA-91-0145, January 1991.

12. XYZ Scientific Applications, Inc. Truegrid: High quality hexahedral grid and mesh generation for fluids and structures. http://www.truegrid.com, October 2006.

13. Pointwise. Gridgen - reliable CFD meshing. http://www.pointwise.com/gridgen, October 2006.

14. ANSYS. ANSYS ICEM CFD, http://www.ansys.com/products/icemcfd.asp, October 2006.

15. David R. White, Mingwu Lai, Steven E. Benzley, and Gregory D. Sjaardema. Automated hexahedral mesh generation by virtual decomposition. In *Proceedings, 4th International Meshing Roundtable*, pages 165–176. Sandia National Laboratories, October 1995.

16. Patrick Knupp. Next-generation sweep tool: A method for generating all-hex meshes on two-and-one-half dimensional geometries. In *Proceedings, 7th In-*

ternational Meshing Roundtable, pages 505–513. Sandia National Laboratories, October 1998.

17. Xevi Roca, Josep Sarrate, and Antonio Huerta. Surface mesh projection for hexahedral mesh generation by sweeping. In *Proceedings, 13th International Meshing Roundtable*, volume SAND 2004-3765C, pages 169–180. Sandia National Laboratories, September 2004.

18. Mingwu Lai, Steven E. Benzley, and David R. White. Automated hexahedral mesh generation by generalized multiple source to multiple target sweeping. *International Journal for Numerical Methods in Engineering*, 49(1):261–275, September 2000.

19. Scott A. Mitchell and Timothy J. Tautges. Pillowing doublets: Refining a mesh to ensure that faces share at most one edge. In *Proceedings, 4th International Meshing Roundtable*, pages 231–240. Sandia National Laboratories, October 1995.

20. Darryl J. Melander, Timothy J. Tautges, and Steven E. Benzley. Generation of multi-million element meshes for solid model-based geometries: The dicer algorithm. *AMD - Trends in Unstructured Mesh Generation*, 220:131–135, July 1997.

21. Ko-Foa Tchon, Julien Dompierre, and Ricardo Camarero. Conformal refinement of all-quadrilateral and all-hexahedral meshes according to an anisotropic metric. In *Proceedings, 11th International Meshing Roundtable*, pages 231–242. Sandia National Laboratories, September 2002.

22. Nathan Harris, Steven E. Benzley, and Steven J. Owen. Conformal refinement of all-hexahedral meshes based on multiple twist plane insertion. In *Proceedings, 13th International Meshing Roundtable*, pages 157–168. Sandia National Laboratories, September 2004.

23. Michael J. Borden, Jason F. Shepherd, and Steven E. Benzley. Mesh cutting: Fitting simple all-hexahedral meshes to complex geometries. In *Proceedings, 8th International Society of Grid Generation Conference*, 2002.

24. Steven R. Jankovich, Steven E. Benzley, Jason F. Shepherd, and Scott A. Mitchell. The graft tool: An all-hexahedral transition algorithm for creating multi-directional swept volume mesh. In *Proceedings, 8th International Meshing Roundtable*, pages 387–392. Sandia National Laboratories, October 1999.

25. Sandia National Laboratories. The CUBIT Geometry and Mesh Generation Toolkit. http://cubit.sandia.gov, 2007.

26. Scientific Computing and Imaging Institute (SCI). SCIRun: A Scientific Computing Problem Solving Environment. http://software.sci.utah.edu/scirun.html, 2002.

27. S.R. Jankovich, S.E. Benzley, J.F. Shepherd, and S.A. Mithcell. The graft tool: an all-hexahedral transition algorithm for creating a mult-directional swep t volume mesh. In *Proceedings, 8th International Meshing Roundtable '99*, pages 387–392. Sandia National Laboratories, 1999.

3A.3

A Selective Approach to Conformal Refinement of Unstructured Hexahedral Finite Element Meshes

Michael Parrish[1], Michael Borden[2], Matthew Staten[3], Steven Benzley[4]

[1] Brigham Young University, Provo, UT U.S.A. mhp7@et.byu.edu
[2] Sandia National Laboratories, Albuquerque, NM, U.S.A. mborden@sandia.gov
[3] Sandia National Laboratories, Albuquerque, NM, U.S.A. mlstate@sandia.gov
[4] Brigham Young University, Provo, UT U.S.A. seb@byu.edu

Summary. Hexahedral refinement increases the density of an all-hexahedral mesh in a specified region, improving numerical accuracy. Previous research using solely sheet refinement theory made the implementation computationally expensive and unable to effectively handle concave refinement regions and self-intersecting hex sheets. The Selective Approach method is a new procedure that combines two diverse methodologies to create an efficient and robust algorithm able to handle the above stated problems. These two refinement methods are: 1) element by element refinement and 2) directional refinement. In element by element refinement, the three inherent directions of a Hex are refined in one step using one of seven templates. Because of its computational superiority over directional refinement, but its inability to handle concavities, element by element refinement is used in all areas of the specified region except regions local to concavities. The directional refinement scheme refines the three inherent directions of a hexahedron separately on a hex by hex basis. This differs from sheet refinement which refines hexahedra using hex sheets. Directional refinement is able to correctly handle concave refinement regions. A ranking system and propagation scheme allow directional refinement to work within the confines of the Selective Approach Algorithm.

1 Introduction

As computing power continues to increase, the finite element method has become an increasingly important tool for many scientists and engineers. An essential step in the finite element method involves meshing or subdividing the domain into a discrete number of elements. Mesh generation has therefore been the topic of much research. Tetrahedral (Tet) or hexahedral (Hex) elements are commonly used to model three dimensional problems. Tet elements have extremely robust modeling capabilities for any general shape while Hex elements provide more efficiency and accuracy in the computational process [1].

Within the realm of hexahedral mesh generation, mesh modification is an area of research that attempts to improve the accuracy of an analysis by locally modifying the mesh to more accurately model the physics of a problem. Hexahedral refinement modifies the mesh by increasing the element density in a localized region.

Several schemes have been developed for the refinement of hexahedral meshes. Methods using iterative octrees[2] have been proposed, however these methods result in nonconformal elements which cannot be accommodated by some solvers. Other techniques insert non-hex elements that result in hybrid meshes or require uniform dicing to maintain a consistant element type[3]. Schneiders proposed an element by element refinement scheme[4] in connection with an octree-based mesh generator, however this technique is limited in that it is unable to handle concavities (see Section 2.2). Schneiders later proposed a sheet refinement method[5] which produces a conformal mesh by pillowing layers in alternating i, j, and k directions but relies on a Cartesian initial octree mesh. Tchon et al. built upon Schneiders' sheet refinement in their 3D anisotropic refinement scheme by expanding the refinement capabilities to unstructured meshes[6][7] however this scheme still has poor scalability inherent in all sheet refinement schemes. Harris et al. further expanded upon Schneiders' and Tchon's work by using templates instead of pillowing to refine the mesh and included capabilities to refine element nodes, element edges, and element faces[8]. While the refinement scheme introduced by Harris is robust in many aspects, it is limited by self-intersecting hex sheets (see Section 2.2), concavities, and poor scalability. The refinement process developed in this paper combines the element by element method proposed by Schneiders and the sheet refinement method proposed by Harris to create a method that overcomes the limitations of using either method alone.

2 Background

A hexahedron, the finite element of interest in this paper, has a dual representation defined by the intersection of three sheets called twist planes[9][10]. Each sheet represents a unique and inherent direction within a hexahedron. Figure 1 shows a hexahedron with its three dual twist planes. Each plane represents a unique direction of refinement.

2.1 Element by Element Refinement

Element by element refinement replaces a single hexahedron with a predefined group of conformal elements effectively refining all three directions of the hexahedron at the same time. As such a nonconformal mesh is temporarily created until all templates have been inserted. Only one template is applied to any initial element thus increasing the efficiency of the refinement process. Figure 2 shows how a mesh is refined using element by element refinement.

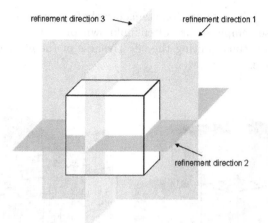

Fig. 1. A Hex with its twist planes representing directions of refinement

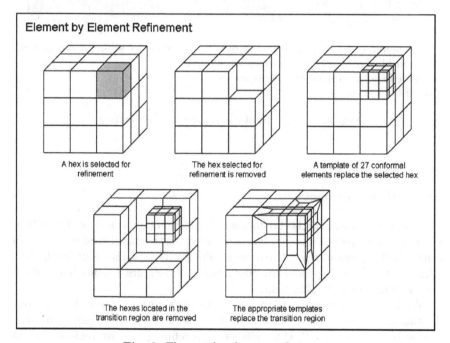

Fig. 2. Element by element refinement

Element by element refinement is limited by its inability to produce a conformal mesh in a concave region. In hexahedral refinement, a concave region refers to any hexahedral element that is not selected for refinement but shares more than one adjacent face with hexahedra that are selected for refinement(see fig. 3(a)). This limitation stems largely from missing or unidentified templates. These templates are often unknown or cannot be created with reasonable quality thus limiting the effectiveness of the element by element refinement scheme.

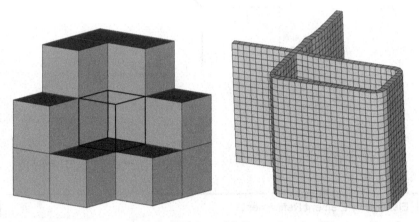

(a) Example of concave region - hex outlined in black is a transition element in a
concave region and shaded elements are selected for refinement

(b) Example of self-intersecting hex
sheet

Fig. 3. Limitations of existing refinement methods

2.2 Sheet Refinement

The sheet refinement method refines a hex one direction at a time. The refinement region is processed in hex sheets allowing unstructured meshes to remain conformal throughout the entire process. Since conformity is maintained, sheet refinement inherently produces a conformal mesh. Figure 4 shows how a mesh is refined using sheet refinement.

While sheet refinement is robust in its capabilities, it has three serious limitations. These limitations are: 1) the inability to effectively treat self-intersecting hex sheets, 2) the inefficiency in refining concave regions, and 3) scalability.

Self-Intersecting Hex Sheets

For conformal, all Hex meshes, a hex sheet must either initiate at a boundary and terminate at a boundary or form a closed surface. Sometimes meshing algorithms will create self-intersecting hex sheets as shown in Figure 3(b).

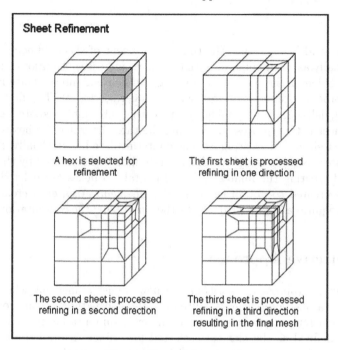

Sheet Refinement

A hex is selected for refinement

The first sheet is processed refining in one direction

The second sheet is processed refining in a second direction

The third sheet is processed refining in a third direction resulting in the final mesh

Fig. 4. Sheet refinement

A self-intersecting hex sheet is defined as any hex sheet that passes through the same stack of elements multiple times (i.e. any dual twist plane that intersects itself). Hexes at the intersection of a self-intersecting hex sheet must be handled as a special case because they need to be processed more than once. Recognizing all the cases where a sheet intersects with itself is a difficult and error prone procedure.

Concavities

Sheet refinement is able to produce a conformal mesh in concave regions however early implementations dealt with these concavities inefficiently. Initially, hexes were added to the concave region until all concavities were removed. While this produces a conformal mesh in a concave region, it leads to excessive refinement. Excessive refinement increases the computational load for both mesh generation and analysis. Templates were later proposed to handle concavities[11] but these templates were never implemented into any sheet refinement scheme.

Scalability

Empirical studies show that the time requirement of sheet refinement grows exponentially as the number of initial elements increases. A major contributor to this problem is the process of creating and deleting intermediate hexes.

The process occurs in the following manner(see fig. 4). The first sheet is processed, deleting the original hex and creating three intermediate hexes. The second sheet is then processed, deleting the three intermediate hexes created by the first sheet and creating nine new intermediate hexes. Finally, the third sheet is processed, deleting the nine intermediate hexes created by the second sheet and creating the final 27 hexes. In total, 13 hexes are deleted and 39 hexes are created to obtain the desired refinement. Also, each creation and deletion requires a data base query further increasing the computational time.

3 A Selective Approach

The Selective Approach Algorithm is a new robust refinement scheme. This procedure (as its name suggests) automatically selects the more appropriate of two different refinement schemes for each hex within a target region. A target region is defined as the elements selected for refinement and the transition elements connecting elements selected for refinement and the coarse mesh. The two refinement schemes used in the Selective Approach Algorithm are element by element (see Section 3.2) and directional (see Section 3.3) refinement. The combination of these two methods allows the Selective Approach Algorithm to overcome the limitations of both element by element and sheet refinement discussed previously.

3.1 Templates

Seven templates[4][11][12] are used within the Selective Approach Algorithm (see fig. 5). Both element by element refinement and directional refinement use templates. The 1 to 27 template and the 1 to 13 template are only used in the element by element refinement scheme while the other five templates are used in both element by element and directional refinement. Figures 5(f) and 5(g) are the templates required to handle any concavity given in a target region. Figure 6 explains how the 1 to 3 template with 1 concavity is constructed. The 1 to 3 template with 2 concavities is constructed in a similar fashion.

3.2 Element by Element Refinement

The general process of performing element by element refinement was discussed in Section 2. Here element by element refinement is discussed in connection with the Selective Approach Algorithm. As stated previously, the element by element refinement method refines all three directions of a hex in

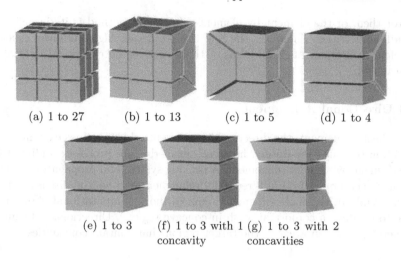

(a) 1 to 27   (b) 1 to 13   (c) 1 to 5   (d) 1 to 4

(e) 1 to 3   (f) 1 to 3 with 1   (g) 1 to 3 with 2
                    concavity   concavities

Fig. 5. Templates used in The Selective Approach Algorithm

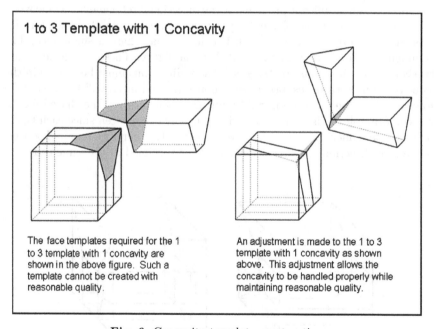

1 to 3 Template with 1 Concavity

The face templates required for the 1 to 3 template with 1 concavity are shown in the above figure. Such a template cannot be created with reasonable quality.

An adjustment is made to the 1 to 3 template with 1 concavity as shown above. This adjustment allows the concavity to be handled properly while maintaining reasonable quality.

Fig. 6. Concavity template construction

one step. A single hex is deleted and the final group of elements is created using one of the seven templates described previously. Since no intermediate hexes are created or deleted, the computational efficiency of element by element refinement is far superior to that of sheet refinement. The limiting factor then, of the element by element refinement method is its inability to handle concavities. Therefore, the Selective Approach Algorithm uses element by element refinement in all areas of the target region except areas local to concavities.

3.3 Directional Refinement

Like sheet refinement, the directional refinement scheme refines each inherent direction of a hex separately, however hexahedra are processed individually like element by element refinement. A ranking system and propagation scheme are new techniques used in directional refinement and will be discussed hereafter. While directional refinement requires more computational effort, it is able to produce a conformal mesh in concave regions. Directional refinement is therefore used in areas of the target region that contain concavities.

The Conformity Problem and Ranking System

Conformity is a significant problem for the directional refinement scheme when hexahedra are processed element by element. An example of the conformity problem is shown in Figure 7 with two hexes that share a single face. The common face for both hexes is shaded in the figure. These two hexes share two common "directions" or "sheets." These directions must be refined in the same order in both hexes, otherwise a nonconformal mesh will be created. In Figure 7, both hexes contain valid refinement schemes yet the shared face is not conformable. This problem could potentially occur often since each hex is refined independently of its neighbors. A method is therefore required so that refinement directions in adjacent hexahedra are refined in the same order.

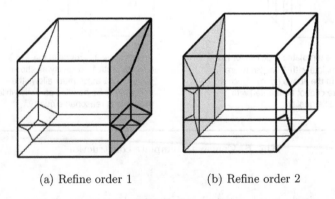

(a) Refine order 1   (b) Refine order 2

Fig. 7. Conformity issues

To solve the conformity problem, the functionality of dual twist planes is used. Twist planes in this refinement scheme represent unique directions of refinement. In the Selective Approach method, connected elements receiving directional refinement are grouped together. Typically there is a single group by each concave region. Since each directional refinement group is confined to a single concavity, the possibility of containing a self-intersecting hex sheet is extremely unlikely. Each group is processed separately by taking an initial arbitrary edge and giving it a rank of 1. All opposite edges of adjacent faces are located for the selected edge. If these new edges need to be directionally refined, they are given the same rank and become selected edges themselves. The rank propagates to all applicable edges intersecting the twist plane defined by the initial edge. The process repeats itself as another unranked edge is arbitrarily selected and given a rank of 2. The ranking scheme is finished when all applicable edges of the entire refinement region are ranked. The ranking system is described graphically in Figure 8. Refinement then occurs on a hex by hex basis starting in the direction with the lowest rank and continuing in ranked order until the hex is completely refined and the algorithm moves onto the next hex.

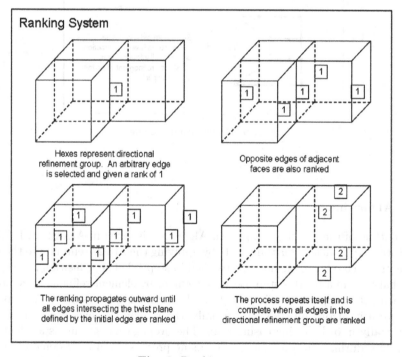

Fig. 8. Ranking system

Propagation Scheme

After a hex is refined in one direction using the directional refinement scheme, new edges exist that may need to be split in order to maintain element quality in the transition region. Only new edges parallel to the direction of refinement are considered in the propagation scheme. Figure 9 graphically shows how the propagation scheme works with a specific example.

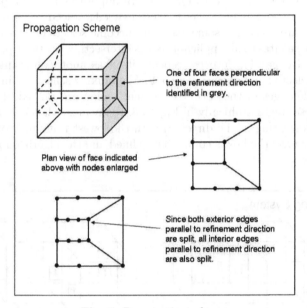

Fig. 9. Propagation scheme

3.4 Algorithm

An outline of the Selective Approach Algorithm is given in Algorithm 1. The Selective Approach Algorithm starts by applying the 1 to 27 template to the elements selected for refinement as specified in step 1.2. The transition hexes are all that remain after this step. Because element by element refinement is more efficient, it is applied first in step 1.4. The remaining hexes are then ranked as shown in algorithm step 1.11. Finally, the remaining hexes are refined directionally in order of increasing rank. The propagation scheme is applied to each hex during the directional refinement process. Figure 10 demonstrates the logic of the algorithm with a simple two-dimensional example.

Algorithm 1 The Selective Approach Algorithm

```
 1: loop target hexes                              ▷ element by element refinement
 2:     apply 1 to 27 template to elements selected for refinement
 3: end loop
 4: loop transition hexes
 5:     if template applies then
 6:         refine hex using template
 7:     else
 8:         add to directional hex list
 9:     end if
10: end loop
11: loop directional hex list
12:     apply ranking system
13: end loop
14: loop directional hex list                      ▷ directional refinement
15:     loop refinement directions in order of increasing rank
16:         apply template
17:         apply propagation scheme
18:     end loop
19: end loop
```

---

4 Results and an Example

The Selective Approach Algorithm solves the sheet refinement limitations of self-intersecting hex sheets, inefficiently handled concavities, and poor scalability. The following section considers the aforementioned limitations individually and discusses how the Selective Approach method eliminates them. Following this discussion, an example will be considered showing the robustness of this algorithm.

4.1 Self-Intersecting Hex Sheets

The Selective Approach Algorithm automatically solves the limitation of self-intersecting hex sheets because both element by element and direction refinement process the target region on a hex by hex basis.

4.2 Concavities

To illustrate the new capabilities of the Selective Approach Algorithm when considering concavities, a simple example problem is presented here. The Selective Approach Algorithm is compared with the sheet refinement scheme implemented by Harris.

The problem involves refining the surfaces composing the right boundary of the model. Figure 11(a) shows the model refined using the sheet refinement scheme implemented by Harris and Figure 11(b) shows the brick refined using

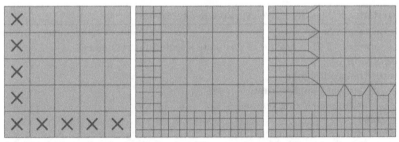

(a) Original mesh where left and bottom hexahedra are selected for refinement

(b) 1 to 27 template applied to elements selected for refinement

(c) Element by element refinement is applied to transition region

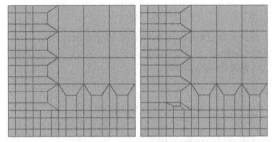

(d) Element is refined in one direction followed by propagation scheme

(e) Element is refined in final directiona resulting in the final mesh

Fig. 10. Example of algorithm

the Selective Approach Algorithm. While sheet refinement could perform the refinement in a similar fashion to the Selective Approach Algorithm, the concave templates were never implemented. The sheet refinement scheme refined the entire bottom right section of the model in an attempt to remove the concavity. Excessive refinement is not a problem with the Selective Approach method. The newly implemented concave templates eliminate the need to add hexes to the target region.

Values for the number of elements, time for both methods, and element quality using a scaled Jacobian metric are given in Table 1. For this example, the Selective Approach method is far superior in both element count and time required to perform the refinement. The Selective Approach Algorithm produced half as many elements and the time requirement was lower as well partially because fewer hexahedra were refined. Solving the mesh using the Selective Appraoch method would also require less time thus lowering the overall time required for a full analysis. The final minimum scaled Jacobian produced by both refinement schemes is the same and adaquate for an accurate analysis.

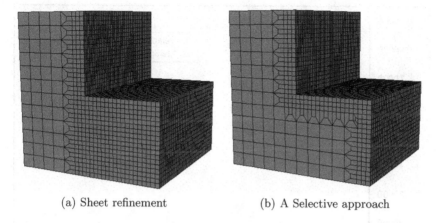

(a) Sheet refinement   (b) A Selective approach

Fig. 11. Simple model where surfaces composing right boundary are refined

Table 1. Results of refining the left and bottom faces of a brick

Measurement	Sheet Refinement	Selective Approach
Initial Elements	1188	1188
Final Elements	16500	8712
Time (sec)	5.359	0.859
Initial Min. Scaled Jacobian	1.0	1.0
Final Min. Scaled Jacobian	0.3143	0.3143

4.3 Scalability

To compare the scalability of the Selective Approach Algorithm to sheet refinement, a simple meshed brick was again used. The number of elements before refinement was increased incrementally by increasing the interval count of the brick. Each meshed brick was completely refined and the required time recorded. The results are shown in Figure 12.

Arguably the greatest advantage of the Selective Approach method over sheet refinement is scalability. Figure 12 decisively shows the exponential increase in time for sheet refinement as the nember of elements before refinement is increased. The scalability of the Selective Approach Algorithm is nearly linear in comparison. The excellent scalability displayed in the Selective Approach Algorithm results from using element by element refinement as the primary refinement scheme.

It should be noted that in the above example, no elements required directional refinement within the Selective Approach Algorithm. A second scalability test was performed where the number of elements of a simple brick was increased incrementally by increasing the interval count as before. However, only elements within a constant radial distance from the top front vertex of

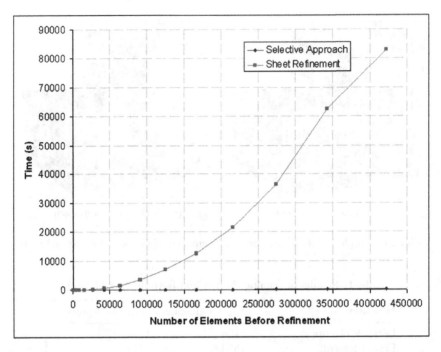

Fig. 12. Comparison of scalability between sheet refinement and the Selective Approach Algorithm

the brick were refined instead of the entire brick as shown in Figure 13. This target region required directional refinement to be used in the refinement process. Using directional refinement will increase the overall computational time of the Selective Approach Algorithm. Figure 14 shows the results of the second scalability test where directional refinement is used. This graph illustrates that while directional refinement may increase the computation time, the Selective Approach Algorithm is still far superior to traditional sheet refinement methods in terms of scalability.

4.4 Example

The example considered is a model of a gear (see Figure 15(a)). All of the teeth of the gear were refined using the Selective Approach Algorithm. Number of elements, speed, and quality using a scaled Jacobian metric were considered in the analysis and the model was smoothed before calculating the final element quality. Figure 15(b) is a closeup of a gear section before refinement. Figure 15(c) shows the same section after refinement. The results are given in Table 2.

In this example, the Selective Approach Algorithm refined the teeth of the gear, adding over 50,000 elements in approximately 20 seconds. The final

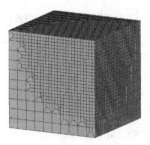

Fig. 13. Brick with constant radius away from top front vertex refined

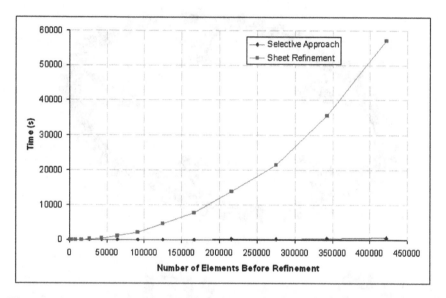

Fig. 14. Comparison of scalability between sheet refinement and the Selective Approach Algorithm with some elements refined using directional refinement

mesh is conformal and the smoothed minimum scaled Jacobian is adaquate for an analysis.

5 Conclusion

The refinement scheme presented in this work is a powerful mesh modification tool. The Selective Approach Algorithm is able to handle self-intersecting hex sheets, concavities, and scalability issues by leveraging the advantages of both element by element and sheet refinement schemes. Directional refinement is a new refinement technique that refines the three inherent directions of a hex sequencially while the target region is processed on a hex by hex basis. A

Table 2. Results of refining the teeth of a gear using the Selective Approach Algorithm

Measurement	Value
Initial Elements	8569
Final Elements	63093
Time (sec)	21.687
Initial Min. Scaled Jacobian	0.4294
Final Min. Scaled Jacobian	0.1580
Final Min. Scaled Jacobian (smoothed)	0.2287

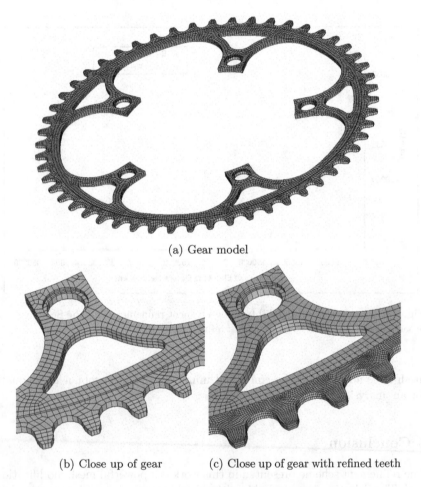

(a) Gear model

(b) Close up of gear   (c) Close up of gear with refined teeth

Fig. 15. Gear Example

ranking system that utilized the dual of the mesh and a propagation scheme allowed directional refinement to work properly within the confines of the Selective Approach Algorithm. The algorithm appears to have a scalability that is nearly linear. Also, the robustness that existed in sheet refinement is not lost within the Selective Approach Algorithm. An Example was also given that provided evidence of this new algorithm's power.

References

1. Benzley S, Perry E, Merkley K, Clark B, and Sjaardema G (1995) A Comparison of All Hexahedral and All Tetrahedral Finite Element Meshes for Elastic and Elasto-plastic Analysis. In: Proceedings 4th International Meshing Roundtable, pages 179-191. Sandia National Laboratories.
2. Tchon K.-F, Hirsch C, and Schneiders R (1997) Octree-Based Hexahedral Mesh Generator for Viscous Flow Simulations. In: 13th AIAA Computational Fluid Dynamics Conference, No. AIAA-97-1980. Snownass, CO.
3. Marechal L (2001) A New Approach to Octree-Based Hexahedral Meshing. In: Proceedings 10th International Meshing Roundtable, pages 209-221. Sandia National Laboratories.
4. Schneiders R (1996) Refining Quadrilateral and Hexahedral Element Meshes. In: 5th International Conference on Numerical Grid Generation in Computational Field Simulations, pages 679-688. Mississippi State University.
5. Schneiders R (2000) Octree-Based Hexahedral Mesh Generation. In: Int Journal Comput Geom Appl. 10, No. 4, pages 383-398
6. Tchon K.-F, Dompierre J and Camerero R (2002) Conformal Refinement of All-Quadrilateral and All-Hexahedral Meshes According to an Anisotropic Metric. In: Proceedings 11th International Meshing Roundtable, pages 231-242. Sandia National Laboratories.
7. Tchon K.-F, Dompierre J and Camerero R (2004) Automated Refinement of Conformal Quadrilateral and Hexahedral Meshes. In: Int Journal Numer Meth Engng 59:1539-1562
8. Harris N (2004) Conformal Refinement of All-Hexahedral Finite Element Meshes. MA Thesis, Brigham Young University, Utah
9. Murdoch P, Benzley S, Blacker T, and Mitchell S (1997) The Spatial Twist Continuum: A Connectivity Based Method for Representing all-Hexahedral Finite Element Meshes. In: Finite Elements in Analysis and Design 28:137-149
10. Murdoch P and Benzley S (1995) The Spatial Twist Continuum. In: Proceedings 4th International Meshing Roundtable, pages 243-251. Sandia National Laboratories.
11. Benzley S, Harris N, Scott M, Borden M, and Owen S (2005) Conformal Refinement and Coarsening of Unstructured Hexahedral Meshes. In: Journal of Computing and Information Science in Engineering 5:330-337

12. Esmaeilian S (1989) Automatic Finite Element Mesh Transitioning with Hexahedron Elements. Doctoral Dissertation, Birgham Young University, Utah
13. Harris N, Benzley S, and Owen S (2004) Conformal Refinement of All-Hexahedral Element Meshes Based on Multiple Twist Plane Insertion. In: Proceedings 13th International Meshing Roundtable, pages 157-167. Sandia National Laboratories.

pCAMAL: An Embarrassingly Parallel Hexahedral Mesh Generator*

Philippe P. Pébay[1], Michael B. Stephenson[2], Leslie A. Fortier[3],
Steven J. Owen[4], and Darryl J. Melander[5]

[1] Sandia National Laboratories
   P.O. Box 969, MS 9051, Livermore CA 94551, U.S.A.
   pppebay@sandia.gov
[2] M.B. Stephenson & Associates
   2005 West 1550 North, Provo, UT 84604-2212, U.S.A
   mbsteph@sandia.gov
[3] Sandia National Laboratories
   P.O. Box 5800, MS 0376, Albuquerque NM 87185-0376, U.S.A.
   laforti@sandia.gov
[4] Sandia National Laboratories
   P.O. Box 5800, MS 0376, Albuquerque NM 87185-0376, U.S.A.
   sjowen@sandia.gov
[5] Sandia National Laboratories
   P.O. Box 5800, MS 0376, Albuquerque NM 87185-0376, U.S.A.
   djmelan@sandia.gov

Summary. This paper describes a distributed-memory, embarrassingly parallel hexahedral mesh generator, pCAMAL (parallel CUBIT Adaptive Mesh Algorithm Library). pCAMAL utilizes the sweeping method following a serial step of geometry decomposition conducted in the CUBIT geometry preparation and mesh generation tool. The utility of pCAMAL in generating large meshes is illustrated, and linear speed-up under load-balanced conditions is demonstrated.

1 Introduction

The requirement for detailed numerical simulation of complex domains has driven the research and development of parallel applications that can take advantage of shared- or distributed-memory machines. This includes tools for numerical simulation, post-processing as well as pre-processing tools for mesh generation. The majority of these techniques for mesh generation have

*This work was supported by the United States Department of Energy, Office of Defense Programs. Sandia is a multiprogram laboratory operated by Sandia Corporation, a Lockheed-Martin Company, for the United States Department of Energy under contract DE-AC04-94-AL85000.

focused on the generation of tetrahedral meshes for computational fluid dynamics. Nikos Chrisochoides [5] provides an extensive survey of parallel mesh generation methods for triangle and tetrahedral methods. While some techniques have been proposed for parallel quadrilateral [16, 8] and block-structured [15, 13] meshing, parallel implementations of unstructured hexahedral meshing techniques such as sweeping [7, 11] have yet to be proposed.

Sweeping is a common tool used for generating quality hexahedral meshes in domains defined by a $2\frac{1}{2}$-dimensional or cylindrical topology. For example, an unstructured meshing algorithm such as paving [4] is applied to one or more source surfaces and then extruded through the volume to generate hexahedral elements ending with a single target surface. While some work has been done to characterize sweepable topologies [17], the user is primarily responsible for developing and executing a strategy for decomposing a CAD model in order to apply the sweeping scheme. Meshing tools such as CUBIT[6] provide extensive geometric tools to assist the user in this process. Nevertheless, the decomposition process can sometimes take days or weeks depending on the complexity of the solid model.

As a result of this inherently user-intensive process, the advantages of parallel techniques for generating sweep meshes has been minimal. This has been because the actual computation time to generate the mesh in serial has been insignificant compared to the overall user time required for sweep meshing. Recently, the granularity of the simulations required by many users within the National Laboratories has been a motivating factor. Meshes of hundreds of millions and eventually billions of elements will be required to respond to the needs of the National Laboratories in the coming years. It is not currently feasible to generate meshes of this size and level of detail using a serial process. Therefore, there is a critical need to develop scalable mesh generators.

This paper outlines the development of a distributed-memory parallel hexahedral mesh generator, pCAMAL (parallel CUBIT Adaptive Mesh Algorithm Library), that utilizes the sweeping method [11] and shows its utility in generating large meshes. It also discusses changes that have been implemented in the geometry preparation and mesh generation tool CUBIT to enable it to prepare the input data required by pCAMAL.

◇

2 Method

Our approach is a two-stage approach that can be summarized as follows:

1. The initial domain (fully described by its watertight boundary) is decomposed into an assembly of sweepable subdomains. Currently, this stage is performed serially and requires direct human intervention. The resulting many-to-one boundary mesh is exported to a file;

2. The pCAMAL application imports the latter file, and subsequently sweeps the interior many-to-one hexahedral meshes in an embarrassingly parallel fashion. The resulting meshes are stored in separate files.

The proposed method for generating a swept hexahedral mesh in parallel is the first phase in a more comprehensive plan of developing a full parallel hexahedral meshing capability. The first phase of this project, described in this paper, is to develop an embarrassingly parallel version of the existing sweeping capability [11] used in CUBIT. Of necessity, this initial version of the parallel sweeping algorithm must not require geometry interactions as it generates the 3D mesh. As a result, the proposed method relies on the two-stage process described above that first depends on the user to provide an initial decomposition and boundary mesh. Because decomposition and surface meshing require significant geometry interaction, these procedures were necessarily left to the traditional user-interactive serial process currently provided in CUBIT.

Geometry interaction, typically managed through the CGM library [14] in the serial CUBIT tool kit, utilizes one or more third party geometry kernels such as ACIS [1] or Granite [2]. In order to develop a hexahedral meshing system that incorporates decomposition and surface meshing, the geometry kernel would also need to reside on each node of a distributed machine, which in their current form would be infeasible. Future work will involve developing a light-weight geometry kernel that will replicate capability used in the serial CUBIT process, but would be more efficient for use in a parallel environment. This will provide the infrastructure for developing a more comprehensive parallel sweeping capability that could potentially include geometry decomposition and surface meshing.

2.1 Serial Domain Decomposition

Serial domain decomposition is performed as an extension to CUBIT, a mesh generation tool kit developed at Sandia National Laboratories. It includes many algorithms to mesh surfaces and solids with triangles, quadrilaterals, tetrahedra, and hexahedra. Some of these algorithms exist in a separate library known as CUBIT Adaptive Meshing Algorithm Library or CAMAL. The algorithms were extracted from CUBIT over several years. CUBIT now uses the algorithms as they are implemented in CAMAL. In addition to mesh generation capabilities, CUBIT also provides geometry manipulation capabilities, including geometry decomposition tools that prepare geometry for the meshing algorithms.

CUBIT Modifications

CUBIT normally attempts to generate a complete hexahedral mesh within each solid. Because CUBIT is a serial application, it is limited in the size of mesh that it can generate by the memory available on a given workstation.

To significantly extend size limitations, the mesh must be generated in pieces rather than in a single, serial session. If the mesh can be generated in pieces, then it can be generated in parallel.

Thus, CUBIT was modified to generate only the boundary mesh for each sweepable volume and then output this mesh, with additional properties, in an Exodus II file format. Exodus II is a file format widely used by Sandia analysts [10].

New commands

Two new commands were added to CUBIT in support of parallel hex meshing.

```
set parallel meshing [on|off]
```

```
export parallel <filename>
```

The first command instructs CUBIT to forgo hexahedra generation when the CUBIT sweeping algorithm is applied to a volume. Instead, meshes are only generated on the volume's boundary surfaces. Note that only one-to-one and many-to-one sweeping currently support this partial sweeping capability.

The second command instructs CUBIT to write an Exodus II file containing the surface meshes of each sweepable volume, organized so that pCAMAL may correctly interpret them to generate the full, hexahedral mesh.

Implementation

PCMLSweeper is a new class derived from the CAMAL class CMLSweepCore. CMLSweepCore is the base-class for all CAMAL sweepers. PCMLSweeper is the interface used by CUBIT and pCAMAL to generate the surface boundary mesh and the hexahedral mesh respectively.

The differences in PCMLSweeper and CMLSweepCore are small but significant. One difference is the signature of the method

```
PCMLSweeper::set_boundary_mesh()
```

The version of this method found in PCMLSweeper has a parameter list that is more like that of the Exodus II routines than its equivalent method in the CMLSweepCore class. This makes it easier to pass data to pCAMAL without need for translation.

The PCMLSweeper class also includes the new method

```
PCMLSweeper::generate_shell()
```

This method generates only the boundary shell of the volume instead of generating hexahedral elements while sweeping the source surface mesh to the target surface. If both source and target surfaces have been meshed before this method is called, then the method only checks that the connectivity of the target mesh matches the surface mesh and that the linking surface meshes are consistent.

Another new method is

```
PCMLSweeper::get_quads_buf()
```

This method retrieves the boundary quadrilateral mesh from the sweeper in a buffer. The output is buffered because of its potential size. It may be impossible to allocate sufficient memory to return the entire mesh in a single call if the surface mesh is huge, i.e., several million elements. The method returns the number of elements in the buffer and is called repeatedly until the number of elements returned is zero.

Another change made to CUBIT is the addition of a new class called PCamalExporter. It is a subclass of ExodusExporter, the Exodus II file writer in CUBIT. Its purpose is to gather all of the default or user specified pCAMAL output blocks and write them to an Exodus II file with additional information necessary for pCAMAL to generate the hexahedral mesh.

PCamalExporter writes each surface mesh for the sweepable volumes as a unique element block. The following element block properties are added to the file.

- Sweep Volume Identifier – Each surface is associated with a sweepable volume with this identifier.
- Surface Type – A surface is either a source, a target or a linking surface.
- Number of Hexahedra – Each source surface has an estimate of the number of hexahedra in the sweepable volume. Surface types other than *source* do not include this property. pCAMAL uses this value in its load balancing scheme.
- User Block Number – The block number assigned to the sweepable volume is recorded here. If the user assigned a block number to the sweepable volume, that block number will be found here. If not, then CUBIT will assign a block number which is the same as the sweep volume identifier. The user may assign many volumes to the same block in the Exodus II format.
- Number of Hex Nodes – The number of nodes to be generated for each hex element. The default number of nodes is eight. This number may increase if the user desires higher-order elements.

A surface may lie on the boundary of two separate volumes. In this case, two instances of the above element properties are added to the Exodus II file for that surface. Because of the different context of each use of the volume, the properties for the surface will differ in each instance.

2.2 Embarrassingly Parallel Sweeping

In the second stage of the parallel sweeping process, the specialized Exodus II file generated in the domain decomposition stage is imported by a corresponding specialized Exodus II reader. The task of pCAMAL is to sweep all interior hexahedral meshes in an embarrassingly parallel fashion on distributed-memory machines. In order to achieve this goal, we take advantage of the fact that mesh consistency is ensured across subdomain boundaries at the domain decomposition stage. As long as the boundary meshes provided to pCAMAL are consistent, the collection of hexahedral meshes are also guaranteed to be consistent across the entire domain. Because mesh consistency is ensured through the surface meshes, each volume can be meshed independently with no communication required between regions.

In traditional serial meshing procedures, the user can display the entire mesh and color code it to visualize mesh quality. With a final mesh distributed across multiple processors, it is not possible to address mesh quality issues in a similar manner. Instead, mesh quality extrema and statistical moments are computed and reported to the user, who can choose to address them within the CUBIT toolkit. In order to offer the user the widest existing choice of mesh element quality functions, we have integrated the Verdict geometric quality library (cf. [12]) in pCAMAL. Since Verdict functions are called on local data by individual compute nodes, this does not affect the embarrassingly parallel nature of pCAMAL: only a global update of the extrema and of the statistical moments must be performed prior to overall termination; this represents an entirely negligible amount of inter-processor communication as only 2 extrema and 4 statistical moments (mean, variance, skewness, and kurtosis) must be reported.

pCAMAL is built on CAMAL and Verdict libraries, and implemented using the single-program, multiple-data (SPMD) paradigm. Specifically, MPI is employed so that:

- the same input data (the file containing the set of meshed surfaces that specify the collection subdomains) is loaded on all processors that participate in the run;
- each processor generates the interior many-to-one sweep meshes that have been attributed to this processor by a load-balancing scheme, saving each mesh as a separate file;
- each processor computes the desired Verdict quality function of each element of its swept meshes, summarizes them in terms of extrema and statistical moments, and collects timing information;
- a single processor collects all local quality and timing statistics via message passing, and aggregates them using appropriate combinations (as constrained by the domain decomposition) prior to reporting them.

Note that currently, only a crude load-balancing scheme has been implemented in pCAMAL. We are currently developing a better strategy that makes use of the numbers of estimated hexahedral elements to be generated for each subdomain, as permitted by the sweeping techniques. In fact, in the case of one-to-one sweeping, the exact number of output hexahedra can even be known prior to mesh generation.

◇

3 Examples

To demonstrate the functionality of pCAMAL, a set of examples ranging from simple (2 subdomains involving the generation of about 7500 hexahedral elements) to very large (1024 subdomains for a maximum of 1024000 hexahedral elements) is presented.

The parallel runs have been executed on Sandia National Laboratories' Catalyst computational cluster, which comprises 120 dual 3.06GHz Pentium Xeon compute nodes with 2GB of memory each. Catalyst has a Gigabit Ethernet user network for job launch, I/O to storage, and users' interaction with their jobs, and a 4X Infiniband fabric high-speed network using a Voltaire 9288 InfiniBand switch. Its operating system has a Linux 2.6.17.11 kernel, and its batch scheduling system is the TORQUE resource manager [3].

3.1 Proof Of Concept: Bisected Cylinder

The first example is that of a split cylindrical model, illustrated in Figure 1, left. The process begins with the geometry, which in this case was created using CUBIT's geometry creation capabilities. In this example, although the original cylindrical volume is sweepable, it was split into two symmetric parts to examplify the use of pCAMAL with a simple geometry. As described in the previous section, the volumes are meshed in CUBIT on their surfaces only, and setting the parallel mesh option on results in exporting them to two files in the modified Exodus II format expected by pCAMAL. Executing pCAMAL with one (in this case, the two sweeping processes are queued by the load-balancing scheme) or two processors, two Exodus II hexahedral mesh files are generated and saved.

These two meshes can be visualized by importing them back into CUBIT or using another visualization tool. Figure 1 shows the model's surface quadrilateral (center) and volume hexahedral (right) meshes as computed by pCAMAL.

3.2 Algorithm Scalability: Similar Cubes

In order to assess speed-up independently of the load-balancing scheme, a series of cubic or rectangular parallelepiped models made up of varying numbers of similar cubes is used. With these synthetic examples, we assess:

1. speed-up at constant total work (as studied for the INL reactor core model above), and
2. speed-up at constant work per processor.

Fig. 1. Original geometry for bisected cylinder, the surface mesh, and the resulting volume mesh.

In both cases, speed-up can be visually inspected by plotting speed-up *versus* number of processors using a log-log scale, as optimal scale-up is revealed by a straight line (more precisely, this line is the angle bisector of the first quadrant). Note however that speed-up is not defined in the same way for the two types of studies. Let $n_{\min}$ denote the smallest utilized number of processors and $T(n)$ the wall clock time measured with n processors. Then, speed-up at constant total work with n_p processors is

$$S_{TW} = \frac{T(n_{\min})}{T(n_p)},$$

whereas speed-up at constant work per processor with n_p processors is

$$S_{WP} = \frac{T(n_p)}{T(n_{\min})}.$$

Fig. 2. Several synthetic models using an increasing number of cubes.

In this demonstration, models were created containing 1, 2, 4, 8, 16, 32, 64, 128, and 1024 identical subdomains. Some of the subdomains are rectangular volumes, others are cubes, though all subdomains in one group of tests will contain the same number of elements when meshed. When the number of subdomains is doubled, the total number of elements is also doubled. Figure 2 illustrates this process for the cubic steps, that is, when the number of cubes between one step and the next differs by a factor of 8.

Table 1. Wall clock time measured on `Catalyst` to sweep the $1,024,000$ elements of a synthetic model as a function of the number of processors used.

Number of processors	1	2	4	8	16	32	64	128
Time (seconds)	140	80	49	35	27	26	24	23

Table 2. Wall clock time measured on `Catalyst` to sweep the $32,000,000$ elements of a synthetic model as a function of the number of processors used.

Number of processors	1	2	4	8	16	32
Time (seconds)	1231	669	339	170	87	48

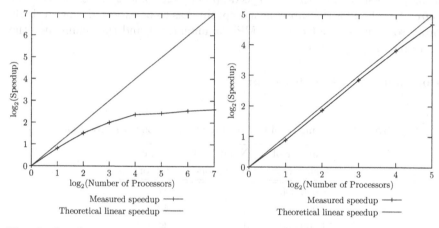

Fig. 3. Speed-up at constant total work when meshing a synthetic model with $1,024,000$ (left) or $32,000,000$ (right) hexahedral elements.

Constant Total Work

In the first series of test runs, at constant load per processor, only the 1024-subdomain model is used. Each cube has a uniform surface mesh consisting of 100 quadrilaterals per face; in other words, each cubic subdomain results in a 1000-element hexahedral mesh. Thus, the swept mesh of the entire model contains $1,024,000$ hexahedral elements. The results obtained on `Catalyst` are provided in Table 1. In this case, speed-up is much more favorable than in the case of the INL reactor core model, being near-optimal for the first 2 or 3 steps. As expected when decreasing the load per processor, speed-up tails off until wall clock time eventually reaches a minimum, which is illustrated in Figure 3, left. This is a normal behavior because a decreasing amount of work per processor ultimately results in a situation where overheads, even small in absolute terms, become dominant as compared to the amount of actual computational work. In this current example, it appears that with 128 processors, minimal wall clock time has been or is almost reached. Note that this corresponds to a per processor load of 8,000 hexahedra to be swept, that is, an almost negligible amount of computational work.

Now if, instead, a much finer level of granularity is required by imposing 10,000 quadrilaterals per face, which in turns results in 1,000,000 hexahedra per subdomain, it is sufficient to consider 32 subdomains to generate a fairly large mesh (32,000,000 hexahedral elements), with which speed-up at constant total work (and thus decreasing work per processor) will remain near-optimal for a while, and thus minimal wall clock time will occur for a larger number of processors. And indeed, as indicated in Table 2 and illustrated in Figure 3, right, speed-up remains next-to-optimal (almost 2) for several steps, and only begins to slightly decrease (while still remaining superior to 1.8) at 32 processors. Note that the speed-up between 1 and 2 processors is lower (*circa* 1.8) than for the subsequent step; this has to be expected, since a single processor run does not create a MPI communicator and thus eliminates the corresponding overhead.

Constant Work Per Processor

Table 3. Timing results of testing constant load per processor.

Number of elements	Number of processors	Time (seconds)
1,000,000	1	41
2,000,000	2	43
4,000,000	4	44
8,000,000	8	47
16,000,000	16	46
32,000,000	32	48

In order to assess speed-up at constant work per processor, the latest model is used, *i.e.*, each subdomain is a cube whose boundary consists of 10,000 quadrilaterals per face and thus its swept mesh contains 1,000,000 hexahedral elements. Increasingly large models with 1, 2, 4, 8, 16, and 32 such subdomains are then meshed with pCAMAL utilizing a number of processors equal to the number of sudomains. Corresponding wall clock times measured on Catalyst are given in Table 3, and illustrated in Figure 4; these clearly exhibit near-optimal speed-up, thus experimentally validating the embarrassingly parallel nature of pCAMAL. Note that the same *caveat* as previously descibed regarding the base (1 processor) case, which does not incur MPI overheads, holds.

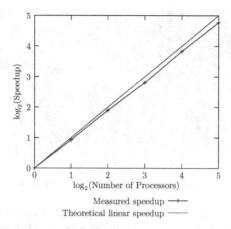

Fig. 4. Speed-up at constant work per processor when meshing a synthetic model with 1, 2, 4, 8, 16, or 32 processors.

3.3 Benefits for a Real-Life Model: Nuclear Reactor Core

A model of a reactor core developed by Idaho National Laboratories (courtesy of Scott Lucas and Glen Hansen), illustrated in Figure 5, is used to demonstrate the benefits of pCAMAL when dealing with real-life multi-volume models.

Fig. 5. Left: the INL reactor core with its casing, right: visualization of the inner portion contained within the casing.

This model contains about a thousand parts, and using serial meshing capabilities would take a considerable amount of time to generate a hexahedral mesh at the desired level of detail. If the element granularity is set to be small enough, mesh generation of this model with CUBIT is not even possible using serial processing due to memory constraints.

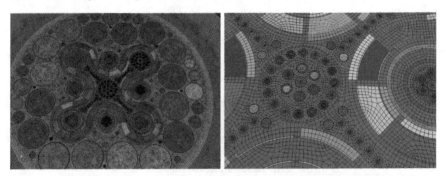

Fig. 6. Details of the mesh of the INL reactor core model.

Table 4. Wall clock time measured on **Catalyst** to sweep a 10, 459, 575-element hexahedral mesh of the INL reactor core model with **pCAMAL** as a function of the number of processors used.

Number of processors	1	3	4	10	20	40	80
Time (seconds)	745	393	363	265	259	246	232

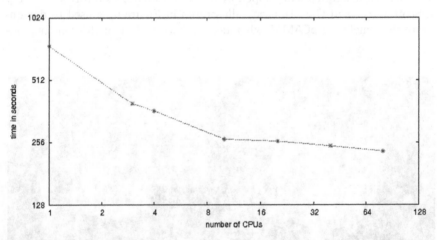

Fig. 7. Wall clock time measured on **Catalyst** to sweep a 10, 459, 575-element hexahedral mesh of the INL reactor core model with **pCAMAL** as a function of the number of processors used.

With a desired level of resolution that results in the creation of 10, 459, 575 hexahedral elements, the wall clock time measured on **Catalyst** to sweep the mesh with **pCAMAL** as a function of the number of processors utilized is indicated in Table 4 and shown in Figure 7. These results allow assessment, by definition, of speed-up at constant global work, and thus with a decreasing load per processor. Although some level of speed-up is achieved, it is evidently sub-optimal as a few of the subdomains are very large in comparison to the other ones, and thus, wall clock time has a lower bound that is constrained by the time it takes to generate the largest swept mesh. This illustrates the need to further develop the domain decomposition step in conjunction with the load-balancing scheme. Nonetheless, with a higher level of detail, it becomes impossible to generate the mesh with a serial mesher, no matter what platform is used, and the capability of pCAMAL becomes a necessity. For instance, Figure 8 illustrate different levels of detail of the interior of a much finer mesh of the same model: this mesh contains 33, 580, 500 hexahedral elements, and cannot be generated by CUBIT on any currently existing platform.

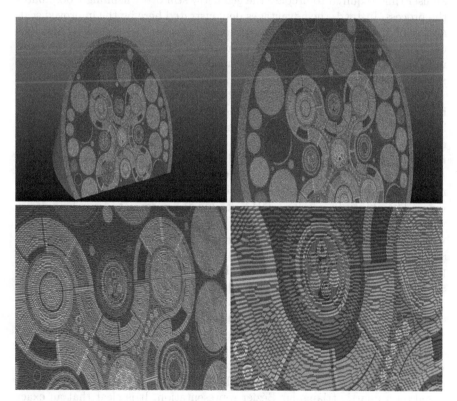

Fig. 8. Different levels of detail of the interior of a 33, 580, 500-element hexahedral mesh of the INL reactor core model generated by pCAMAL.

◇

4 Conclusions and Perspectives

An approach to embarrassingly parallel hexahedral sweep mesh generation has been proposed. While the objective of improving computational efficiency for generating large models has clearly been met, the main objective for this work has been to address the scalability problem. The sweep meshing problem continues to be an inherently user-intensive procedure, requiring detailed user knowledge of the CAD model in order to apply an appropriate decomposition strategy. While the advantages of improved computational efficiency demonstrated here are dramatic compared to the traditional serial process, the overall user time required to prepare the geometry still overwhelmingly dominates the process. Instead, the advantages demonstrated here by the improved scalability of the hexahedral meshing process represent the main contributions of this work, permitting complex simulations where meshes of hundreds of millions and potentially billions of elements are required to understand local physical phenomena.

The presented work is an initial phase of a more comprehensive plan to build a general parallel code for hexahedral meshing. Future work will include improving geometric load balancing, integration of element quality metrics, addition of a geometry query kernel, implementation of a parallel surface meshing process and integration with simulation tools as a component library.

The geometric load balancing solution proposed here for parallel sweep meshing takes advantage of the natural sweep direction of the individual volumes. The ability to decompose the volume at appropriate layer boundaries will address most of the load balancing problem. The problem arises, however when the number of processors will require a decomposition at a resolution smaller that a single layer. This implies a decomposition strategy that will impose a subdivision of individual layers. Future work will address this issue and should provide a more general load balancing strategy for parallel meshing.

The ability to mesh the geometric surfaces in a parallel environment is one of the significant issues to be addressed in the future. This will necessarily involve the integration of a geometry query engine within pCAMAL. The serial solution currently used by CUBIT[1, 2] will likely prove to be overly heavyweight to include on every compute node. Instead a light-weight facet-based geometry kernel is proposed. [9] describes a G1 continuous surface definition based on a quartic triangular Bézier representation. It is clear that an exact NURBS representation of the original CAD model would provide a more accurate representation; however, for the applications that are currently intending to utilize pCAMAL, an interpolated definition will likely be sufficient.

Once a geometry kernel is available within pCAMAL, surface meshing algorithms [4] may then be incorporated within the parallel meshing procedure. This will initially involve meshing each geometric surface on separate compute nodes prior to invoking the sweep meshing procedure. Inter-processor communication will be necessary to distribute the shared surface meshes to the appropriate processors for hexahedral meshing. An open question, yet to be resolved is a method for load balancing for individual surface domains. For example, the ability to effectively decompose a surface for quadrilateral meshing in a manner where the final quad mesh will be independent of the number of subdivisions will need to be addressed.

pCAMAL is currently implemented as a stand-alone application that accepts input of a boundary quad mesh and exports a full hexahedral mesh via Exodus II files. Future work will involve integrating pCAMAL as a component library callable from a simulation code. The objective will be to eliminate the need to export the full finite element mesh to disk, but rather provide an API to objects in memory. Future work will include exploring the necessity to generate the entire mesh. Instead, a method where only the required portions of the mesh are generated real-time as they are requested by the API would be developed.

Finally, future work will involve more targeted mesh quality assessment: in addition to reporting global extrema and statistical moments, pCAMAL will offer the ability to define lower and upper bounds, below and above which elements IDs will be stored for inspection and/or modification of the input geometry prior to re-meshing.

◇

Acknowledgments

The authors would like to thank:

- David Evensky (Sandia National Laboratories) for his help with the Catalyst computational cluster, and for having allowed us to use much time as needed on this system;
- David C. Thompson (Sandia National Laboratories) for having set up parallel visualization capabilities to post-process pCAMAL's outputs.
- Scott Lucas and Glen Hansen (Idaho National Laboratories) for the INL reactor core model.

References

1. 3D ACIS modeler. URL http://www.spatial.com/products/acis.html.
2. GRANITE interoperability kernel. URL http://www.ptc.com/appserver/mkt/products/home.jsp?k=369.
3. TORQUE resource manager. URL http://www.clusterresources.com/pages/products/torque-resource-manager% .php.
4. T.D. Blacker and M.B. Stephenson. Paving: A new approach to automated quadrilateral mesh generation. *International Journal for Numerical Methods in Engineering*, 32:811–847, 1991.
5. Nikos Chrisochoides. A survey of parallel mesh generation techniques. Technical report, Brown University. URL http://www.andrew.cmu.edu/user/sowen/pmesh_survey.pdf.

6. Computational Modeling Science Dept., Sandia National Laboratories. *CU-BIT 10.2 User Documentation*, October 2006. URL `http://cubit.sandia.gov/help-version10.2/cubithelp.htm`.
7. P. M. Knupp. Next-generation sweep tool: A method for generating all-hex meshes on two-and-one-half dimensional geometries. In *Proceedings of the 7th International Meshing Roundtable*, pages 505–513. Dearborn, MI, October 1998. URL `http://www.imr.sandia.gov/papers/imr7/knupp_sweep98.ps.gz`.
8. Randy Lober, T.J. Tautges, and R.A. Cairncross. The parallelization of an advancing front, all-quadrilateral meshing algorithm for adaptive analysis. In *Proceedings of the 4th International Meshing Roundtable*, pages 59–70. Albuquerque, NM, October 1995.
9. Steven J. Owen, David R. White, and Timothy J. Tautges. Facet-based surfaces for 3d mesh generation. In *Proceedings of the 11th International Meshing Roundtable*, pages 297–312. Ithaca, NY, September 2002. URL `http://www.imr.sandia.gov/papers/imr11/owen.pdf`.
10. Larry A. Schoof and Victor R. Yarberry. Exodus II: A finite element data model. Technical Report SAND92-2137, Sandia National Laboratories, November 1995. URL `http://endo.sandia.gov/SEACAS/Documentation/exodusII.pdf`.
11. Michael A. Scott, Matthew N. Earp, Steven E. Benzley, and Michael B. Stephenson. Adaptive sweeping techniques. In *Proceedings of the 14th International Meshing Roundtable*, pages 417–432. San Diego, CA, September 2005. URL `http://www.imr.sandia.gov/papers/imr14/scott.pdf`.
12. C. Simpson, C. D. Ernst, P. Knupp, P. P. Pébay, and D. C. Thompson. *The Verdict Library Reference Manual*. Sandia National Laboratories, April 2007. URL `http://www.vtk.org/Wiki/images/6/6b/VerdictManual-revA.pdf`.
13. Moitra Stuti and Anutosh Moitra. Considerations of computational optimality in parallel algorithms for grid generation. In *Proceedings of the 5th International Conference on Numerical Grid Generation in Computational Field Simulations*, pages 753–762, 1996.
14. Timothy J. Tautges. The common geometry module (CGM). In *Proceedings of the 9th International Meshing Roundtable*, pages 337–348. New Orleans, LA, October 2000. URL `http://www.andrew.cmu.edu/user/sowen/abstracts/Ta758.html`.
15. Joe F. Thompson, Bharat K. Soni, and Nigel Weatherill. *Handbook of Grid Generation*. CRC Press, 1999.
16. B.H.V. Topping and B. Cheng. Parallel adaptive quadrilateral mesh generation. *Computers and Structures*, (73):19–536, 1999.
17. David R. White and Timothy J. Tautges. Automatic scheme selection for toolkit hex meshing. *International Journal for Numerical Methods in Engineering*, 49 (1):127–144, 2000.

Surface Meshing

3B.1

An Extension of the Advancing Front Method to Composite Geometry

G. Foucault[1], J-C. Cuillière[1], V. François[1], J-C. Léon[2], R. Maranzana[3]

(1) Université du Québec à Trois-Rivières, CP 500, Trois-Rivières
(Québec) G9A-5H7, Canada Gilles.Foucault, Jean-Christophe.Cuilliere,
Vincent.Francois @uqtr.ca
(2) Institut National Polytechnique de Grenoble, Laboratoire G-SCOP,
France, Jean-Claude.Leon@hmg.inpg.fr
(3) Ecole de Technologie Supérieure de Montréal, GPA, Laboratoire
LIPPS, Canada, Roland.Maranzana@etsmtl.ca

Abstract
  This paper introduces a new approach to automatic mesh generation over
composite geometry. This approach is based on an adaptation of advancing
front mesh generation techniques over curved surfaces, and its main
features are :
- elements are generated directly over multiple parametric surfaces:
  advancing front propagation is adapted through the extension to
  composite geometry of propagation direction, propagation length, and
  target point concepts,
- each mesh entity is associated with sets of images in each reference
  entity of the composite geometry,
- the intersection tests between segments are performed in the parametric
  domain of their images.

Keywords: Adaptivity, advancing front, mesh generation, composite
geometry, virtual topology, defeaturing.

1. Introduction

The recent integration of FEA in CAD has greatly reduced the median time requirements to prepare FE models [1].

However, the preparation of FE models from CAD models is still a difficult task when it contains many shape details, and when its Boundary Representation (B-Rep) is composed of a large number of faces, many of them being much smaller than the desired FE size. Such configurations are often at the origin of poorly-shaped elements and/or over-densified elements, not only increasing the analysis time, but also producing poor simulation results.

Several efforts have been made to avoid poorly-shaped elements and over-densified mesh elements generated from an unprepared (for analysis) CAD model [2-5]. In these approaches, well-known mesh topological transformations perform mesh element removal, e.g. decimation of surface meshes, or collapsing the faces of a tetrahedron in order to remove it. One limitation of these operators is that entity collapsing operations are not intrinsically suited for through hole removal, and need more complex extensions. Therefore, hole details may require specific treatments such as removing these details in the initial CAD model.

Other approaches consist in adapting CAD models directly. For example, feature recognition and extraction processes can be used to simplify details like fillets and blends [6, 7], bosses, pockets [8], and holes. These approaches generate tree-structured simplified models, where each simplification is identified as a feature.

Lee et al. [9] propose a feature removal technique which starts from a feature tree and provides the ability to suppress and subsequently, reinstate features independently from the order in which they were suppressed, within defined limitations [9]. Most of these approaches manage a restricted set of feature types, and interactions between features remains a major issue. Moreover, even recognized features are often difficult to suppress, which makes these approaches non-robust and very restrictive when used alone. Nevertheless, these approaches are efficient when used to remove holes and bosses prior to mesh simplification. However, feature suppression does not guarantee that the object's boundary decomposition can be directly used for meshing. In this context, virtual topology techniques can contribute afterwards to adapt the boundary decomposition.

Virtual topology approaches proposed by Sheffer et al. [10] and Inoue et al. [11] aim at editing the B-Rep definition of a CAD model in order to produce a new topology that is more suited to mesh generation constraints. These approaches implement split and merge operators aimed at clustering adjacent surfaces into nearly planar regions in order to generate a new B-

Rep topology that is more suited to mesh generation, while preserving its geometry. However, face clustering algorithms proposed in [10, 11] show limitations in the context of FE models preparation: they do not support the definition of edges and vertices interior to faces, while these non-manifold surface configurations are required for various needs (modelling boundary conditions, taking into account specific features and sharp curves lying inside faces).

In a previous work, we have introduced alternate virtual topology concepts, dedicated to mesh generation requirements, designated as Meshing Constraints Topology (*MCT*) [12]. One of the basic and most important features of the MCT is enabling non-manifold surface transformations: edge deletion, vertex deletion, edge splitting, edge to vertex collapsing, vertices merging. The *MCT* preparation algorithm is based on local analysis of faces and edges regarding mesh generation constraints (local face width, normal vector deviation across edges, etc.) enabling the definition of interior edges and vertices on faces when required.

Virtual topology models are basically defined using composite geometry. For instance, a composite face (designated below as a *MC face*) is defined as a set of adjacent faces in original B-Rep structure, each of which associated with a bi-parametric surface. Consequently, using virtual topology for mesh generation requires the ability to automatically generate finite elements across composite geometry. At this point, mesh generation techniques aiming at this ability are based on the following concepts :

- global parameterization of a composite surface [13, 14]: this method defines a bijective projection between any point inside sets of surface patches and a global parametric domain. The bijective transformations proposed in [13, 14] are both based on a cellular decomposition of each reference (non-composite) surface mapped into the global parametric space. Each cell is related to a reference surface image, and a global parametric space image. Any point of a cell in the global parametric space is represented using its barycentric coordinates, and projected in the equivalent cell of the corresponding reference surface using the same barycentric coordinates. This new parameterization enables a transparent use of parametric meshing schemes. Unfortunately, this type of approach is limited by:

  - open and non-periodic surface requirements: this method can only be applied to open surfaces (homeomorphic to a disc), but not to periodic surfaces and closed surfaces homeomorphic to a n-torus.
  - the determination of surfaces outer loops is not automatic,
  - smoothness and planarity requirements: planar projection of distorted surfaces may cause high variations in the global parameterization

metrics, and often results, either in failures during mesh generation or highly distorted mesh results [13].
- using direct 3D advancing front techniques on a tessellated (triangulated) representation of composite surfaces [15]:
  - the mesh accuracy depends on the tessellation accuracy,
  - more sophisticated discrete representations such as subdivision surfaces and higher order triangulations allow curved mesh generation but still generate approximation errors.

In order to overcome these weaknesses, we are introducing, in this paper, a new approach to automatic mesh generation over composite geometry. This approach is based on an adaptation of advancing front mesh generation techniques over curved surfaces and its main features are :
- elements are generated directly over multiple parametric surfaces: advancing front propagation is adapted through the extension to composite geometry of propagation direction, propagation length, and target point concepts,
- each mesh entity is associated with sets of images in each reference entity of the composite geometry,
- the intersection tests between segments are performed in the parametric domain of their images.

2. Preparing CAD models for meshing

This section presents the FEA context, proposes a FE model preparation procedure, then sets up the *MCT* model (major input of the mesh generation process presented in this paper).

2.1. The FEA context

Prior to FEA itself, the analyst specifies mechanical hypotheses and boundary conditions on the CAD model, i.e. materials, loads and restraints. The analyst also specifies an analysis accuracy objective with regard to the engineering quantities (such as stresses) he is trying to compute. Based on his FEA skills, the analyst also specifies a priori a size map adapted to the component's shape and to the analysis accuracy objective. This size map can either be rough to quickly obtain an approximate solution or refined, generally for accuracy needs.

This size map is a central issue in our automatic feature removal and topology adaptation processes because it represents the analyst intent with regard to the size of shape and topology features that can be neglected for analysis purposes. In fact, the size map is the principal input on which automatic feature removal and topology adaptation processes are based.

2.2. A three step approach to CAD automatic adaptation for FEA

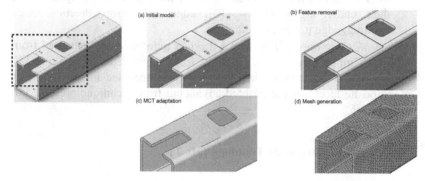

Fig. 1. Preparation of a CAD model for mesh generation

A fully automatic adaptation process has been designed to prepare CAD models for FEA. The automatic simplification criteria are based on the imposed size map and the process takes place through the following three steps:

- *Step1- Feature removal:* CAD design features (holes, fillets, pockets and protrusions) that are too small (with regard to the imposed size map) to affect analysis results are automatically identified as shape details (see fig. 1(b)). Two types of operations are used to automatically remove these details:
  - (a) Suppressing the feature directly in the feature-based model: this applies for details that have been designed as features, and for which suppression does not affect any other feature
  - (b) Performing a *delete-face* operation: this operation deletes detail faces and reconstructs a closed solid envelope by filling holes [8, 9].
- *Step2- MCT adaptation:* the B-Rep topology obtained after feature removal often requires additional preparation for mesh generation : small edges, narrow faces, must be transformed. *MCT* operators and criteria, based on adjacency hypergraphs, have been designed for automatic topology adaptation (see   fig. 1(c)):

° *MCT* criteria, based on the size map and on boundary conditions zones, automatically identify *MCT* operations required [16]:
  – irrelevant edges located in narrow faces or planar surfaces are removed by edge deletion,
  – irrelevant vertices located in small edges or smooth curves are removed by vertex deletion or edge contraction,
  – constricted sections of faces are collapsed by vertices merging.
° *MCT* operators then automatically edit the topology hypergraphs and their underlying geometry [12]: edge and vertex deletion, edge splitting, edge to vertex collapsing,

- *Step3-Meshing the MCT model:* a mesh is automatically generated from the *MCT* model (see   fig. 1(d)): the front is initialized by meshing *MC edges* (composite edges), then the mesh is propagated inside *MC faces* (composite faces) by the adapted advancing front technique presented in this paper.

2.3. Meshing Constraints Topology (*MCT*)

The *MCT* is represented as a B-Rep structure, providing a full description of orientation and topological links between entities, as shown in fig. 2. In this structure, **MCT entities** are defined versus **reference entities** as a outlined below:

Definitions
*The **reference model** is the B-Rep model obtained after performing step 1 in the process presented above (after feature removal and prior to topology adaptation).*
***Reference entities** (reference face, reference edge, reference vertex) are topological entities of the reference model. Their underlying geometry is represented through a single mathematical definition:*
- *the surface underlying a reference face is a Riemannian surface: plane, sphere, torus, NURBS, etc*
- *the curve underlying a reference edge is a Riemannian curve: line, circle, ellipse, NURBS, etc*

MCT entities *(MC face, MC edge, MC vertex) are composite topological entities created for mesh generation requirements. Their geometry is defined as sets of adjacent reference entities:*
- *The composite surface underlying a MC face is designated as a PolySurface, defined as the union of reference faces,*

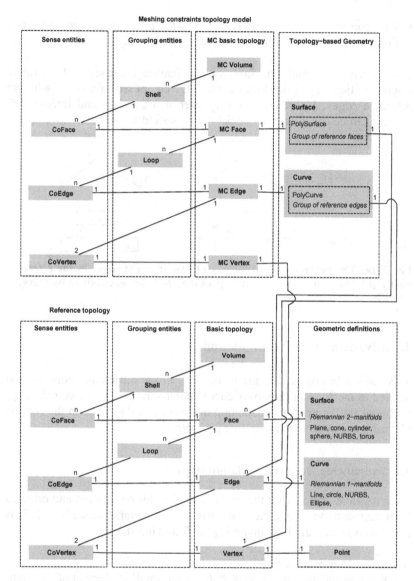

Fig. 2. (top) the *MCT* composition diagram (bottom) reference topology composition diagram

- *The composite curve underlying a MC edge is designated as a PolyCurve, defined as the union of reference edges.*

Thus, *PolyCurves* and *PolySurfaces* can feature tangency and curvature discontinuities. Fig. 3 illustrates a *MC edge* composed of a set of adjacent reference edges. *MC faces* featuring interior *MC edges* and interior *MC vertices* can be considered as special non-manifold faces.

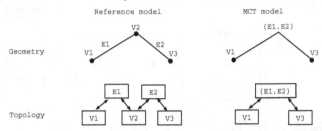

Fig. 3. (a) The geometry and topology of two adjacent edges in the *reference model*. (b) The geometry and topology of the *MCT* model obtained by merging edges *E₁* and *E₂*.

3. Advancing front triangulation

This section briefly recalls the main steps in the advancing front method applied to mesh generation over curved parametric surfaces (see ref [17]). Section 4 will present the adaptation of this classical scheme in the context of mesh generation over composite geometry.

3.1. Advancing front triangulation steps

After its initialization with the generation of nodes on vertices and oriented front segments on edges, the triangulation of a parametric surface [17] is based on a procedural loop involving the following steps:

While the front is not empty:
  Let the candidate front element $P_A P_B$ be the smallest element of the front,
  Compute α_1 the angle between the *candidate front element* and its *previous neighbour,* and α_2 the angle between the *candidate front* and its *next neighbour,*
  Identify the front configuration using α_1 and α_2, among the six front configurations illustrated in fig. 4.
  If the front configuration type lies between 1 and 5 (reference to fig. 4),

Then triangles are generated using neighbouring fronts (see fig. 5) and continue the loop.

Else,

> Compute the optimal candidate node location P_{OPT} to generate a triangle that fulfils both element's shape and size requirements (see §3.2),
>
> Search a node P_F of an existing front segment inside the area determined by two circles C_M and C_{OPT} (see Fig. 5):
>
> > C_M being centred on P_M (middle of the candidate front), and which radius is $1.5 \square \|P_M P_{OPT}\|$
> >
> > C_{OPT} being centred on P_{OPT} and which radius is $\dfrac{\sqrt{3}}{2} \|P_M P_{OPT}\|$

If P_F is found, then $P_C = P_F$, **else** $P_C = P_{OPT}$

Verify that the triangle (P_A, P_C, P_B) does not overlap existing triangles, if so, create the triangle (P_A, P_C, P_B) and replace candidate front element $P_A P_B$ with front elements $P_A P_C$ and $P_C P_B$.

	1	2	3	4	5	6
Configuration of the candidate front	△	▱	$\alpha_1\ \alpha_2$	$\alpha_1\quad\alpha_2$	$\alpha_1\quad\alpha_2$	$\alpha_1\qquad\alpha_2$
Generated triangles	◬	◺	◹	◿	◺	▲

Fig. 4. The 6 front configurations and their specific triangulation process

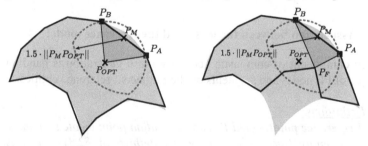

Fig. 5. (a) Triangle created from the candidate node at the optimal location (b) Triangle created from an existing node located inside the search area

3.2. Computing the optimal candidate node location

Given a candidate front element $P_A P_B$ and an isotropic size map function $H(x,y,z)$, the optimal candidate node is located by following a surface path starting from $P_M = (P_A + P_B)/2$ and progressing in the orthogonal direction to $P_A P_B$, until reaching a distance d from P_M.

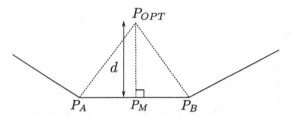

Fig. 6. Creation of the optimal point location

The distance d is calculated as a compromise between elements shape and size requirements (see fig. 6):

$$d = \frac{\sqrt{3}}{2}\left(w_T \cdot H(\boldsymbol{P}_M) + w_S \cdot \| \boldsymbol{P}_A \boldsymbol{P}_B \|\right)$$

w_T controls the size map respect, whereas w_S controls the elements shape quality. Past experience has shown that $(w_T, w_S) = (0.65, 0.35)$ is a good compromise between these two contradictory requirements.

4. Extension to composite geometry

This section presents mesh generation data-structures and algorithms aiming at the extension of advancing front triangulation techniques to the context of composite geometry.

4.1. Association between the mesh and the reference model

A key component of composite geometry mesh generation is handling the images of FE nodes and segments on their adjacent reference entities.

Definitions:
*A **reference point**, noted $\boldsymbol{P}_i$, is an Euclidian point located on one single reference entity (face, edge, or vertex, defined in §2.3), which is not necessarily a node of the final mesh.*
*A **reference sub-segment**, noted $\boldsymbol{P}_i\boldsymbol{P}_{i+1}$, is a curvilinear segment defined by two reference points and located on one single reference edge or face. The shape of a reference sub-segment lies on the surface of the reference B-Rep.*

In our data-structure, a **node** of the mesh is associated with (see Fig. 7(a)):
• its underlying *MCT* entity (defined in §2.3),

- one reference entity (defined in §2.3),
- its images on its adjacent reference entities (see fig. 7(a)), each of them defined by their parametric coordinates.

And a **segment** of the mesh is associated with:
- its underlying *MCT* entity,
- P_A and P_B, its two extremity nodes,
- an image curve defined as a sequence of N reference sub-segments $P_A P_1$, $P_1 P_2$, ..., $P_{N-1} P_B$ (see fig. 7(b), fig. 8(c), and fig. 9), each of them being associated with :
  - a reference entity,
  - its image on this reference entity,
  - a set images on its adjacent reference entities.

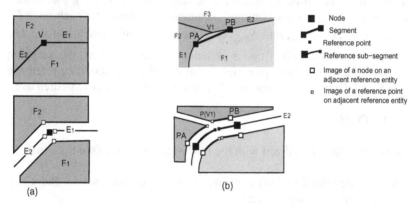

Fig. 7. Links between mesh elements and reference topology (a) a mesh node lying on a *MC Vertex* (b) a mesh segment lying on a *MC Edge* (*PolyCurve*)

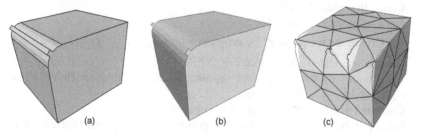

Fig. 8. (a) reference topology (b) *MCT* model (c) image curves of segments lying on reference entities

4.2. Meshing *MC edges*

Conventions:
x_i^{Fj} *represents parametric coordinates (u_i, v_i) underlying reference point P_i
, on reference face F_j.*
S_j *represents the parametric surface equation underlying face F_j. Thus $P_i =$
$S_j(x_i^{Fj})$.*
*Variables t designate curve parameter associated with a reference
geometry, while variables t' designate curve parameter associated with a
composite geometry.*

The *MCT* meshing process begins with the generation of a set of nodes on
MC vertices. Then nodes are intercalated on *MC edges*. Given an *MC edge*
parametric curve *PolyC(t')* and the size map function $H(x,y,z)$, the method
presented in [18] generates optimal parametric coordinate for each node P_i:
t'_i with $i = 0, 1, .., N_{seg}$, and N_{seg} being the number of segments of the *MC*
edge discretization. When nodes parameters t'_i are defined, the mapping
function of the *PolyCurve* provides, for any t', the corresponding reference
edge E and the parametric coordinate (on this reference edge) t written as :

$(E, t) = PolyC(t')$

Each node is then initialized with the following procedure (see fig. 7(a)):

A process described in [18] generates t'_i nodes coordinates on *PolyC(t')*,
respecting the size map $H(x,y,z)$
For each parameter t'_i:
   $(E_i, t_i) = PolyC(t'_i)$
   Create the node $P_i = E_i(t_i)$,
   For each face F_j adjacent to E_i:
      Create the image of P_i by projecting it on F_j : $x_i^{Fj} = proj(F_j, P_i)$
For each segment (t'_i, t'_{i+1}) (see Fig. 7(b)):
Let P_A and P_B be the current segment nodes, located on *PolyC(t')* at
$t'_A = t'_i$ and $t'_B = t'_{i+1}$ respectively
Construct segment $P_A P_B$ on the *MC edge*
Let $P[t']$ be the list of reference points of the segment sorted by t'
parameter
   $P[t'_A] = P_A$ and $P[t'_B] = P_B$
   For each reference vertex V_j of *PolyC(t')*:
      Let $t'(V_j)$ be the parameter of the reference vertex V_j on *PolyC(t')*

If $t'(V_j) \in \,]t'_A; t'_B[$ **then** create a reference point $P(V_j)=E_j(t'(V_j))$
and add $P[t'(V_j)] = P(V_j)$
For $j=0$ **to** size $(P[t'])-1$:
    Let P_j and P_{j+1} be the j-th and $j+1$-th nodes of $P[t']$
    Create a reference sub-segment $P_j P_{j+1}$ on reference edge E_j=$PolyC($
    $(t'(P_j)+ t'(P_{j+1}))/2$)
    For each reference face F_k adjacent to the reference edge E_j:
        Create the image of $P_j P_{j+1}$ on F_k

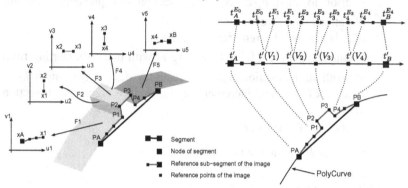

Fig. 9. Images of a segment lying on multiple reference edges and faces

4.3. Creating the optimal candidate node from a candidate segment

Conventions:

$C_{uv}^{Fj}(t)$ *represents a curve in the parametric space of reference face F_j.*

The equivalent curve in the Euclidian space is $C^{Fj}(t) = S_j\left(C_{uv}^{Fj}(t)\right)$.

$N(S, P)$ *represents the normal vector to surface S at point P, where:*
    S can be a MC face, a PolySurface, a reference face, or a parametric surface.
    P can be an Euclidian point of S or, a parametric coordinate of S.

Given a candidate front element starting at P_A and ending at P_B, constructing the optimal candidate node P_{OPT} on the *PolySurface* S_{Poly} of a MC face F_{MCT} can be achieved by creating :
- P_M the middle point of the curvilinear segment $P_A P_B$
- plane Φ, orthogonal to $P_A P_B$ and containing P_M
- Path $C(t')$ defined as the intersection curve between plane Φ and S_{Poly}, oriented with regard to the orientation of the advancing front.

The main part of this procedure is the creation of the path $C(t')$, illustrated in fig. 10 and fig. 11, and it is processed as follows:

Initialize $j=0$, $P_0 = P_M$, $L_{max} = d$ (where d is defined in section §3.2)
Do
  Set $A_j = N(S_{Poly}, P_j) \wedge (P_B\text{-}P_A)$ (*PolySurface* normal vector is defined in §4.4 see fig. 11)
  Find face F_j adjacent to P_j and contained in S_{Poly}, such as A_j is interior to face F_j at x_i^{Fj} (see section 4.4 presenting interior directions on a face)
  If F_j is not found, **then** break the loop,
  Create the path $C_{uv}^{Fj}(t)$ intersecting F_j with plane Φ, starting from P_i, with start direction $A_{3D,0}{=}A_j$, accuracy δ_{max}, target length $L_{Fj} = L_{max} - \| C(t') \|$ and increment length ds (see algorithm in section 4.5.4)
  Add curve $C^{Fj}(t)$ to $C(t')$
  initialize P_{j+1} with the last point of $C^{Fj}(t)$
  $j=j+1$
While $\| C(t') \| < L_{max}$
P_{OPT} is located at parameter t'_{OPT} of the curve $C(t')$ such as $\| C(t'_{OPT}) \|= L_{max}$

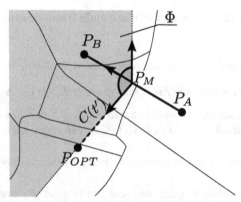

Fig. 10. The optimal location of candidate node P_{OPT} from candidate segment $P_A P_B$ is determined using path $C(t')$ intersecting the *PolySurface* with a plane

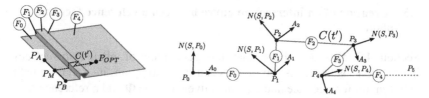

Fig. 11. The *PolySurface* **path** $C(t')$ **is constituted by multiple curves lying on reference faces of the** *PolySurface*

4.4. Definition of *PolySurface* normal vectors for discontinuous configurations

Section 4.3 underlined that the generation of paths on *PolySurfaces* requires the definition of normal vectors at locations where tangency and curvature discontinuities occur:

- on a reference edge shared by two reference faces of the *PolySurface* (see Fig. 11),
- on a reference vertex shared by multiple reference faces of the *PolySurface* (see Fig. 12).

Then, we define the normal vector at sharp locations P_i of a *PolySurface* S_{Poly} by a simple weighted average of adjacent faces normal vectors. The weights used are the normalized spanning angles β_j of faces F_j adjacent to point P_i and contained in S_{Poly} (see Fig. 12):

$$N(S_{Poly}, P_i) = \frac{\sum_{j=1}^{n} \beta_j \cdot N(F_j, P_i)}{\sum_{j=1}^{n} \beta_j}$$

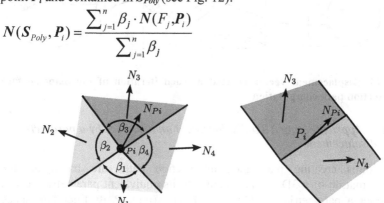

Fig. 12. Spanning angle and normal vectors along *PolySurface* **discontinuities**

4.5.   Creation of an intersection curve between a reference face and a plane

Section 4.3 also underlined that the generation of plane-*PolySurfaces* intersection paths relies on a procedure creating an intersection path between a reference face and a plane. Given a plane Φ and a reference face F_j, the problem is to create a path starting from a given point P_0, and following the intersection curve $C^{Fj}(t)$ between Φ and F_j. A target point P_n can be optionally prescribed as the ending point of the path. The path $C^{Fj}(t)$ is represented as a polyline $C_{uv}^{Fj}(t)$ defined in the parametric space of face F_j. The polyline $C_{uv}^{Fj}(t)$ is constructed by:

- calculating a small displacement Δ_{3D} in the Euclidian space,
- calculating Δ_{uv} equivalent to Δ_{3D} in the parametric space (detailed in §4.5.1),
- correcting the displacement error δ (detailed in §4.5.1) by projecting the incremented point on the intersection plane Φ.

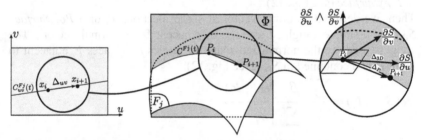

Fig. 13. Displacement vector created at each iteration of the plane-surface intersection path computation

4.5.1.   Approximating a 3D displacement on a surface by a parametric displacement

Algorithms creating plane-face intersection paths are based on the transformation of a 3D displacement into its equivalent parametric vector. Consider a parametric point x_i on surface $S(u,v)$, P_i its Euclidian space equivalent and Δ_{3D} an infinitesimal 3D displacement tangent to S at P_i:

$$\Delta_{3D} = \frac{\partial S}{\partial u} du + \frac{\partial S}{\partial v} dv$$

Let us consider the tangent (to S at point P_i) plane frame $\left(\dfrac{\partial S}{\partial u} \quad \dfrac{\partial S}{\partial v} \quad \dfrac{\partial S}{\partial u} \wedge \dfrac{\partial S}{\partial v}\right)_{xi}$ and Δ_{uv} with coordinates $(du, dv, 0)$ in this tangent plane frame. We can state that (see Fig. 13):

$$\Delta_{3D} = T_P(x_i) \cdot \Delta_{uv}$$

where $T_p(x_i)$ is the transformation matrix relating the 3D space frame and the tangent plane frame :

$$T_P(x_i) = \left(\dfrac{\partial S}{\partial u} \quad \dfrac{\partial S}{\partial v} \quad \dfrac{\partial S}{\partial u} \wedge \dfrac{\partial S}{\partial v}\right)_{xi} = \begin{pmatrix} \dfrac{\partial x}{\partial u} & \dfrac{\partial x}{\partial v} & \dfrac{\partial y}{\partial u}\dfrac{\partial z}{\partial v} - \dfrac{\partial z}{\partial u}\dfrac{\partial y}{\partial v} \\ \dfrac{\partial y}{\partial u} & \dfrac{\partial y}{\partial v} & \dfrac{\partial z}{\partial u}\dfrac{\partial x}{\partial v} - \dfrac{\partial x}{\partial u}\dfrac{\partial z}{\partial v} \\ \dfrac{\partial z}{\partial u} & \dfrac{\partial z}{\partial v} & \dfrac{\partial x}{\partial u}\dfrac{\partial y}{\partial v} - \dfrac{\partial y}{\partial u}\dfrac{\partial x}{\partial v} \end{pmatrix}_{xi}$$

First order approximation

Let P_{i+1} be the surface point obtained by the displacement (du, dv) from point P_i :

$$x_{i+1} = (u + du, v + dv) \text{ and } P_{i+1} = S(x_{i+1})$$

The 1st order Taylor expansion of the surface $S(x)$ equation about a point x_i gives:

$$\Delta_p = P_{i+1} - P_i = \left(\dfrac{\partial S}{\partial u}\right)_{xi} du + \left(\dfrac{\partial S}{\partial v}\right)_{xi} dv + R_1 = T_P(x_i) \cdot \Delta_{uv} + R_1 = \Delta_{3D} + R_1$$

Where R_1 is the remainder term of the first-order Taylor expansion. Therefore, $\Delta_{uv} = T_p{}^{-1} \square \Delta_{3D}$ can be considered as a first order approximation of the parametric space displacement that is equivalent to Δ_{3D}.

Second order approximation

Remainder R_1 of the first order approximation can estimated by a 2nd order Taylor expansion :

$$R_1 = \dfrac{du^2}{2}\left(\dfrac{\partial^2 S}{\partial u^2}\right)_{xi} + \dfrac{dv^2}{2}\left(\dfrac{\partial^2 S}{\partial v^2}\right)_{xi} + du \cdot dv\left(\dfrac{\partial^2 S}{\partial u \partial v}\right)_{xi} + R_2$$

Where R_2 is the remainder term of the second-order Taylor expansion.

Now, correcting Δ_{uv} by projecting R_1 in the tangent plane frame reduces the approximation error:

$$\Delta_{uv}'=\Delta_{uv} - T_P^{-1}(x_i)R_1 \text{ with } R_1 \approx \frac{du^2}{2}\left(\frac{\partial^2 S}{\partial u^2}\right)_{xi} + \frac{dv^2}{2}\left(\frac{\partial^2 S}{\partial v^2}\right)_{xi} + du \cdot dv\left(\frac{\partial^2 S}{\partial u \partial v}\right)_{xi}$$

In the remaining of the text, Δ_{uv} is used as a 2-dimensional vector (the third component of Δ_{uv}, being equal to 0 is neglected).

4.5.2.   Verifying interior directions on a reference face

The creation of a path in a reference face F_j requires to verify that the path remains interior to F_j when advancing in a given direction Δ_{uv}. Given an initial point x_i^{Fj} in the parametric space of F_j, the parametric direction $\Delta_{uv}=(du, dv)=(\rho \cos \beta, \rho \sin \beta)$ is interior only if ε exists so that $\varepsilon > 0$ and $x_i^{Fj} + \varepsilon \square \Delta_{uv}$ is interior to F_j.

Three topological configurations of the initial point x_i^{Fj} must be handled differently:

- If x_i^{Fj} is on an edge, then Δ_{uv} is interior to F_j if $\square\square$ $[\beta_1 ; \beta_1+\pi]$ where β_1 is the angular direction of coedge tangent t_1 at x_i^{Fj} in the parametric space (see fig. 14(a))
- If x_i^{Fj} is on a vertex, then Δ_{uv} is interior to F_j if $\beta\square[\beta_2 ; \beta_1+\pi]$ where β_1 and β_2 are the angular direction of coedges tangent vectors t_1 and t_2 at x_i^{Fj} preceding and succeeding the vertex (see fig. 14(b))
- Otherwise, Δ_{uv} is interior to F_j (see fig. 14 (c))

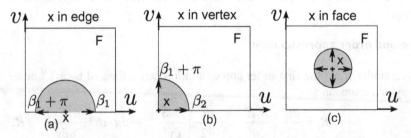

Fig. 14. interior directions of a point x on a face F for various topological configurations of x

Verifying if an Euclidian direction Δ_{3D} is interior to F_j at a point P_i is achieved by transforming Δ_{3D} to its equivalent bi-parametric vector Δ_{uv} at parameter x_i^{Fj} using equation (3) presented in section 4.5.1.

4.5.3. Verifying if a parametric segment crosses outside the face domain

During the construction of the path intersecting face F_j and plane Φ with accuracy δ_{max} (see section 4.5), it is necessary to verify for each offset P_iP_{i+1}, if it crosses outside the domain of F_j (see fig. 13). If so, the *PolySurface* path should be continued on next face F_{j+1}. This is achieved as follows (see fig. 15):

Let P_k be the list of intersection points between plane Φ and edges bounding F_j

For each P_k

Let $P'_k \; \square \; P_iP_{i+1}$ be the closest point of P_iP_{i+1} to P_k

If $\|P_k\text{-}P'_k\| < \delta_{max}$ and $x_{i+1}^{Fj} - x_i^{Fj}$ is not interior to F_j, then P_iP_{i+1} crosses outside F_j

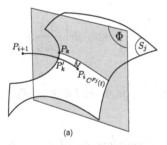 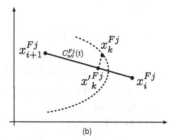

Fig. 15. Segment crossing an edge of the face F_j (a) Euclidian space (b) parametric space of S_j

4.5.4. Creating the path

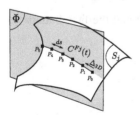 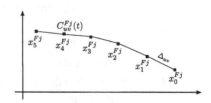

Fig. 16. Construction of the path following the plane / reference face intersection curve

In this section, we describe the algorithm creating the path $C^{Fj}(t)$ on surface S of reference face F_j intersecting the plane Φ while respecting the following constraints:

- Let P_0 be the initial point of $C^{Fj}(t)$,
- Let $A_{3D,0}$ be the initial tangent vector
- The distance between any point of $C^{Fj}(t)$ to the plane Φ must respect the accuracy constraint δ_{max}
- The length of each new segment of the polyline must be limited to ds
- The path computation is stopped when :
  - The segment intersects any edge or vertex of the face
  - Its length reaches the target length L^{Fj}_{max}
  - The segment pass through the target point P_n

i=0 ; stop = false
Do

$$\Delta_{3D} = ds \frac{A_{3D,i}}{\| A_{3D,i} \|}$$

$$\Delta_{uv} = T_P^{-1}(x_i^{Fj}) \cdot \Delta_{3D}$$

$$\Delta_{uv}' = \Delta_{uv} - T_P^{-1}(x_i^{Fj}) \cdot \left(\frac{du^2}{2} \left(\frac{\partial^2 S}{\partial u^2} \right)_{xi} + \frac{dv^2}{2} \left(\frac{\partial^2 S}{\partial v^2} \right)_{xi} + du \cdot dv \left(\frac{\partial^2 S}{\partial u \partial v} \right)_{xi} \right)$$

$x_{i+1}^{Fj} = x_i^{Fj} + \Delta_{uv}'$ and $P_{i+1} = S_j(x_{i+1}^{Fj})$

Let P_{proj} be the normal projection of point P_{i+1} on Φ, and $\delta = P_{proj} - P_{i+1}$ be the error vector,

While $\|\delta\| > \delta_{max}$

$\delta_{uv} = T_P^{-1}(x_{i+1}^{Fj}) \cdot \delta$ (1st order approximation of δ, see § 4.5.1)

$x_{i+1}^{Fj} = x_{i+1}^{Fj} + \delta_{uv}$.

If $x_i^{Fj} \, x_{i+1}^{Fj}$ leaves the domain of F_j, **then:**

$P_{i+1} = P_k$ (intersection between Φ and the face boundary, see § 4.5.3),

$x_{i+1}^{Fj} = x_k^{Fj}$

stop = true

If $x_i^{Fj} \, x_{i+1}^{Fj}$ reaches the target point P_n, **then:**

$P_{i+1} = P_n$

$x_{i+1}^{Fj} = x_n^{Fj}$

stop = true

Add x_{i+1}^{Fj} to the parametric polyline $C_{uv}^{Fj}(t)$

$A_{3D,i+1} = N(S_j, x_i^{Fj}) \square N_\Phi$

If $A_{3D,i+1} \square A_{3D,i} < 0$ **then** $A_{3D,i+1} = - A_{3D,i+1}$
i=i+1;
While $(\|C^{Fj}(t)\| < L^{Fj}{}_{max}$ **and** stop=false)

The number of parametric points constituting the polyline $C_{uv}^{Fj}(t)$ is equal to $L^{Fj}{}_{max}/ds+1$, but only few points are really required to respect the accuracy constraint δ_{max} of the plane/surface intersection. In order to reduce the number of polyline points, an additional procedure is included during path creation in order to keep the minimum number of points that is necessary to respect δ_{max}.

4.6. Image of a segment on a *PolySurface*

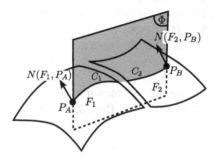

Fig. 17. Construction of the image of segment $P_A P_B$ on a *PolySurface* (F_1, F_2)

Sections 4.3, 4.4, and 4.5 presented optimal candidate node generation on a *PolySurface*. This section presents another algorithm needed to extend advancing front mesh generation to composite geometry: generating a segment on a *PolySurface*.

Constructing the image curve of a segment $P_A P_B$ on a *MC face F* consists in projecting it on the underlying *PolySurface* S_{poly}. The projection is calculated as a plane-surface intersection:

• The normal vector to *PolySurface* S_{Poly} along the segment is averaged by:
$N_{avg} = (N(S_{Poly}, P_A) + N(S_{Poly}, P_B))/2$

- The plane Φ, parallel to N_{avg} and containing points P_A and P_B, is defined by its normal $N_\Phi = N_{avg} \wedge (P_B - P_A)$ and its origin $P_\Phi = P_A$
- Create a path $C(t')$ intersecting S_{poly} with plane Φ with:
  - starting point $P_0 = P_A$ and target point $P_n = P_B$,
  - start direction $A_0 = N(S_{poly}, P_A) \wedge N_\Phi$,
  - $ds = \| P_B - P_A \| / n_{seg}$, where n_{seg} is the number of sub segments of the polyline.
  - accuracy $\delta_{max} = \varepsilon \square \| P_B - P_A \|$

$j=0$,
Find face F_0 adjacent to P_0 such as A_0 is interior to face F_0 at P_0 (see section 4.4 presenting interior directions on a face)
Do
Create the path $C^{Fj}(t)$ by intersecting face F_j with plane Φ, starting from P_j, with start direction $A_{3D,0} = A_j$, and (if P_n lies on F_j) ending at P_n (algorithm section 4.5.4)
Add curve $C^{Fj}(t)$ to $C(t)$
initialize P_{j+1} with the last point of $C^{Fj}(t)$
$A_{j+1} = N(S_{Poly}, P_{j+1}) \wedge N_\Phi$
F_{j+1} = face adjacent to P_{j+1} such as:
$F_{j+1} \neq F_j$,
$F_{j+1} \in S_{Poly}$,
A_{j+1} or $-A_{j+1}$ is interior to F_j
If A_{j+1} is not interior to F_j, **then** $A_{j+1} = -A_{j+1}$
$j = j+1$
While $P_j \neq P_n$

4.7. Calculating intersection points between segments

The creation of a triangle during the advancing front method mainly relies on the intersection test between newly created segments and existing segments. The intersection point of two segment is determined by the intersection of their image curves in the parametric space of reference faces (see fig. 18).

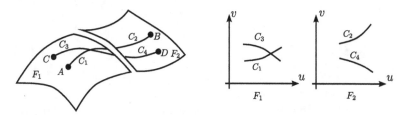

Fig. 18. Segments *AB* and *CD* are intersecting on reference face *F₁*

5. Examples and results

This section illustrates results obtained applying the *MCT* preparation process and the mesh generation algorithm described in this paper. We start with Fig. 19. to illustrate that our contribution overcomes the main limitations of meshing composite surfaces using a single parametric description. The contribution presented in this paper is:

° *independent of surface topology*: closed surfaces homeomorphic to a sphere or to a n-torus can be meshed as well as open surfaces.

° *not limited by stretched geometries and steep metrics variations*: highly stretched surfaces, such as the one representing a glove, result in good quality meshes.

Fig. 20. illustrates the quality distribution of meshes, evaluated by a classical shape criterion (inscribed radius divided by the maximum triangle edge length), and by the geodesic size criterion. The geodesic size criterion (defined in [17]) is related to the difference between the actual area of a triangle and its optimal area with regard to the size map.

Cutter: results presented in Fig. 21 represent an automatically generated *MCT* generated using our adaptation process (detailed in [12, 16]) : the B-Rep entities representing the cutter size label, and other irrelevant edges have been removed. Fig. 20 shows that shape and size quality distributions are both very good (few triangles are significantly smaller than the expected size).

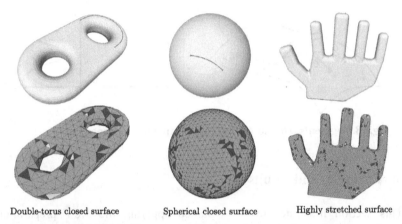

Double-torus closed surface   Spherical closed surface   Highly stretched surface

Fig. 19. Models after topology adaptation and meshes obtained on closed composite surfaces. The remaining *MC edge* (colored in blue) has been used for the advancing front initialization. For these configurations, approaches based on a mapping into a unique parameterization would likely fail.

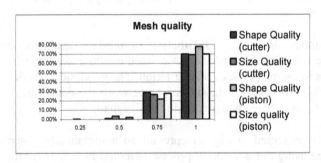

Fig 20. Quality distribution of the Cutter and Piston meshes

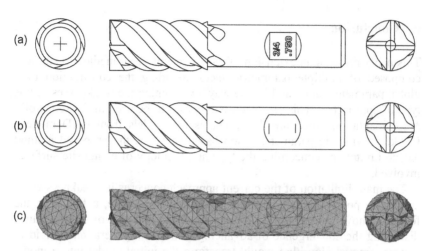

Fig. 21. Mesh generated on a CAD model representing a cutter.

Quarter of piston: fig. 22 shows a piston CAD model, split into one quarter due to symmetry considerations. The original CAD model features many narrow faces and small edges, which are irrelevant for mesh generation. On this sample part, the *MCT* simplification process [12, 16] operated 60 edges deletions, 62 vertex deletions, and collapsed one edge to a vertex. The number of faces has been reduced from 71 to 21, the number of edges from 182 to 76, and the number of vertices from 113 to 63. Again, size and shape quality distributions are quite satisfying as illustrated in fig. 20.

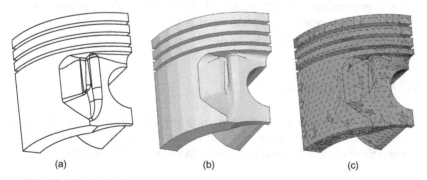

(a)   (b)   (c)

Fig. 22. Mesh obtained on a CAD model representing a quarter of piston

6. Conclusion

This paper presents an extension of the advancing front method to surfaces composed of multiple parametric faces, avoiding the construction of a global parameterization and by the way, overcoming the weaknesses of this type of approaches. This extension is intended to be used in the scope of a *MCT* adaptation procedure that prepares FE models from CAD models. Unlike previous re-parameterization based schemes, the proposed method has no limitations concerning the type and topology of composite surfaces involved.

The main limitation of the current approach concerns its weakness when applied to poorly prepared feature models. For example, a failure in the feature-removal preparation step often cause the presence of small features disturbing the convergence of advancing front mesh generation. The ideal feature-removal algorithm would transform the initial model into a model that conforms to the specified size map. Further work could overcome this weakness by improving the robustness of feature-removal and advancing front mesh generation processes.

Future potential directions in this research include:

- *Extension to elements with higher degree*: quadratic triangles (T6) can be easily generated on the exact geometry by inserting a middle node on each segment's image curve. This method should include a quality criterion to avoid squeezed elements in highly curved zones. The segment-swapping and node moving optimization steps will require quality criteria adapted to curved mesh elements.
- *Extension to mixed-dimensional models*: extending feature-removal and *MCT* preparation algorithms to 3D geometric models mixing curves (meshed with beam elements), surfaces (meshed with shell elements), and solids (meshed with solid elements).

7. Acknowledgements

This study was carried out as part of a project supported by research funding from Québec Nature and Technology Research Fund and by the Natural Sciences and Engineering Research Council of Canada (NSERC).

8. References

1. Halpern, M., *Industrial requirements and practices in finite element meshing: a survey of trends*, in *6th International Meshing Roundtable*. 1997. p. 399--1997.

2. S. Dey, S.S. Mark, and K.G. Marcel, *Elimination of the Adverse Effects of Small Model Features by the Local Modification of Automatically Generated Meshes.* Engineering with Computers, 1995. **13**(3): p. 134--152.

3. Shephard, M.S., M.W. Beall, and R.M.O. Bara, *Revisiting the Elimination of the Adverse Effects of Small Model Features in Automatically Generated Meshes*, in *7th International Meshing Roundtable*. 1998. p. 119-132.

4. Mark, W.B., W. Joe, and S.S. Mark, *Accessing CAD Geometry For Mesh Generation*, in *Proceedings of 12th International Meshing Roundtable, Sandia National Laboratories*. 2003.

5. L. Fine, L. Remondini, and J.C. Leon, *Automated generation of FEA models through idealization operators.* International Journal for Numerical Methods in Engineering, 2000. **49**(1): p. 83--108.

6. S. Venkataraman, M. Sohoni, and R. Rajadhyaksha, *Removal of blends from boundary representation models*, in *Proceedings of the seventh ACM symposium on Solid modeling and applications*. 2002. p. 83--94.

7. S. Venkataraman, M. Sohoni, and G. Elber, *Blend recognition algorithm and applications*, in *Proceedings of the sixth ACM symposium on Solid modeling and applications*. 2001. p. 99--108.

8. S. Venkataraman and M. Sohoni, *Reconstruction of feature volumes and feature suppression*, in *Proceedings of the seventh ACM symposium on Solid modeling and applications*. 2002. p. 60--71.

9. K. Y. Lee, et al., *A small feature suppression/unsuppression system for preparing B-rep models for analysis*, in *SPM '05: Proceedings of the 2005 ACM symposium on Solid and physical modeling*. 2005: New York, NY, USA. p. 113--124.

10. A. Sheffer, *Model simplification for meshing using face clustering.* Computer Aided Design, 2001. **33**(13): p. 925--934.

11. K. Inoue, et al., *Clustering a Large Number of Faces For 2-Dimensional Mesh Generation*, in *8th International Meshing Roundtable*. 1999. p. 281-292.

12. G. Foucault, et al., *A Topological Model for the Representation of Meshing Constraints in the Context of Finite Element Analysis*, in *Proceedings of ASME 2006 International Design Engineering Technical Conferences and Computers and Information in Engineering Conference*. 2006: Philadephia, USA.

13. DL. Marcum, J.G. *Unstructured Surface Grid Generation Using Global Mapping and Physical Space Approximation.* in *8th International Meshing Roundtable*. 1999.

14. F. Noel, *Global parameterization of a topological surface defined as a collection of trimmed bi-parametric patches: Application to automatic*

mesh construction. International Journal for Numerical Methods in Engineering, 2002. **54**(7): p. 965--986.

15. S. H. Lo and T. S. Lau, *Mesh generation over curved surfaces with explicit control on discretization error*. Engineering Computations, 1998. **15**(3): p. 357--373.

16. G Foucault, J-C Cuillière, V François, J-C Léon, R Maranzana,, *Towards CAD models automatic simplification for finite element analysis*, in *2nd World Congress of Design and Modelling of Mechanical Systems*. 2007: Monastir.

17. J-C Cuilliere, *An adaptative method for the automatic triangulation of 3D parametric surfaces*. Computer Aided Design, 1998. **30**: p. 139--149.

18. J-C Cuilliere, *Direct method for the automatic discretization of 3D parametric curves*. Computer Aided Design, 1997. **29**(9): p. 639 - 647.

A MESH MORPHING TECHNIQUE FOR GEOMETRICALLY DISSIMILAR TESSELLATED SURFACES

Radhika Vurputoor[1], **Nilanjan Mukherjee[2], Jean Cabello[3] & Michael Hancock[4]

Meshing & Abstraction Group,
Digital Simulation Solutions
UGS PLM Software
SIEMENS
2000 Eastman Dr, Milford, OH 45150, USA

Email: 1 [radhika.vurputoor@siemens.com]
2 [mukherjee.nilanjan@siemens.com]
3 [jean.cabello@siemens.com]
4 [hancock.michael@siemens.com]

** Author to whom all correspondence should be addressed

Abstract. Many industrial structural mechanics applications require topologically similar meshes on topologically similar surfaces that might differ in shape or size. Rotationally symmetric mesh models requiring cyclic symmetry boundary conditions, contact meshes that connect large structural assemblies and automotive and space panels with minor structural variations are some common challenges. The present paper discusses a generic technique that assembles several simple mapping /morphing tools to map a mesh from a source surface to a topologically similar target surface that might vary in shape or size. A loop/edge based relationship is used to define the dependency. The boundary discretization method adopted maps m nodes on a loop with n edges to a target loop with p edges. A two-dimensional boundary constrained sample mesh is constructed on the related surfaces, which is efficiently used to map the mesh on the source surface to the target using linear iso-parametric mapping techniques. Local distortions of the target mesh are checked for and corrected by a mesh relaxation procedure. Although the algorithm works for both parametric NURBS and discrete surfaces, the present discussion focuses on tessellated surfaces. Results include some key examples of practical interest.

1. INTRODUCTION

In the finite element analysis world, several structural mechanics problems require the user to create topologically similar meshes on topologically similar surfaces for more than one purpose. An elementary sub-structuring approach involves scooping out a radial pie-piece of a large rotationally symmetric structure for various analyses. In order to simulate the global environment, these mesh models require cyclic symmetry boundary conditions to be created on the cut-faces. The mesh on the cut-faces must be topologically similar such that a node-to-node/element-to-element correspondence exists.

Another interesting problem deals with large assembly structures where contact meshes are used to connect faces of different bodies. These faces are usually topologically congruent but might vary in shape or size, thus implying geometric incongruency. Topologically identical meshes need to be generated on these faces such that contact elements can be created between matching nodes and elements.

In the automotive and aerospace industry, similar problems could be found. It is a standard design procedure today to reuse legacy FE models of car-body panels with minor structural variations. This produces surfaces that have a legacy quad mesh on them but have been minutely modified geometrically. The modifications mostly imply repositioning or resizing of existing features. Thus, the modified surface is a topological twin of the original surface although its geometry is now different. If the same mesh needs to fit the modified geometry, it needs to be morphed to reflect the geometric change. The scope of the present paper is to lay down a technique to morph triangular and quadrilateral meshes to surfaces that are topologically similar but geometrically dissimilar.

2. PAST RESEARCH

Many investigations have been reported on 2D and 3D mesh morphing especially in the area of multiresolution mesh morphing [2-5]. Similarly, many investigations exist in the area of moving boundary meshes [6-7]. However, in most of these proposed methods, the required topological proximity between the source and the target is stringent (faces need to have the same number of loops and edges). Furthermore, the need to preserve the topology of the original mesh is not as critical as the present case. The quality of the finite element mesh is also vastly important, as opposed to meshes for graphic use, as these meshes need to solve. Staten et al [1], in their BM Sweep technique, presented an attractive method to locate interior nodes during sweep mesh generation. The method is used in a completely different context – Hexahedral sweep meshing and does not aim at morphing a surface mesh to another geometrically dissimilar surface. Accordingly, it reveals certain bottlenecks when applied to solve the present problems. One limitation of the method is that it morphs nodes from the source face to volume interior layers, which are usually planar. In addition to that, it could not be used to move nodes or morph meshes across periodic surfaces. To solve similar problems in the hex mesh generation area, Shih & Sakura [8] and Lai & Benzley [9] provided solutions to place volume interior nodes. These algorithms suffer from very similar limitations.

3. PROBLEM STATEMENT

The present paper aims at laying down a robust, general-purpose algorithm for morphing 2D meshes across surfaces that are topologically identical but geometrically incongruent. The approach presented works for both NURBS surfaces as well as discrete, tessellated geometry. However, with discrete geometry several analytical disadvantages could be overcome. This paper highlights some of those virtues. The method proposed is general-purpose and particularly addresses three classes of structural problems, namely

1. Cyclic symmetry situations where the source and target faces are part of the same rotationally symmetric body and are congruent. The morphed mesh on the target face must be a pure transformation of the source face and requires stringent positional accuracy. Cyclic symmetry boundary conditions are then created on these meshes. A typical example is shown in Figure 1.

 Cyclic
Sym metry
source face,
the target face is
hidden on the
opposite side

Fig. 1. A sector of a rotor showing faces where cyclic symmetry boundary conditions need to be applie.

2. Mesh assembly situations where contact elements are created between two bodies through certain faces. These faces are usually geometrically incongruent but topologically identical. An example is shown in Fig. 11.

3. Nearly identical sections of automotive body or space panels that contain large multi-featured faces that have some minute geometric dissimilarity. Either certain dimensions of these faces change or some features are modified to a minor degree. These classes of problems also extend to handling legacy FE data while minutely updating old automotive designs to create new ones. Car doors or excavator carriage frames are perfect examples. Figures 2. & 3 exemplify a simplified situation where a hole has been translated on the target face to a different location. t^S and t^T ($t^S \neq t^T$) are the margin width of the holes on the source and the target faces respectively. The loops on the source and the target face are defined in the opposite directions. The morphed mesh on the target face is shown in Fig. 3.

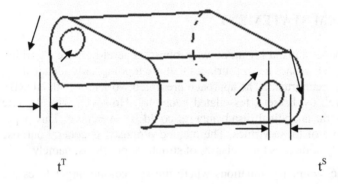

t^T $\qquad\qquad\qquad\qquad\qquad\qquad\qquad\qquad$ t^S

Fig. 2. Topologically similar source and target faces

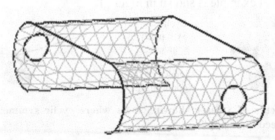

Fig. 3. Triangular mesh on source is morphed where a feature (hole) is displaced on to the target.

4. MORPHING THE FACE BOUNDARIES

Given a source face and a corresponding target face that are topologically congruent i.e. that have the same number of loops, we first define corresponding pairs of loop between the source and the target as well as, for each pair, a start vertex and direction on the source (similarly a corresponding start vertex and direction on the target). The goal is to map with the least distortion the boundary node distribution on the source to a corresponding boundary node distribution on the target.

Let us use the following notations:

E_i : edge bounded by vertices V_i, V_{i+1}.

E_i^S : edge on the source, resp. E_i^T edge on the target.

L_i^S : loop on the source, resp. L_i^T loop on the target.

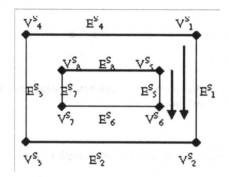

Fig 4a. Source face, loop, edges, vertices

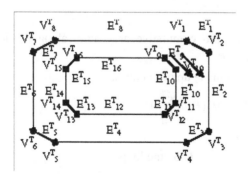

Fig 4b. Target face, loop, edges, vertices

In the case depicted by Fig.4a & Fig 4b., we have :

$$L^S{}_1 = (E_1, E_2, E_3, E_4) \rightarrow L^T{}_1 = (E_1, E_2, E_3, E_4, E_5, E_6, E_7, E_8)$$
$$L^S{}_2 = (E_5, E_6, E_7, E_8) \rightarrow L^T{}_2 = (E_9, E_{10}, E_{11}, E_{12}, E_{13}, E_{14}, E_{15}, E_{16})$$

where the upperscripts have been dropped for the edges. Let us consider the first loop, L_1 for both source and target.
Defining :
AL_i : total arc length for edge E_i
$AL_i(P)$: relative local arc length to a point P along edge E_i .

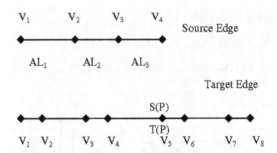

Fig.5 : **Normalized perimeter based parameterization (middle) of the source (top) and target (bottom) loop.**

$$AL_i = AL_i(V_{i+1}) \tag{1}$$

$NAL_i(P)$: normalized relative local arc length of a point P belonging to edge E_i .

$$NAL_i(P) = AL_i(P) / AL_i$$
$$NAL_i(V_i) = 0$$
$$NAL_i(V_{i+1}) = 1$$

$AL(P)$: Normalized cumulative arc length along the loop at point P belonging to edge E_i
Assume the loop is made up of m edges.

$$AL(P) = (\sum_{j=1}^{j=i-1} AL_j + AL_i(P)) / (\sum_{j=1}^{j=m} AL_j)$$
$$\forall P \in E_i \tag{2}$$
$$0 \leq AL(P) \leq 1 \ \forall \ P \text{ on the loop}$$

$$L = \bigcup_{j=1}^{j=m} E_j \tag{3}$$

The affine function AL(P) defines a "normalized perimeter based" parameterization of a loop made up of m edges obtained by accumulation of the arc length while traveling along the loop. Such a parameterization can be defined both for the source and target loop that are in correspondence. The parameterization for the source loop is denoted $S(P)$ and for the target loop $T(P)$ and provides a common reference parameterization to distribute the nodes along their respective loops. This is an extension of the normalized arc length parameterization along an edge to a loop. Using this common parameterization for both source and target loops, the vertices of the target loop can be positioned along the source loop. For example, V_2^T falls on edge E_1^S.

Next, the boundary node discretization of the source edge is performed. Each boundary node N_j has parameter location $S(N_j)$ along the source loop given by :

$$N_j \in E_i \tag{4}$$

$$S(N_j) = AL(N_j) \tag{5}$$

Assuming that the number of edges that make up the target loop is smaller than the number of boundary nodes on the source, the algorithm proceeds by first snapping the target vertices locations $T(V_j)$ to the closest existing boundary node. This operation defines a piecewise decomposition of the source loop. A target edge E_i^T (parameter location $[T(V_i) : T(V_{i+1})]$) is then mapped to a segment $[S(N_j), S(N_k)] \quad k > j$.

$$T(V_i) \cong S(N_j) \quad \text{and} \quad T(V_{i+1}) \cong S(N_k) \tag{6}$$

The k-j-1 interior nodes on the source are then created on the target edge according to their relative local arc length. For example, an interior source node N_{jo} with parameter location $S(N_{j0})$ on the source satisfies the following relations:

$$j < j_0 < k \tag{7}$$

$$0 \le S(N_j) < S(N_{j0}) < S(N_k) \le 1 \tag{8}$$

$$S(N_{j0}) = AL^S(N_{j0}) \tag{9}$$

Its corresponding target node N_{jo} verifies the equations:

$$T(N_{jo}) = S(N_{j0}).$$

(10)

$$T(N_{j0}) = AL^T(N_{j0})$$

(11)

$$0 \le T(V_i) < T(N_{j0}) < T(V_{i+1}) \le 1$$

(12)

The parameter location of source node N_{j0} is computed using equations (9) and (2). Equation (11) provides the parameter location of the corresponding target node location. The inverse of the function defined by equation (2) is used to derive the normalized relative local arc length from the normalized perimeter based arc length i.e.

$$AL_{Loop}^T = \sum_{j=1}^{j=m} AL_j$$

(13)

$$NAL_i^T(N_{j0}) = [S(N_{j0}) * AL_{Loop}^T - \sum_{j=1}^{j=i-1} AL_j]/ AL_i^T$$

(14)

In the special case when the number of edges on the source and the target is equal, i.e. $n = m$ the algorithm is modified so that the step that involves finding the closest boundary node to snap to a target vertex is replaced by mapping target vertices to their corresponding source vertices. A source node belonging to an edge E_i^S is then mapped to a target node belonging to an edge E_i^T using equations (13) and (14) above. Notice than in this special case, equation (10) does not hold when the edges on the source and target have different arc lengths. Piecewise identity, contraction or dilation are uniformly applied to each source edge nodal distribution to get the corresponding target edge nodal distribution.

It is important to state here that the orientation and the start nodes of the source and target loops are critical for obtaining a valid, good quality map.

5. MORPHING THE MESH TO THE TARGET FACE

Once the boundary of the source and target faces are discretized, the next major task is to morph the mesh to the target face. The essential steps involved in the process are –

a. Flatten both the source and target faces into a non-intersecting planar domain using a procedure described by Beatty [11].
b. Generate a boundary constrained 2D triangular mesh on the source face. This mesh is boundary conforming and serves as a background mesh over which the 2D mesh needs to be generated.
c. Create a topologically conforming background template mesh on the target face by mapping the triangle connectivity of the source face.
d. Generate a 2D tria/quad/quad-dominant mesh (as desired by the user) of a given size on the planar domain on the source face.
e. Use the background mesh to map the interior nodes of the source face to the target face similar to the method described by Staten et al [1].
f. Complete the mesh on the target face by constructing the elements by mapping their connectivity from the source face.

The details of the procedure will vary for periodic faces over non-periodic faces.

5.1 Mesh Morphing On Non-periodic faces

a. The source and the target faces dealt with for this case are open tessellated faces that are developable [11]. Two such faces are shown in Fig. 3. Several domain development techniques for discrete surfaces could be used. For NURBS surfaces, the parameter space can also be used for this purpose. We have used the following techniques [10-12] the details of which are outside the scope of the present paper.

b. Using the boundary discretization discussed in section 2 (Fig. 3), a boundary constrained background mesh is now generated on the source face. The TriaQuaMesher algorithm, previously discussed in detail [13], is employed for the purpose in a boundary-constrained mode. The resultant mesh is depicted in Fig.6a.

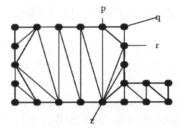

Fig. 6a Boundary-constrained background template mesh on source surface

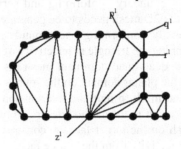

Fig. 6b Boundary constrained background template mesh on target surface

c. The boundary constrained background mesh on the target surface can be generated by mapping the element connectivity information from the source template as shown in Fig. 6b. For example,

$$\Delta(p,q,r) \equiv \Delta_T(p^1, q^1, r^1) \quad \text{and}$$
$$\Delta_S(p,r,z) \equiv \Delta_T(p^1,r^1,z^1) \tag{15}$$

Likewise, the source and target meshes can be correlated as

$$\bigcup_{i=1}^{N} \Delta_{Si}(n_1,n_2,n_3) \equiv \bigcup_{i=1}^{N} \Delta_{Ti}(n_1^1, n_2^1, n_3^1) \tag{16}$$

where N represents the number of triangles in the mesh

Δ_{Si} represents the i-th triangle of the source template mesh whose connectivity is defined by 3 nodes (n_1,n_2,n_3).

Δ_{Ti} represents the i-th triangle of the target template mesh whose connectivity is defined by 3 nodes (n_1^1, n_2^1, n_3^1) which correspond to (n_1,n_2,n_3).

If any triangle $_{Ti}$ gets inverted, the domains and thus the source and target faces are deemed largely dissimilar. This explains the limit of geometric dissimilarity that can be handled by the algorithm.

d. The mesh on the source face is generated by the TriQuaMesher
    (embedded with transfinite meshing) algorithm first proposed decades
    back by Sluiter et al [13] and later modified by Cabello [14]. TQM is a
    recursive domain subdivision algorithm that starts with discretized
    loops defining the mesh area. The split line locations are governed by
    the included angles of the loop segments. The splitting algorithm aims
    at recursively decimating the area into convex regions. Hermite
    polynomials (and alternatively a discrete background sample triangular
    mesh) are used to represent the split lines. A boundary-blended
    transfinite meshing algorithm, discussed by Mukherjee [15], is
    optionally employed in the sub-areas when they become convex and 4-
    sided.

e. In this stage, the mesh on the source face is done and is morphed to the
    target. To accomplish this, each face-interior node of the source face
    and the triangle containing it, is used to map its location on the
    corresponding triangle of the target. The following equation family (17-
    18) explain the mapping procedure [refer Fig. 7(a-b)].

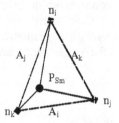

Fig 7a. Source triangle $\triangle_{Sm}$ containing node P_{Sm}

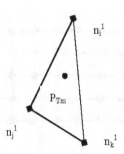

Fig 7b. Corresponding target triangle $\triangle_{Tm}$ containing node P_{Tm}

$$P_{Tm}(u,v) = N_i.n_i^1(u,v) + N_j.n_j^1(u,v) + N_k.n_k^1(u,v) \qquad (17)$$

where

$P_{Sm}(u,v)$ represent the location of the m-th interior node P_{Sm} on the source

$P_{Tm}(u,v)$ represent the location of the m-th interior node P_{Tm} on the target

N_i, N_j, N_k represent the shape functions of source-face background triangle Δ_{Sm} that contain the source node P_{Sm}

n_i^1, n_j^1, n_k^1 are three nodes of the target background triangle Δ_{Tm}^1 (corresponds to Δ_{Sm}) containing the target node P_{Tm} .

The shape functions N_i, N_j, N_k can be expressed as

$$N_i = A_i/A_{Sm}; \quad N_j = A_j/A_{Sm}; \quad N_k = A_k/A_{Sm} \qquad (18)$$

where A_i, A_j, A_k are the sub-areas as shown in Fig 5a.

A_{Sm} represents the area of the m-th source triangle Δ_{Sm}

After all the interior nodes are morphed, the mesh elements are constructed on the target face using the connectivity information of the source face. Fig. 6(a, b) depict a structured quad mesh on the source morphed to a topologically compatible mesh on a geometrically dissimilar target face. The i-th quad element of the target face can be constructed using the relation given below

$$\square_{Ti}(p^1, q^1, r^1, s^1) \equiv \square_{Si}(p, q, r, s) \qquad (19)$$

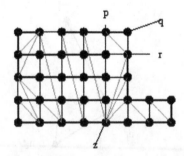

Fig. 8a Final mesh created on the source

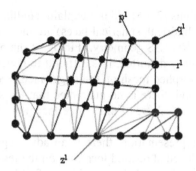

Fig. 8b Final mesh morphed on the target face face over the background mesh

Depending on the geometric dissimilarity of the target face from the source, the mesh could suffer distortion (Fig. 8b). A mesh relaxation or smoothing procedure that does not alter the mesh topology could be used in such situations to reduce mesh distortion. To measure the mesh distortion caused on the target face, an element quality metric, λ, is used, where

λ_i (for the i-th triangle)   = α_{iT}/α_{iS};

where $\alpha = 2\sqrt{3}\ (\overline{BC}\ X\ \overline{AC})/\ (\|AB\|^2 + \|BC\|^2 + \|CA\|^2)$ as proposed by Lo [16]

for a triangle ABC   (20)

λ_i (for the i-th quadrilateral)   =

$(\ |J_{min}|/|J_{max}|\ )_{iT}/\ (\ |J_{min}|/|J_{max}|\ )_{iS}$ (21)

$(\ |J_{min}|/|J_{max}|\ )_{iT}$ denotes the Jacobian determinant ratio for the ith element of the target face

$(\ |J_{min}|/|J_{max}|\ )_{iS}$ denotes the Jacobian determinant ratio for the ith element of the source face

If λ_i falls below a certain critical value λ_{cr} (which may differ for element topology), it indicates that at least one element in the target mesh is severely distorted and the smoother kicks in. A variational smoothing algorithm discussed by Mukherjee [17] that selectively combines numerous iterative solutions based on nodal valency is used for this purpose.

5.2 Mesh Morphing On Periodic Faces

For general structural mechanics applications like contact etc. mesh morphing of periodic faces have to be often dealt with. The BM Sweep [1]

and other past algorithms [2,4,5] did not explain handling of periodic faces. Periodic faces (e.g cylindrical, spherical faces) are more challenging to deal with, especially with NURBS geometry. Firstly, if the parameter spaces of these surfaces are overtly distorted, it is difficult to generate a good quality mesh; let alone the morphed mesh on the target face. Secondly, even if the parameter space is good, there is no guarantee that the seam/pole(s) of the target face will align with those of the source face, thus distorting the mesh on the target.

With a discrete representation, there is an advantage. Arbitrary seams (cuts) could be introduced at desired locations on the tessellated faces (both source and target) as shown below in Figure 9a-9b. The generated 2D domain is thus geometrically more uniform and congruent resulting in good quality meshes on both the source and the target face.

The algorithmic details for a cylinder are given below –

- Cylindrical faces are cut so that the facets can be unzipped to get a flattened domain. Essentially a two-loop cylinder becomes a one-loop cylinder with the cut edge being repeated twice in the loop (edge e3 as shown in Fig. 9a).
- Each cut edge in the loop is represented as a "dummy" edge (edge e4 as shown in Fig. 9a) and an edge dependency relationship is established between these edges.
- For the purpose of mapping the source nodes on the target, duplicate nodes (as shown in Fig. 10a) are introduced in the master mesh along the cut edge. This enables a one-to-one mapping with the source and the target edges.
- At this point, a periodic surface is treated just like a non-periodic face and the mapping of boundary and interior nodes is done as described in 5.1(e).
- At the point of creation of elements on the target face, duplicate nodes are eliminated based on the edge-dependency relation ship between the cut edges of the target loop.
- The details of how the best location for the seam is chosen is discussed by Beatty & Mukherjee [10]. The algorithm for generating a flattened 2D domain from a facetted 3D surface is discussed in detail by Beatty [11] and Hormann [12]. We avoid a repetition of these techniques as they are outside the scope of this paper.

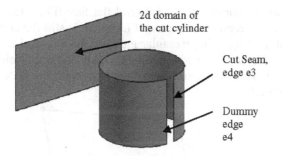

Fig. 9a. A cylindrical surface cut along the seam

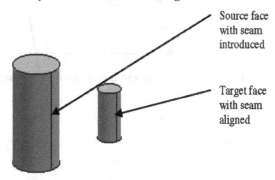

Fig. 9b. Topologically aligned seams introduced with its 2D domain in the background source and target cylindrical surfaces.

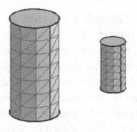

Fig. 10b. Triangular mapped mesh generated on the source is morphed to the target.

8. EXAMPLES AND DISCUSSION

Figs. 11 (a & b) presents a simple case of topological congruency with geometric dissimilarity. A triangular mesh is created using the TQM

algorithm on a four-edged quadrilateral flat face (Fig. 11a). The elliptical
face in Fig. 11b, represents the target face that is still flat and 4-edged, but
is elliptical in shape. The triangular mesh on the elliptical face is morphed
from the mesh on the quadrilateral face. All nodes and elements of the
source face have a one-to-one correspondence with the nodes/elements on
the target face.

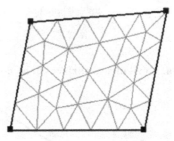

Fig. 11a. Triangular mesh on a four-edged source face

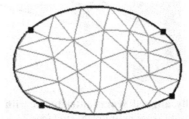

Fig. 11b. Morphed contact mesh on a four-edged target face

A radial sector of an Aluminium impeller as shown in Fig.1, is sub-
structured with cyclic symmetry boundary conditions. Two radially cut
faces, as shown in Fig. 12, represent the cyclic symmetry faces. A boundary
condition needs to be applied to these faces that equate the degree of
freedom of each node on these faces in the Hoop direction (θ-direction). In
order to achieve this, it is necessary to have the same number of
nodes/elements on these faces. The proposed mesh morphing technique is
used to morph the mesh on the source face (any one of the symmetry faces)
to the target. The meshes are shown in Fig. 12. Cyclic symmetry boundary
conditions are applied on the nodes of the source and target faces. Next, this
radial sector of the impeller, is analyzed for centrifugal load (rotating at
3000 RPM). Table 1.0 lists the maximum displacement, maximum Von
Mises and maximum Octahedral shear stress for the radial substructure in
comparison with those obtained from a full model solve. The cyclic
symmetry model has about a fifth of the DOFs of the full model. The
results indicate the accuracy of the sub-structured finite element model.

This kind of sub-strictured analysis would not be possible without a surface mesh morphing functionality. The displacement plots are shown in Fig 13(a-b).

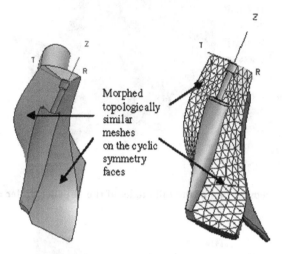

Morphed topologically similar meshes on the cyclic symmetry faces

Fig. 12. Morphed mesh on cyclic symmetry faces. The entire rotationally symmetric model is shown in Fig. 1.

TABLE 1.0 Stress-Deflection Result Comparison

Type of Model	Max. Disp. (in)	Max. Von Mises Stress (psi)	Max. Tresca (Octahedral Shear) Stress (psi)	DOFs
Full-Model	1.014e-5	13.13	6.19	75,238
Cyclic Symmetry Model	1.141e-5	15.28	6.25	14,970

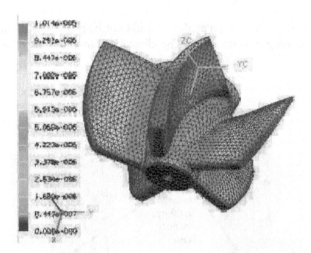

Fig. 13a. Displacement plot of the full model of the impeller under centrifugal load.

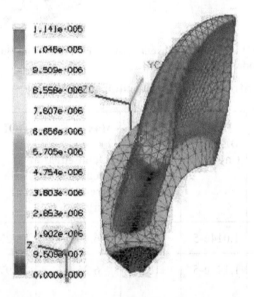

Fig. 13a. Displacement plot of the sub-structured model of the impeller (with cyclic symmetry boundary conditions on the mesh-morphed source and target faces) under centrifugal load.

Fig. 14 shows an interesting contact-mesh problem involving two hollow cylinders. The outer and the inner cylinders share a common contact face. This is represented by a pair of faces - the outer cylindrical face of the inner cylinder and the inner cylindrical face of the outer cylinder. A geometry abstraction operator, that is intrinsic to this algorithm, cuts the cylindrical faces by introducing real seams that align with each other. Any one face of the pair could be used as the source and is first meshed with a transfinite meshing algorithm. This mesh is morphed on the target face using the proposed morphing strategy. The mesh nodes that correspond are used to define a contact mesh.

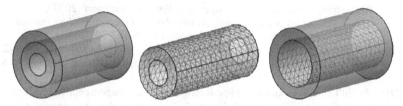

Fig 14. A contact mesh problem showing two concentric hollow cylinders. The cylindrical (seamed) surfaces on the mating bodies define the contact surfaces.

9. CONCLUSION

The present paper focuses on identifying a robust, general-purpose algorithm for morphing 2D meshes across surfaces that are topologically identical but geometrically incongruent. The approach presented is discussed for discrete, tessellated geometry although it will work well for spline geometry too. The paper presents three different categories of industrial problems that could be solved by the proposed algorithm. These are- a) rotational substructures with cyclic symmetry constraints; b) surface contact situations across finite element assemblies and c) legacy finite element models of automotive body panels with minor modification of face-interior features. The proposed method can also be elegantly applied to periodic surfaces. Several examples show the range, robustness and versatile strength of the algorithm. The algorithm is not limited by the geometric difference between topologically identical surfaces. However, the quality of the morphed mesh is limited by the degree of geometric deformation between the source and the target surfaces.

REFERENCES

[1]   M.L. Staten, S. A. Canann and S.J. Owen, "BMSweep: Locating Interior Nodes During Sweeping" *Proceedings, 7th International Meshing Roundtable*, pp.7-18 (1998).

[2] A.W.F. Lee, D.Dobkin, W. Sweldens and P. Schroder, "Multiresolution Mesh Morphing" *Proceedings of SIGGRAPH 99*, pp343-350.(1999).

[3] E. Praun, W. Sweldens and P. Schroder, "Consistent Mesh Parameterizations" *Proceedings of SIGGRAPH 2001.* pp.179-184 (2001).

[4] K. Hormann and G. Greiner "MIPS: An Efficient Global Parameterization Method" *Curve and Surface Design-Saint Malo 1999.* pp.153-162 Vanderbilt University Press (2000).

[5] P. Alliez, P. C. de Verdiere, O. Devillers and M. Isenburg, "Isotropic Surface Remeshing" *Proceedings of Shape Modelling International* (2003).

[6] T. J. Baker, "Mesh movement and metamorphosis". *Proceedings, 10th International Meshing Roundtable*, Sandia National Laboratories pp. 387- 396, (2001).

[7] T.J. Baker, "Interpolation from a cloud of points", *Proceedings, 12th International Meshing Roundtable*, Sandia National Laboratories, pp.55-63, (2003).

[7] B. Shih and H. Sakurai, "Shape Recognition and Shape-Specific Meshing for Generating All Hexahedral Meshes" *Proceedings, 6th International Meshing Roundtable*, Sandia National Laboratories, pp.197-209, (1997).

[9] Mingwu Lai and S.E. Benzley, "Automated Hexahedral Mesh Generation by Virtual Decomposition" *Proceedings, 5th International Meshing Roundtable,*, pp 217-225, (1996).

[10] Kirk Beatty and Nilanjan Mukherjee, "Mesh Generation On Cylindrical Geometric Abstractions" *Internal Report, Digital Simulation Solutions, UGS* (2004).

[11] Kirk Beatty, "Strategies for Generating 2D Domains for Facetted Surfaces" *Internal Report, Digital Simulation Solutions, UGS* (2005).

[12] Kai Hormann, "Theory and Applications of Parameterizing Triangulations" *PhD Thesis,* Department of Computer Science, University of Erlangen (2001).

[13] A.J.G. Schoof, L.H.Th. M. van Buekering and M.L.C. Sluiter, "A General Purpose Two-Dimensional Mesh" *Advances in Engineering Software,* Vol.1(13) pp 131-136, (1979).

[14] Jean Cabello, "Towards Quality Surface Meshing", *Proceedings, 12th International Meshing Roundtable,*, pp 201-213, (2003).

[15] Nilanjan Mukherjee, "High Quality Bi-Linear Transfinite Meshing with Interior Point Constraints", *Proceedings, 15th International Meshing Roundtable*, Ed. P.P.Pebay, Springer, pp 309-323, (2006).

[16] S.H. Lo, "A New Mesh Generation Scheme For Arbitrary Planar Domains", *International Journal For Numerical Methods in Engineering*, No. 21, pp.1403-1426, (1985).

[17] Nilanjan Mukherjee, "A Hybrid, Variational 3D Smoother For Orphaned Shell Meshes", *Proceedings, 11th International Meshing Roundtable*, pp 379-390, (2002).

Mesh Sizing with Additively Weighted Voronoi Diagrams

Lakulish Antani[1], Christophe Delage[2], and Pierre Alliez[2]

[1] IIT Bombay lakulish@cse.iitb.ac.in
[2] INRIA Sophia-Antipolis firstname.lastname@sophia.inria.fr

Summary. We address the problem of generating mesh sizing functions from a set of points with specified sizing values. The sizing functions are shown to be maximal and K-Lipschitz, with arbitrary parameter K provided by the user. These properties allow generating low complexity meshes with adjustable gradation.

After constructing an additively weighted Voronoi diagram, our algorithm provides fast and accurate answers to arbitrary point queries. We have implemented our mesh sizing technique as a sizing criterion for the 2D triangle meshing component from the CGAL library. We demonstrate the performance advantages of our technique through experimental results on various inputs.

1 Introduction

Mesh sizing is one important aspect of the quality mesh generation problem. The size of the elements may be either a consequence of the meshing algorithm when no sizing constraints are provided as input, or may be explicitly controlled by the user or by the physical simulation carried on over the mesh. This paper addresses the latter case, where fine control over the mesh sizing is sought after.

When generating a mesh from an input domain boundary and a set of constraints, it is desirable for the output mesh to satisfy quality, geometry, as well as sizing constraints. Quality herein refers to the shape of the elements. Geometry refers to the preservation (or faithful approximation) of domain boundary and constraints. Sizing refers to the maximum size of the elements defined at each point in the domain to be discretized. The corresponding sizing function may, for example, be derived from the accuracy of the simulation. Obviously, sizing and geometric constraints are related as preserving or approximating the domain boundary and constraints influences the final sizing. Similarly, quality constraints over the shape of the elements may influence the mesh grading, and hence the final mesh sizing.

The current mesh generation algorithms may take as input a sizing function provided, for example, as a so-called background mesh (sizing values are specified at the nodes of this mesh and interpolated). Although specifying values at some nodes of this mesh is certainly easy, in some cases the values are known only on the domain boundary, and/or at very few points. Altering these values so as to reflect global mesh gradation constraints, or interpolating sizing values everywhere in the domain from a sparse set of values is a difficult task. In general, the features that a good sizing function must possess are the following:

- In the approximating case, the domain boundary and constraints are approximated to a high degree of accuracy.
- The mesh grading is adjustable by the user.
- The number of elements in the final mesh is as small as possible.

Alliez et al. [ACSYD05] propose a default *sizing function* $\mu(x)$ which provides the element size at each point x. In this case the approximating requirement listed above is met by basing the sizing function upon the estimated local feature size (LFS, described later) of the domain boundary. However, computing this function directly requires iterating over each point on the boundary, which turns out to be intractable for large point sets.

In this paper we propose a method to compute sizing functions over arbitrary domains using additively weighted Voronoi diagrams, which eliminates the need for iterating over each point where the sizing is specified. Although we illustrate the potential of our technique through the example of LFS-based sizing functions, our technique applies to arbitrary point sets and sizing values. More generally, the main message of this paper is that the additively weighted Voronoi diagram is a powerful data structure for interpolating scattered values. We now briefly review the background used in the narrative.

2 Background

2.1 Basic Definitions

In this section, we review a few basic definitions that we will use throughout the rest of the paper.

Definition 1 *The Local Feature Size of a point x on a shape S (denoted by $LFS(x)$) is the distance from x to the medial axis of S.*

Intuitively, the local feature size captures local thickness, separation and curvature altogether. For example, regions of the shape which are thin or have high curvature are indicated by low values of LFS. To compute the local feature size from a point set P sampled from a shape S, we need to compute a good approximation of the medial axis of S. This can be done if the shape S is "densely sampled". Before discussing how the LFS can be computed, we need a few definitions.

Definition 2 *A point sample P of a shape S is an ϵ-sample if no point x on S is farther than $\epsilon LFS(x)$ from a point in P.*

Definition 3 *The poles of a sample point x on a surface S are the two farthest vertices of the Voronoi cell of x, one on each side of S.*

It can be shown [AB98] that for an ϵ-sampling P of a shape S, the medial axis of S can be approximated by the set of poles of all points in P.

2.2 Sizing Function

It has been shown [ACSYD05] that if $\mu(x)$ is the maximal K-Lipschitz function that is nowhere greater than $LFS(x)$, then $\mu(x)$ possesses the features described in the previous section. This sizing function is defined as:

$$\mu(x) = \min_{y \in \partial \Omega} (Kd(x,y) + LFS(y)), \tag{1}$$

and is shown to be one maximal such K-Lipschitz function. Here, $\partial \Omega$ denotes the boundary point set, and $d(x,y)$ denotes the usual Euclidean distance between x and y. K is a "gradation parameter", which determines, effectively, the relative levels of detail required at or near the boundary of the object, as compared to the details around regions of high curvature and/or low thickness or separation. Previous methods have used a somewhat special case of the above sizing field function, such as [MTT99] and [ORY05], which use a 1-Lipschitz version of the above function (i.e., $K = 1$).

Note that this is only one form of the sizing function. Since it is based on the Euclidean distance to boundary points and local feature size, it is purely geometric in nature. However, we could use other functions in place of the local feature size to capture other properties of the data. Once the values of such a function are known for all points on the boundary, it can be used without any modifications to later stages of the algorithm.

2.3 Additively Weighted Voronoi Diagrams

Consider a point set $P = \{p_1, \ldots, p_n\}$, and a set of *weighted sites* $S = \{s_1, \ldots, s_n\}$ such that $s_i = (p_i, w_i)$ where w_i is the *weight* of the point. The *additively weighted distance* from a point x to a site $s_i = (p_i, w_i)$ is defined as:

$$d_+(s_i, x) = d(p_i, x) - w_i. \tag{2}$$

The additively weighted Voronoi cell of s_i, $V(s_i)$ is defined as:

$$V(s_i) = \{x \in \mathbb{R}^d \mid \forall j \; d_+(s_i, x) \leq d_+(s_j, x)\}. \tag{3}$$

The additively weighted Voronoi diagram $V(S)$ is the cell complex whose d-cells are the additively weighted Voronoi cells $V(s_i)$.

3 Computing the Sizing Function

We can now recast the problem of computing the sizing function $\mu(x)$ for a point x in a point set in terms of some operations on an additively weighted Voronoi diagram. Suppose $P = \{p_1, \ldots, p_n\}$ is the set of sample points using which the mesh is to be reconstructed. Then we consider the set of sites $S = \{(p_i, w_i)\}$ with centers at the sample points and weights given by $w_i = -\frac{LFS(p_i)}{K}$. We see that:

$$\mu(x) = \min_i (Kd_+(s_i, x)). \tag{4}$$

By definition, point x belongs to the Voronoi cell $V(s_k)$ of the site s_k if and only if there is no site s_j such that $d_+(s_j, x) < d_+(s_k, x)$. In other words, the minimum in the above definition of $\mu(x)$ occurs with $i = k$. (Since $K > 0$.) The LOCATE operation of an additively weighted Voronoi diagram does exactly this: It computes the following quantity:

$$s^* = \arg\min_s (Kd_+(s, x)). \tag{5}$$

Therefore, evaluating $\mu(x)$ simply involves locating x in the additively weighted Voronoi diagram, and computing the additively weighted distance to the corresponding site s^*. Thus, a broad outline of the procedure to evaluate $\mu(x)$ given a point set P would be:

1. Compute the Voronoi diagram of P.
2. Extract the poles from the set of Voronoi vertices. Assuming an ϵ-sampling, the poles approximate the medial axis.
3. For each boundary point p_i, compute $LFS(p_i)$ by finding the distance to the closest pole to p_i.
4. Compute the additively weighted Voronoi diagram of S, the set of sites derived from P [BD05].
5. For each query point x, compute $\mu(x)$ by finding the additively weighted Voronoi cell it belongs to and computing the additively weighted distance to the corresponding site.

Further optimizations can be made, based on the insertion algorithm for additively weighted Voronoi diagrams [BD05]. The algorithm spends less time on "hidden" sites as compared to the time spent on other sites. Therefore, we first sort the sites in descending order of their weights. As a result, a site s which would finally (when the algorithm terminates) be hidden by some other site s' would never be processed by the algorithm, since it would already have encountered s' and would simply be marked as hidden.

4 Experiments

We have implemented the above algorithm in C++, using the CGAL library [FGK+00]. Our test machine is a Pentium 4 PC running Linux at 1.8GHz. Our implementation is illustrated with the data sets shown by Fig. 1(a) and 1(b).

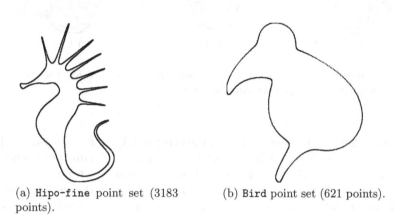

(a) `Hipo-fine` point set (3183 points).

(b) `Bird` point set (621 points).

Fig. 1. Example 2D point sets.

The sizing functions generated by the algorithm for the above two data sets are shown by Fig. 2 and 3. In both sets of figures, red indicates lower values, and blue indicates higher values of the sizing function.

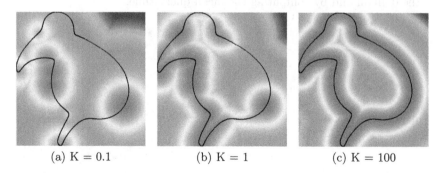

(a) K = 0.1   (b) K = 1   (c) K = 100

Fig. 2. Sizing functions generated by our implementation for the point set `bird`, with three different parameters K.

As illustrated, for $K = 0.1$ (smooth grading), the sizing function takes its lowest values (equivalently, indicates highest detail) in regions which are

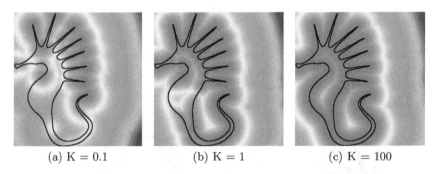

(a) K = 0.1   (b) K = 1   (c) K = 100

Fig. 3. Sizing functions generated by our implementation for the point set hipo-fine, with three different parameters K.

thinner or have higher curvature. On the other hand, for $K = 100$, the sizing function takes its lowest values all around the boundary. This is as expected, as higher values of K give a higher weight to the distance term $d(x, y)$ with respect to the $LFS(y)$ term, and vice-versa. To analyze the performance of our implementation, we measure the following quantities (results are shown by Figs 4 and 5):

- Time required to build the data structures. This includes steps 1 through 4 of the algorithm above, i.e., computing the Voronoi diagram, extracting the poles, computing LFS and constructing the additively weighted Voronoi diagram.
- Average time to query the sizing function at a point. This quantity was measured by querying 100 random points inside a loose bounding box of the domain and by computing the mean query time.

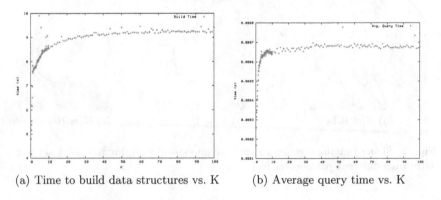

(a) Time to build data structures vs. K   (b) Average query time vs. K

Fig. 4. Performances of our implementation for the point set hipo-fine.

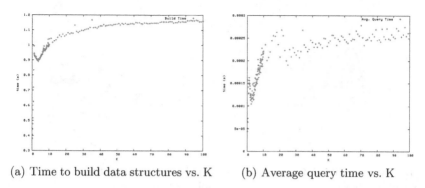

(a) Time to build data structures vs. K   (b) Average query time vs. K

Fig. 5. Performances of our implementation for the point set bird.

The shape of the graphs occurs due to the following reason. While inserting sites into the additively weighted Voronoi diagram, a lower value of K leads to overall lower weights, and consequently a smaller number of sites contained within other sites. Higher values of K increase the number of such sites, and hence more processing time is required to handle this. This same issue results in a slowdown with increasing K while querying the value of the sizing function.

To measure how the performance of our implementation depends upon the input size, we use an ellipse data set. We consider the ellipse defined by the implicit equation $\frac{x^2}{a^2} + \frac{y^2}{b^2} = 1$ with $a = 10$ and $b = 4$. The ellipse was generated with a varying number of points and the performance was measured for each case. Results are depicted by Fig. 6. For any given number of points, the shape is as in the previous cases. The time increases with the number of points as expected.

The performance of the implementation was also tested on a 3D point set sampled on the bimba model, obtained by scanning a physical statue. We used several versions of the point set with increasing numbers of points (from 100 to 55,000) randomly sampled from the original point set. Results are depicted by Fig. 7.

We also compare the performance of our method with the naive approach which examines every boundary point when computing the sizing function. The comparison was performed using the ellipse point set with $K = 0.1$. Results are shown by Fig. 8.

Figures 9, 10 and 12 illustrate our sizing function when provided as input to a 2D mesh generator based upon Delaunay refinement. A plot of the number of vertices in the butterfly mesh versus K is shown in Fig. 11. The number of vertices decreases with increasing K and then stabilizes. This is due to the fact that the amount of details given to areas with smaller LFS values decreases

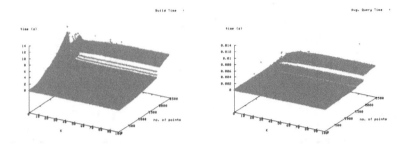

(a) Time to build data structures vs. (b) Average query time vs. No. of
No. of points vs. K   points vs. K

Fig. 6. Performance of our implementation for the point set `ellipse`.

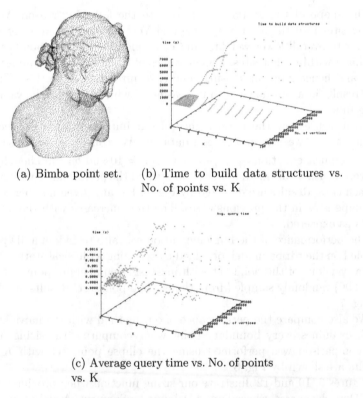

(a) Bimba point set.   (b) Time to build data structures vs.
                         No. of points vs. K

(c) Average query time vs. No. of points
vs. K

Fig. 7. Performance of our implementation for the 3D point set `bimba`.

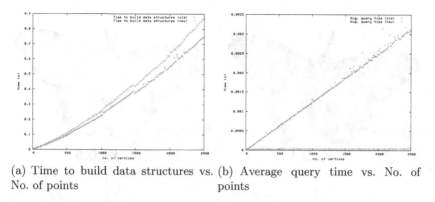

(a) Time to build data structures vs. No. of points

(b) Average query time vs. No. of points

Fig. 8. Comparison between the naive exhaustive method for obtaining the sizing function and ours. (The older method is in red, ours is in green.)

with increasing K, and the sizing function becomes increasingly uniform in the interior of the mesh.

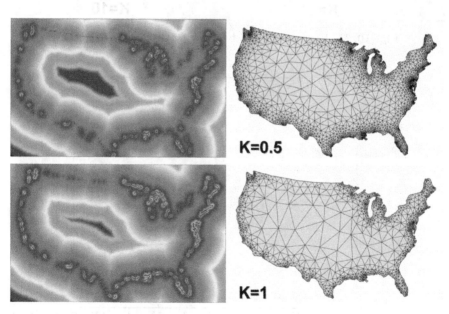

Fig. 9. Mesh generation of the USA map by Delaunay refinement.

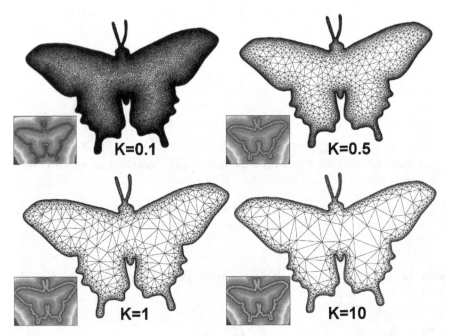

Fig. 10. Mesh generation of a butterfly by Delaunay refinement.

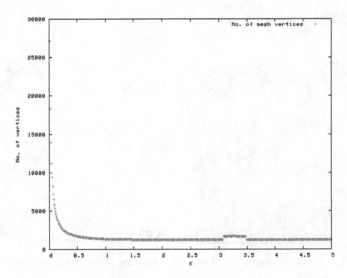

Fig. 11. No. of vertices in the `butterfly` mesh against K.

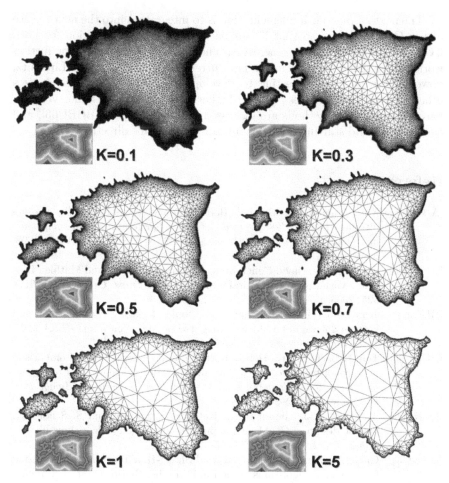

Fig. 12. Mesh generation of the Estonia map by Delaunay refinement.

5 Conclusions and Future Work

In this paper we show how the additively weighted Voronoi diagram can be used as a powerful data structure for speeding up nearest-point queries. By incorporating the additively weighted Voronoi diagram into an algorithm for computing mesh sizing functions, we avoid iterating over every input point at each point query. The resulting implementation is much faster than previous methods in practice, and the sizing field computed by our method is not an approximation, so we do not end up sacrificing accuracy for speed. Although we used LFS as an example point-wise sizing, it could be replaced by some other (application-specific) sizing values.

The next step in our implementation is to integrate it into the next release of the CGAL library, so that it can be used as a sizing criterion for both CGAL's 2D and 3D mesh generators. There is also much scope for further work on the sizing function itself. The current sizing function is K-Lipschitz, however it would be interesting to develop sizing functions which possess other interesting properties, such as higher-order continuity. Some Voronoi-based interpolation methods are discussed and compared in [BHBU06], and can make a good starting point for further work in this direction.

References

[AB98]   Nina Amenta and Marshall Bern. Surface reconstruction by voronoi filtering. In *SCG '98: Proceedings of the fourteenth annual symposium on Computational geometry*, pages 39–48, New York, NY, USA, 1998. ACM Press.

[ACSYD05]   Pierre Alliez, David Cohen-Steiner, Mariette Yvinec, and Mathieu Desbrun. Variational tetrahedral meshing. *ACM Trans. Graph.*, 24(3):617–625, 2005.

[BD05]   Jean-Daniel Boissonnat and Christophe Delage.   Convex hull and voronoi diagram of additively weighted points. *Algorithms - ESA 2005, 13th Annual European Symposium*, pages 367–378, 2005.

[BHBU06]   Tom Bobach, Martin Hering-Bertram, and Georg Umlauf. Comparison of Voronoi based scattered data interpolation schemes. In *Proceedings of International Conference on Visualization, Imaging and Image Processing*, pages 342–349, 2006.

[FGK+00]   A. Fabri, G.-J. Giezeman, L. Kettner, S. Schirra, and S. Schönherr. On the design of CGAL, a computational geometry algorithms library. *Softw.-Pract. Exp.*, 30(11):1167–1202, 2000.

[MTT99]   Gary L. Miller, Dafna Talmor, and Shang-Hua Teng. Data generation for geometric algorithms on non-uniform distributions. *Int. J. Comput. Geometry Appl*, 9(6):577–599, 1999.

[ORY05]   Steve Oudot, Laurent Rineau, and Mariette Yvinec. Meshing volumes bounded by smooth surfaces. In *Proc. 14th International Meshing Roundtable*, pages 203–219, 2005.

3B.4

Overlaying Surface Meshes: Extension and Parallelization

Ankita Jain* and Xiangmin Jiao**

College of Computing, Georgia Institute of Technology, Atlanta, GA, 30332

Summary. Many computational applications involve multiple physical components and require exchanging data across the *interface* between them, often on parallel computers. The interface is typically represented by surface meshes that are non-matching, with differing connectivities and geometry. To transfer data accurately and conservatively, it is important to construct a *common refinement* (or *common tessellation*) of these surface meshes. Previously, Jiao and Heath developed an algorithm for constructing a common refinement by overlaying the surface meshes. The original algorithm was efficient and robust but unfortunately was complex and difficult to implement and parallelize. In this paper, we present a modified algorithm for overlaying surface meshes. Our algorithm employs a higher-level primitive, namely *face-face intersection*, to facilitate easy parallelization of mesh overlay while retaining the robustness of the original algorithm. We also introduce a safeguarded projection primitives to improve the robustness against non-matching features and potential topological inconsistencies. We present numerical examples to demonstrate the robustness and effectiveness of the new method on parallel computers.

Key words: Computational geometry, surface mesh overlay, common refinement, robustness, parallel algorithms

1 Introduction

Many emerging scientific and engineering applications involve interactions of multiple surface meshes. An example is multiphysics simulations, such as simulations of foil flutter of airplane wings in aerodynamics or of blood flows in biomedical applications. These multiphysics systems nowadays constitute one of the greatest challenges in computational science, and they are also very computational demanding and often can be solved only on parallel computers. In these applications, the different

---

* Current affiliation: Cisco Systems, Inc.

** Corresponding author. Current affiliation: Department of Applied Mathematics and Statistics, Stony Brook University. Email: jiao@ams.sunysb.edu

sub-systems have their own computational domains, which are discretized into volume meshes. A sub-system communicates at the boundaries with other sub-systems, and an integrated simulation of the entire system requires accurate and conservative data exchange at these common boundaries or interfaces. An interface between these components has more than one realization, one for each subdomain abutting the interface. These surface meshes are in general non-matching with differing connectivities and geometry. Many computer graphics applications, such as texture mapping and shape morphing, also involve correlating multiple surface meshes. Correlating these non-machines meshes is an important and challenging task.

There are at least two different types of techniques for correlating surface meshes. The first and the simpler type is *mesh association* [6], which maps the vertices of one mesh onto a discrete surface defined by another mesh. The second and the more sophisticated type is *mesh overlay* [8, 9, 10], which computes a finer mesh, called the *common refinement* or *common tessellation*, whose vertices are the "intersections" of the edges of the input meshes and whose faces are the "intersections" of the faces of the input meshes. While mesh association is useful sometimes, such as for pointwise interpolation, the mesh overlay provides a more comprehensive correspondence of the meshes and can enable more accurate data transfers across different meshes in scientific applications [7].

The subject of this paper is to develop a robust algorithm for mesh overlay that can be used in large-scale scientific simulations, especially on parallel computers. In [8], Jiao and Heath proposed an algorithm for overlaying two surface meshes. Albeit the original algorithm was efficient and robust, it was difficult to implement and parallelize. The goal of this paper is to overcome the complexity and the inherently sequential nature of the original algorithm for mesh overlay.

The first contribution of this paper is the simplification of the serial algorithm while maintaining the accuracy and robustness. The new algorithm introduces a high-level primitive, *face-face intersection*, which not only simplifies the implementation but also streamlines the logic for analyzing potential inconsistencies. Secondly, we introduce a new safeguarded primitive for projecting a vertex onto a discrete surface, and also make mesh association a preprocessing step of mesh overlay for improved robustness and flexibility. Thirdly, our new algorithm can be applied to partially overlapping meshes. Even for meshes that have disparate topologies, the new algorithm can build correspondence whenever possible, unlike the original algorithm, which required the input meshes to have the same topology and matching features.

The remainder of the paper is organized as follows. Section 2 defines some basic terminology and reviews the original algorithm for surface mesh overlay and some other related work. Section 3 describes the modified primitives and the modified serial algorithm. Section 4 describes the parallel algorithm. Section 5 contains some preliminary results with our algorithms. Section 6 concludes the paper with a brief discussion on future research directions.

2 Background

2.1 Terminology

We first define some terms that are used throughout this paper. A *surface mesh* is a discretization or tessellation of the geometry of a surface. We refer to the topological objects of a mesh as *cells*. The 0-D cells are called *vertices*, the 1-D cells *edges*, and the 2-D cells *faces*. Each d-dimensional cell σ has a *geometric realization* in $\mathbb{R}^3$, denoted by $|\sigma|$. The realization of an edge is the line segment between its two vertices, and that of a triangle or quadrilateral is a linear or a bilinear patch bounded by their edges. The *realization* of a mesh M, denoted by $|M|$, is the union of the realizations of its faces. We refer the $|M|$ as a *discrete surface*.

A mesh R is a *refinement* of a mesh M if their geometric realizations $|R|$ and $|M|$ are homeomorphic (i.e., there is a mapping between them that is bijective and continuous and whose inverse is also continuous) and every cell of M is partitioned into one or more cells of R. Given two meshes whose realizations are homeomorphic, a *common refinement* (or *common tessellation*) of them is a mesh that is a refinement of both given meshes. For meshes that are not homeomorphic globally, a *common refinement* may be defined for the subsets of the input surfaces that are homeomorphic. We refer to the 0-D, 1-D, and 2-D cells of R as *subvertices*, *subedges*, and *subfaces*, respectively, and refer to them collectively as *subcells*.

An *overlay* of two meshes is special type of common refinement. A subvertex of an overlay is either a vertex of the input meshes or an "intersection" of a blue and a green edge. The subedges of the overlay are the intervals in blue and green edges between the subvertices. A subface corresponds to a 2-D "intersection" of a blue and a green face. Figure 1 shows a blue and a green mesh and their overlay. The subfaces in the overlay are polygons with up to eight edges, which are inconvenient for many applications. A common-refinement mesh may further subdivide these subfaces into triangles. A cell in the blue or green mesh *hosts* one or more subcells of a common refinement. The lowest dimensional blue or green cell that contains a subcell is called the blue or green *parent* of the subcell, respectively. Each subcell has two realizations, one in each of its parents. Through the subcells, the common refinement defines a piecewise one-to-one correspondence between the input discrete surfaces, which can be used in applications such as exchanging data between the surface meshes accurately and conservatively.

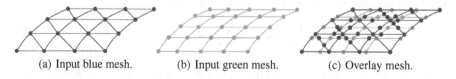

(a) Input blue mesh.   (b) Input green mesh.   (c) Overlay mesh.

Fig. 1. Sample blue (a) and green (b) meshes and their overlay mesh (c).

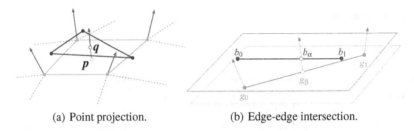

(a) Point projection.	(b) Edge-edge intersection.

Fig. 2. Illustrations of projection and intersection primitives.

2.2 Basic Primitives

The surface mesh overlay has two basic primitives. The first is *point projection*, which maps a point on one discrete surface onto a nearby point on another, and vice versa. Different types of mapping can be used, such as nearest-point projection. In [8], a nearly orthogonal projection was defined through interpolating the normals at green vertices using the shape functions of the green mesh, where the vertex normals can be given as an input or be estimated numerically. Let us first define the projection from $|G|$ to $|B|$ and then use its inverse to define the projection from $|B|$ to $|G|$. For any point $\mathbf{p} \in |G|$, let its projection $\mathbf{q} \in |B|$ be $\mathbf{p} + \gamma\mathbf{d}(\mathbf{p})$, where $\mathbf{d}$ is the unit projection vector at $\mathbf{p}$ obtained from interpolating the green vertex normals (c.f. Fig. 2(a)). Let $\mathbf{g}_i$ denote the ith green vertex and $\mathbf{d}_i$ its associated projection direction. Let ϕ_i denote the shape function corresponding to the ith vertex of the green mesh. For any point $\mathbf{p} = \sum_i \phi_i \mathbf{g}_i$, its projection direction $\mathbf{d}$ is $\sum_i \phi_i \mathbf{d}_i / \| \sum_i \phi_i \mathbf{d}_i \|$. Let $\mathbf{b}_j$ denote the jth vertex of a blue face $b \in B$, and ψ_j the corresponding basis function associated with the vertex. The projection of $\mathbf{p}$ onto b is then the solution of the system of equations

$$\sum_j \psi_j \mathbf{b}_j - \gamma\mathbf{d}(\mathbf{p}) = \mathbf{p}. \tag{1}$$

Because ψ_j is a function of the two local coordinates of b, this system has three equations (one for each component of the physical coordinates) and three unknowns (γ and the two local coordinates in b). This system of equations is linear for triangular meshes and bilinear for quadrilateral meshes. Its solution gives the local coordinates of the projection, which determines whether $\mathbf{q}$ is in the interior, on an edge, at a vertex, or in the exterior, of b. The projection from a point $\mathbf{q} \in |B|$ to a green face $g \in G$ is computed by solving the system of equations

$$\sum_i \phi_i \mathbf{g}_i - \gamma \sum_i \phi_i \mathbf{d}_i = \mathbf{q}, \tag{2}$$

which also has three equations and three unknowns. This system is in general non-linear and can be solved using Newton's method.

The second primitive is *edge-edge intersection*. Given a blue edge b and a green edge g, the primitive computes the "intersection" of them. Again, the intersection has two realizations, one in $|b|$ and one in $|g|$. Let the blue edge be $b = \mathbf{b}_0\mathbf{b}_1$ and the green edge $g = \mathbf{g}_0\mathbf{g}_1$. We parameterize them as $\mathbf{b}_0 + \alpha(\mathbf{b}_1 - \mathbf{b}_0)$ and $\mathbf{g}_0 + \beta(\mathbf{g}_1 - \mathbf{g}_0)$, respectively (c.f. Fig. 2(b)). Let $\mathbf{d}_0$ and $\mathbf{d}_1$ denote the unit normal vectors at $\mathbf{g}_0$ and $\mathbf{g}_1$, respectively. The green realization of the intersection is contained in the plane containing b with normal direction $\mathbf{n} = (\mathbf{b}_1 - \mathbf{b}_0) \times (\mathbf{d}_0 + \beta(\mathbf{d}_1 - \mathbf{d}_0))$. Therefore, the parameter β of the intersection in g can be obtained by solving

$$\mathbf{n}^T(\mathbf{g}_0 + \beta(\mathbf{g}_1 - \mathbf{g}_0) - \mathbf{b}_0) = 0, \tag{3}$$

which is a quadratic equation with the unknown variable β, and we in general choose the positive solution with the smaller magnitude. After solving for β, the parameter α of the intersection in b is the solution to the linear equation

$$\mathbf{m}^T(\mathbf{b}_0 + \alpha(\mathbf{b}_1 - \mathbf{b}_0) - \mathbf{g}_0) = 0, \tag{4}$$

where $\mathbf{m} = (\mathbf{g}_1 - \mathbf{g}_0) \times (\mathbf{d}_0 + \beta(\mathbf{d}_1 - \mathbf{d}_0))$. A solution corresponds to an actual edge intersection if $\alpha \in [0, 1]$ and $\beta \in [0, 1]$ simultaneously. The intersection is in the interior of b if $0 < \alpha < 1$ and at a vertex if $\alpha = 0$ or 1; similarly for g. Because of the nonlinearity of some of the primitives, extra care must be taken due to the inevitable numerical errors. We refer readers to [8, 9] for more detail.

2.3 Original Edge-Based Algorithm

The original mesh-overlay algorithm in [8] proceeds in two phases. The first phase determines the subvertices of the overlay and their parents using the aforementioned primitives and sorts the subvertices on their parent edges. Specifically, it traverses the edges of B in a breadth-first order. For each edge $b \in B$ it locates the subvertices within b using edge-edge intersection and determines the parents of the end-vertices of b using point projection. This operation is repeated until all the blue edges have been processed. To start the algorithm, it is necessary to know the green parent of a *seed* blue vertex within each connected component of B, and the parent of the seed can be obtained by brute-force. Thereafter, the subvertices are sorted in their green parent edges, and the green vertices are projected onto blue faces. During the first phase, robustness is achieved through thresholding and a set of consistency rules [9].

The second phase of the algorithm uses the parent information of the subvertices to create the subfaces of the overlay using divide-and-conquer. In particular, it first creates a list of the subvertices in the edges of a blue face in counterclockwise order. It then identifies a sequence of green subedges to subdivide the face into two parts. The subdivision procedure is then applied recursively to each part until no part is further divisible.

2.4 Other Related Work

There are various algorithms to construct mesh association and mesh overlay. For mesh association, a few methods were reviewed in [6], including a locality-based search called *vine search*. The key primitive in mesh association is point projection, which commonly uses a nearest-point projection or orthogonal projection. Recently, van Brummen proposed a smoothed normal field for the projection to improve accuracy [17]. A related work in computer graphics is the construction of homeomorphisms. In particular, Chazal et al. [2] defined a mapping (Orthomap) between two (n-1)-dimensional manifolds S and S' in $\mathbb{R}^n$, which associates a point $\mathbf{x} \in S$ with the closest point on S' lying on the line passing through the normal to S at $\mathbf{x}$. This mapping is homeomorphic if the minimum feature sizes of S and S' both exceed $h/(2 - \sqrt{2})$, where h is the Hausdorff distance between S and S'. More recently, Chazal et al. [3] proposed a ball mapping that relaxed the requirements of the minimum feature sizes. However, these mappings are presently limited to smooth surfaces and are not directly applicable to discrete surfaces.

For mesh overlay, some algorithms have also been developed in the computer graphics community. Some of them use a virtual reference surface, such as a plane or sphere, onto which the input meshes are first projected. Such methods have sparked extensive research on spherical and planar parameterizations of surfaces. An algorithm to compute the overlay of planar convex subdivisions was proposed by Guibas and Seidel in [5]. Alexa [1] proposed a method to compute overlay on a unit sphere, which is limited to overlaying genus-0 polyhedral meshes. More recently, Schreiner et al. proposed a method to construct a common refinement (which they referred to as a *meta-mesh*) of two triangle meshes [15]. The method partitions the two surfaces into a corresponding set of triangular patches by tracing a set of corresponding paths and then creates progressive mesh representations of the input meshes. A number of researchers have also developed other types of inter-surface correspondences. For example, the zippering algorithm of Turk and Levoy [16] establishes a correspondence between nearby overlapping surface patches to "zipper" them into one mesh. In [11], Kraevoy and Sheffer developed an inter-mesh mapping by constructing a common base mesh, which in general is not a common refinement. These methods are advantageous for their specific application domains but are not well suited for scientific applications that require accurate correspondence for physically meaningful and numerically accurate data transfers.

In the scientific community, parallelization of mesh correlation have drawn some attention for large-scale simulations, but they have thus far mostly focused on mesh association instead of mesh overlay. In particular, Farhat et al. proposed a parallel algorithm for mesh association in [12]. Plimpton et al. proposed a so-called rendezvous algorithm for mesh association [14], which constructs a separate partitioning of the input meshes for better load balancing. The MpCCI (www.mpcci.com) is a commercial software package that supports parallel mesh association and data interpolation. Different from these previous efforts, the focus of this paper is on mesh overlay for accurate and conservative data transfer and their effective parallelization.

3 Simplified Serial Algorithms

The original algorithm in [8] was efficient and robust, but it is difficult to implement and parallelize. Part of the complexity is due to its edge-based traversal of the meshes. In this section, we propose a modified algorithm to traverse the blue mesh face-by-face, which simplifies the implementation, streamlines the logic, and will also ease its parallelization. We also propose some enhancement to the primitives to improve robustness.

3.1 Modified Primitives

Safeguarded Point Projection

Our first modification is to introduce a safeguard for point projection to improve robustness. The point-projection primitive constructs a nearly orthogonal projection by interpolating the green vertex normals. If the normals change rapidly, such as near features or large curvatures, this primitive may project a vertex onto more than one face and lead to ambiguity.

To address this issue, we define a *volume of influence* for each green face g, denoted by $V(g)$, so that the blue vertex b projects onto g only if b falls within $V(g)$. We define $V(g)$ in a way consistent with the point projection primitive. Specifically, the interpolated projection directions along an edge form a bilinear surface, which subdivide $\mathbb{R}^3$ into two half-spaces, and $V(g)$ contains all the points that lie in the same half-space as the face with respect to the bilinear surfaces of all the edges of g. If the edges of the face are in counter-clockwise order, then a point in $V(g)$ must lie to the left of bilinear surfaces for all the edges of g.

Whether a particular point is to the left of the bilinear surface can be determined by the sign of the vector from the point to its orthogonal projection onto the bilinear surface. Given a point $\mathbf{p}$, let its projection be $\mathbf{q} = \mathbf{p} + \gamma\mathbf{n}$, where $\mathbf{n}$ is normal to the bilinear surface (c.f. Fig. 3(a)). We parameter $\mathbf{q}$ by α and β, i.e.,

$$\mathbf{q} = \alpha\mathbf{g}_1 + (1 - \alpha)\mathbf{g}_0 + \beta(\alpha\mathbf{d}_1 + (1 - \alpha)\mathbf{d}_0), \tag{5}$$

where α corresponds to its parameterization in the green edge $\mathbf{g}_0\mathbf{g}_1$ and β is the parameterization along the green normal directions. Therefore, $\mathbf{t}_1 = (\mathbf{g}_1 + \beta\mathbf{d}_1) - (\mathbf{g}_0 + \beta\mathbf{d}_0)$ and $\mathbf{t}_2 = \alpha\mathbf{d}_1 + (1 - \alpha)\mathbf{d}_0$ correspond to two tangent vectors of the bilinear surface at $\mathbf{q}$, and the normal to the bilinear surface at $\mathbf{q}$ is

$$\mathbf{n} = \mathbf{t}_1 \times \mathbf{t}_2 = (\mathbf{g}_1 - \mathbf{g}_0 + \beta(\mathbf{d}_1 - \mathbf{d}_0)) \times (\alpha\mathbf{d}_1 + (1 - \alpha)\mathbf{d}_0). \tag{6}$$

Plugging (5) and (6) into $\mathbf{q} = \mathbf{p} + \gamma\mathbf{n}$, we obtain a system of three equations for three unknowns. A negative γ would indicate that $\mathbf{q}$ is to the left of the green edge. To avoid degeneracy, we scale $\mathbf{d}_0$ and $\mathbf{d}_1$ by a factor c so that

$$(\mathbf{g}_1 - \mathbf{g}_0 + c(\mathbf{d}_1 - \mathbf{d}_0))^T (\mathbf{g}_1 - \mathbf{g}_0) > 0. \tag{7}$$

This scaling does not change the sign of γ. In addition, the resulting β must also satisfy

$$(\mathbf{g}_1 - \mathbf{g}_0 + \beta c(\mathbf{d}_1 - \mathbf{d}_0))^T (\mathbf{g}_1 - \mathbf{g}_0) > 0. \tag{8}$$

If this condition is violated, then the point lies outside the volume of influence of g. If a blue vertex falls into the volumes of influence of multiple green faces, then

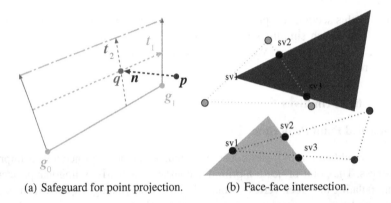

(a) Safeguard for point projection.   (b) Face-face intersection.

Fig. 3. Illustrations of safeguarded point projection and face-face intersection.

we project it onto the closer face. If a blue vertex falls outside of the volumes of influence of all the green faces, then the surfaces mismatch locally and we will not project the vertex onto the green mesh.

Face-Face Intersection

To facilitate face-based traversal, we define a new primitive *face-face intersection*, which identifies the subvertices hosted by a pair of green and blue faces. Specifically, this primitive computes the intersections of the blue and green edges of the faces and determines the projections of the blue vertices onto the green face and vice versa. Face-face intersection therefore is higher level than the point-projection and edge-edge-intersection primitives. Naively, given a blue face b and a green face g, one can independently compute the pairwise intersections of the edges of b and g, compute the projections of the vertices of b onto g, and vice versa. However, such an approach may potentially not only introduce redundant computations but also lead to inconsistencies due to numerical errors.

To avoid these potential redundant computations and inconsistencies, we compute the intersection in two passes (c.f. Fig. 3(b)). First, we determine the projections of the vertices of b onto $|g|$ and the projections of the vertices of g onto $|b|$. In general, this is done using the safeguarded point-projection primitive outlined earlier. If the projection has already been determined previously, then this computation is skipped for efficiency. Thereafter, we compute the intersection of a pair of blue and green edge by checking the barycentric coordinates of the vertices and compute the intersections only if the two blue vertices fall onto different sides of the green edge and the two green vertices fall onto different sides of the blue edge. In addition, we skip the computation of the intersection of a pair of edges if it has already been computed previously. For robustness, we also compare the normals of the blue and green faces and reject the intersection if their face normals point opposite to each other (i.e., their dot product is negative).

3.2 Face-Based Serial Algorithm

We now present our new face-based algorithm for overlaying surface meshes using the modified primitives. The algorithm is composed of three steps, as we explain in the following subsections. During the process, we pay special attention to the robustness in the presence of numerical errors.

Projecting blue vertices onto green surface

The projection and intersection primitives of mesh overlay in general assume the surface meshes are close to each other. This assumption is valid for most physical applications, but it may be violated sometimes near large curvatures. For best generality and robustness, we first project the blue vertices onto the green discrete surface. We use the safeguarded point projection to project the blue vertices onto $|G|$. For each blue vertex b, we identify a unique point b' on $|G|$ and store the green face that b projects onto and the local coordinates of b' in that green face. The projected points along with the connectivity of B compose a mesh B' and a discrete surface $|B'|$. If a blue vertex projects outside $|G|$, we omit the blue vertex in later steps so that partially overlapping meshes can be overlaid.

Essentially, this projection step is *mesh association*, which we can solve using a number of different techniques, such as *vine search* [6]. For ease of implementation, our present implementation uses a kd-tree [4]. The kd-tree is a multidimensional search tree for points in k-dimensional space, where $k = 3$ in our case. We insert the centroids of the green faces into a kd-tree. We then iterate through the blue vertices. For each blue vertex b we extract from the kd-tree the green faces that are nearby and project b onto them. For better efficiency, an optimized implementation can take advantage of locality similar to vine search. In particular, starting from a blue face with a known intersection with a green face, one can visit its incident blue face c and attempt to project the blue vertices of c onto nearby green faces. This procedure constitutes a face-based version of the vine search.

Separating out the projection step from the remainder of the overlay algorithm has a number of advantages. First, since this step is mesh association, which may suffice for some simpler applications such as point-wise interpolation, the later steps of the algorithm may be skipped for those applications. More importantly, this modification allows enhancing the robustness of later steps in a number of ways. During the process, we can perform additional pre-processing steps, such as aligning the features, smoothing the normal directions of the green vertices, or optimizing the projections of the blue vertices. Furthermore, for applications with moving meshes, the projections of the blue vertices can be updated efficiently from the previous projections. Finally, after the projection step, the discrete surface $|B'|$ will in general be closer to $|G|$. We therefore will use the mesh B' instead of B in the later steps, which will substantially reduces the chances of inconsistencies.

Locating subvertices

This step identifies the subvertices and their blue and green parents. A subvertex is either a vertex in one of the meshes or is the intersection of a blue edge and a green edge. We identify the subvertices by traversing the blue faces and for each blue face apply the face-face intersection primitive to the nearby green faces. To identify the nearby green faces of a blue face b (contained in $|B'|$), we first make a list of the green faces that are projected onto by the vertices of b. For each green face g that intersects with b, we add a neighbor face of g, say h, into the list if $h \cap g$ intersects with b. When applying the face-face intersection primitive, we minimize redundant computations as much as possible as discussed previously. In particular, we never compute the intersection of a blue and a green edge more than once.

The non-linear primitives for mesh overlay can be solved only approximately. This may lead to potential inconsistencies of the subvertices. Such inconsistencies must be identified and resolved properly for a valid overlay of the input meshes. As noted in [8], there are three common types of inconsistencies that may arise. The inconsistencies and their remedies are shown in Fig. 4. In the first two cases, numerical computations return two intersections between the blue edge b and green edges, with the difference that in Case (a) one of the intersection is close to (or at) the origin of b, say b_0, whereas in Case (b) both the intersections are far away from b_0. Case (a) occurs only when b_0 is too close to one of the green edges. We perturb b_0 onto its nearby green edge (by reassigning the green parent of b_0). Case (b) occurs only when b is too close to the common vertex of the green edges that intersect b. We resolve this inconsistency by perturbing the intersection onto the green vertex. In Case (c), numerical errors cause b to fall into an artificial gap at a given green vertex between the green edges so that no edge intersection is found. Case (c) happens only when b is too close to the green vertex, and we also perturb the intersection to the green vertex in this case.

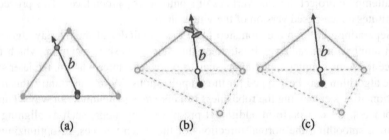

(a)   (b)   (c)

Fig. 4. Different types of inconsistencies and their remedies.

Creating and triangulating subfaces

We enumerate the subfaces for the blue mesh and green mesh simultaneously. This step is broken down into three sub-steps. First, we sort the subvertices on their host edges in both input meshes. Sorting is done based on the distance of a subvertex from the starting vertex of the face in counter-clockwise direction, with the subvertices lying farther away from the starting vertex listed earlier in the list. During the second sub-step, we create polygonal subfaces. Unlike the original algorithm, we directly construct the subface for each pair of intersecting faces from their shared subvertices and subedges. Starting from a known subvertex, we traverse the subvertices hosted by the blue and green faces in counter-clockwise order, as illustrated in Fig. 5. The sorted lists of the subvertices within their parent edges are used to identify the ordering. The subfaces obtained from face-face intersections are polygons with up to eight edges, and they are inconvenient to use for most applications. In the third substep, we further triangulate each subface. We use the ear-removal algorithm [13] to triangulate the subfaces formed above.

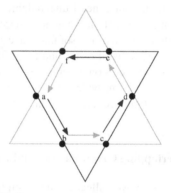

Fig. 5. Illustration of construction of subface.

3.3 Analysis and Comparison

Let N_b and N_g be the numbers of faces in the blue and green mesh, respectively. The first step traverses the nodal graph of B to locate the subvertices in its incident edges. With the simple implementation using kd-trees, the cost of building the kd-tree is $O(N_g log N_g)$ and the cost of each query is $O(N_g^{2/3} + s)$, where s is the maximum number of returned points. The later steps intersect each blue face b only with the green faces that are projected onto by b, so they take linear time in input plus output size. Therefore, the total cost of using the kd-tree is $O(N_b(N_g^{2/3} + s))$. However, when an optimized locality-based search is used during point projection, the total computational cost can be reduced to linear complexity in input plus output size, which is expected to be linear in the input size for well-shaped input meshes.

A main difference of the original and modified algorithms is that they use edge-based and face-based traversals, respectively. Face traversal allows replacing the lower-level primitive by the higher-level primitive, face-face intersection, which simplified the data structures and also simplified the resolution of inconsistencies. When using this new primitive, the inconsistencies could be resolved using the knowledge of the location of the subvertices relative to their blue and green parents. Introducing a higher-level primitive, as explained before, may potentially introduce some redundant computations, which we minimized through additional bookkeeping. The modified algorithm can also be more easily applied to partially overlapping meshes, which allows parallelization of the algorithm.

4 Parallel Algorithm

The main motivation behind modifying the original algorithm was to make mesh overlay easily parallelizable. The key challenges in the parallelization come from the fact that the input meshes are often partitioned independently of each other, so inter-processor communication is in general unavoidable. In particular, the resolution of inconsistencies along the partition boundaries requires coordinations across different processors. In addition, each processor may have a different numbering system for the elements and vertices that it owns, and the different numbering systems on different processors introduce additional implementation complexity. We present a relatively simple parallel algorithm, which follows the same control flow as the serial algorithm, with one additional communication step at the beginning and one at the end.

4.1 Communicating Overlapping Green Patches and Blue Ghost Layers

During this communication step, we collect all the green faces that may potentially intersect with the blue faces that are local to a processor. A naive approach is for each processor to broadcast its green partition to all the processors. Such an approach leads to too much communication overhead as well as memory requirement. Instead, we let each process compute the bounding box of each connected component of its blue partition and send the bounding boxes to all the other processors. Thereafter, each processor p_i sends to processor p_j the green faces that intersect with the bounding boxes of the blue partition on p_j. Note that each vertex along the partition boundary of the green mesh may be shared by multiple processors. Each instance of a shared vertex may be identified by the two-tuple of its partition ID and its local vertex ID within the partition. We assign a unique global ID to each shared vertex, taken to be the two-tuple ID of the instance with the smallest partition ID. When sending the green faces, we also send the global IDs of the shared vertices along the green partition boundaries to facilitate the identification of shared vertices.

Besides sending the green faces, each processor also collects the blue faces from the other processors that are adjacent to the faces of the blue mesh. This builds up a blue mesh with a layer of "ghost elements," which will substantially simplify the resolution of inconsistencies as explained shortly. We implement the above communications using non-blocking sends and receives.

4.2 Applying Face-Based Serial Algorithm

After receiving the green faces from all the processors, we then apply the serial algorithm to compute the intersections. One potential strategy to compute the subvertices is to use the serial algorithm for every set of green faces received from the other processors, one at a time. The advantage of this method is that a processor can then overlap computation with communication. However, such an approach makes it very difficult to identify and resolve inconsistencies along partition boundaries and hence defeats the goal of the new algorithm, namely, to simplify the implementation.

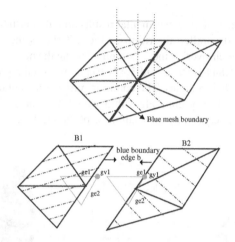

Fig. 6. Inconsistencies along blue partition boundary.

To overcome this disadvantage, we gather all the green faces from different processors first. We number the received vertices and faces using the local numbering scheme of the sender process, and compute the subvertices and the subfaces using the serial algorithm. For the green faces from the same sender, we resolve the inconsistencies in the same way as in the serial algorithm. The resolution of the inconsistencies along partition boundaries deserve special attention. If a blue face produces different intersections with the green faces on two different processors, as shown in Fig. 6, then some significant amount of communication would be required to exchange data back and forth to resolve them. In our algorithm, if two processes share a blue vertex or edge, then they would contain all the incident blue faces and the intersecting green faces. In the implementation, we further ensure that the primitives are invoked consistently on different processors to avoid inconsistent numerical results due to round-off errors. Therefore, the algorithm would produce identical results along partition boundaries on different processors, eliminating the needs of additional communication for resolving inconsistencies across different processors.

4.3 Redistributing Subfaces

After all the subvertices have been computed and the subfaces are created, a processor sends all the subvertices and subfaces to the senders of their green parent faces during the pre-communication step. Therefore, at the end of the algorithm every processor has a list of the subvertices and subfaces lying on its blue and green mesh partitions.

5 Experimental Results

We present some preliminary experimental results and demonstrate the performance of the new serial and parallel algorithms. Figure 7 shows the two meshes of the Stanford bunny and their overlay mesh computed using the new serial algorithm. The blue mesh contains 1,844 vertices and 3,471 triangles, and the green mesh contains 3,546 vertices and 6,944 triangles. The common refinement contains 20,853 vertices and 40,945 triangles. The original algorithm would have failed for this test due to non-matching features caused by the roughness of the meshes.

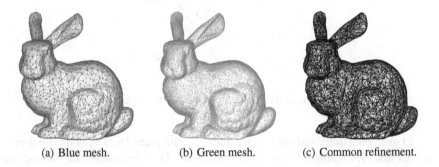

(a) Blue mesh.   (b) Green mesh.   (c) Common refinement.

Fig. 7. Blue and green meshes of Stanford bunny and their common refinement

To demonstrate our parallel algorithm, we used a relative simple geometry of two cylinders connected by conic patches. The blue mesh contains 5,903 vertices and 11,806 triangles, and the green mesh contains 22,730 vertices and 45,460 triangles. The two meshes are shown in Fig. 8. Figures 9 and 10 show the different partitions of the common refinement, where the colored edges indicate the input meshes and black edges correspond to the common refinement. Figure 11 shows a zoomed-in view of a blue partition. We conducted the experiments on the IA-64 Linux cluster at San Deigo Super Computing Center (SDSC), which has 262 nodes with dual 1.5 GHz Intel Itanium 2 processors. Figure 12 shows the speed-ups of our algorithm up to 16 processors. The achieved speed-ups were nearly linear. On a single processor, the current unoptimized implementation using kd-trees took about 24 seconds, which

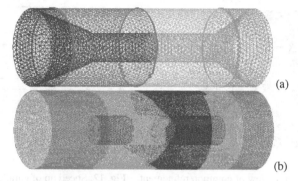

(a)

(b)

Fig. 8. Blue (a) and green (b) meshes for testing of parallel mesh overlay. Each mesh is decomposed into four partitions, indicated by the different colors.

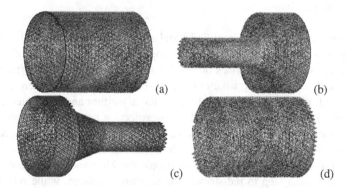

(a)

(b)

(c)

(d)

Fig. 9. Partitions of blue meshes and the common refinement.

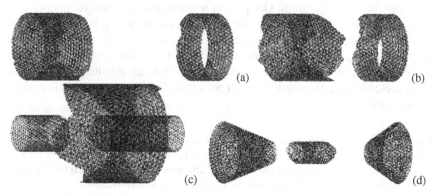

(a)

(b)

(c)

(d)

Fig. 10. Partitions of green meshes and their common refinement.

is relatively slow compared to the original algorithm. A primary cause of this slow-down of the serial algorithm is due to the use of kd-trees during the point-projection step. We are presently working on optimizing this step for better performance.

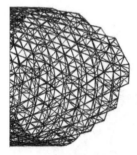
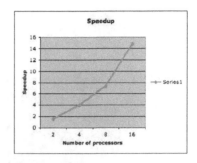

Fig. 11. Zoomed in view of common refinement.   **Fig. 12.** Speed up of parallel algorithm.

6 Conclusion

In this paper, we presented a simplified serial algorithm and a parallel algorithm for overlaying two surface meshes, which has applications in many areas of computational science, such multi-physics or multi-component simulations. Different from the previous algorithm in [8], which used edge-based traversals, the new algorithms are face-based, which simplified the logic of the algorithm and its implementation. We also introduced a new safeguarded point-projection primitive for improved robustness of mesh overlay, and also defined a higher-level primitive, namely, face-face intersection. Our parallel algorithm uses the serial algorithm for the kernel, requiring only a pre-communication step to distribute the blue and green meshes and a post-communication step to redistribute the resulting subfaces, while retaining the robustness of the serial algorithm. We presented some preliminary experimental results with the new algorithms. Our present implementation is not yet optimized in terms of performance, but the experimental results are promising in terms of robustness and speed-ups. Our future research directions will focus on further improving the projection step of the algorithm, i.e., to project the blue vertices on the green discrete surface, in terms of more accurate treatment of singularities, smoothing of the projected blue vertices and the green projection directions, as well as improving its efficiency.

Acknowledgement. This work was supported by a subcontract from the Center for Simulation of Advanced Rockets of the University of Illinois at Urbana-Champaign funded by the U.S. Department of Energy through the University of California under subcontract B523819 and in part by the National Science Foundation grants ATM-0326431 and 0325046.

References

1. M. Alexa. Merging polyhedral shapes with scattered features. *Vis. Comput.*, 16(1):26–37, 2000.

2. F. Chazal, A. Lieutier, and J. Rossignac.   Orthomap: Homeomorphism-guaranteeing normal-projection map between surfaces. In *ACM Symposium on Solid and Physical Modeling (SPM)*, pages 9–14, 2005.
3. F. Chazal, A. Lieutier, J. Rossignac, and B. Whited. Ball-map: Homeomorphism between compatible surfaces. Technical Report GIT-GVU-06-05, Georgia Institute of Technology, 2006.
4. M. de Berg, M. van Kreveld, M. Overmars, and O. Schwarzkopf. *Computational Geometry – Algorithms and Applications*. Springer, second edition, 2000.
5. L. Guibas and R. Seidel. Computing convolutions by reciprocal search. In *SCG '86: Proceedings of the Second Annual Symposium on Computational geometry*, pages 90–99. ACM Press, 1986.
6. X. Jiao, H. Edelsbrunner, and M. T. Heath. Mesh association: formulation and algorithms. In *Proceedings of 8th International Meshing Roundtable*, pages 75–82, 1999.
7. X. Jiao and M. Heath.   Common-refinement-based data transfer between nonmatching meshes in multiphysics simulations. *Int. J. Numer. Methods Engr.*, 61(14):2402–2427, 2004.
8. X. Jiao and M. Heath. Overlaying surface meshes, part i: algorithms. *Int. J. Comput. Geom. Appl.*, 14(6):379–402, 2004.
9. X. Jiao and M. Heath. Overlaying surface meshes, part ii: topology preservation and feature matching. *Int. J. Comput. Geom. Appl.*, 14(6):379–402, 2004.
10. X. Jiao and M. T. Heath. Efficient and robust algorithms for overlaying surface meshes. In *Proceedings of 10th International Meshing Roundtable*, 2001.
11. V. Kraevoy and A. Sheffer. Cross-parameterization and compatible remeshing of 3d models. *ACM Trans. Graph.*, 23(3):861–869, 2004.
12. N. Maman and C. Farhat. Matching fluid and structure meshes for aeroelastic computations: a parallel approach. *Comput. Struct.*, 54(4):779–785, 1995.
13. J. O'Rourke. *Computational Geometry in C*. Cambridge University Press, second edition, 1998.
14. S. J. Plimpton, B. Hendrickson, and J. R. Stewart. A parallel rendezvous algorithm for interpolation between multiple grids. *J. Parallel Distrib. Comput.*, 64(2):266–276, 2004.
15. J. Schreiner, A. Asirvatham, E. Praun, and H. Hoppe. Inter-surface mapping. In *SIGGRAPH '04*, pages 870–877. ACM Press, 2004.
16. G. Turk and M. Levoy. Zippered polygon meshes from range images. In *SIGGRAPH '94*, pages 311–318. ACM Press, 1994.
17. E. van Brummelen. Mesh association by projection along smoothed-normal-vector fields: association of closed manifolds. Technical Report 1574-6992, Delft Aerospace Computational Science(DACS), Kluyverweg 1, 2629HS Delft, The Netherlands, 2006.

Applications and Three-dimensional Techniques

4.1

Automatic 3D Mesh Generation for a Domain with Multiple Materials[*]

Yongjie Zhang[1], Thomas J.R. Hughes[1], and Chandrajit L. Bajaj[1,2]

[1] Institute for Computational Engineering and Sciences, The University of Texas at Austin, United States.

[2] Department of Computer Sciences, The University of Texas at Austin, United States.

Abstract: This paper describes an approach to construct unstructured tetrahedral and hexahedral meshes for a domain with multiple materials. In earlier works, we developed an octree-based isocontouring method to construct unstructured 3D meshes for a single-material domain. Based on this methodology, we introduce the notion of *material change edge* and use it to identify the interface between two or several materials. We then mesh each material region with conforming boundaries. Two kinds of surfaces, the boundary surface and the interface between two different material regions, are distinguished and meshed. Both material change edges and interior edges are analyzed to construct tetrahedral meshes, and interior grid points are analyzed for hexahedral mesh construction. Finally the edge-contraction and smoothing method is used to improve the quality of tetrahedral meshes, and a combination of pillowing, geometric flow and optimization techniques is used for hexahedral mesh quality improvement. The shrink set is defined automatically as the boundary of each material region. Several application results in different research fields are shown.

Key words: Unstructured 3D meshes, multiple materials, conforming boundaries, material change edge, pillowing, geometric flow.

1 Introduction

With the recent development of finite element analysis in various active research fields, such as computational medicine and computational biology, geometric modelling and mesh generation become more and more important. In scanned Computed Tomography (CT) and Magnetic Resonance Imaging (MRI) data, multiple materials are typically present. For example, as shown in Figure 1, the MRI brain data has been segmented into about 40 areas, and each area has a specific functionality. In analysis, multiple material regions correspond to various material data or regions of different physical phenomena. For this purpose, quality meshes for all material regions with

---

[*]http://www.ices.utexas.edu/~jessica/paper/meshmaterial

Email addresses: jessica@ices.utexas.edu (Yongjie Zhang), hughes@ices.utexas.edu (Thomas J.R. Hughes), bajaj@cs.utexas.edu (Chandrajit L. Bajaj). Yongjie Zhang's new address will be Department of Mechanical Engineering, Carnegie Mellon University after August 15, 2007.

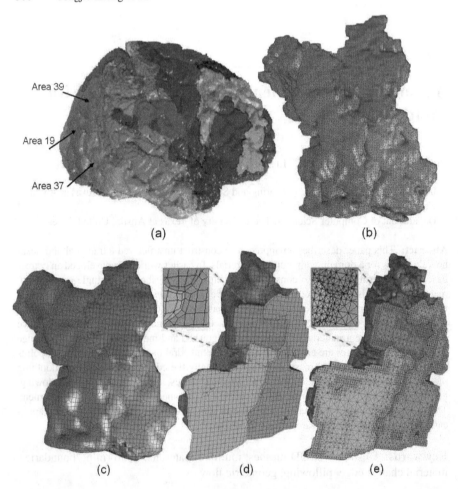

Fig. 1. Mesh generation for the segmented brodmann brain atlas with about 40 areas or materials. In (b)-(e), only three areas (Area 19, 37, and 39) are shown. (a) - smooth shading of the constructed brain model, each color represents one material; (b) - a triangular mesh; (c) - a quadrilateral mesh; (d) - one cross-section of a hexahedral mesh; (e) - one cross-section of a tetrahedral mesh. Red windows show details.

conforming boundaries are required. However, it is still very challenging to construct good geometric models for each of the materials. This issue is addressed in this paper in which we propose an approach to automatically mesh a domain with multiple materials.

We have developed an octree-based isocontouring method [1, 2] to construct adaptive and quality tetrahedral/hexahedral meshes from imaging data with conforming boundaries defined as level sets, but it only works for a domain with a single material. In order to construct 3D meshes for multiple material domains, we introduce and analyze a so-called *material change edge* to relocate all boundaries, including the boundaries of each material domain and the interfaces between two materials. All the boundaries are meshed into triangles or quadrilaterals. Besides material change edge, we also analyze each interior edge for each material domain to construct tetrahedral meshes. Each interior grid point is analyzed for hexahedral mesh construction.

Mesh adaptivity can be controlled in different ways: by a feature sensitive error function, by regions that users are interested in, by finite element solutions, or by a user-defined error function. The feature sensitive error function measures topology and geometry changes between isocontours at two neighboring octree levels. Adaptive tetrahedral and hexahedral meshes are generated by balancing the above four criteria and mesh size. Edge contraction and geometric flows [3] are used to improve the quality of tetrahedral meshes. A combination of pillowing, geometric flow, and optimization techniques are chosen for quality improvement of hexahedral meshes. The shrink set is defined in an automatic way, and the pillowing technique guarantees for each element in hexahedral meshes that at most one face lies on the boundary. This provides us with considerable freedom to further improve the element aspect ratio, especially for elements along the boundary.

We have applied our meshing method on a segmented human brain and rice dwarf virus (RDV) data, both containing multiple materials. Quality tetrahedral and hexahedral meshes are generated automatically with conforming boundaries, and some quantitative statistics such as area and volume for each domain are computed. Our results provide useful information to check the anatomy of the human brain, or identify and understand the RDV structure.

The remainder of this paper is organized as follows: Section 2 summarizes related previous work. Section 3 reviews the octree-based unstructured mesh generation techniques we developed. Section 4 discusses the detailed algorithm of mesh generation for a domain with multiple materials. Section 5 explains how to improve the mesh quality using various techniques. Section 6 presents the results. Section 7 draws conclusions and outlines planned future work.

2 Previous Work

Octree-based Mesh Generation: The octree technique [4, 5], primarily developed in the 1980s, recursively subdivides the cubes containing the geometric model until the desired resolution is reached. Irregular cells are formed along the geometry boundary, and tetrahedra are generated from both the irregular cells on the boundary and the interior regular cells. Unlike Delaunay triangulation and advancing front techniques, the octree technique does not preserve a pre-defined surface mesh. The resulting meshes also change as the orientation of octree cells changes. In order to generate high quality meshes, the maximum octree level difference during recursive subdivision is restricted to be one. Bad elements may be generated along the boundary, therefore quality improvement is necessary after mesh generation.

The grid-based approach generates a fitted 3D grid of structured hex elements on the interior of the volume [6]. In addition to regular interior elements, hex elements are added at the boundaries to fill gaps. The grid-based method is robust, however it tends to generate poor quality elements along the boundary. The resulting meshes are also highly dependent upon the grid orientation, and all elements have similar sizes. A template-based method was developed to refine quad/hex meshes locally [7].

Quality Improvement: Algorithms for mesh improvement can be classified into three categories [8] [9]: local refinement/coarsening by inserting/deleting points, local remeshing by face/edge swapping and mesh smoothing by relocating vertices. Laplacian smoothing generally relocates the vertex position at the average of the nodes connecting to it. Instead of relocating vertices based on a heuristic algorithm, an optimization technique to improve the mesh quality can be utilized. The optimization algorithm measures the quality of the surrounding elements to a node and attempts to optimize it. The optimization-based smoothing yields better results, but is more expensive. Therefore, a combined Laplacian/optimization-based approach was recommended [10, 11], which uses Laplacian smoothing whenever possible and only uses optimization-based smoothing when necessary.

Pillowing Techniques: A 'Doublet' is formed when two neighboring hexes share two faces, which have an angle of at least 180 degrees. In this situation, it is practically impossible to generate a reasonable Jacobian value by relocating vertices. The pillowing technique was developed to remove doublets by refining quad/hex meshes [12, 13, 14, 15]. Pillowing is a sheet insertion operation, which provides a fairly straight forward method to insert sheets into existing meshes. The speed of the pillowing technique is largely dependent upon the time needed to figure out the shrink set. The number of newly introduced hexahedra equals to the number of quads on the inserted sheet.

We have developed octree-based isocontouring methods to construct tetrahedral and hexahedral meshes from gridded imaging data [1, 2]. In this paper, we extend these methods to automatic tet/hex mesh generation for a domain with multiple materials. In addition, we will also discuss how to automatically define the shrink set and use the pillowing technique to improve the quality of hex meshes.

3 A Review of The Octree-based Isocontouring Method for Mesh Generation

There are two main isocontouring methods, primal contouring (or Marching Cubes) and Dual Contouring. The Marching Cubes algorithm (MC) [16] visits each cell in a volume and performs local triangulation based on the sign configuration of the eight vertices. MC and its variants have three main drawbacks: (1) the resulting mesh is uniform, (2) poor quality elements are generated, and (3) sharp features are not preserved. By using both the position and the normal vectors at each intersection point, the Dual Contouring method [17] generates adaptive isosurfaces with good aspect ratio and preserves sharp features. In this section, we are going to review the Dual Contouring method, and the octree-based algorithms we developed for 3D mesh generation.

3.1 Dual Contouring Method

The octree-based Dual Contouring method [17] analyzes each *sign change edge*, which is defined as one edge whose two end points lie on different sides of the isocontour. For each octree cell, if it is passed by the isocontour, then a minimizer point

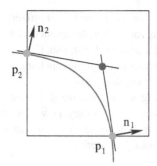

Fig. 2. A minimizer point (the red one) is calculated within an octree cell as the intersection point of two tangent lines. The two green points are intersection points of the red curve with cell edges. p_i and n_i are the position and unit normal vectors at a green point, respectively.

is calculated within this cell by minimizing a predefined Quadratic Error Function (QEF) [18, 19],

$$QEF(x) = \sum_i (n_i \cdot (x - p_i))^2 \qquad (1)$$

where p_i and n_i represent the position and unit normal vectors of the intersection point, respectively. For example, in Figure 2, the red curve is the true curve inside an octree cell, and the two green points are intersection points of the red curve with cell edges. The calculated minimizer point (the red one) is actually the intersection point of the two tangent lines.

In uniform case, each sign change edge is shared by four cells, and one minimizer point is calculated for each of them to construct a quadrilateral. In an adaptive case, each sign change edge is shared by either four cells or three cells, and we obtain a hybrid mesh, including quadrilateral and triangular elements.

3.2 Tetrahedral Mesh Generation

We have extended the Dual Contouring method to tetrahedral mesh generation [1, 20]. Each sign change edge belongs to a boundary cell, which is an octree cell passed through by the isocontour. Interior cells only contain interior edges. In order to tetrahedralize boundary cells, we analyze not only sign change edges but also interior edges. Only interior edges need to be analyzed for interior cell tetrahedralization. Each sign change edge is shared by four or three cells, and we obtain four or three minimizer points. Therefore those minimizers and the interior end point of this sign change edge construct a pyramid or tetrahedron. For each interior edge, we can also obtain four or three minimizers. Those minimizers and the two end points of this interior edge construct a diamond or pyramid. A diamond or pyramid can be split into four or two tetrahedra. Finally, the edge contraction and smoothing method is used to improve the quality of the resulting meshes.

3.3 Hexahedral Mesh Generation

Instead of analyzing edges, we analyze each interior grid point to construct hexahedral meshes from volumetric data [2]. In a uniform case, each grid point is shared by eight octree cells and we can obtain eight minimizers to construct a hexahedron. There are three steps to construct adaptive hexahedral meshes: (1) a starting octree level is selected to construct a uniform hex mesh, (2) templates are used to refine the uniform mesh adaptively without introducing any hanging nodes, (3) an optimization method is used to improve the mesh quality.

The octree-based meshing algorithms we developed are very robust, and they work for complicated geometry and topology. However, they only work for a domain with a single material. In the following section, we will discuss how to construct 3D meshes for a domain with multiple materials.

4 Mesh Generation for A Domain with Multiple Materials

4.1 Problem Description

Given a geometric domain Ω consisting of N material regions, denoted as Ω_0, Ω_1, ..., and Ω_{N-1}, it is obvious that $\cup_{i=1}^{N-1}\Omega_i$ is the complement of Ω_0 in Ω. Suppose B_i is the boundary of Ω_i, we have $\Omega_i \cap \Omega_j = B_i \cap B_j$ when $i \neq j$. $\cup B_i$ may not always be manifold, it can also be non-manifold curves or surfaces. Figure 3 shows two examples in 2D. There are three materials in Figure 3(a) denoted as Ω_0, Ω_1 and Ω_2, and we can observe that $\cup B_i$ are manifold curves. In Figure 3(b), there are four materials, but $\cup B_i$ consists of non-manifold curves and a square outer boundary. Non-manifold boundaries $\cup B_i$ cannot be represented by isocontours because each data point in a scalar domain can only have one function value. Therefore isocontouring methods do not work for a domain with non-manifold boundaries.

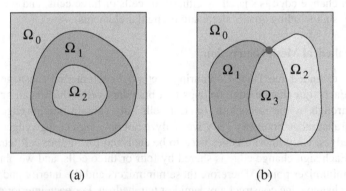

(a)   (b)

Fig. 3. A domain with multiple materials. (a) - $\cup B_i$ consists of manifold curves; (b) - $\cup B_i$ consists of non-manifold curves and a square outer boundary.

One possible way to mesh a domain with non-manifold boundaries is to only consider one material region at a time using the method of function modification and isocontouring. After meshes for all the material regions are obtained, we merge them together. However, there are four problems in this method:

1. During the whole process, we have to choose the same octree data structure, including interfaces between two different materials. Otherwise the resulting meshes are not conforming to the same boundary.
2. When we mesh each material domain, we detect all boundaries surrounding this material. Part of the boundaries may be shared by more than two materials, for example, the red point in Figure 3(b) is shared by four materials, Ω_0, Ω_1, Ω_2, and Ω_3. When we process each material, we may obtain four different points to approximate the red one. In other words, the meshes obtained may not conform to each other around the interface shared by more than two materials.
3. It is difficult to find the corresponding points on the interface shared by two materials if only position vectors are given.
4. Since we analyze only one material region at a time, we need to process the data N times for a domain with N materials. This is very time consuming.

In this section, we are going to present an approach to automatically detect all boundaries and mesh a domain with multiple materials simultaneously. Here are some definitions used in the following algorithm description:

Boundary Cell: A boundary cell is a cell which is passed through by the boundary of a material region.

Interior Cell: An interior cell is a cell which is not passed through by the boundary of any material region.

Material Change Edge: A material change edge is an edge whose two end points lie in two different material regions. A material change edge must be an edge in a boundary cell.

Interior Edge: An interior edge is an edge whose two end points lie inside the same material region. An interior edge is an edge in a boundary cell, or an interior cell.

Interior Grid Point: An interior grid point of one material is a grid point lying inside this material region.

4.2 Minimizer Point Calculation

In our octree-based method, only one minimizer point is calculated for each cell, and each octree cell has an unique index. This property provides us a lot of convenience to uniquely index the minimizer point of octree cells without introducing any duplicates.

There are two kinds of octree cells in the analysis domain: interior cells and boundary cells. For each interior cell, we simply choose the center of the cell as the minimizer point. For each boundary cell, we cannot separately calculate the minimizer point using Equation (1) for each material region because different minimizers may be obtained within this cell. For example in Figure 4(a), the red curve is the boundary which is shared by two materials Ω_1 and Ω_2. The same minimizer (the red point) is obtained when we mesh each material separately. However in Figure 4(b), there are three materials inside the octree cell, Ω_1, Ω_2 and Ω_3, and the blue point is shared by all three materials. Three different minimizers (the red points) are obtained for this cell when we mesh each material separately. Therefore instead of meshing each material separately, we include all intersection points (green points) in the QEF and calculate one identical minimizer point within this cell no matter how many materials are contained in it. This guarantees conforming meshes around boundaries.

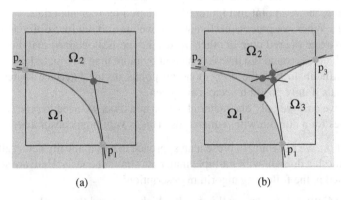

(a)   (b)

Fig. 4. Minimizer point calculation within a boundary octree cell. The red curves are boundaries between materials, green points are intersection points of red curves with cell edges, and red points are calculated minimizers. (a) - the boundary is shared by two materials, and the same minimizer is obtained when we mesh each material separately; (b) - the octree cell contains three materials, and the blue point is shared by all the three materials. Three different minimizers are obtained within this cell when we mesh each material separately.

Our meshing algorithm assumes that only one minimizer is generated in a cell. When complicated topology appears in the finest cell, non-manifold surface may be constructed. Schaefer et al. extended the dual contouring method to manifold surface generation [21], but multiple minimizers were introduced within a cell. This cannot be further extended to 3D hexahedral mesh generation. Here we prefer a heuristic subdivision method. If a cell contains two components of the same material boundary, we recursively subdivide the cell into eight identical sub-cells until each sub-cell contains at most one component of the same material boundary. The subdivision method can be easily extended to hex mesh generation, but it may introduce a lot of elements. Fortunately, complicated topology rarely happens in a cell, for example in the segmented brain data (Figure 1) and the RDV data (Figure 14), this situation does not exist.

4.3 2D Meshing

4.3.1 2D Triangulation

Material change edges and interior edges are analyzed to construct triangular meshes for each material region as shown in Figure 5:

1. Material change edge: each material change edge is shared by two boundary octree cells in a uniform case, and one minimizer point is calculated by minimizing the quadratic error function defined in Equation (1). We can obtain two

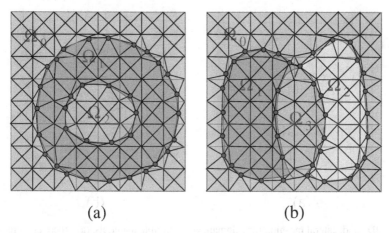

(a)   (b)

Fig. 5. 2D triangulation for a domain with multiple materials. (a) - there are three materials, and $\cup B_i$ constructs manifold curves; (b) - there are four materials, and $\cup B_i$ constructs non-manifold curves and a square outer boundary.

minimizer points. The two minimizer points and each end point of the material change edge construct a triangle. Therefore two triangles are obtained. In an adaptive case, each material change edge is also shared by two boundary octree cells, but the two octree cells may have different sizes. We can also get two minimizer points and construct two triangles.

2. Interior edge: each interior edge is shared by two octree cells. If the octree cell is a boundary cell, then we use Equation (1) to calculate a minimizer point. Otherwise we choose the center of this cell to represent the minimizer point. This interior edge and one minimizer point construct a triangle, therefore two triangles are obtained. In a uniform case, the two ocree cells have the same size, while in an adaptive case, the two octree cells have different sizes. However, we use the same method to construct triangles.

4.3.2 2D Quadrilateral Meshing

Instead of analyzing edges such as material change edges and interior edges, we analyze each interior grid point to construct a quadrilateral mesh. In a uniform case, each interior grid point is shared by four octree cells, and we can calculate four minimizer points to construct a quad. Six templates defined in [2] are used to refine the mesh locally. As shown in Figure 6(a, b), some part of the domain is refined using templates. The mesh quality improvement will be discussed later.

4.4 3D Meshing

4.4.1 Tetrahedral Meshing

We analyze each edge in the analysis domain, which contains multiple material regions. The edge can be a material change edge or an interior edge. In a 3D uniform

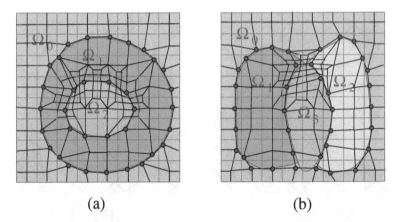

(a)   (b)

Fig. 6. 2D quadrilateral meshing for a domain with multiple materials. (a) - there are three materials, and $\cup B_i$ constructs manifold curves; (b) - there are four materials, and $\cup B_i$ constructs non-manifold curves and a manifold outer boundary.

case, each edge is shared by four cells. We obtain a total of four minimizers, which construct a quad. The quad and the two end point of the edge construct two pyramids as shown in Figure 7(a), and each pyramid can be divided into two tetrahedra. In an adaptive case, each edge is shared by four or three octree cells. Therefore we can obtain four or three minimizers, which construct a quad or a triangle. The quad/triangle and the two end point of the edge construct a diamond or a pyramid as shown in Figure 7(a, b). Finally we split them into tetrahedra.

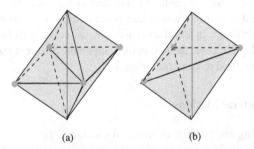

(a)   (b)

Fig. 7. Material change edges and interior edges are analyzed in 3D tetrahedralization. (a) - the red edge is shared by four cells, and two pyramids are constructed; (b) - the red edge is shared by three cells, and two tetrahedra are constructed. The green points are minimizer points.

We have applied our approach to an example with three materials. In Figure 8, a wireframe of the domain is shown in (a), (b) shows the constructed triangular mesh for the surface, and (c) shows one cross-section of the tetrahedral mesh for all three material regions. Note the presence of conforming boundaries.

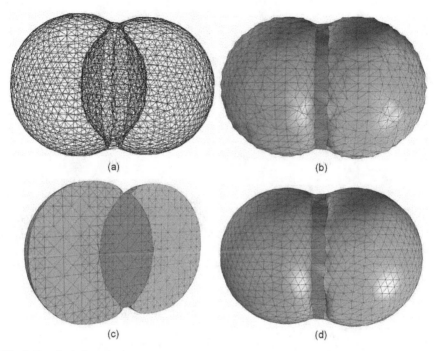

Fig. 8. Tetrahedral mesh generation for a domain with three materials. (a) - a wireframe visualization; (b) - the constructed triangular surface mesh; (c) - one cross-section of the constructed tetrahedral mesh; (d) - the mesh is improved by edge-contraction and geometric flow.

4.4.2 Hexahedral Meshing

For hexahedral mesh generation, we first choose a starting octree level and analyze each interior grid point in a uniform case. In 3D, each grid point is shared by eight octree cells, and we can obtain eight minimizer points to construct a hexahedron. An error function is calculated for each octree cell and compared with a threshold to decide the configuration of the minimizer point for this cell. All configurations can be converted into five basic cases defined in [2, 7], which are chosen to refine the uniform mesh adaptively without introducing any hanging nodes. The templates satisfy one criterion: in all templates, the refinement around any minimizer points/edges/faces with the same configuration is the same. This criterion guarantees that no hanging nodes are introduced during the process of mesh refinement. Figure 9(a) shows quadrilateral and hexahedral meshes constructed for a domain with three materials. Figure 9(c) shows one cross-section of the hexahedral mesh.

4.5 Mesh Adaptivity

The mesh adaptivity is controlled flexibly by different techniques:

1. A feature sensitive function defined in [1] is based on a trilinear function $f^i(x,y,z)$ within an octree cell,

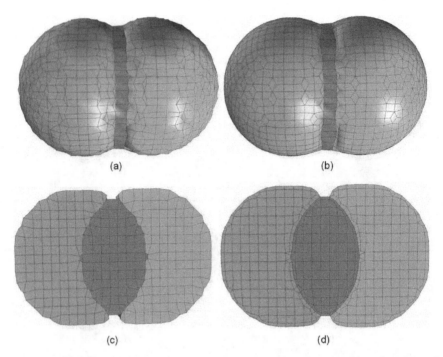

Fig. 9. Hexahedral mesh construction for a domain with three materials. (a) - surface quad mesh; (b) - surface mesh after quality improvement with geometric flow; (c) - one cross-section of the original hex mesh; (d) - one cross-section of the improved hex mesh.

$$F = \sum \frac{|f^{i+1} - f^i|}{|\nabla f^i|} \tag{2}$$

where

$$
\begin{aligned}
f^i(x,y,z) &= f_{000}(1-x)(1-y)(1-z) + f_{001}(1-x)(1-y)z \\
&+ f_{010}(1-x)y(1-z) + f_{100}x(1-y)(1-z) \\
&+ f_{011}(1-x)yz + f_{101}x(1-y)z \\
&+ f_{110}xy(1-z) + f_{111}xyz.
\end{aligned} \tag{3}
$$

f_{lmn} (where $l, m, n = 0$ or 1) is the function value at a grid point of the octree cell. The feature sensitive error function measures the isocontour difference between two neighboring octree levels, Level i and Level $(i+1)$.

2. Regions that users are interested in: According to various application requirements, location can be included in the error function to control the mesh adaptivity.
3. Finite element solutions: The physical domain is approximated by finite element solutions. We can also use them to efficiently and dynamically control the mesh adaptivity.
4. User-defined error function: Our algorithm is very flexible, and a user-defined error function can be substituted into the code for the mesh adaptivity control.

5 Quality Improvement

Mesh quality is a very important factor influencing the convergence and stability of finite element solvers. In the meshes generated from the above algorithm, most elements have good aspect ratio, especially the interior elements, but some elements around the boundaries may have poor aspect ratio, therefore the mesh quality needs to be improved.

5.1 Tetrahedral Mesh

First we choose three quality metrics to measure the quality of tetrahedral meshes, then use edge-contraction and geometric flow smoothing to improve it. The three quality measures are: (1) edge-ratio, which is defined as the ratio of the longest edge length to the shortest edge length in a tetrahedron; (2) Joe-Liu parameter $\frac{2^{\frac{4}{3}} \times 3 \times |V|^{\frac{2}{3}}}{\sum_{0 \le i < j \le 3} |e_{ij}|^2}$, where $|V|$ denotes the volume, and e_{ij} denotes the edge vectors representing the 6 edges [22] ; (3) Minimum volume bound.

Edge-contraction: We detect the element with the worst edge-ratio, and use the edge-contract method to remove it until the worst edge-ratio arrives at a predefined threshold, e.g., 8.5. A special case is shown in Figure 10. When one vertex P is embedded in a triangle T_{tri} or a tetrahedron T_{tet}, this vertex and each edge of T_{tri} construct a triangle in 2D, or this vertex and each face of T_{tet} construct a tetrahedron in 3D. If we contract any edge of T_{tri} or T_{tet} before removing the vertex P, then we will generate two duplicated and useless elements. This special case needs to be detected, and duplicated vertices/elements and useless vertices need to be removed after edge-contraction.

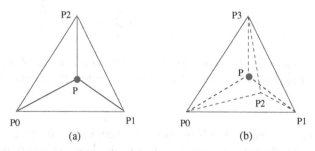

(a)　　　　　　　　　　(b)

Fig. 10. A special case for edge-contraction. In (a), a point P is embedded in a triangle $P_0 P_1 P_2$. P and each edge of the triangle $P_0 P_1 P_2$ construct a triangle. In (b), a point P is embedded in a tetrahedron $P_0 P_1 P_2 P_3$. P and each face of the tetrahedron $P_0 P_1 P_2 P_3$ construct a tetrahedron. When we contract any edge of triangle $P_0 P_1 P_2$ or tetrahedron $P_0 P_1 P_2 P_3$, duplicated and useless elements are generated.

Geometric flow smoothing: There are two kinds of vertices in 3D meshes, boundary vertices and interior vertices. For each boundary vertex, we use geometric flow to denoise the surface and improve the aspect ratio. For each interior vertex, we use the weighted averaging method to improve the aspect ratio, e.g., volume-weighted averaging. During the smoothing process, the Joe-Liu parameter and minimum volume bound are chosen as the quality metrics.

Geometric flow or geometric partial differential equations (PDEs) have been intensively used in surface and image processing [23, 24]. Here we choose surface diffusion flow to smooth the surface mesh because it preserves volume. A discretization scheme for the Laplace-Beltrami operator over triangular meshes is given in [24] and so we do not go to detail here.

The main aim of edge-contraction is to improve the element with the worst edge-ratio for each iteration. However, the edge-contraction method cannot remove slivers, therefore we should couple it with the smoothing scheme. Geometric flow smoothing tends to improve the mesh globally. We repeat running the two steps until a threshold or an optimized state is reached.

Figure 8 shows the difference of the mesh before and after the quality improvement. Figure 8(b) shows the original mesh, and Figure 8(d) shows the improved mesh. It is obvious that after quality improvement, the surface mesh is more regular and has better aspect ratio.

5.2 Hexahedral Mesh

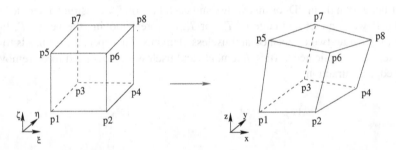

Fig. 11. A hexahedron [p1p2...p8] is mapped into a trilinear parametric volume in terms of ξ, η, and ζ. The eight basis functions in Equation (4) correspond to the eight vertices of the hexahedron.

Knupp et al. defined the Jacobian matrix of a vertex using its three edge vectors [25, 26], here we choose the usual definition of the Jacobian matrix in the Finite Element Method [27, 28]. Given a hexahedron with eight vertices as shown in Figure 11, there is a basis function ϕ_i in terms of ξ, η and ζ corresponding to each of them in the parametric domain. The eight basis functions are:

$$\phi_1 = (1-\xi)(1-\eta)(1-\zeta),$$
$$\phi_2 = \xi(1-\eta)(1-\zeta),$$
$$\phi_3 = (1-\xi)\eta(1-\zeta),$$
$$\phi_4 = \xi\eta(1-\zeta), \tag{4}$$
$$\phi_5 = (1-\xi)(1-\eta)\zeta,$$
$$\phi_6 = \xi(1-\eta)\zeta,$$
$$\phi_7 = (1-\xi)\eta\zeta,$$
$$\phi_8 = \xi\eta\zeta.$$

For any point inside this parametric hexahedral element, its coordinates can be calculated as $x = \sum x_i \phi_i$, $y = \sum y_i \phi_i$, and $z = \sum z_i \phi_i$. The Jacobian matrix is constructed as follows:

$$J = \begin{pmatrix} \frac{\partial x}{\partial \xi} & \frac{\partial x}{\partial \eta} & \frac{\partial x}{\partial \zeta} \\ \frac{\partial y}{\partial \xi} & \frac{\partial y}{\partial \eta} & \frac{\partial y}{\partial \zeta} \\ \frac{\partial z}{\partial \xi} & \frac{\partial z}{\partial \eta} & \frac{\partial z}{\partial \zeta} \end{pmatrix}. \tag{5}$$

The determinant of the Jacobian matrix is called the *Jacobian*. An element is said to be *inverted* if its Jacobians ≤ 0 somewhere in the element. We use the *Frobenius norm* as a matrix norm, $|J| = (tr(J^T J)^{1/2})$. The condition number of the Jacobian matrix is defined as $\kappa(J) = |J||J^{-1}|$, where $|J^{-1}| = \frac{|J|}{det(J)}$. Therefore, the three quality metrics for a vertex x in a hex are defined as $Jacobian(x) = det(J)$, $\kappa(x) = \frac{1}{3}|J^{-1}||J|$, and $Oddy(x) = \frac{(|J^T J|^2 - \frac{1}{3}|J|^4)}{det(J)^{\frac{4}{3}}}$ [29]. A combination of pillowing, geometric flow [3] and optimization techniques is used to improve the quality of hexahedral meshes.

Pillowing technique: The pillowing technique was developed to remove 'doublets' as shown in Figure 12(a-c), which are formed when two neighboring hexahedra share two faces. The two faces have an angle of ≥ 180 degrees, and generally they only appear along the boundaries between two materials. There is another similar situation in our meshes as shown in Figure 12(d-f). Two faces of a hexahedron lie on the boundary but they are shared by two other different elements. Since the meshes have to conform to the boundary, it is practically impossible to generate reasonable Jacobian values by relocating vertices. Here we use the pillowing technique to remove the two situations. As shown in Figure 12, first we identify the boundary for each material region, if there is a 'doublet' or an element with two faces on the boundary, then we create a parallel layer/sheet and connect corresponding vertices to construct hexes between the inserted sheet and the identified boundary. The number of newly generated hexes equals the number of quads on the boundary for each material region. For the boundary shared by two material regions, one or two parallel sheets may be inserted. Finally we use geometric flow to improve the mesh quality.

The speed of the pillowing technique is largely decided by the time needed to figure out the shrink set, therefore the main challenge is how to automatically and efficiently find out where to insert sheets. In our octree data structure, each grid point belongs to one material region. Material change edge is analyzed to construct and detect boundaries, which are shared by two different materials. For example, the

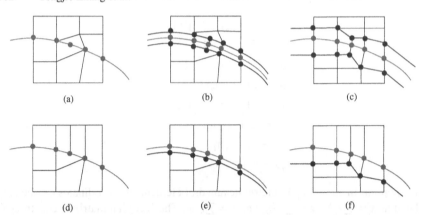

Fig. 12. The pillowing technique. (a-c) show a 'doublet' and (d-f) show an element whose two faces lie on the boundary, but the two faces are shared by two other elements. (a, d) - the original mesh, the red layer is the boundary shared by two materials; (b, e) - a parallel layer (the blue one) is created for the material with an element whose two faces lie on the boundary. Two layers are created in (b), but only one layer is created in (e); (c, f) - geometric flow is used to smooth the resulting mesh. The red layer is still on the boundary.

two end points of a material change edge are in two material regions, Ω_1 and Ω_2. Therefore the constructed boundary is shared by these two materials. In this way, the boundary for each material region is detected, and is defined as the shrink set automatically.

Geometric flow: First boundary vertices and interior vertices are distinguished. For each boundary vertex, there are two kinds of movement: one is along the normal direction to remove noise on the boundary, the other one is on the tangent plane to improve the aspect ratio of the mesh. The surface diffusion flow is selected to calculate the movement along the normal direction because the surface diffusion flow preserves volume. A discretized Laplace-Beltrami operator is computed numerically [3]. For each interior vertex, we choose the volume-weighted averaging method to relocate it.

Optimization method: After applying pillowing and geometric flow techniques on the meshes, we use the optimization method to further improve the mesh quality. For example, we choose the Jacobian equation as our object function, and use the conjugate gradient method to improve the worst Jacobian value of the mesh. The condition number and the Oddy number are also improved at the same time.

We have applied our quality improvement techniques on some hexahedral meshes. Figure 9(c) shows one cross-section of the original mesh, and Figure 9(d) shows the improved mesh. It is obvious that the hexahedral mesh is improved and each element has at most one face lying on the boundary. Figure 13 shows some statistics of quality metrics for the Brodmann brain atlas (Figure 1) and the segmented RDV data (Figure 14).

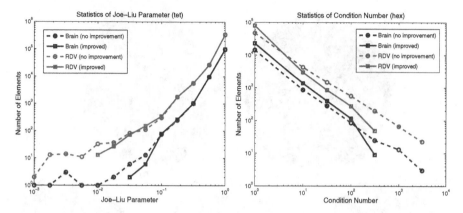

Fig. 13. The histogram of Joe-Liu parameter (left) of tet meshes, and the condition number (right) of hex meshes for the human brain and the RDV data. It is obvious that the mesh quality, especially the worst parameter, is improved

6 Results

In this section, we present applications of our meshing approach to two datasets: the brodmann brain atlas and a segmented rice dwarf virus (RDV) volumetric data.

Brodmann brain atlas: The brodmann brain atlas is a segmented volume map with about 40 different areas or materials, and each area controls a different functionality (http://www.sph. sc.edu/comd/rorden/mricro.html). We apply our meshing algorithms on the atlas to construct meshes for all material regions, and provide some statistics, such as the surface area and the volume of each region. The results are shown in Figure 1, and three areas are selected to show details of the constructed meshes, including the peristriate area (Area 19, surface area 144.1 cm^2 and volume 96.9 cm^3), the occipitotemporal area (Area 37, surface area 128.9 cm^2 and volume 88.7 cm^3) and the angular area (Area 39, surface area 47.0 cm^2 and volume 41.4 cm^3). The total volume of the brain is 1373.3 cm^3 (the normal volume of the human brain is 1300-1500 cm^3). Some statistics of quality metrics are shown in Figure 13.

Rice dwarf virus: Another main application of our algorithms is on segmented biomolecular data, for example, the cryo-electron magnetic data (cryo-EM) of a rice dwarf virus (RDV) with a resolution of 6.8 angstrom. In Figure 14, (a) shows the segmentation result [30], and each color represents four 1/3 unique trimers. Each trimer is further segmented into three P8 monomers as shown in (b-e). Different kinds of meshes are constructed for a trimer which has three materials. Our method provides a convenient approach to visualize the inner structure of RDV. Figure 13 shows some statistics of quality metrics.

7 Conclusions and Future Work

We have developed an automatic 3D meshing approach to construct adaptive and quality tetrahedral or hexahedral meshes for a volumetric domain with multiple

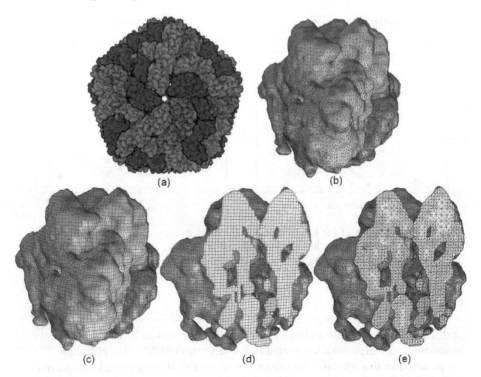

Fig. 14. Mesh generation for the segmented rice dwarf virus data (RDV). (a) - smooth shading of the segmented RDV model [30], each color represents four $1/3$ trimers; (b) - a triangular mesh of one trimer consisting of three monomers; (c) - a quadrilateral mesh of one trimer; (d) - one cross-section of a hexahedral mesh; (e) - one cross-section of a tetrahedral mesh.

materials. Boundaries between materials are detected, and each material region is meshed with conforming boundaries. Edge-contraction and geometric flow schemes are used to improve the quality of tetrahedral meshes, while a combination of pillowing, geometric flow and optimization techniques is employed for the quality improvement of hexahedral meshes. We also provide an automatic way to define the shrink set for the pillowing technique.

We have successfully applied our method to three volumetric imaging data, which involve a segmented human brain model and the rice dwarf virus. In the future, we hope to apply the techniques described here to more applications.

Acknowledgement

We would like to thank Yuri Bazilevs for proofreading. Y. Zhang was partially supported by a J. T. Oden ICES Postdoctoral Fellowship and NSF-DDDAS-CNS-054033. C. Bajaj was supported in part by NSF grants EIA-0325550, CNS-0540033, and NIH grants P20-RR020647, R01-GM074258, R01-GM073087.

References

1. Zhang Y., Bajaj C., Sohn B.S. "3D Finite Element Meshing from Imaging Data." *The special issue of Computer Methods in Applied Mechanics and Engineering (CMAME) on Unstructured Mesh Generation*, vol. 194, no. 48-49, 5083–5106, 2005
2. Zhang Y., Bajaj C. "Adaptive and Quality Quadrilateral/Hexahedral Meshing from Volumetric Data." *Computer Methods in Applied Mechanics and Engineering (CMAME)*, vol. 195, no. 9-12, 942–960, 2006
3. Zhang Y., Bajaj C., Xu G. "Surface Smoothing and Quality Improvement of Quadrilateral/Hexahedral Meshes with Geometric Flow." *Proceedings of 14th International Meshing Roundtable*, pp. 449–468. 2005
4. Yerry M.A., Shephard M.S. "Three-Dimensional Mesh Generation by Modified Octree Technique." *International Journal for Numerical Methods in Engineering*, vol. 20, 1965–1990, 1984
5. Shephard M.S., Georges M.K. "Three-Dimensional Mesh Generation by Finite Octree Technique." *International Journal for Numerical Methods in Engineering*, vol. 32, 709–749, 1991
6. Schneiders R. "A grid-based algorithm for the generation of hexahedral element meshes." *Engineering With Computers*, vol. 12, 168–177, 1996
7. Schneiders R. "Refining quadrilateral and hexahedral element meshes." *5th International Conference on Grid Generation in Computational Field Simulations*, pp. 679–688. 1996
8. Teng S.H., Wong C.W. "Unstructured mesh generation: Theory, practice, and perspectives." *International Journal of Computational Geometry and Applications*, vol. 10, no. 3, 227–266, 2000
9. Owen S. "A survey of unstructured mesh generation technology." *7th International Meshing Roundtable*, pp. 26–28. 1998
10. Canann S., Tristano J., Staten M. "An approach to combined Laplacian and optimization-based smoothing for triangular, quadrilateral and quad-dominant meshes." *7th International Meshing Roundtable*, pp. 479–494. 1998
11. Freitag L. "On combining Laplacian and optimization-based mesh smoothing techniques." *AMD-Vol. 220 Trends in Unstructured Mesh Generation*, pp. 37–43, 1997
12. Mitchell S., Tautges T. "Pillowing Doublets: Refining a Mesh to Ensure That Faces Share at Most One Edge." *4th International Meshing Roundtable*, pp. 231–240. 1995
13. Borden M., Shepherd J., Benzley S. "Mesh Cutting: Fitting Simple All-Hexahedral Meshes to Complex Geometries." *Proceedings of the 8th International Socity of Grid Generation conference*. 2002
14. Shepherd J., Tuttle C., Silva C., Zhang Y. "Quality Improvement and Feature Capture in Hexahedral Meshes." *SCI Institute Technical Report UUSCI-2006-029, University of Utah*. 2006
15. Shephard J. "Topologic and Geometric Constraint-Based Hexahedral Mesh Generation." *University of Utah*, 2007
16. Lorensen W.E., Cline H.E. "Marching Cubes: A High Resolution 3D Surface Construction Algorithm." *Proceedings of SIGGRAPH*, pp. 163–169. 1987
17. Ju T., Losasso F., Schaefer S., Warren J. "Dual Contouring of Hermite Data." *SIGGRAPH*, pp. 339–346. 2002
18. Garland M., Heckbert P.S. "Simplifying Surfaces with Color and Texture using Quadric Error Metrics." *IEEE Visualization '98*, pp. 263–270. 1998. URL citeseer.nj.nec.com/garland98simplifying.html
19. Hoppe H., DeRose T., Duchamp T., McDonald J., Stuetzle W. "Mesh optimization." *SIGGRAPH*, pp. 19–26. 1993

20. Zhang Y., Xu G., Bajaj C. "Quality Meshing of Implicit Solvation Models of Biomolecular Structures." *The special issue of Computer Aided Geometric Design (CAGD) on Applications of Geometric Modeling in the Life Sciences*, vol. 23, no. 6, 510–530, 2006

21. Schaefer S., Ju T., Warren J. "Manifold Dual Contouring." *IEEE Transactions on Visualization and Computer Graphics, accepted*, 2007

22. Liu A., Joe B. "Relationship between tetrahedron shape measures." *BIT*, vol. 34, no. 2, 268–287, 1994

23. Xu G. "Discrete Laplace-Beltrami Operators and Their Convergence." *Computer Aided Geometry Design*, vol. 21, 767–784, 2004

24. Xu G., Pan Q., Bajaj C. "Discrete Surface Modelling Using Partial Differential Equations." *Computer Aided Geometric Design, in press*, 2005

25. Knupp P. "Achieving finite element mesh quality via optimization of the jacobian matrix norm and associated quantities. Part II - a framework for volume mesh optimization and the condition number of the jacobian matrix." *International Journal of Numerical Methods in Engineering*, vol. 48, 1165–1185, 2000

26. Kober C., Matthias M. "Hexahedral Mesh Generation for the Simulation of the Human Mandible." *9th International Meshing Roundtable*, pp. 423–434. 2000

27. Oden J.T., Carey G.F. "Finite Elements: A Second Course." *Prentice Hall, Englewood Cliffs, NJ*, 1983

28. Hughes T.J. "The Finite Element Method: Linear Static and Dynamic Finite Element Analysis." *Prentice Hall, Englewood Cliffs, NJ*, 1987

29. Oddy A., Goldak J., McDill M., Bibby M. "A distortion metric for isoparametric finite elements." *Transactions of CSME, No. 38-CSME-32, Accession No. 2161*, 1988

30. Yu Z., Bajaj C. "Automatic Ultra-structure Segmentation of Reconstructed Cryo-EM Maps of Icosahedral Viruses." *IEEE Transactions on Image Processing: Special Issue on Molecular and Cellular Bioimaging*, vol. 14, no. 9, 1324–1337, 2005

4.2

Mixed-element Mesh for an Intra-operative Modeling of the Brain Tumor Extraction

Claudio Lobos[1], Marek Bucki[1], Nancy Hitschfeld[2], and Yohan Payan[1]

[1] TIMC-IMAG Laboratory, UMR CNRS 5525, University J. Fourier, 38706 La Tronche, France [claudio.lobos|marek.bucki|yohan.payan]@imag.fr
[2] Departamento de Ciencias de la Computación, FCFM, Universidad de Chile, Santiago, Chile nancy@dcc.uchile.cl

Summary. This paper presents a modified-octree technique that generates a mixed-element mesh. The final output mesh consider cubes, prisms, pyramids and tetrahedra. This technique is optimized for brain tumor extraction simulation in a real-time application. The proposed technique is based on the octree algorithm with a specific constraint: elements will be split only if they intersects a certain region of interest. With this approach we pursued a refined mesh only in the path from the skull opening point to the tumor. Fast computation by the Finite Element Method (FEM) is achieve thanks to the local refinement. Examples are given and comparison with other approaches are presented.

Key words: Modified-Octree, Mixed-elements, Finite Elements, Real-time Application, Patient-specific Simulation, Region Of Interest.

1 Clinical Background

Accurate localization of the target is essential to reduce morbidity during a brain tumor removal intervention. Image guided neurosurgery nowadays faces an important issue for large skull openings, with intra-operative changes that remain largely unsolved.

Once the skull is open a deformation of the brain naturally occurs. This phenomena is known as "the brain-shift". The causes of this deformation can be grouped by:

- physical changes (dura opening, gravity, loss of cerebrospinal fluid, actions of the neurosurgeon, etc) and
- physiological phenomena (swelling due to osmotic drugs, anesthetics, etc).

As a consequence of this intra-operative brain-shift, pre-operative images no longer correspond to reality. Therefore the neuro-navigation system based on those images doesn't necessarily represent the current situation.

In order to face this problem, various teams have proposed to incorporate into existing image-guided neurosurgical systems, a biomechanical modeling to compensate the brain deformations by updating the pre-operative images and planning according to intra-operative brain shape changes. For this, such measured changes (for example the changes of the external shape of the brain tissues in the opening skull region) are given as new boundary conditions to the biomechanical model of the brain tissues that infers the new position of the tumor. Such intra-operative use of a biomechanical model implies that a strong modeling effort must be carried out. Our group tries to provide some clues towards such a modeling effort. Three steps are followed to design the brain model:

- The segmentation of pre-operative images (MRI) to build the external surface mesh of the brain, including the tumor.
- The generation of a volume mesh optimized for real-time simulation.
- The creation of a model of the brain-shift with Finite Elements (FE).

The focus of this paper is the second point. The proposed meshing technique starts from a surface representation of the brain with the tumor and produces a volume mesh of it. Some results about the third point are also shown in section 6.2.

2 Meshing constraints

Time is crucial in surgery thus, to produce an intra-operative real-time modeling of the brain-shift the FEM computation must be fast. The speed of the FEM directly depends on the number of degrees of freedom the system has, thus an optimal mesh in terms of quantity of nodes must be provided.

A good representation of the tumor as well as the Skull Opening Point (SOP) and the path between them is mandatory because here is where a greater deformation is expected [16]. This path will from now on be referred to as the Region of Interest (RoI). A mesh without quality elements can lead to errors in the computation of the FE thus quality is also an important issue in this problem. Therefore the constraints to model the brain-shift in a real-time application are:

1. The final mesh must be refined enough in the RoI and coarse elsewhere.
2. Achieve surface representation for the input FE mesh.
3. Guarantee element quality throughout the entire mesh.

3 Meshing Background

This section gives a description of several meshing techniques in order to find the most suitable one to solve the brain-shift simulation problem.

3.1 Advancing Front

The Advancing Front (AF) technique [11, 7] uses as input a closed surface. All the faces that describe the input domain are treated as fronts and are expanded into the volume in order to achieve a final 3D representation. The selection of points to create the new faces encourages the use of existing points. Figure 1 shows some steps of the AF technique where only a portion of the surface has been taken for clarity purposes. Additional process to improve the quality of the elements can be made.

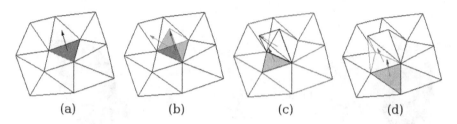

(a)   (b)   (c)   (d)

Fig. 1. The advancing front technique: (a) A portion of a surface mesh with one front to expand, (b) the tetrahedron is created and the new faces can be treated as fronts, (c) another expansion using a recently inserted front and (d) another expansion using already inserted points.

This technique is strongly dependent on the input surface mesh. If the input is accurate, it will have a high number of faces producing a refined volume mesh in the entire input domain. This is contradictory with our first constraint: the mesh must be refined in the RoI and coarse elsewhere. On the other hand, if the input mesh is coarse, this technique should be combined with others strategies of local refinement to respect the first constraint.

In relation to the second constraint, this technique achieves a perfect surface representation because it uses the input surface to produce the final volume mesh. However a little remark can be made: If the input surface mesh is coarse and the RoI is constrained to have a large quantity of points, the resulting final volume mesh will continue to be coarse in the portion of the RoI that intersects the input surface mesh.

Two solutions are proposed: (1) pre-treatment of the input surface mesh to be more refined in the RoI or (2) local refinement in the surface portion of the RoI once the volume mesh is generated.

The third constraint is not managed by this technique. However different algorithms can be applied to improve the quality of the resulting volume mesh, for example in the case of tetrahedral AF meshes, constraint the new tetrahedra to fulfill the Delaunay [5] property as long as the fronts continues their expansion.

3.2 Mesh Matching

Mesh matching is an algorithm that starts with a generic volume mesh and tries to match it onto the input domain [4, 13]. The generic volume can be obtained from an interpolation of several sample models. To generate a new mesh, in our case of the brain, the problem is reduced to finding a 3D nonuniform transformation function that will be applied to the entire generic mesh (atlas) and produce in this way the final volume mesh for the current patient. This is shown in figure 2 in a maxillofacial meshing example, from Chabanas et al. [2]

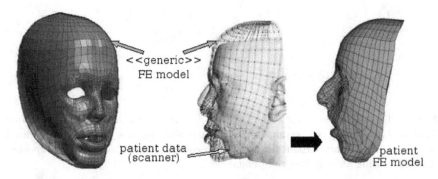

Fig. 2. The mesh matching algorithm in a maxillofacial simulation.

This technique does not satisfy the first constraint. The atlas does not consider the tumor information thus the mesh is not refined in the RoI. It would be necessary to combine this technique with others strategies of local refinement in order to respect the first constraint. In other words, a generic solution cannot be applied to a problem that is patient dependent, i.e. with positions and sizes of tumor and the SOP that change from one patient to another.

The second constraint is satisfied because the atlas normally resembles the input domain. This can be not true for very specific cases where the brain is malformed.

The third constraint like the second one, is also satisfied because the atlas has a great quality coefficient in the entire mesh. It is a quite perfect model that sometimes has modifications usually made by hand in order to produce the best starting point. A loss of quality can occur only in cases where the input domain and the atlas are not alike.

3.3 Regular Octree

The octree meshing technique starts from the bounding box of the surface [17, 14]. This basic cube or "octant" is split into eight new octants. Each octant is then iteratively split into eight new ones, unless it resides outside the input surface mesh, in which case it is deleted. The algorithm stops when a predefined maximum level of iterations is reached or when a condition of surface approximation is satisfied.

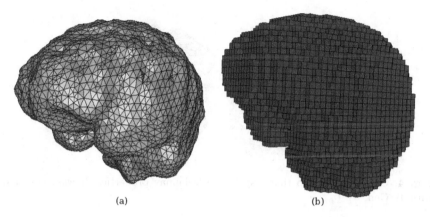

(a)   (b)

Fig. 3. (a) Input surface triangle mesh of 3152 points and (b) output hexahedral mesh of 20,000 points.

The problem with a regular octree mesh for our problem is that it result in a high number of points even in regions where very few deformation of the brain is expected [9]. Therefore, a non-optimal mesh would be the input for the FEM producing unnecessary time consumption for the entire simulation. For example the mesh shown in figure 3 is unacceptable for the brain-shift modeling due to excessive quantity of points. Therefore the first constraint is not fulfilled.

The octree by itself does not consider a surface approximation algorithm once the split process is done. Therefore it has to be combined with other techniques in order to achieve a surface representation. Two main approaches are considered:

- Marching cubes [12]: this algorithm crops the cubes that lie within the surface and produces, in most cases, tetrahedra.
- Surface projection: this technique projects the points of those elements that intersect the surface, onto it. The main problem is that this can produce degenerate elements unless a minimal displacement is needed.

The quality of the mesh is normally acceptable. The elements with a bad quality can be found only at the surface of the input domain because they have changed in order to achieve the correct representation of the domain.

3.4 Delaunay Mesh

In the literature a Delaunay [5] mesh is said to be a mesh of tetrahedra that respects the Delaunay constraint. An example mesh of the Delaunay property can be seen in figure 4

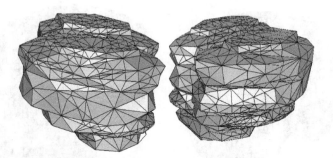

Fig. 4. One output mesh that respect the Delaunay properties. It was generated with TetGen.

A Delaunay mesh can be useful to solve the brain-shift (meshing) problem, however it must be combined with a point insertion strategy that consider the constraints presented in section 2. Current implementations, like TetGen [19], do not offer as much control over point quantity and quality as the constraints needs. One alternative implementation would be the following work flow:

- generate a basic tetrahedralization of the input domain (that doesn't necessarily respect the Delaunay constraint).
- improve the quality of the elements in the RoI and constrain them to a certain size.
- without inserting an excessive number of points, increase the quality of the others elements (outside the RoI).

What is critical in this process is to control the number of nodes. A classical problem of Delaunay meshes is to control the insertion of new nodes when going from one refined zone into another not so refined.

Constraint two can be satisfied by the constrained Delaunay meshes that consider the representation of a domain by using certain predefined points. Those points would be the input surface domain in the brain-shift case. And finally, the third constraint requires the insertion of points to fulfil the underline quality requirement. See also [1] for a Delaunay based approach to isotropic tetrahedral meshing.

4 Discussion Regarding Presented Techniques

The mesh matching and the advancing front techniques are ruled out because it is not easy to incorporate a RoI i.e. to generate a mesh that is refined in a specific region and coarse elsewhere. These techniques need major adjustments in order to achieve the desired mesh.

The Delaunay approach is closer to achieving the desired features, however its emphasis is on the quality of the tetrahedra (i.e. the achievement of the Delaunay property) and not on the quantity of nodes.

Even though the octree technique doesn't accomplish the constraints by itself it allows the refinement of some elements directly by the splitting process. This is a very powerful feature that easily help to fulfill the constraint on local refinement in the RoI. However a process to manage the transition between regions of different refinement levels is needed. A method for surface representation is also needed for the octree in order to achieve an optimal mesh for real-time simulation.

The following section proposes some adaptations provided to the octree technique in order to fulfill all the desired constraints. We consider that this is the best starting point to respect the constraints mentioned in section 2.

5 Meshing Technique for the Brain Shift

The basic octree algorithm is applied but with two main modifications:

- The condition to stop the refinement is the number of nodes the mesh has. The quantity of nodes is provided as input to the algorithm.
- One element will be refined only if it is inside the RoI.

The first modification corresponds to a basic constraint of real-time simulation. Even with high quality elements, a mesh that would have too many nodes would increase the FE time response during surgery and this is not acceptable.

The second modification is due to the fact that the RoI is the area where the predictions of the developed model should be the most accurate one. Consequently only those elements are refined.

This basic octree mesh is not suitable for the FEM. Figure 5 shows an example of different refinement levels for neighbor elements. The implemented algorithm deals with the different possible cases by adding mixed-elements: pyramids, prisms and tetrahedrons. At the end of this subprocess the mesh is valid for the FEM.

The final step is to achieve a surface representation regarding the quality of the elements. Next part details each step of our algorithm.

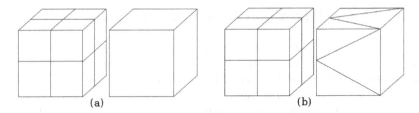

Fig. 5. An example of transition management between different levels of refinement. (a) Two neighbor octants, the left one is more refined. (b) The same two octants but now the right one has a subdivision: 5 pyramids and 4 tetrahedra, making the mesh congruent and suitable for the FEM.

5.1 Inputs and basic mesh

In order to produce a final mesh several inputs are requested:

- A surface mesh of the input domain (S_d).
- A desired quantity of points. This input will be used as the condition to stop the octree subdivision.
- A surface mesh that describes the RoI (S_{RoI}). This mesh has to intersect the input domain.

The first step is to obtain the bounding box of the domain. A global search is performed to detect the maximum and minimum coordinates of each axis. The result is an hexahedron that will correspond to the root octant of the octree. Then a point (P_{out}) that is outside this bounding box is calculated to be used as a reference in several tests that are described later.

The algorithm continues with the classical octree subdivision but with one difference. An octant will be split only if it intersects or is completely inside the RoI (defined by S_{RoI}). The process continues until the maximum number of nodes, provided as input, is reached.

When an element is created the P_{out} is used to check if the element is completely outside the domain. A virtual segment is created between the P_{out} and each vertex of the element. If one of those segments crosses an odd quantity of S_d faces, this element is said to be inside. If not, the element is removed from the mesh. For each new element there is also a test to check if it intersects the S_{RoI}. If not it won't continue to split.

Figure 6 plots the mesh generated for a given quantity of points, the input S_d and the S_{RoI}. Note that the mesh is strongly refined in the RoI.

At this step, the mesh is not suitable to carry out a classical Finite Element Analysis (FEA) because of incongruent elements (an element is said to be incongruent when it has one or more points inserted in each face or edge of it). Some modifications of the FE basis functions (see [15] for example) could enable this mesh to be used for an FEA. However, it was chosen in this paper to remain with a classical FEM approach, which means that the mesh has to be modified in order to manage transitions between incongruent elements.

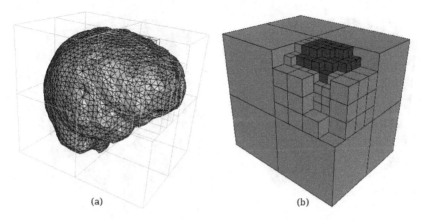

(a)   (b)

Fig. 6. Octree: only the elements that intersects the RoI are split in order to achieve a higher density in that region: (a) the input and the generated mesh and (b) the generated mesh in solid: dark elements corresponds to the RoI. This mesh has 416 points and cannot be used for classical FEA.

5.2 Managing transitions

The developed application has several patterns [8] that detect and split the incongruent elements. In particular, we can address to incongruent elements that has at most one point inserted in each face or edge of it. This is a property called "one-irregular". In order to produce a one-irregular mesh, all the octants that do not respect this property are split. In this manner, regions outside the RoI have more points to represent the volume. In other words, the density of points outside the RoI is a consequence of the RoI itself. The resulting mesh can be seen in figure 7.

For each new element generated from this subdivision a test is run to check if it is outside the contour of the brain, as described in the previous section.

It is important to keep the faces as planar as possible and the triangle faces are the only ones that preserve this property all the time. In order to represent the surface the points outside the S_d must be projected into it. For this reason, all the elements that intersects the input domain are tetrahedralized. In this manner the outside points can be moved without losing the planar property of the faces.

The octree hierarchical structure is used to assign a default subdivision orientation for each type of element. The neighbor face information is also used to test if there are faces that were already triangulated in which case this orientation is used.

An hexahedron will be split into five tetrahedra as shown in figure 8a. The prism will be divided into three tetrahedra like in figure 8b and the pyramid into only two as figure 8c shows.

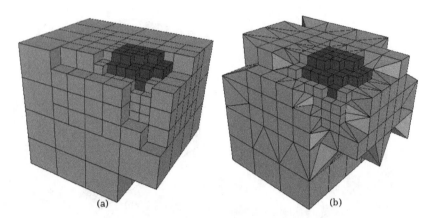

Fig. 7. (a) A one-irregular mesh: each cube has at most one point inserted in each face or edge of it. This mesh has 594 points. (b) A mixed-element mesh that handle the transitions. This mesh has 660 points and can be used for classical FEA.

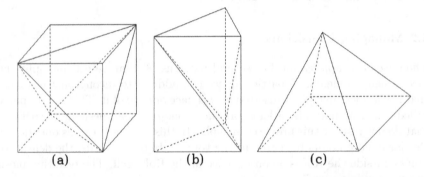

Fig. 8. (a) The subdivision of an hexahedron into five tetraedra, (b) The subdivision of a prism into three tetrahedra and (c) The subdivision of a pyramid into two tetrahedra.

After the local partition of a surface element into tetraedra, it is possible that some neighbors (such as an hexahedron) get a diagonal in their square face. This diagonal must be incorpored and the neighbor element must be tessellated. This new tessellation depends on the type of the element with a diagonal in their square face. For example, if the element is a pyramid: the base face is triangulated congruently with the neighbors and the propagation (of the triangulation) stops. If the element is a prism it depends on the number of already triangulated faces it has:

- 0: the prism is split in one tetrahedra and one pyramid. The face that wasn't with a diagonal before is updated with the information to be handled by the neighbor element.

- 1: if it is possible to build one pyramid and one tetrahedra this option is preferred and the propagation of diagonalizations stops. If not, one diagonal is added and the corresponding face is updated.
- 2: there is only one option and the propagation stops.

In the case of the hexahedron the inner middle point of it is inserted. With this new point the hexahedron is replaced by 6 pyramids. Then the faces that have diagonals trigger the split of the corresponding pyramid and the propagation stops.

5.3 Mesh Quality

An extraordinary number of measures has been proposed, ranging from bounds on solid angles to more complex geometric ratios (See [10, 18, 20] for more details). To our knowledge, only one team [6] dealing with the brain-shift problem through a biomechanical FE modeling of the brain have proposed a mesh quality measure. We decided to use the same measure: the aspect-ratio coefficient (ARC). This coefficient is obtained in different manners for each type of element.

The ARC is normalized so that ARC = 1 corresponds to an ideal element and ARC $\rightarrow \infty$ as the element becomes increasingly distorted.

The ARC of the tetrahedra is obtained as follows [6]:

$$\mathrm{ARC} = \frac{(\frac{1}{6}\sum_{i=1}^{6}(l_i^2))^{3/2}}{8.47867 V^{el}}$$

Where $l_i(i = 1, \ldots, 6)$ are its edge lengths and V^{el} is the volume of the tetrahedron. The value 1 corresponds in this case to an equilateral tetrahedra.

For the Hexahedron the segments between oppose faces middle points are obtained. The ratio between the longest and the shortest of the three segments will be used as the ARC of this element.

Three segments are also used in the case of the pyramid. The two firsts are constructed using the middle points of opposite edges in the base face. The third segment is the one that represents the height of the pyramid. Like in the case of the hexahedron the coefficient will be the ratio between the longest and shortest segment.

To obtain the ARC of the prism a selection of a square face of it is made. Two segments are obtained as in the case of the pyramid base face. The third segment is obtained using the height of the triangle face that corresponds to the average of the two original triangle faces of the prism. This can be seen in figure 9. The ARC, as in previous cases, is obtained from the ratio between the longest and shortest segment.

Measures over the quality of the mesh are described in section 6.1.

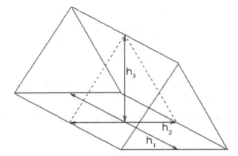

Fig. 9. The aspect-ratio coefficient for the prism is obtained from the ratio between the longest and shortest segment of the three heights presented.

5.4 Surface representation

We propose a novel mesh adaptation technique. The goal is to achieve an acceptable surface representation after a quality adaptative octree-based mesh has been generated. The nodes of the mesh obtained until this point of the algorithm do not necessarily lie on the object surface, as described in figure 7. It is thus necessary to improve the surface representation of the mesh while keeping an acceptable level of quality of its elements.

The inputs to our algorithm are:

- The source mesh: a generic volumetric mesh comprising quality elements (in our case the octree-based mixed-element mesh generated above).
- Destination mesh: the surface we want our source mesh to be adapted to (in our case S_d.

The key idea of our algorithm is to use a mechanical simulation to constraint the deformation between the source and the destination mesh. We drive the surface nodes of the source mesh with a step-by-step displacement towards the destination mesh, leaving a mechanical model perform the inner nodes relaxation throughout the deformation. It is also important to mention that it was chosen to implement a compressible material behavior (poisson $\nu = 0.3$), in order to allow a mechanical compression of the 3D mesh.

Our approach, unlike direct projection of surface nodes that disregards inner nodes position, is guaranteed to not produce element inversions. To preserve the mesh overall quality we check the elements quality at each step of the deformation. If necessary we artificially increase the mechanical resistance, or stiffness, of those elements that suffer the greatest quality loss before proceeding to the next deformation step. This can result in oscillations between neighboring elements. Therefore in order to guarantee the algorithm termination, a constraint was used: we stop increasing the stiffness of an element if it's $Young$'s modulus value reaches a predefined threshold. Note that our goal is not to compute a realistic deformation of the mesh. The virtual mechanical medium is merely used to compute the inner relaxation of the

---

Algorithm 1 Surface representation

Require: source mesh and destination mesh.
  Let E be the set of elements in the source mesh, defining the mechanical model.
  Let S be the set of surface nodes in the source mesh.
  Let I be the set of inner nodes in the source mesh.
  Let D be the destination surface mesh.
  for all surface nodes P in S **do**
    compute the projection vector of P on D: $U(P)$.
  end for
  Let $Step$=0.
  repeat
    for all surface nodes P in S **do**
      Compute displaced node P': $P' = P + U(P)/\text{MAX\_STEP}$.
    end for
    Let S' be the set of resulting surface nodes positions
    Compute the inner nodes deformations using the mechanical model E, constrained by the new surface nodes positions S'
    Let I' be the resulting inner nodes positions
    for all elements in E **do**
      Let Q be the quality of E given the new nodes positions S' and I'
      Let Y be the stiffness of the current element
      if Q is not acceptable and $Y < \text{Y\_MAX}$ **then**
        increase Y in E: $Y = 2 * Y$
      end if
    end for
    if no change in element stiffness have occurred in E **then**
      accept the deformation and proceed to next step: $S = S'$, $I = I'$ and $Step = Step + 1$
    end if
  until $Step = \text{MAX\_STEP}$

---

nodes at each deformation step: the $Youngs$ modulus arbitrary incrementation prevents the most exposed elements from being excessively deformed and possibly degraded.

The entire process can be seen in algorithm 1. Note that in the Do-While loop, if some elements need to be stiffened, the deformation for a given step is redone using the same initial node positions S and I along with the updated elements E. Another very important remark is that this method do not insert new points to the input mesh.

6 Results

6.1 Quality of the mesh

Measurements of the elements quality are performed before and after the projection onto the surface. Before the projection the quality of each element ranges between 1.3 and 2.7, which is quite normal since the octree-based algorithm can generate high quality elements. After the projection, 97% of the elements has a quality value between 1.3 and 7. This can be seen in figure 10. In addition, table 1 shows the quality measurement before and after the projection of the nodes onto the input domain. The average quality values don't drastically change from one to another.

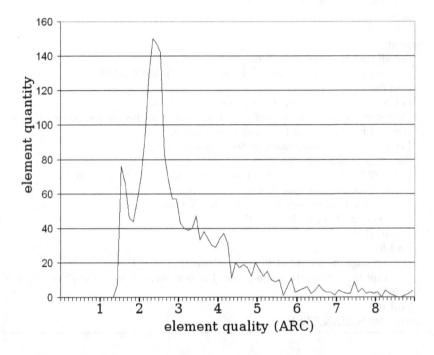

Fig. 10. Quality measurement of the elements after the projection.

The region of interest is mostly represented by hexahedron due to the local refinement of this zone as explained in section 5.2. After projection, the aspect-ratio of the hexahedra is not drastically increased thus the deformation of these elements is reasonable. The same happens with the pyramids and the situation only changes for the tetrahedra, although the aspect-ratio average is still acceptable. We can conclude that the quality in the RoI remains good enough for the simulation. Note that in the presented example no prism was generated. This is because, in our experience, even though they can be found in a mesh, these elements are very rare (just a few patterns consider this type of element).

	Element type	Aspect ratio average	quantity of elements
before projection	Hexahedra	1.27056	72
	Prism	0	0
	Pyramid	2.1387	386
	Tetrahedra	1.58253	1418
	Total	1.68499	1876
after projection	Element type	Aspect ratio average	quantity of elements
	Hexahedra	1.35187	72
	Prism	0	0
	Pyramid	1.7168	386
	Tetrahedra	3.27791	1418
	Total	2.88278	1876

Table 1. Aspect-ratio per element type average before and after the projection onto the surface.

6.2 Modeling

The final mesh is used to simulate the brain-shift with the help of MRI images. Figure 11 shows how the output mesh is finally put together with the initially scanned images.

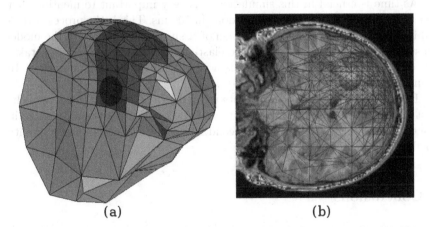

(a)   (b)

Fig. 11. MRI images with the final mesh.

The tumor is simulated by a sphere and it can be seen how the mesh is congruently more refined in the RoI, i.e. the path between the Opening Skull Point and the tumor.

The FEM deforms the mesh and with this information, the images are updated to show the tissue displacement. A linear elastic small deformation framework is chosen for this computation ($E = 1kPa, \nu = 0.45$). The set of pictures in figure 12 shows how this is done. The first picture shows the initial shape of the brain following the skull aperture. The four points in the picture are control points used to change the model boundary conditions assuming some measurements of the intra-operative changes. The second picture shows the brain deformations induced by such changes in the model boundary conditions, using the control points. And the third picture simulates the resection (removal) of the tumor.

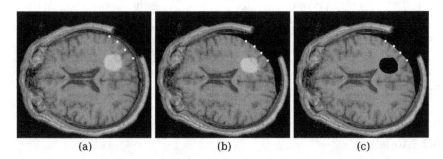

(a)   (b)   (c)

Fig. 12. (a) the initial input with skull aperture, (b) brain-shift simulation using 4 control points and (c) update of the model with tumor resection

As time is crucial in this simulation it is very important to mention that the computation of a deformation is done in 100 ms. This is the process (a) → (b) in figure 12. When there is resection of tissue a main update to the model must be done (re-computation of the elasticity matrix), this operation takes 12 sec for the presented case which is a very acceptable time for surgery. In figure 12 this corresponds to (b) → (c).

Finally figure 13 shows the outputs using different RoIs. In the case of the first one, at the end it has 628 points and the quality coefficient average in the entire mesh is 2.52882. In the second one it has 617 points and a quality coefficient average of 2.9747.

7 Conclusions

This paper aimed at proposing a method to automatically generate a 3D mesh of the brain adapted to the constraints of an intra-operative use, i.e. a good representation of the area between the targeted tumor and the Opening Skull Point, with a coarser mesh elsewhere in order to allow a fast Finite Element computation. The method was successfully evaluated on a given brain geometry with different simulations for tumor and SOP locations. The algorithm generates a mixed-element mesh that achieves correct surface representation. The overall mesh quality is preserved in the final mesh although the termination rule can lead to excessive quality degradation for some elements. In our

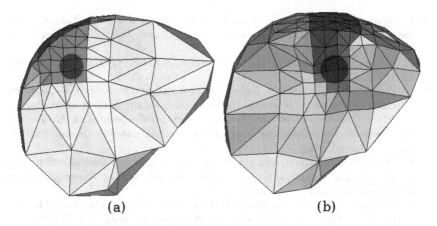

(a)   (b)

Fig. 13. (a) a refinement with a RoI in the top-back section of the brain and (b) a refinement with a RoI in the top-middle section of the brain.

experience the degraded elements quantity represents less than 0.1% of the total elements count and a subsequent sliver removal procedure [3] should be applied with little impact on the final mesh topology.

This method could also be used to adapt a mesh in an inter-subject context for fitting a generic mesh on specific patient data. In particular, the issues encountered in the Mesh-Matching procedure as presented in [4] and [13] should be overcome by the mechanical approach to the displacement of inner nodes.

8 Acknowledgement

Francisco Galdames and Fabrice Jaillet for the input surface mesh of the brain used in the explanation of the developed algorithm. This project has been financially supported by FONDECYT 1061227, ALFA IPECA project, FONDEF D04-I-1237 and ECOS-Sud C06E04.

References

1. P. Alliez, D. Cohen-Steiner, M. Yvinec, and M. Desbrun. Variational tetrahedral meshing. *ACM Transactions on Graphics*, 24:617–625, 2005. SIGGRAPH '2005 Conference Proceedings.
2. M. Chabanas, V. Luboz, and Y. Payan. Patient specific finite element model of the face soft tissue for computer-assisted maxillofacial surgery. *Medical Image Analysis*, 7:131–151, 2003. Issue 2.
3. S.W. Cheng, T.K. Dey, H. Edelsbrunner, M.A. Facello, and S.H. Teng. Sliver exudation. *Journal of the ACM*, 47:883–904, 2000.

4. B. Couteau, Y. Payan, and S. Lavalle. The mesh-matching algorithm: an automatic 3d mesh generator for finite element structures. *Journal of Biomechanics*, 33:1005–1009, 2000.
5. B. Delaunay. Sur la sphère vide. *Bull. Acad. Sci. USSR(VII)*, pages 793–800, 1934.
6. M. Ferrant, S.K. Warfield, A. Nabavi, F.A. Jolesz, and R. Kikinis. Registration of 3d intraoperative mr images of the brain using a finite element biomechanical model. In *Proceedings of the Third International Conference on Medical Image Computing and Computer-Assisted Intervention*, pages 19–28, London, UK, 2000. Springer-Verlag.
7. P.J. Frey, H. Borouchaki, and P.L. George. Delaunay tetrahedralization using an advancing front approach. In *5th International Meshing Roundtable*, pages 31–46. Sandia National Laboratories, 1996.
8. N. Hitschfeld. Generation of 3d mixed element meshes using a flexible refinement approach. In *Engineering with Computers*, volume 21, pages 101–114, 2005.
9. J. Hu, X. Jin, and L. et al Zhang. Intraoperative brain shift prediction using a 3d inhomogeneous patient-specific finite element model. *Journal of Neurosurgery*, 106:164–169, 2007.
10. X. Li, J-F. Remacle, N. Chevaugeon, and M. S. Shephard. Anisotropic mesh gradation control. In *Thirteenth International Meshing Roundtable*, pages 401–412. Sandia National Laboratories, September 2004.
11. C. Lobos and N. Hitschfeld. 3d noffset mixed-element mesh generator approach. In *14th International Conference in Central Europe on Computer Graphics, Visualization and Computer Vision*, pages 47–52, 2006.
12. W. Lorensen and H. Cline. Marching cubes: A high resolution 3d surface construction algorithm. In *Proceedings of the 14th annual conference on Computer graphics and interactive techniques*, volume 21, pages 163–169. ACM Press, July 1987.
13. V. Luboz, M. Chabanas, P. Swider, and Y. Payan. Orbital and maxillofacial computer aided surgery: Patient-specific finite element models to predict surgical outcomes. *Computer Methods in Biomechanics and Biomedical Engineering*, 8:259–265, 2005.
14. L. Marechal. A new approach to octree-based hexahedral meshing. *10th International Meshing Roundtable*, pages 209–221, October 2001.
15. M. Nesme, F. Faure, and Y. Payan. Hierarchical multi-resolution finite element model for soft body simulation. *Lecture Notes in Computer Science*, 4072:40–47, july 2006.
16. I. Reinertsen, M. Descoteaux, K. Siddiqi, and D.L. Collins. Validation of vessel-based registration for correction of brain shift. *Medical Image Analysis*, 11:374–388, 2007. doi: http://dx.doi.org/10.1016/j.media.2007.04.002.
17. R. Schneiders. Octree-based hexahedral mesh generation. *Int. J. of Comp. Geom. & Applications*, 10:383–398, 2000.
18. J. R. Shewchuk. What is a good linear element? interpolation, conditioning, and quality measures. In *Eleventh International Meshing Roundtable*, pages 115–126. Sandia National Laboratories, 2002.
19. TetGen. *A Quality Tetrahedral Mesh Generator*. http://tetgen.berlios.de.
20. Y. Zhang and C.L. Bajaj. Adaptive and quality quadrilateral/hexahedral meshing from volumetric imaging data. In *Thirteenth International Meshing Roundtable*, pages 365–376. Sandia National Laboratories, 2004.

4.3

Geometric Algorithms for 3D Interface Reconstruction

Hyung Taek Ahn[1] and Mikhail Shashkov[2]

[1] Los Alamos National Laboratory `htahn@lanl.gov`
[2] Los Alamos National Laboratory `shashkov@lanl.gov`

Summary. We describe geometrical algorithms for interface reconstructions for 3D generalized polyhedral meshes. Three representative piece-wise linear interface calculation methods are considered, namely gradient based method, least squares volume-of-fluid interface reconstruction algorithm, and moment-of-fluid method. Geometric algorithms for the 3D interface reconstructions are described. Algorithm for the intersection of a convex polyhedron with half-space is presented with degenerate cases. Fast iterative methods for volume matching interface computation are introduced. The numerical optimization method for interface normal computation is presented, and its super-linearly convergence is demonstrated. Finally, actual reconstruction of complex geometry is demonstrated.

Key words: Interface reconstruction, polyhedral mesh, volume of fluid, moment of fluid

1 Introduction and Background

There are several well established methods for dealing with interfaces in fluid flow: the volume of fluid (VoF) method, [8, 16, 2]; the front tracking, [22, 6, 21]; and level set method,[20, 19, 13]. For modeling three-dimensional, high-speed compressible, multi-material flows, on general meshes with the interface topology changing in time, and when exact conservation is critical, VoF seems to be method of choice, [2]. The typical VoF method consists of two steps: interface reconstruction (using volume fractions) and updating the volume fractions in time. Excellent reviews of VoF methods and, in particular, general interface reconstruction methods can be found in the following papers [16, 2, 17]. In this paper we are only interested in the *geometric algorithms* for interface reconstruction.

The most common interface representation used by interface reconstruction methods consists of a single linear interface (a line in 2D and a plane in

3D) per cell composed of multi-materials. This class of interface representation is commonly called Piecewise-Linear Interface Calculation (PLIC). The location of the linear interface, for a given volume at a cell, is uniquely defined by the interface normal. There are a number of ways to define the direction of the normal. We will describe three representative methods in Section 3.

Most of the papers related to interface reconstruction deal with the two dimensional case. There are only several papers which describe interface reconstruction in 3D; almost all of them deal with a cartesian mesh and the case of two materials, e.g., [11], and references therein. Descriptions of 3D interface reconstruction methods on distorted and unstructured meshes are very rare and are usually in unpublished reports related to ALE methods, e.g., [10, 4].

The goal of this paper is to explain geometric algorithms for 3D interface reconstructions on generalized polyhedral meshes. We consider three representative PLIC methods: (i) gradient based method (GRAD), (ii) the least squares volume-of-fluid interface reconstruction algorithm (LVIRA) [15, 14], and (iii) most recent moment of fluid (MoF) method [3, 1].

The outline of the rest of this paper is as follows. In Section 2, we describe the concept of generalized polyhedral mesh. In Section 3, we describe the three representative interface reconstruction methods. In Section 4, we introduce the intersection algorithm of convex polyhedron with half-space with degenerate cases. In section 5, we introduce two iterative volume matching interface computation algorithms. In section 6, we describe optimization method for interface normal computation. In section 7, we give a demonstration of interface reconstruction. Finally, we conclude in Section 8.

2 Multi-material interface representation in generalized polyhedral meshes

In this paper, we are interested in a mesh that consists of generalized polyhedra - *generalized polyhedral mesh - (GPM)*. A generalized polyhedron can be thought as a 3D solid obtained from a polyhedron by perturbing positions of its vertices, which makes its faces non-planar as well as the polyhedron itself non-convex.

The geometry of a face whose vertices are not all in a single plane, however, is not unique. Therefore, we adopt a faceted representation to obtain a consistent definition of its geometry (see [5] for more detail). As illustrated in Fig. 1, first we define the "face center" by averaging vertex coordinates associated with the face. Next, the faces of the generalized polyhedral cell are triangulated by using the face center and two vertices of each edge. In the second step, the triangulated generalized polyhedral cell is decomposed into sub-tetrahedra by using triangulated surfaces and one additional vertex inside the cell, called the "cell center", defined by averaging coordinates of all vertices of original polyhedral cell.

Fig. 1. Generalized polyhedral representation of a hexahedron with non-planar faces. The left figure shows the initial hexahedral cell with a non-planar top face (the wire frame of a cube is overlapped to emphasize the non-planar top face), the middle figure shows the surface triangulation of the hexahedron, and the right figure shows a sub-cell decomposition of the hexahedron.

Hence, an m-vertexed polygonal face is divided into m-triangles, and an n-faced polyhedron is further decomposed into $n \times m$ sub-tetrahedra provided that each face is composed of m vertices. For example, the generalized polyhedral representation of a hexahedral cell results in 6×4 sub-tetrahedra as displayed in Fig. 1.

This generalization of polyhedral cells has two advantages. First, the planar face restriction of a polyhedral cell is relaxed so that it can have vertices not always in a plane. Second, it allows us to deal with non-convex cells, as long as the cell can be decomposed into valid sub-cells (that is, the cell center can "see" all the vertices). These features are advantageous for dealing with meshes in ALE methods.

Three different types of generalized polyhedral mixed cells are represented in Fig. 2. Each cell includes three materials (colored red, green, and blue). Each interface is reconstructed by the intersection of the polyhedral cell with half-spaces. The sub-cells, initially tetrahedra, evolve to convex polyhedra as they intersect with their corresponding half-spaces. A wire frame view of these sub-cells, shown in the bottom row of Fig. 2, reveals these sub-cell structures.

3 Interface reconstruction methods

Three representative piece-wise linear interface calculation (PLIC) methods are discussed: the gradient based method, the least squares volume-of-fluid interface reconstruction algorithm, and the moment-of-fluid method.

3.1 Review of representative PLIC methods

In PLIC methods, each mixed cell interface between two materials is represented by plane. It is convenient to specify this plane in *Hessian normal form*

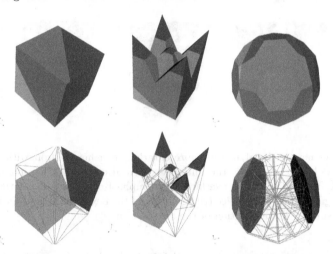

Fig. 2. Generalized polyhedral cells with multi-materials (red, green, and blue). The left column shows a hexahedral cell, the middle column shows a *non-convex* enneahedron (obtained by subdividing the top face of a hexahedron and disturbing the vertices on the faces), and the right column represents a truncated icosahedron. The top row shows the solid view, and the bottom row shows the wire-frame view of the solid which reveals the sub-cell structure.

$$\mathbf{n} \cdot \mathbf{r} + d = 0 \,, \tag{1}$$

where $\mathbf{r} = (x, y, z)$ is a point in the plane, $\mathbf{n} = (n_x, n_y, n_z)$ are components of the unit normal to the plane, and d is the signed distance from the origin to the plane. The principal reconstruction constraint is local volume conservation, i.e. the reconstructed interface must truncate the cell, c, with a volume equal to the reference volume $\mathcal{V}_c^{ref}$ of the material (or equivalently, the volume fraction $f_c^{ref} = \mathcal{V}_c^{ref}/\mathcal{V}_c$, where $\mathcal{V}_c$ is the volume of the entire cell c).

Since a unique interface configuration does not exist, the interface geometry must be inferred based on local data and the assumptions of the particular algorithm. PLIC methods differ in how the normal $\mathbf{n}$ is computed. For a given normal, d is uniquely defined from the reference volume $\mathcal{V}_c^{ref}$.

In the remainder of this section, we briefly describe the main ideas of GRAD, LVIRA and MoF methods, give some details relevant to 3D extensions of these methods, and present a summary of algorithms for their implementation.

3.2 Gradient based interface reconstruction (GRAD)

In the gradient based method, the interface normal, $\mathbf{n}$, is computed by approximating the gradient of the volume fraction function f as

$$\mathbf{n} \sim -\left(\frac{\partial f}{\partial x}, \frac{\partial f}{\partial y}, \frac{\partial f}{\partial z} \right) . \tag{2}$$

In the case of a 3D unstructured mesh consisting of generalized polyhedra, it is convenient to use a least squares procedure (see, for example, [5]) to estimate the gradient of the volume fraction.

To define the interface, one needs to find the distance d in Eq. (1) such that intersection of the corresponding half-space and cell has volume $\mathcal{V}^{ref}$. To find d we need to solve the equation

$$\mathcal{V}(d) = \mathcal{V}^{ref}. \tag{3}$$

The volume $\mathcal{V}(d)$ is a continuous and monotone function of d, which guarantees that Eq. (3) always has a unique solution. Let us note that all PLIC methods require solving Eq. (3) many times. The geometrical algorithms for the intersection of a half-space and a convex polyhedron, and computation of the volume of a polyhedron are presented in the following sections.

3.3 Least squares volume-of-fluid interface reconstruction algorithm (LVIRA)

In the LVIRA interface reconstruction method introduced by Puckett [15, 14], the interface normal is computed by minimizing the following error functional:

$$E_c^{LVIRA}(\mathbf{n}) = \sum_{c' \in \mathcal{C}(c)} (f_{c'}^{ref} - f_{c'}(\mathbf{n}))^2 \tag{4}$$

where $f_{c'}^{ref}$ is the reference volume fraction of neighbor c', and $f_{c'}(\mathbf{n})$ is the actual (reconstructed) volume fraction of neighbor c' taken by extending the interface of central cell-c, under the constraint that the corresponding plane exactly reproduces the volume fraction in the cell under consideration.

The stencil for the error computation in LVIRA is illustrated in Fig. 3, where 2D meshes are employed for simplicity. The neighboring cells around a central cell-c are referenced with index j. The stencil is composed of immediate vertex neighbors. The picture on the left of Fig. 3 represents a structured quadrilateral mesh, and picture on the right shows a stencil on an unstructured polygonal mesh.

Like the GRAD method, LVIRA also requires information about the volume fractions from all immediate neighboring cells. In contrast with the GRAD method, LVIRA requires minimization of the non-linear objective function, as shown in Eq. (4). In 3D, the normal can be described by polar angles, and therefore implementation of LVIRA requires an algorithm for the minimization of a non-linear function of two variables.

3.4 Moment-of-fluid interface reconstruction (MoF)

The *moment-of-fluid* method was introduced by Dyadechko and Shashkov [3, 1] for interface reconstruction in 2D. The MoF method uses information about

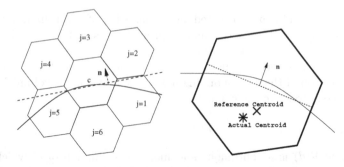

Fig. 3. Left – stencil for LVIRA error computation. The solid curved line represents the true interface, and the dashed straight line represents the extension of a piecewise linear volume fraction matching interface at the central cell-c. Right – stencil for MoF error computation. The stencil is composed of only the cell under consideration.

the volume fraction, f_c^{ref} and centroid, $\mathbf{x}_c^{ref}$ of the material, but *only* from the cell c under consideration. No information from neighboring cells is used, as illustrated in Fig. 3.

In the MoF method, the computed interface is chosen to match the reference volume exactly and to provide the best possible approximation to the reference centroid of the material. That is, in MoF, the interface normal, $\mathbf{n}$, is computed by minimizing (under the constraint that the corresponding pure sub-cell has exactly the reference volume fraction in the cell) the following functional:

$$E_c^{MoF}(\mathbf{n}) = \| \mathbf{x}_c^{ref} - \mathbf{x}_c(\mathbf{n}) \|^2 \tag{5}$$

where $\mathbf{x}_c^{ref}$ is the reference material centroid and $\mathbf{x}_c(\mathbf{n})$ is the actual (reconstructed) material centroid with given interface normal $\mathbf{n}$.

Similar to the LVIRA, the implementation of MoF method requires the minimization of the non-linear function of two variables. The computation of $E_c^{MoF}(\mathbf{n})$ requires the following steps. The first step is to find the parameter d of the plane such that the volume fraction in cell c exactly matches f_c^{ref}; this is also performed in the GRAD and LVIRA algorithms. Secondly we compute the centroid of the resulting polyhedron. This is a simple calculation, described in [12]. Finally, one computes the distance between actual and reference centroids.

4 Intersection of convex polyhedron with half-space

Convex polyhedron intersection with a half-space is the base operation for interface reconstruction in 3D. First the algorithm of intersection is presented for regular case (no vertices on cutting plane), and later issues and strategies for degenerate cases (vertices on cutting plane) will be addressed.

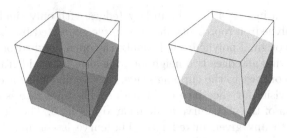

Fig. 4. Convex polyhedron intersection by clipping and capping. The left shows an open hexahedron by clipping the hexahedron with a given plane, and the right shows the closed polyhedron by clipping as well as capping.

4.1 Regular case

Algorithm for intersection of convex polyhedron with a half-space is composed of two different conceptual stages; clipping and capping [18]. The idea of clipping and capping is delineated in Fig. 4 with an example of hexahedron and plane intersection. In clipping stage, each face of polyhedron is visited and polygonal intersection is performed if the cutting plane passes through it. Depending on the distance and orientation of the given plane, it may be no intersection and the face is considered as a *pure* face, *i.e.* the face is completely above or below with respect to the given plane. In the clipping, stage, no specific order is necessary for visiting polyhedron faces, and each face visit can be considered as a polygon and plane intersection in 3D.

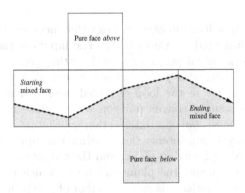

Fig. 5. Unfolded faces of hexahedron. Clipped hexahedron faces are displayed in gray color, and continuation of slice curve (polylines on unfolded plane) for capping is illustrated with dashed arrow lines.

In the latter, capping stage, the polygonal slice face has to be constructed. Without capping, the merely clipped polyhedron will result in an *open* polyhedron as shown in Fig. 4. The boundary (edges with only single neighbor) of the open polyhedron represents the slice curve generated by the given cutting plane and original polyhedron. To make the open polyhedron be a closed polyhedron having all edges two neighbors, the slice face is identified by capping stage. In contrast to the clipping stage, the capping stage needs a proper orders of face-visits for slice face construction. The slice face is constructed by continuation of the slice curve as delineated in Fig. 5. The slice curve can get started with any given mixed face. The curve is continued by looping the adjacent mixed faces until it returns back to the initial mixed face and completing closed slice curve.

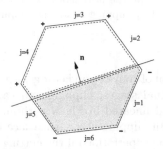

Fig. 6. Convex polygon intersection with a plane of interface normal (**n**) in 3D. Polygon intersection routine returns two sub-polygons indicated by dashed lines: first sub-polygon which is below to the cutting plane (gray part) and second sub-polygon above (void part). New vertices are generated by intersection of the plane and edges with different signs ($j = 2, 5$).

The convex polyhedron intersection algorithm incorporating both clipping and capping is illustrated in Algorithm 1. The inputs of the polyhedron intersection routine is initial *polyhedron* and cutting *plane*, and the outputs are two closed sub-polyhedra; *phed1* below the given cutting plane and *phed2* above the plane. Inside of the loop of mixed faces, polygon intersection is performed. Intersection of convex polygon with a plane in 3D is illustrated in Fig. 6. In the convex polygon intersection subroutine as described in Algorithm 2, like polyhedron intersection routine, the inputs are a *polygon* (a face of *polyhedron*) and cutting *plane*, and the outputs are two closed polygons; *pgon1* which is below the plane and *pgon2* which is above the plane. Polygon intersection algorithm is similar to that of polyhedron intersection as presented in Algorithm 1. The main difference of polygon intersection is edgewise loop is carried out instead of face-wise loop in polyhedron intersection.

4.2 Degenerate cases

Cutting plane does not always intersect edges (break the edge into two parts), and either or both of the vertices can be exactly on the cutting plane. For

Input: *polyhedron, plane*
Output: *phed1, phed2*
foreach *face of polyhedron* **do**
    if *unvisited face* **then**
        if *face below to plane* **then**   /* pure face */
            add this face to *phed1* ;
            mark this face *visited* ;
        else if *face above to plane* **then**   /* pure face */
            add this face to *phed2* ;
            mark this face *visited* ;
        else if *face gets intersection* **then**   /* mixed face */
            $face_{start}$ ← this face ;
            repeat
                perform polygon/plane intersection ;
                add *face below* to *phed1* ;
                add *face above* to *phed2* ;
                mark this face *visited* ;
                $face_{next}$ ← next face ;
            until $face_{start} = face_{next}$;
        end
    end
end

Algorithm 1: Convex polyhedron intersection

Input: *polygon, plane*
Output: *pgon1, pgon2*
foreach *edge of polygon* **do**
    if *edge below to plane* **then**   /* both vertices (-) */
        add this edge to *pgon1* ;
    else if *edge above to plane* **then**   /* both vertices (+) */
        add this edge to *pgon2* ;
    else if *edge gets intersection* **then**   /* vert's sign mixed */
        perform segment/plane intersection ;
        add *edge below* to *pgon1* ;
        add *edge above* to *pgon2* ;
    end
end

Algorithm 2: Convex polygon intersection

example, the vertices of polyhedron can be exactly (or within some tolerance) on the given plane. This results in degenerate cases, as delineated in Fig. 7. Degenerate cases requires two additional considerations. First, in polygon intersection routine, additional vertex may not be generated by intersection (of plane and edge in 3D), instead an existing vertex is used for it. Second, for the continuation of the slice curve as delineated in Fig. 8, next adjacent face should be found carefully because not only the mixed faces but also the pure

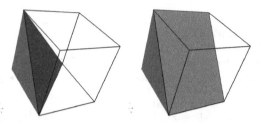

Fig. 7. Degenerate cases in polyhedron intersection: vertices on the intersecting plane (left), and vertices as well as edges on the intersection plane (right).

faces (if an edge is on the cutting plane) can be the candidate for the next face.

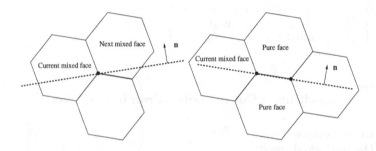

Fig. 8. Degenerate cases in polygon intersection: one vertex on the intersecting plane (left), and two vertices (edge) on the intersection plane (right). Cutting plane is delineated with dashed line, and vertices on the plane is marked with •.

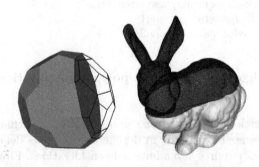

Fig. 9. Intersection of complex polyhedra. Left shows 32 faced truncated icosahedron, and right shows 725,000 faced bunny mesh. Both surface mesh represents a polyhedron.

4.3 Test cases

The convex polyhedron intersection algorithm is applied for more general cases in Fig. 9. Two polyhedra intersected by the present algorithms are displayed. First the intersection of truncated icosahedron (*a.k.a* soccer ball geometry with 12 pentagons and 20 hexagons) is presented, and next the polyhedral representation of bunny (725,000 triangular surface mesh) is intersected. The bunny geometry is not an example of convex polyhedron, but as long as the slice face is *simply connected* and the polyhedron is convex-faced (triangulated surface here) the current algorithm can be applied.

4.4 Volume and centroid computation

For each polyhedron intersection, volume and centroids of the intersected sub-cell have to be computed for measuring the error of interface reconstructed. For this purpose, fast and accurate computation of moment data of general polyhedron is indispensable, and our implementation is based on [12]. The algorithm is based on multi-step reduction of the volume integral to successively lower dimensions by using Divergence and Green's theorems.

5 Volume matching interface computation

In this section, the target volume fraction matching interface calculation, with *given normal*, methods are presented. The primary mechanism of volume preserving interface reconstruction requires cutting appropriate volume fraction of the cell, as expressed in Eq. (3). The equation can also be expressed as follows

$$f(d) = f^{ref}$$

where $f(d) = \mathcal{V}(d)/\mathcal{V}_{cell}$ is volume fraction defined by d and $f_{ref} = \mathcal{V}^{red}/\mathcal{V}_{cell}$ is reference (target) volume fraction, which are both normalized by cell volume $\mathcal{V}_{cell}$. Since the normal (orientation) of cutting plane is given, the volume of intersection is purely function of distance, $d \in [d_{min}, d_{max}]$. For example, $f(d_{min}) = 0$ and $f(d_{max}) = 1$.

Several approaches are proposed, but they are mainly described in 2D. These methods, *e.g.* analytical method [3] and semi-iterative method [16], require two pre-processing: first vertex-wise volume fraction evaluation ($O(n)$ volume fraction evaluation) and then another vertex-wise volume fraction sorting ($O(n \log n)$ operations in sorting), where n is number of vertices.

The analytical approach can be extended for tetrahedral cell in 3D [23]. For cells with small number of vertices, such as triangles in 2D and tetrahedra in 3D, this pre-processing and analytical approach could save CPU time. As the cells contains more vertices, typical for 3D polyhedral cells, these pre-processing demand considerable amount of CPU time as well as extra memory space besides the implementation efforts.

In order to cut target volume fraction accurately as well as efficiently two fully iterative schemes are employed, namely secant method and bisection method. The algorithm for the iterative methods is described in Algorithm 3.

Input: f^{ref}, **n**, d_{min}, d_{max}
Output: d
$d_1 = d_{min}$;
$d_2 = d_{max}$;
$f_1 = 0$;
$f_2 = 1$;
repeat

> **if** *Secant method used* **then**
> > | $secant = (f_2 - f_1)/(d_2 - d_1)$;
>
> **end**
> update d by Secant or Bisection method;
> intersect cell with defined interface $(\mathbf{n}, d)$;
> compute $f(d)$;
> $\Delta f = |f^{ref} - f(d)|$;
> update d_1, d_2, f_1, f_2;

until *($\Delta f < tol$)* ;

Algorithm 3: Iterative volume fraction matching interface computation

---

These iterative schemes have too distinctive advantages:

1. no pre-processing: vertex-wise volume fraction evaluation or sorting
2. fixed number of iteration regardless of number of vertices

First, no vertex-wise volume fraction evaluation or sorting is needed. For the start of the iteration, only the minimum and maximum distances with respect the the given interface normal are required. This is because of the monotonically increasing behavior (actually C^1 if the cell is convex and C^0 if not) of volume fraction with respect to the distance. Second, both iterative schemes are converging in almost fixed number of iterations regardless of cell size. In bisection method, with unit interval of distance $[0,1]$ the number of iterations required to achieve distance error tolerance of $tol = 10^{-10}$ is

$$\log_2 \frac{1}{10^{-10}} = 33.2193$$

regardless of function behavior [7]. Due to monotonicity of the function, the volume fraction error tolerance of $tol = 10^{-10}$ is also achieved with approximately same number of bisection iterations as shown in Fig. 10.

Fig. 10 shows the volume fraction convergence history of the two iterative methods applied to three polyhedral cells shown in Fig. 2. First, the secant method shows super linear convergence in volume fraction error. Less than 10

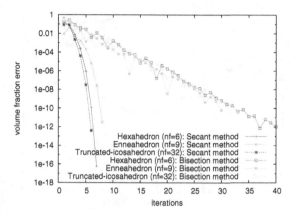

Fig. 10. Volume fraction convergence of secant and bisection methods with three polyhedral cells shown in Fig. 2.

Fig. 11. Spherical cell containing two materials (blue and red) with successive refinement. From the coarsest (left) mesh to the finest (right), the numbers of faces are 80, 320, 1280, and 5120.

iterations are required to achieve the volume fraction error $< 10^{-10}$. Bisection method shows linear convergence, but it guarantees that only fixed number of iteration is required regardless of the number of vertices, n, for the cell.

The efficiency of iterative methods are further demonstrated with large size spherical cells as displayed in Fig. 11. Four levels of successively refined spherical surface meshes are used as a single polyhedral cell representation. In Fig. 12, using the four levels of spherical cells, the number of secant iteration to achieve $err(f) < 1.e - 10$ is measured with target volume fraction between [0,1]. The number of iteration required is irrespective to the size of polyhedral cell, and the target volume fraction is achieved almost less than 10 iterations.

6 Numerical optimization

The second-order accurate interface reconstruction methods, LVIRA and MoF, require additional optimization process minimizing the error function-

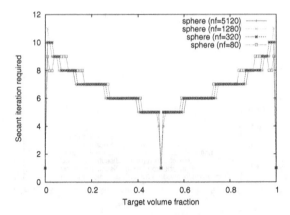

Fig. 12. Number of secant iterations required for appropriate volume fraction cutting measured with four levels of spherical cells displayed in Fig. 11

als. In both case optimization has to be performed with respect to the normal **n** in the Eq. (1). In 3D normal **n** is defined by two polar angles (ϕ, θ).

Fig. 13. Interface configuration by a sphere, centered at $(-0.1, -0.2, -0.3)$ with radius $r = 1.3$, and equispaced 3^3 hexahedral mesh covering the domain of $[0, 1]^3$. For visualization purpose, a fraction of transparent sphere and the central cell of the mesh are displayed.

By using the spherical interface configuration delineated in Fig. 13, typical behavior of the objective functions are displayed in Fig. 14. For the interface configuration in Fig. 13, sphere centered at $(-0.1, -0.2, -0.3)$ with radius $r = 1.3$ is used and equispaced $3 \times 3 \times 3$ hexahedral mesh covering the domain of $[0, 1] \times [0, 1] \times [0, 1]$ is used. For visualization purpose, a transparent fraction of the sphere and the cell centered at $(0.5, 0.5, 0.5)$ are displayed.

The behavior of objective functions for the center cell shown in Fig. 13 are displayed in the top row of Fig. 14. General trends of these objective functions

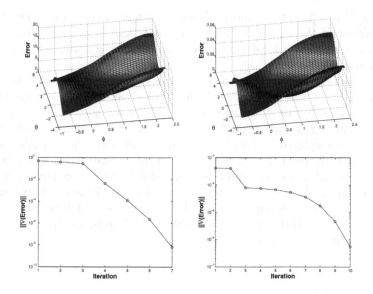

Fig. 14. Behavior of objective function and optimization of it: Left — LVIRA, and right — MoF. Top row is 3D view of the objective functions, and the bottom row shows the convergence history of optimization process.

are similar for both LVIRA and MoF. However, the scale of absolute values of the functions are different. This is because LVIRA uses accumulated volume fraction difference from neighbors and MoF uses normalized distances between centroids as the objective function.

For the above multi-dimensional minimization, Broyden-Fletcher-Goldfarb-Shanno (BFGS) method [9] is used. It is a quasi Newton method, approximating Hessian matrix with a set previous of gradients. The gradients of the objective function are computed by finite differences. For each search direction, a quadratic or cubic polynomial line search is performed for sufficient decrease in the error with the Armijo rule for step size control. Detailed discussion of the BFGS method can be found in [9]

For the initial guess of the optimization, the gradient of volume fraction computed as in GRAD is utilized for LVIRA.

$$\mathbf{n}_0(\phi_0, \theta_0)_{LVIRA} = -\mathbf{GRAD}(f). \tag{6}$$

For MoF, the unit vector along the given material centroid to the cell centroid is used as follows

$$\mathbf{n}_0(\phi_0, \theta_0)_{MoF} = \frac{\mathbf{x}_c^{cell} - \mathbf{x}_c^{ref}}{\| \mathbf{x}_c^{cell} - \mathbf{x}_c^{ref} \|} \tag{7}$$

where $\mathbf{x}_c^{cell}$ is cell centroid and $\mathbf{x}_c^{ref}$ is the reference material centroid. Once the gradient of the objective function becomes less than a given tolerance, it is considered that a local minimum is found and the optimization process terminates.

The convergence history of the LVIRA and MoF are displayed in the bottom row of Fig. 14. Both LVIRA and MoF show super-linear convergence rate, and $\parallel \nabla(Error) \parallel < 10^{-6}$ are achieved with 10 iterations.

The final reconstructed interface for the mixed cell configuration shown in Fig. 13 is displayed in Fig. 15. Fraction of original spherical interface is overlapped with transparency. Depending on the reconstruction methods, the interface normal is different and this results different interface reconstruction as shown in the figure. The volumes of symmetric difference, between the original and reconstruction, at the cell are measured as follows: 6.1616e-04 (GRAD), 5.9999e-04 (LVIRA), 5.5676e-04(MoF). This result strengthens that MoF gives the best accuracy.

Fig. 15. Interface reconstruction of the configuration shown Fig. 13. Left – GRAD, middle – LVIRA, and right – MoF.

7 Reconstruction of complex interfaces

The effectiveness of the geometric algorithms is demonstrated by reconstructing complex interfaces, as shown in Fig. 16. The initial geometry of the object is given by a surface mesh, and then a tetrahedral volume mesh is generated based on it. Tetrahedron-tetrahedron intersection is performed to compute the volume fraction and moment data exactly. The reconstructed object by MoF method is displayed in Fig. 16.

8 Conclusion

We have described geometric algorithms for 3D interface reconstructions with 3D generalized polyhedral meshes. Three different reconstruction methods are

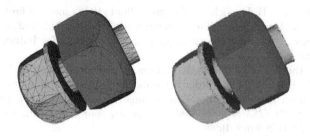

Fig. 16. Multi-material ($n_{mat} = 3$, i.e. bolt, nut, and background) interface reconstruction with MoF. The left represents original material region represented by tetrahedral volume meshes, and the right shows its reconstruction with the MoF method on an unstructured tetrahedral mesh.

considered, namely GRAD, LVIRA, and MoF. Intersection algorithm for convex polyhedron with a half-space is presented including degenerate cases. Fast iterative methods for volume fraction matching interface calculation are presented. Optimization algorithm for second order accurate interface reconstruction methods (LVIRA and MoF) is explained and super-linear convergence of it is demonstrated. Performance of the methods is demonstrated with actual reconstruction of complex geometry.

Acknowledgement. This work was supported by the Advanced Simulation and Computing (ASC) program at the Los Alamos National Laboratory.

References

1. H. T. Ahn and M. Shashkov. Multi-material interface reconstruction on generalized polyhedral meshes. *Journal of Computational Physics*, In press, 2007.
2. D. J. Benson. Volume of fluid interface reconstruction methods for multi-material problems. *Applied Mechanics Reviews*, 55:151–165, 2002.
3. V. Dyadechko and M. Shashkov. Moment-of-fluid interface reconstruction. Technical Report LA-UR-05-7571, Los Alamos National Laboratory.
4. D. M. Gao. A three-dimensional hybrid finite element-volume tracking model for mould filling in casting processes. *International Journal for Numerical Methods in Fluids*, 29:877–895, 1999.
5. R. Garimella, M. Kucharik, and M. Shashkov. An efficient linearity and bound preserving conservative interpolation (remapping) on polyhedral meshes. *Computers and Fluids*, 36:224–237, 2007.
6. J. Glimm, J. W. Grove, X. L. Li, K. Shyue, Y. Zeng, and Q. Zhang. Three-dimensional front tracking. *SIAM Journal on Scientific Computing*, 19:703–727, 1998.
7. Michael T. Heath. *Scientific Computing: An Introductory Survey, Second Edition*. McGraw-Hill, New York, 2002.

8. C. W. Hirt and B. D. Nichols. Volume of fluid (VOF) method for the dynamics of free boundaries. *Journal of Computational Physics*, 39:201–225, 1981.
9. C.T. Kelly. *Iterative methods for optimization*. Society for Industrial and Applied Mathematics, 1999.
10. D. B. Kothe, M. W. Williams, K. L. Lam, D. R. Korzekwa, P. K. Tubesing, and E. G. Puckett. A second-order accurate, linearity-preserving volume tracking algorithm for free surface flows on 3-d unstructured meshes. In *Proceedings of the 3rd ASME/JSME joint fluid engineering conference, San Francisco, CA, USA, fEDSM99-7109*, 1999.
11. P. Liovic, M. Rudman, J.-L. Liow, D. Lakehal, and D. Kothe. A 3d unsplit-advection volume tracking algorithm with planarity-preserving interface reconstruction. *Computers & Fluids*, 35:1011–1032, 2006.
12. B. Mirtich. Fast and accurate computation of polyhedral mass properties. *Journal of Graphics Tools*, 1:31–50, 1996.
13. S. Osher and R. P. Fedkiw. Level set methods: An overview and some recent results. *Journal of Computational Physics*, 169:463–502, 2001.
14. J. E. Pilliod and E. G. Puckett. Second-order accurate volume-of-fluid algorithms for tracking material interfaces. *Journal of Computational Physics*, 199:465–502, 2004.
15. E.G. Puckett. A volume-of-fluid interface tracking algorithm with applications to computing shock wave refraction. In H. Dwyer, editor, *Proceedings of the Fourth International Symposium on Computational Fluid Dynamics*, pages 933–938, 1991.
16. W. J. Rider and D. B. Kothe. Reconstructing volume tracking. *Journal of Computational Physics*, 121:112–152, 1998.
17. R. Scardovelli and S. Zaleski. Direct numerical simulation of free-surface and interfacial flow. *Annual Review of Fluid Mechanics*, 31:567–603, 1999.
18. Michael B. Stephenson and Henry N. Christiansen. A polyhedron clipping and capping algorithm and a display system for three dimensional finite element models. *ACM SIGGRAPH Computer Graphics*, 9:1–16, 1975.
19. M. Sussman, A. S. Almgren, J. B. Bell, P. Colella, L. H. Howell, and M. L. Welcome. An adaptive level set approach for incompressible two-phase flows. *Journal of Computational Physics*, 148:81–124, 1999.
20. M. Sussman, E. Fatemi, P. Smereka, and S. Osher. Improved level set method for incompressible two-phase flows. *Computers & Fluids*, 27:663–680, 1998.
21. G. Tryggvasona, B. Bunnerb, A. Esmaeelic, D. Juricd, N. Al-Rawahic, W. Tauberc, J. Hanc, S. Nase, and Y.-J. Jan. A front-tracking method for the computations of multiphase flow. *Journal of Computational Physics*, 169:708–759, 2001.
22. S. O. Unverdi and G. Tryggvason. A front-tracking method for viscous, incompressible, multi-fluid flows. *Journal of Computational Physics*, 100:25–37, 1992.
23. Xiaofeng Yang and Ashley J. James. Analytic relations for reconstructing piecewise linear interfaces in triangular and tetrahedral grids. *Journal of Computational Physics*, 214:41–54, 2006.

κ-Compatible Tessellations*

Philippe P. Pébay and David Thompson

Sandia National Laboratories
P.O. Box 969, Livermore CA 94551, U.S.A.
{pppebay,dcthomp}@sandia.gov

Summary. The vast majority of visualization algorithms for finite element (FE) simulations assume that linear constitutive relationships are used to interpolate values over an element, because the polynomial order of the FE basis functions used in practice has traditionally been low – linear or quadratic. However, higher order FE solvers, which become increasingly popular, pose a significant challenge to visualization systems as the assumptions of the visualization algorithms are violated by higher order solutions. This paper presents a method for adapting linear visualization algorithms to higher order data through a careful examination of a linear algorithm's properties and the assumptions it makes. This method subdivides higher order finite elements into regions where these assumptions hold (κ-compatibility). Because it is arguably one of the most useful visualization tools, isosurfacing is used as an example to illustrate our methodology.

1 Introduction

People have been approximating solutions to partial differential equations (PDEs) ever since PDEs were conceived. Much more recently, a class of techniques known as *hp*-adaptive methods have been developed in an effort to converge to a solution faster than previously possible, or to provide more accurate approximations than traditional finite element simulation within the same amount of computational time. These new techniques can increase both the hierarchical (h) and polynomial (p) levels of detail – or degrees of freedom – during a simulation.

Once these solution approximations have been computed, they must be characterized in some way so that humans can understand and use the results. This paper develops a technique for partitioning higher-order cells in order to

*This work was supported by the United States Department of Energy, Office of Defense Programs. Sandia is a multiprogram laboratory operated by Sandia Corporation, a Lockheed-Martin Company, for the United States Department of Energy under contract DE-AC04-94-AL85000.

characterize the behavior of their geometric and scalar field curvatures during post-processing. We say that such partitions are κ-*compatible*. While a past paper [8] has presented our software framework for creating these partitions, this paper presents a full description of the algorithm and a rigorous proof of the conditions under which it will work and terminate.

Currently, visualization techniques for quadratic and higher order FE solutions are very limited in scope and/or cannot guarantee that all topological features are captured [1, 3, 7, 8, 6]. These techniques are also limited in their applicability to a subset of visualization techniques. Moreover, although some production-level tools currently offer support for quadratic elements, they do not always do so correctly (*cf.* [8] for a discussion); in the case of isocontouring, consider for example the following scalar field:

$$\Phi_1 : \begin{array}{l} [-1,1]^3 \longrightarrow \mathbb{R} \\ (x,y,z)^T \mapsto x^2 + y^2 + 2z^2, \end{array}$$

interpolated over a single Q2 Lagrange element (hexahedron with degree 2 Lagrange tensor-product interpolation over 27 nodes), with linear geometry, where one is interested in the isocontours $\Phi_1^{-1}(1)$ and $\Phi_1^{-1}(2)$.

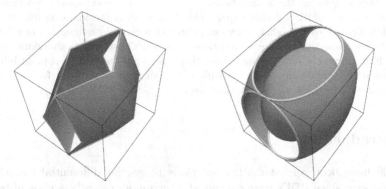

Fig. 1. Isocontours of Φ_1 for the isovalues 1 (cyan) and 2 (green): linear isocontouring approach (left), and our new topology-based approach (right).

As shown in Figure 1, left, the linear isocontouring approach (implemented here in ParaView) completely misses $\Phi_1^{-1}(1)$, because it is entirely contained within the cell. Meanwhile, the topology of $\Phi_1^{-1}(2)$ is correct, but its geometry is poorly captured. On the other hand, our new method (Figure 1, right) captures the correct topologies of both isocontours, and provides a much better geometric approximation of $\Phi_1^{-1}(2)$ than linear isocontouring does.

The lack of tools that are applicable to most visualizations of higher order element simulations, and that can guarantee correctness of the results, prevents analysts from exploiting such simulations. In this paper, we propose a solution to this problem.

1.1 Higher Order Finite Elements

Here we briefly review the FE method to develop notation used throughout the paper. Recall that the FE method approximates the solution, $f : \Omega \mapsto \mathbb{R}$, of some PDE as a set of piecewise functions over the problem domain, $\Omega \subset \mathbb{R}^d$. Although Ω may be any general d-dimensional domain, we'll assume it is 3-dimensional. The fact that we have a piecewise approximant divides Ω into subdomains $\Omega_e \subseteq \Omega$, $e \in E_\Omega$ that form a closed cover of Ω. Each Ω_e is itself a closed set that is parametrized with a map (usually polynomial in form) from parametric coordinates $\mathbf{r} = (r, s, t) \in R \subset \mathbb{R}^3$ to geometric coordinates $\mathbf{x} = (x, y, z) \in \Omega_e$:

$$\Xi_e(\mathbf{r}) = B_{0,0,0} + B_{1,0,0}r + B_{0,1,0}s + B_{0,0,1}t + B_{1,1,1}rst + \cdots$$

where $B_{i,j,k} \in \mathbb{R}^3$ such that Ξ_e is invertible for all points of Ω_e. Therefore, the approximate solution may be written in terms of parametric coordinates (as Φ_e) or in terms of geometric coordinates: $f_e(\mathbf{x}) = \Phi_e \circ \Xi_e^{-1}(\mathbf{x})$, where

$$\Phi_e : R \subset \mathbb{R}^3 \longrightarrow \mathbb{R}$$
$$(r, s, t)^T \mapsto \Phi_e(r, s, t)$$

is the approximating function over Ω_e expressed in terms of parametric coordinates. Furthermore, we require that each Φ_e is polynomial in the parameters:

$$\Phi_e(\mathbf{r}) = A_{0,0,0} + A_{1,0,0}r + A_{0,1,0}s + A_{0,0,1}t + A_{1,1,1}rst + \cdots$$

These coefficients, $A_{i,j,k} \in \mathbb{R}$, are known as *degrees of freedom* (DOFs). Each one corresponds to a particular modal shape, and sets of modal shapes can be grouped together into nodes by the regions of parameter-space over which they have an effect (a given corner, edge, face, or the interior volume).

A global approximant $f(\mathbf{x})$ can then be constructed from the piecewise elemental approximants. This leaves only the matter of what to do where Ω_e and $\Omega_{j,j \neq e}$ intersect. Usually, these subdomains intersect over $(d - 1)$-dimensional or lower regions (2-dimensional faces, 1-dimensional edges, and "0"-dimensional vertices in our case). In these regions, Φ is not well-defined since Φ_e and Φ_j may disagree. Usually, the FE method constrains Φ_e and Φ_j to be identical in these regions, however some methods such as the discontinuous Galerkin method do not require this and subsequently have no valid approximant in these regions.

For a given decomposition of Ω, the FE method may not converge to the correct (or indeed, any) solution. When Φ_e and Ξ_e are trilinear polynomials for all e, a technique called h-adaptation is often used to force convergence and/or increase solution accuracy. In this technique, some subdomains $E \subseteq E_\Omega$ are replaced with a finer subdivision E' such that $\bigcup_{e \in E} \Omega_e = \bigcup_{e \in E'} \Omega_e$ but $|E'| > |E|^2$. Similarly, p-adaptation is the technique of increasing the

---

[2]In temporal simulations, we do allow $|E'| < |E|$ in regions where Ω has been adapted to some time-transient phenomenon.

order of polynomials Φ_e and/or Ξ_e rather than the number of finite elements. hp-adaptation is then some combination of h- and p-adaptation over Ω. In the end, the FE method provides an approximation to $f = \Phi \circ \Xi^{-1}$ by solving a collection of equations for coefficients (A and B in the examples above). Our task is then to characterize the maps Φ_e and Ξ_e in a way that aids human understanding of the solution.

The rest of this paper presents a method for partitioning higher-order finite elements into regions were visualization assumptions are satisfied.

2 Partitioning Finite Elements

Using these notation, we now examine what requirements must be satisfied by κ-compatible tessellations. A common assumption made by linear visualization algorithms is that critical points (points where all partial derivatives of f_e vanish) may only occur at vertices. This one assumption can show up in many different ways. Algorithms that iterate over an element's corner vertices to identify extrema (*e.g.*, thresholding) all make this assumption. Other examples include (but are not limited to) isosurfacing, cutting, and clipping.

This paper is concerned with tessellating finite elements into regions where the critical-points-at-vertices assumption holds as part of an overall strategy to adapt existing techniques to work with higher order elements. Because we use it as a running example of an important application driving the development of κ-compatible tessellation, we will focus on the prerequisites of the linear tetrahedral isosurfacing algorithm, keeping in mind that other visualization techniques result in the same set of constraints. The linear tetrahedral isosurfacing algorithm assumes:

(C1^0) each tetrahedron edge intersects a particular isocontour at most once,
(C2^0) no isocontour intersects a tetrahedron face without intersecting at least two edges of the face,
(C3^0) no isocontour is completely contained within a single tetrahedron, and
(C4^0) the map from parametric to geometric coordinates must be bijective.

Remark 1. Note that (C4^0) only regards Ξ_e and can be restated as:

(C4) $\forall \mathbf{x} \in \Omega_e, \exists! \mathbf{r} \in R$ such that $\Xi_e(\mathbf{r}) = \mathbf{x}$.

In this paper, it is satisfied by hypothesis as we focus on the Φ_e map. However, in the context of higher order element mesh generation and modification, our methodology can be applied to Ξ_e to verify the correctness of each element.

Let's examine how changing to a higher-order interpolant affects these assumptions. For instance, the $\Phi_1^{-1}(1)$ isocontour in the example of §1 violates (C3^0), which explains why it is entirely missed by linear isocontouring.

We now translate the criteria into requirements on Φ_e, that are slightly stronger for reasons that are discussed later on.

Proposition 1. *($C1^0$), ($C2^0$) and ($C3^0$) are respectively implied by:*

(C1) for each edge E of R, with direction vector (a_x, a_y, a_z),

$$a_x \frac{\partial \Phi_e}{\partial r} + a_y \frac{\partial \Phi_e}{\partial s} + a_z \frac{\partial \Phi_e}{\partial t} \neq 0 \quad \text{over the interior of } E.$$

(C2) for each face F of R, with vector basis $((a_x, a_y, a_z), (b_x, b_y, b_z))$,

$$\begin{cases} a_x \dfrac{\partial \Phi_e}{\partial r} + a_y \dfrac{\partial \Phi_e}{\partial s} + a_z \dfrac{\partial \Phi_e}{\partial t} \neq 0 \\[2mm] b_x \dfrac{\partial \Phi_e}{\partial r} + b_y \dfrac{\partial \Phi_e}{\partial s} + b_z \dfrac{\partial \Phi_e}{\partial t} \neq 0 \end{cases} \quad \text{over the interior of } F.$$

(C3) $\nabla \Phi_e \neq 0$ over the interior of R.

Proof. By definition, ($C1^0$), ($C2^0$) and ($C3^0$) are equivalently stated as:

($C1^0$) No restriction of Φ_e to an element edge has extrema interior to the edge unless the restriction is constant on the edge.

($C2^0$) No restriction of Φ_e to an element face has extrema interior to the face unless the restriction is constant on the face.

($C3^0$) Φ_e has no extrema inside the interior of the element unless the interpolant is constant over the entire element.

Being a polynomial function, Φ_e is C^1, and so are its restrictions to the edges and faces of R. Let E be an edge of R that passes through the point (p_x, p_y, p_z), with direction vector (a_x, a_y, a_z). This edge can be parametrized as follows:

$$\eta : I \subset \mathbb{R} \longrightarrow \mathbb{R}^3$$
$$t \longmapsto (a_x t + p_x, a_y t + p_y, a_z t + p_z)^T,$$

from which we obtain the restriction of Φ_e to E, $\Phi_{e|E} = \Phi_e \circ \eta : I \longrightarrow \mathbb{R}$. Its derivative $\mathrm{d}\Phi_{e|E} \in \mathcal{L}(\mathbb{R})$ at any arbitrary point $t_0 \in I$ is thus

$$\mathrm{d}\Phi_{e|E}(t_0) = \mathrm{d}\Phi_e(\eta(t_0)) \circ \mathrm{d}\eta(t_0) = \left(\frac{\partial \Phi_e}{\partial r}, \frac{\partial \Phi_e}{\partial s}, \frac{\partial \Phi_e}{\partial t} \right)\Bigg|_{\eta(t_0)} (a_x, a_y, a_z)^T \mathrm{d}t. \quad (1)$$

As $\Phi_{e|E}$ is C^1 over I, a necessary condition for $\Phi_{e|E}$ to have an extremum in t_0, interior to I, is that $\mathrm{d}\Phi_{e|E}(t_0) = 0$, *i.e.*,

$$a_x \frac{\partial \Phi_e}{\partial r}(t_0) + a_y \frac{\partial \Phi_e}{\partial s}(t_0) + a_z \frac{\partial \Phi_e}{\partial t}(t_0) = 0.$$

Similarly, let F be a face of R that passes through the point (p_x, p_y, p_z) and has basis $((a_x, a_y, a_z), (b_x, b_y, b_z))$. This face can be parametrized with two variables as follows:

$$\eta : U \subset \mathbb{R}^2 \longrightarrow \mathbb{R}^3$$
$$(u, v)^T \longmapsto (a_x u + b_x v + p_x, a_y u + b_y v + p_y, a_z u + b_z v + p_z)^T.$$

The derivative $d\Phi_{e|F} \in \mathcal{L}(\mathbb{R}^2, \mathbb{R})$ of $\Phi_{e|F} = \Phi_e \circ \eta : U \longrightarrow \mathbb{R}$ is

$$d\Phi_{e|F}(u_0, v_0) = d\Phi_e(\eta(u_0, v_0)) \circ d\eta(u_0, v_0) \tag{2}$$

$$= \left(\frac{\partial \Phi_e}{\partial r}, \frac{\partial \Phi_e}{\partial s}, \frac{\partial \Phi_e}{\partial t}\right)\Bigg|_{\eta(u_0, v_0)} \begin{pmatrix} a_x & b_x \\ a_y & b_y \\ a_z & b_z \end{pmatrix} \begin{pmatrix} du \\ dv \end{pmatrix}. \tag{3}$$

As $\Phi_{e|F}$ is C^1 over U, a necessary condition for $\Phi_{e|F}$ to have an extremum in (u_0, v_0), interior to U, is that $d\Phi_{e|F}(u_0, v_0) = 0$, which may be restated as

$$a_x \frac{\partial \Phi_e}{\partial r}(\eta(u_0, v_0)) + a_y \frac{\partial \Phi_e}{\partial s}(\eta(u_0, v_0)) + a_z \frac{\partial \Phi_e}{\partial t}(\eta(u_0, v_0)) = 0$$

$$b_x \frac{\partial \Phi_e}{\partial r}(\eta(u_0, v_0)) + b_y \frac{\partial \Phi_e}{\partial s}(\eta(u_0, v_0)) + b_z \frac{\partial \Phi_e}{\partial t}(\eta(u_0, v_0)) = 0.$$

Finally, as Φ_e is C^1 over R, a necessary condition for Φ_e to have extremum in (r_0, s_0, t_0), interior to R, is that $d\Phi_e(r_0, s_0, t_0) = 0$, i.e., $\nabla \Phi_e(r_0, s_0, t_0) = 0$. □

Note that the proof mainly relies on the fact that, for a C^1 function over an open domain, an extremum is a critical point. However, the converse is not true and thus each (Ci) is stronger than the corresponding (Ci^0). Therefore, using (Ci) rather than (Ci^0) can yield non-extremal critical points – just consider $r \mapsto r^3$ in 0. In order to eliminate such "false positives", we would need to evaluate second- and higher-order derivatives, which would incur further computational costs, and even this would not always suffice as degenerate cases may occur. Therefore, rather than degrading computational performance, the trade-off we make is to accept the stronger (Ci) conditions and extra points that are not required by the (Ci^0) conditions. This may even be beneficial in terms of geometric approximation, as non-extremal critical points can be the locus of important geometric features (e.g., saddle points). In short, (Ci) conditions mean that all of the differences between the linear and higher order isocontouring implementations can be attributed to critical points of Φ_e.

2.1 Creating the Partition

In §2, we presented the requirements a tetrahedral element must meet for isocontouring algorithm to work. As we noted earlier, higher order elements that have non-simplicial domains (such as hexahedra, pyramids, etc.) will have to be decomposed into tetrahedra $\mathcal{T}$. However, these tetrahedra must additionally meet the requirements (C1) to (C4). As explained in Remark 1, (C4) is satisfied by hypothesis; therefore the partition is subdivided until it meets criteria (C1), (C2), and (C3). Because not only a single scalar field, but a set κ of such fields may be of interest, these criteria must be satisfied for all fields in κ. Once this is achieved, the final partition is said to be κ-compatible. For the sake of legibility, we only discuss here the case where $\kappa = \{\Phi\}$, but the method remains the same when κ is not a singleton.

From now onwards, **it is assumed that all critical points are isolated**. This requirement is necessary so that the set of all critical points is finite, since a polynomial function can only have a finite number of isolated critical points; its implications are discussed at the end of this section. In this context, the general scheme of our method applied to the input parameters M (initial mesh) and Φ (field interpolated over M) is

PARTITION-MESH(M, Φ)

1   $C \leftarrow$ DOF-CRITICALITIES(M, Φ)
2   $(T_0, S) \leftarrow$ TRIANGULATE-BOUNDARIES(M, C)
3   CORRECT-TRIANGLE-TOPOLOGY(M, Φ, S, T_0)
4   $(T_1, S) \leftarrow$ TETRAHEDRALIZE-INTERIOR(M, T_0, C)
5   CORRECT-TETRAHEDRAL-TOPOLOGY(M, Φ, S, T_1)
6   **return** T_1

The output of the scheme is a tetrahedral subdivision T_1 of M. We now discuss each step in details,and theoretically establish the validity of the approach.

DOF-CRITICALITIES

This algorithm locates the critical points of Φ inside each element and of the restrictions of Φ to all element faces and edges. It takes M and Φ as inputs and yields a set C of critical points. BDY$^1(R)$ and BDY$^2(R)$ respectively denote the set of 1- and 2-dimensional boundaries of the parametric domain R.

DOF-CRITICALITIES(M)

1   **for** $e \leftarrow |M|$
2   **do** Find critical points of Φ_e in R
3   Store critical points indexed by volumetric DOF node.
4   **for** $f_i^2 \in$ BDY$^2(R)$
5   **do if** $\nabla \Phi_{e|f_i^2} = 0$ not marked,
6   **then** Find critical points of $\Phi_{e|f_i^2}$
7   Store critical points of $\Phi_{e|f_i^2}$ in $C_{f_i^2}$
8   Mark $\Phi_{e|f_i^2}$ as done
9   **for** $f_i^1 \in$ BDY$^1(R)$
10   **do if** $\nabla \Phi_{e|f_i^1} = 0$ not marked,
11   **then** Find critical points of $\Phi_{e|f_i^1}$
12   Store critical points of $\Phi_{e|f_i^1}$ in $C_{f_i^1}$
13   Mark $\Phi_{e|f_i^1}$ as done
14   **return** $C = \left(\cup_i C_{f_i^1} \right) \cup \left(\cup_i C_{f_i^2} \right)$

Note that, because finding critical points is a time-consuming process, we do not wish to process shared edges or faces multiple times. This extra work is avoided by storing critical points indexed by the DOF with which they are associated – critical points on a face are stored with the index used to retrieve the coefficients for that face's degrees of freedom, and likewise for edges. Therefore, DOF-CRITICALITIES operates on the mesh a whole, and not independently on each element. The issue of how to actually find the critical points is a complex and challenging problem of its own. We have not specifically researched this issue, and we handle it as follows:

- for edge critical points, where the problem amounts to finding all roots of a polynomial in a bounded interval, we have implemented exact solvers for up to quartic equations (and hence, quintic interpolants), and a Lin-Bairstow solver for higher order equations;
- for face and body critical points, where the problem amounts to finding all roots of a polynomial system within a bounded domain, we solve exactly if the system is linear, and otherwise make use of the PSS package [4, 5]. However, we think that this aspect deserves much further investigation.

If one assumes that the polynomial system solver always terminates in finite time, then DOF-CRITICALITIES does as well. As mentioned earlier, because restrictions of the field to edges and faces are marked as they are done, each is processed only once even when it is shared by several elements.

Remark 2. The methodology we present requires the ability to detect all critical points on arbitrary line segments and triangular faces in the domain of an element. Most polynomial system solvers require a power-basis representation of a system to be solved and that is not usually how finite elements are represented. Given that we wish to perform this change of basis as infrequently as possible, it behooves us to find a way to derive the restriction of Φ_e to a line or face from the full representation of Φ_e.

TRIANGULATE-BOUNDARIES

Once the critical points have been located, the second step of our scheme consists in triangulating the two-dimensional boundaries of all elements. This ensures that all volumetric elements that reference a particular face use the same triangulation – otherwise our model could have cracks along element boundaries[3]. In order to satisfy (C2) on a given element e, a restriction of Φ_e to a face of the tessellation of e is not permitted to have a critical point; therefore, we "eliminate" the critical points of the restriction of Φ_e to the faces of e by inserting them in the list of points to be triangulated. The algorithm TRIANGULATE-BOUNDARIES thus takes the mesh M and its related set of critical points C as inputs, and returns a triangulation T_0 of the set of faces in M. In this algorithm, the method FACE-CENTER takes a face as its input and returns its parametric center, and $\text{STAR}^2(c, Q_1, \ldots, Q_n)$ creates a

---

[3]Discontinuous Galerkin elements can be accommodated by using different indices (as opposed to a shared index i) for edges and faces of adjoining elements. Cracks would occur, but they would be faithful representations of the interpolant discontinuity.

triangulation composed of triangles $cQ_1Q_2,..., cQ_nQ_1$ (with the requirement that c is contained in the interior of the convex hull of $\{Q_1, \ldots, Q_n\}$).

TRIANGULATE-BOUNDARIES(M, C)

```
 1   for f_i^2 ← each 2-boundary of every 3-D finite element
 2       do if |C_{f_i^2}| > 0
 3           then c ← C_{f_i^2,0}
 4           else  c ← FACE-CENTER(BDY_i^2(ℝ))
 5       T_i ← ∅
 6       Q ← corner points of face i ⋃ isolated critical points
             of all bounding edges of face f_i^2, ordered in a
             counterclockwise loop around f_i^2.
 7       for j ∈ {0,..., |Q| − 1}
 8           do Insert STAR^2(c, Q_j, Q_{(j+1)mod|Q|}) into T_i
 9       C_{f_i^2}' ← {C_{f_i^2} \ C_{i,0}}
10       for c ∈ C_{f_i^2}'
11           do Find t ∈ T_i such that c ∈ t
12               Remove c from C_{f_i^2}'
13               Remove t from T_i
14               Subdivide t into 2 or 3 triangles t_k
15               Insert t_k into T_i
16   return T_0 = ∪_i T_i
```

All sets involved in TRIANGULATE-BOUNDARIES are finite, and no recursion is involved. Therefore, this procedure terminates in finite time. In addition,

Proposition 2. *Upon completion of* TRIANGULATE-BOUNDARIES, *(C1) is satisfied on the edges of the triangulation T_0 that either belong to M or are subdivisions of edges or faces of M, and (C2) is satisfied across T_0.*

Proof. By construction, TRIANGULATE-BOUNDARIES inserts in T_0 all critical points of restrictions of Φ to edges of M. Since this process cannot create novel critical points on these edges, the first part of the proposition ensues. Compliance with (C2) across T_0 is ensured because all critical points of restrictions of Φ to faces of M have been inserted as vertices of T_0, and no new such critical point may have been created. $\square$

Note that that there is no guarantee that (C1) is satisfied across T_0 upon completion of TRIANGULATE-BOUNDARIES, as shown in the following example:

Example 1. Consider a face f_i, illustrated in Fig. 2(a), such that the restriction of the field to the interior of f_i has 3 critical points (denoted a, b, and c), and each of the restrictions of the field to the edges of f_i has at least one critical point (denoted d, f, h, i, j, and k). The triangulation displayed in Fig. 2(b) is obtained once all STAR^2 procedures have performed by taking a as the first internal critical point to be inserted. The final tessellation, shown in Fig. 2(c), has eliminated the remaining interior critical points b and c by making them

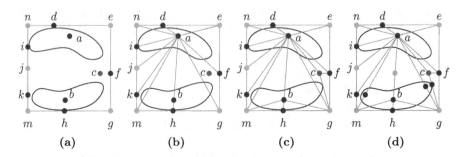

Fig. 2. An example face f_i^2 with critical points shown as blue dots (maxima), red dots (saddles), and green dots (minima). **(a)** The input to TRIANGULATE-BOUNDARIES. **(b)** Resulting triangulation of f_i^2 after all STAR^2 procedures of TRIANGULATE-BOUNDARIES have been performed. **(c)** The triangulation at the completion of TRIANGULATE-BOUNDARIES. **(d)** The new critical points introduced by the first stage of CORRECT-TRIANGLE-TOPOLOGY.

nodes of the triangulation; however, as illustrated in Fig. 2(d), new critical points have appeared, on the restrictions of Φ_e to edges ab, ag, am, and cg.

Remark 3. Line 14 of TRIANGULATE-BOUNDARIES allows for 2 different ways to subdivide a triangle, depending on whether the face critical point lies within or on the boundary of the triangle; for instance, in Fig. 2(c), triangles ahm and afg are split in, respectively, 2 and 3 triangles. In practice, to avoid unnecessary creation of quasi-degenerate triangles, the implementation of TRIANGULATE-BOUNDARIES uses a predefined threshold (that can be related to the distance of the critical point to the closest triangle edge, or to a triangle quality estimate of the subdivided triangles) below which a critical point is moved to the appropriate edge; for instance, in the example illustrated in Fig. 2(c), point b is considered as belonging to ah, even if it slightly off.

CORRECT-TRIANGLE-TOPOLOGY

This algorithm searches for critical points in the restriction of Φ_e to each unmarked edge of T_0. When points are found, they are inserted into T_0, and iteratively repeats the procedure until (C1) and (C2) are satisfied throughout the entire triangulation of the faces of M. Fig. 3(b) shows this procedure applied to Example 1.

CORRECT-TRIANGLE-TOPOLOGY(M, Φ, S, T_0)

```
1    while S not empty
2        do Pop t from S
3            C ← ∅
4            for e ← marked edges of t
5                do Insert critical points of e into C
6            for c ← C
7                do Find t ∈ T₁ s. t. c ∈ t
8                    Remove t from T₁ and S
9                    U ← STAR²(c, t)
10                   for t' ← U
11                       do if MARK-TRIANGLE(t')
12                           then Push t' onto S
13                       Insert t' into T₀
```

Proposition 3. *If, for all e in M, all critical points of the restriction of Φ_e to any face of e are isolated, then* CORRECT-TRIANGLE-TOPOLOGY *terminates. In addition, upon termination, (C1) and (C2) are satisfied across T_0.*

Proof. For brevity, the proof is not provided here: it can readily be obtained by reducing the proof of Proposition 5 to the 2-dimensional case.□

Remark 4. Because there is no need to refine below a certain size for visualization purposes, our implementation uses a tolerance $\varepsilon \in \mathbb{R}_+$ so that, in line 5, a new critical point p of the restriction of Φ to a marked edge is inserted only if there is no p' in C such that $|p - p'| < \varepsilon$. Therefore, the procedure may terminate before (C1) and (C2) are fully satisfied, but are satisfied up to ε.

TETRAHEDRALIZE-INTERIOR

Each element interior is now tetrahedralized, and although we treat the whole mesh, there is here no issue of inter-element consistency, as this part of the scheme only regards element interiors. As in TRIANGULATE-BOUNDARIES, we "eliminate" critical points of Φ_e that are interior to e by adding them to the set of points to be tetrahedralized. Therefore, the tetrahedralization of each element e is constrained by the triangulations of the faces of e that result from CORRECT-TRIANGLE-TOPOLOGY, and by the critical points of Φ_e that are interior to e. Additionally, when the finite element is starred into a set of tetrahedra, we know that the triangular base of each tetrahedron and its 3 bounding edges will not have any critical points since those have already been identified and inserted into the triangulation of the two-dimensional boundary of the element. However, the remaining 3 faces and 3 edges must be marked because Φ_e restricted to their domain may contain critical points. This is accomplished by MARK-TETRAHEDRON, which sets a bit code for each edge and face not on the base of the given tetrahedron (which must be properly oriented when passed to the subroutine). The algorithm is then as follows:

TETRAHEDRALIZE-INTERIOR(M, T_0, C)

```
 1   S ← ∅
 2   T₁ ← ∅
 3   for e ← |M|
 4         do Let T ⊆ T₀ be all triangles on BDY²(R)
 5            if |Cₑ| > 0
 6               then c ← Cₑ,₀
 7               else  c ← ELEMENT-CENTER(R)
 8            V ← STAR³(c, T)
 9            for t ← V
10                do if MARK-TETRAHEDRON(t)
11                      then Push t onto S
12            for c ∈ {Cₑ \ Cₑ,₀}
13                do Find t ∈ V s. t. c ∈ t
14                   Remove t from V and S
15                   U ← STAR³(c, t)
16                   for t' ← U
17                       do if MARK-TETRAHEDRON(t')
18                             then Push t' onto S
19                          Insert t' into V
20            Insert V into T₁
21   Return (T₁, S)
```

All sets involved in TETRAHEDRALIZE-INTERIOR are finite, and no recursion is involved. Therefore, this procedure terminates in finite time. In addition,

Proposition 4. *(C3) is satisfied across the tetrahedralization T_1 upon completion of* TETRAHEDRALIZE-INTERIOR. *Moreover, any sub-tetrahedralization of T_1 satisfies (C3) as well.*

Proof. Let p be a critical point of the field Φ; then, 2 cases may occur:

1. p is contained in the interior of an element $e \in M$. In this case, p belongs to C_e (and to no other $C_{e'}$) and thus, thanks to STAR³, p is a tetrahedron vertex in T_1.
2. p is contained on the boundary of an element $e \in M$. In this case, it is also a lower-dimensional critical point, *i.e.*, a critical point for the restriction of Φ_e to one of its faces or edges, because the fact $d\Phi_e$ vanishes in p ensures that the left hand side of (1) (if p is on an edge) or (2) (if p is on a face) vanishes as well. Therefore, p belongs to C_{f_i} for an edge f_i^1 or a face f_i^2 and hence has been made a triangle vertex of T_0 by TRIANGULATE-BOUNDARIES. Since all triangles of T_0 become tetrahedral faces in T_1, then p is a tetrahedron vertex in T_1.

In both cases, upon completion of TETRAHEDRALIZE-INTERIOR, p is a tetrahedron vertex in T_1. Therefore, for all $t \in T_1$, p is not a critical point of Φ_t interior to t. As this is true for any critical point p of the field Φ, it follows that T_1 satisfies (C3). Finally, as (C3) is satisfied on any tetrahedron $t \in T_1$, it is also satisfied on any tetrahedral subdivision of t: indeed, if one could find a tetrahedron $t' \subset t$ and a critical point p of Φ contained in the interior of t', then p would also be in the interior of t, which would contradict the hypothesis; *ad absurdum*, the result ensues.□

CORRECT-TETRAHEDRAL-TOPOLOGY

A this point, a brief summary of what has been obtained through the 3 first stages of our scheme will most likely be useful to the reader. A tetrahedralization T_1 of the initial mesh M has been obtained, such that

- (C3), and (C4) by hypothesis, are satisfied;
- (C1) and (C2) are satisfied on all edges and faces of T_1 that either belong to M or are subdivisions of edges or faces of M.

However, there is no guarantee that (C1) and (C2) are satisfied on all edges and faces of T_1 that are not included (*stricto* or *lato sensu*) in M.

Example 2. Consider the scalar field defined by $\Phi_2(x, y, z) = x^2 - y^2 + z$ over a single Lagrange Q2 element with linear geometry and coordinates in $[-1, 1]^3$, as illustrated in Fig. 3(c): each restriction of Φ_2 to the faces perpendicular to the z-axis has a critical point at the corresponding face center (labeled 7 and 11), and each restriction of Φ_2 to the edges perpendicular to the z-axis has a critical point at the corresponding edge midpoint (labeled 6, 23, 20, 24, 22, 14, and 26 for those that are visible), whereas Φ_2 proper does not have any critical point. Upon completion of TETRAHEDRALIZE-INTERIOR, all of these points have been inserted in T_1 which contains, among others, edges from the element center to points 6, 20, 22, and 26. It is easy to check that the restrictions of Φ_2 to each of these edges have a critical point at the edge midpoint, and thus (C1) is not satisfied across T_1, that therefore requires further modification.

We must therefore examine how, in general, T_1 can be modified so in order to satisfy (C1) and (C2). A natural question is to wonder whether it is possible to perform a series of edge-face flips on the T_1 so that the final tessellation satisfies the desired properties.

Example 3. Given the triangulation of a face in Fig. 2(c) for which additional critical points (of the restrictions of the field to some of the new edges) have appeared, one can easily perform a series of edge flips so that the final connectivity satisfies (C1), as shown in the resulting tessellation in Fig. 3(a).

Nevertheless, it is unclear whether it is always possible to retrieve a tessellation that satisfies (C1) using only edge flips. Although this may be the case, we have not further explored this path, because the matter is more complicated as both (C1) and (C2) must be satisfied in a problem that is intrinsically three-dimensional: for instance, although any 2-D triangulation can always be converted to a Delaunay triangulation by a finite sequence of edge flips, but

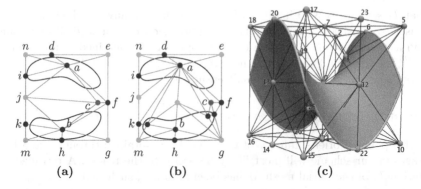

Fig. 3. (a) A triangulation satisfying (C1) with edge flips from Fig. 2(c). (b) Another triangulation satisfying (C1) using CORRECT-TRIANGLE-TOPOLOGY and the same initial tessellation. (c) Topology-based tetrahedralization of a single Lagrange Q2 element with linear geometry for the field $\Phi_2(x, y, z) = x^2 - y^2 + z$.

this result does **not** extend to 3-D tetrahedralizations [2], making us skeptical of the flipping approach. We therefore took a different route, guided by Proposition 4, as not only T_1 satisfies (C3), but we also know that will not be altered by subsequent subdivisions of T_1. Rather than attempting to identify a set of "problem" entities and prove that they may always be flipped into a satisfactory configuration, we introduce each critical point of a restriction of Φ to a new entity into the tessellation. Any edges and faces created by this operation must then be examined as well for critical points of Φ restricted to their respective domains. However, it is only necessary to focus on critical points of the restrictions of Φ to the previously marked edges and faces, as others already satisfy (C1) and (C2), thanks to TRIANGULATE-BOUNDARIES.

CORRECT-TETRAHEDRAL-TOPOLOGY(M, Φ, S, T_1)

```
1   while S not empty
2       do Pop t from S
3           C ← ∅
4           for e ← marked edges of t
5               do Insert critical points of e into C
6           for f ← marked faces of t
7               do Insert critical points of f into C
8           for c ← C
9               do Find t ∈ T₁ s. t. c ∈ t
10                  Remove t from T₁ and S
11                  U ← STAR³(c, t)
12                  for t' ← U
13                      do if MARK-TETRAHEDRON(t')
14                          then Push t' onto S
15                              Insert t' into T₁
```

Remark 5. The same approach as that explained in Remark 4 is used here, using a tolerance to avoid refining below a certain size. Therefore, and for the same reason, this allows the algorithm to possibly terminate faster than normally possible if all critical points created on new edges and faces and the refinement proceeds had to be inserted without regard to the distance to an already inserted vertex.

Proposition 5. *If, for all e in M, all critical points of the restriction of Φ_e to any arbitrary face are isolated, then* CORRECT-TETRAHEDRAL-TOPOLOGY *terminates. In addition, upon termination, (C1), (C2) and (C3) are satisfied.*

Proof. To establish this result, it is sufficient to make sure the algorithm terminates, starting from any arbitrary face of an arbitrary element in M. So, let f_i be one of the faces of an arbitrary $e \in M$, and we then shall prove that CORRECT-TETRAHEDRAL-TOPOLOGY terminates.

First, remark that if the restriction $\Phi_{e|f_{ij}}$ of Φ_e to one edge f_{ij} of f_i has a non-isolated critical point, then the derivative of $\Phi_{e|f_{ij}}$ vanishes along a nonempty open segment of f_{ij}, and therefore has an infinity of zeros. Because this derivative is itself a univariate polynomial function, it can thus only be zero everywhere, and thus $\Phi_{e|f_{ij}}$ is constant along the edge. Therefore, the only case when non-isolated critical points along a bounding edge of f_i arises is when the interpolant is constant along that edge, and therefore no other points than its endpoints are contained in P. P is indeed a finite set, as polynomials can only have a finite number of isolated critical points.

Now assume the restriction $\Phi_{e|f_i}$ of Φ_e to the interior of f_i has $n \in \mathbb{N}^*$ critical points. The innermost loop of CORRECT-TETRAHEDRAL-TOPOLOGY will insert these n points, and yield a triangulation of f_i in $N \in \mathbb{N}^*$ triangles $t_{i,k}$, such that $\cup_{k=1}^N t_{i,k} = f_i$ and

$$(\forall\, 1 \leq k, k' \leq N) \quad k \neq k' \iff \overset{\circ}{t}_{i,k} \cap \overset{\circ}{t}_{i,k'} = \emptyset,$$

where all the points of C are vertices of some of $t_{i,k}$ ($\overset{\circ}{t}$ is the interior of t in the sense of the natural topology induced on t by embedding it in $\mathbb{R}^2$). Therefore, none of the restrictions of Φ_e to $t_{i,k}$ has an internal critical point (otherwise this point would belong to C, which is impossible because all points of C are vertices of some of $t_{i,k}$).

However, the restriction of Φ_e to some edges of this triangulation of f_i may have critical points[4]. Denote η such an edge. If the restriction of Φ to η has any non-isolated critical point, then the same argument as above holds and thus the corresponding edges do not need to be further refined. On the contrary, if such an edge critical point is isolated (in this case, the edge must be internal to f_i, as all isolated critical points along the edges of f_i have been inserted priorly), then CORRECT-TETRAHEDRAL-TOPOLOGY recursively proceeds on

---

[4]In other words, the subdivision of f_i cannot create new face critical points, but it can create new edge critical points.

η. However, the process terminates because all face critical points are supposed to be isolated according to the hypothesis. Therefore, for each critical point p_i of the restriction of Φ_e to f_i, there exists a neighborhood of p_i in which all directional derivatives of Φ_e are nonzero and thus, there exists a finite number of triangle subdivisions after which no edge critical points are left (because such a critical point implies one directional derivative is equal to 0).

Finally, upon completion of the algorithm, for the same reasons as for TETRAHEDRALIZE-INTERIOR, (C3) is satisfied for each tetrahedron of the final partition (the final tetrahedra are subdivisions of initial tetrahedra that all satisfied (C3), according to Proposition 4).□

Example 4. Consider the same case as Example 2: the execution of CORRECT-TETRAHEDRAL-TOPOLOGY results in the insertion of the 4 previously mentioned edge midpoints (3 of which are visible in Fig. 3(c), labeled 3, 19, and 25) in a sub-tetrahedralization of T_1 that becomes the final tessellation, across which conditions (C1) through (C4) are satisfied. In this case, only one level of refinement was necessary, as none of the edges and faces created upon insertion of the 4 aforementioned edge midpoints violates (C1) or (C2).

Remark 6. Note that the hypothesis of Proposition 5 is very stringent, as it is not limited to faces of the mesh, but extends to all possible faces. In fact, it is sufficient that the restrictions of Φ_e to any face of the successive partitions of e only has isolated critical points. However, as this condition depends on the particular subdivision path, it is, albeit weaker, more difficult to prove.

3 Application to Isocontouring

We now illustrate the application of our method to isocontouring, and the discuss an additional benefit of κ-tessellation that comes as a side-effect. Consider the following scalar field:

$$\Phi_3 : [-1, 1]^3 \longrightarrow \mathbb{R}$$
$$(x, y, z)^T \mapsto x^2 + y^2 + z^2(z - 1),$$

interpolated over a single Q3 Lagrange element with linear geometry, i.e., a tricubic (hence with total degree 9) cell obtained by tensorization of cubic Lagrange interpolants. The test consists of representing the 0-isocontour of Φ_3, which is tricky because an entire lobe of the resulting isosurface is contained within the element.

Figure 4, left, shows that if a linear isocontouring marching cubes technique is applied (after uniform subdivision of the hexahedron into 48 tetrahedra), then a substantial part of $\Phi^{-1}(0)$ is missing. This example is interesting because the missing part of $\Phi^{-1}(0)$ is not a disconnected component and, therefore, (C3) is *not* violated for this particular isovalue. While intuition may suggest that a linear isocontouring technique should retrieve the correct

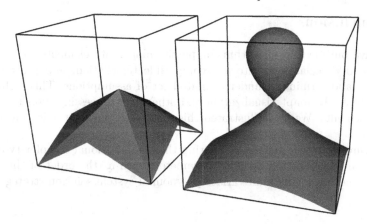

Fig. 4. Tricubic isocontouring of Φ_3 for the isovalue 0, using a linear isocontouring technique (left), and using our approach (right).

topology of $\Phi^{-1}(0)$ in the interior of the hexahedral element, in fact $\Phi^{-1}(0)$ is not a 2-dimensional submanifold of $\mathbb{R}^3$ at (and only at) point $(0,0,0)$ (where $\Phi^{-1}(0)$ is not simply connected). Therefore, $\Phi^{-1}(0)$ is not a surface and this causes the isocontouroing algorithm to fail, as illustrated. Moreover, the Implicit Function Theorem shows that for any value of α in $]-2,0[$, $\Phi^{-1}(\alpha)$ is a surface, but can also easily check that for such values of α, it is not connected; in fact, it has 2 disjoint connected components, one of which is entirely contained in the interior of the element. Hence, in this case, (C3) is indeed violated; which also causes the linear isocontouring technique to fail (by missing this connected component).

From what we already know of our topology-based technique, we can expect it to properly retrieve $\Phi^{-1}(\alpha)$ for these problematic values of α in $]-2,0[$, and it does indeed. However, what was not initially expected, is that it could fix the case when $\alpha = 0$, as we do not expect the technique to handle non-manifold isocontours. Nonetheless, as shown in Figure 4, right, our topology-based method is able to construct the correct topology of the non-simply connected $\Phi^{-1}(0)$. And in fact, this not an anecdotal effect valid for this example only, but we can see that this is always the case: by "eliminating" critical points from the final tessellation, the method ensures that (thanks to the Implicit Function Theorem), the isocontour is locally a 2-dimensional manifold within the interior of each element. In other words, the scheme produces a tessellation that not only satisfies criteria (C1) through (C4), but additionally ensures that the isocontour is indeed a surface inside each element. Note that this extends to the lower-dimensional case, for the same reason: the isocontours inside the faces of the final tessellation are simply connected curves.

4 Conclusions

We have outlined an algorithm for partitioning finite elements to which is associated a scalar field into a κ-compatible tessellation, and proved that it works and terminates under a limited set of assumptions. This technique allows to easily adapt visualization and other post-processing tools to higher order elements. We have illustrated this methodology with the isocontouring operation.

Future work will include estimating the computational complexity of the algorithm as a function of input parameters such as the order of the interpolant, and improving multivariate polynomial system solution strategies.

References

1. Michael Brasher and Robert Haimes. Rendering planar cuts through quadratic and cubic finite elements. In *Proceedings of IEEE Visualization*, pages 409–416, October 2004.
2. Barry Joe. Three dimensional triangulations from local transformations. *SIAM Journal on Scientific and Statistical Computing*, 10:718–741, 1989.
3. Rahul Khardekar and David Thompson. Rendering higher order finite element surfaces in hardware. In *Proceedings of the first international conference on computer graphics and interactive techniques in Australasia and South East Asia*, pages 211–ff, February 2003.
4. Gregorio Malajovich. PSS 3.0.5: Polynomial system solver, 2003. URL http://www.labma.ufrj.br:80/~gregorio.
5. Gregorio Malajovich and Maurice Rojas. Polynomial systems and the momentum map. In *Proceedings of FoCM 2000, special meeting in honor of Steve Smale's 70th birthday*, pages 251–266. World Scientific, July 2002.
6. M. Meyer, B. Nelson, R.M. Kirby, and R. Whitaker. Particle systems for efficient and accurate finite element subdivision. *IEEE Trans. Visualization and Computer Graphics*, 13, 2007.
7. J.-F Remacle, N. Chevaugeon, E. Marchandise, and C. Geuzaine. Efficient visualization of high-order finite elements. *Intl. J. Numerical Methods in Engineering*, 69:750–771, 2006.
8. W. J. Schroeder, F. Bertel, M. Malaterre, D. C. Thompson, P. P. Pébay, R. O'Bara, and S. Tendulkar. Framework and methods for visualizing higher-order finite elements. *IEEE Trans. on Visualization and Computer Graphics, Special Issue Visualization 2005*, 12(4):446–460, 2006.

Tetrahedral Meshing 2

Meshing 3D Domains Bounded by Piecewise Smooth Surfaces[*]

Laurent Rineau[1] and Mariette Yvinec[2]

[1] INRIA, BP 93 06902 Sophia Antipolis, France.
  Laurent.Rineau@sophia.inria.fr
[2] INRIA, BP 93 06902 Sophia Antipolis, France.
  Mariette.Yvinec@sophia.inria.fr

Summary. This paper proposes an algorithm to mesh 3D domains bounded by piecewise smooth surfaces. The algorithm may handle multivolume domains defined by non connected or non manifold surfaces. The boundary and subdivision surfaces are assumed to be described by a complex formed by surface patches stitched together along curve segments.

The meshing algorithm is a Delaunay refinement and it uses the notion of restricted Delaunay triangulation to approximate the input curve segments and surface patches. The algorithm yields a mesh with good quality tetrahedra and offers a user control on the size of the tetrahedra. The vertices in the final mesh have a restricted Delaunay triangulation to any input feature which is a homeomorphic and accurate approximation of this feature. The algorithm also provides guarantee on the size and shape of the facets approximating the input surface patches. In its current state the algorithm suffers from a severe angular restriction on input constraints. It basically assumes that two linear subspaces that are tangent to non incident and non disjoint input features on a common point form an angle measuring at least 90 degrees.

1 Introduction

Mesh generation is a notoriously difficult task. Getting a fine discretization of the domain of interest is the bottleneck of many applications in the areas of computer modelling, simulation or scientific computation. The problem of mesh generation is made even more difficult when the domain to be meshed is bounded and structured by curved surfaces which have to be approximated as well as discretized in the mesh. This paper deals with the problem of generating unstructured tetrahedral meshes for domains bounded by piecewise smooth surfaces. A common way to tackle such a problem consists in building

---

[*]Work partially supported by the IST Programme of the EU as a Shared-cost RTD (FET Open) Project under Contract No IST-006413 (ACS - Algorithms for Complex Shapes) and by the European Network of Excellence AIM@SHAPE (FP6 IST NoE 506766).

first a triangular mesh approximating the bounding surfaces and then refining the volume discretization while preserving the boundary approximation. The meshing of bounding surfaces are mostly performed through the highly popular marching cubes algorithm [LC87]. The marching cubes algorithm provides an accurate discretization of smooth surfaces but the output surface mesh includes poor quality elements and fails to recover sharp features. The marching cubes may be followed by some remeshing step to improve the shape of the elements and to adapt the sizing of the surface mesh to the required density, see [AUGA05] for survey on surface remeshing. Once a boundary surface mesh is obtained, the original surface is replaced by its piecewise linear approximation. The 3-dimensional mesh is then obtained through a meshing algorithm which either conforms strictly to the boundary surface mesh (see e.g. [FBG96, GHS90, GHS91]) or refine the surface mesh within the geometry of the piecewise linear approximation [She98, CDRR04]. See e.g. [FG00] for a survey on 3-dimensional meshing. In both cases, the quality of the resulting mesh and the accuracy of the boundary approximation highly depends on the initial surface mesh.

This paper proposes an alternative to the marching cubes strategy. In this alternative, the recovery of bounding curves and surfaces is based on the notion of restricted Delaunay triangulations and the mesh generation algorithm is a multi-level Delaunay refinement process which interleaves the refinement of the curve, surface and volume discretization.

Delaunay refinement is recognized as one of the most powerful method to generate meshes with guaranteed quality. The pioneering work of Chew [Che89] and Ruppert [Rup95] handles the generation of 2-dimensional meshes for domains whose boundaries and constraints do not form small angles. Shewchuk improves the handling of small angles in two dimensions [She02] and generalizes the method to generate 3-dimensional meshes for domains with piecewise linear boundaries [She98]. The handling of small angles formed by constraints is more puzzling in three dimensions, where dihedral angles and facet angles come into play. Using the idea of protecting spheres around sharp edges [MMG00, CCY04], Cheng and Poon [CP03] provide a thorough handling of small input angles formed by boundaries and constraints. Cheng, Dey, Ramos, and Ray [CDRR04] turn the same idea into a simpler and practical mesh generation algorithm for polyhedral domains. In 3-dimensional space, Delaunay refinement produces tetrahedral meshes free of all kind of degenerate tetrahedra except slivers. Further work [CDE+00, CD03, LT01, CDRR05] was needed to deal with the problem of sliver exudation.

Up to now, little work has been dealing with curved objects. The early work of Chew [Che93] tackles the problem of meshing curved surfaces and Boivin and Ollivier-Gooch [BOG02] consider the meshing of 2-dimensional domains with curved boundaries. In [ORY05], we propose a Delaunay refinement algorithm to mesh a 3-dimensional domain bounded by a smooth surface. This algorithm relies on recent results on surface mesh generation [BO05, BO06]. It involves Delaunay refinement techniques and the notion of restricted Delaunay triangulations to provide a nice sampling of both the volume and the bounding surface. The restriction to the boundary surface of the Delaunay triangulation of the final mesh vertices forms an homeomorphic and accurate approximation of this surface. The present paper extends this mesh generation algorithm to handle 3-dimensional domains defined by piecewise smooth surfaces, i. e. patches of smooth surfaces stitched together along 1-dimensional smooth curved segments. The main idea is to approximate the 1-dimensional sharp features of boundary surfaces by restricted Delaunay triangulations. The accuracy of this approximation is controlled by a few additional refinement rules in the Delaunay refinement process. The algorithm produces a controlled quality mesh in which each surface patch and singular curved segment has a homeomorphic piecewise linear approximation at a controlled Hausdorff distance. The algorithm can handle multi-volume domains defined by non connected or non manifold piecewise smooth surfaces. The algorithm relies only on a few oracles able to detect and compute intersection points between straight segments and surface patches or between straight triangles and 1-dimensional singular curved segments. Therefore it can be used in various situations like meshing CAD-CAM models, molecular surfaces or polyhedral models. The only severe restriction with respect to the input features is an angular restriction. Roughly speaking, tangent planes at a common point of two adjacent surface patches are required to make an angle bigger than 90°.

Our work is very closed to a recent work [CDR07] where Cheng, Dey and Ramos propose a Delaunay refinement meshing for piecewise smooth surfaces. This algorithm suffers no angular restriction but uses topology driven refinement rules which involve computationally intensive and hard to implement predicates on the surface.

The paper is organized as follows. Section 2 precises the input of the algorithm and provides a few definitions. In particular we define a local feature size adapted to the case of piecewise smooth surfaces. The mesh generation algorithm is described in section 3. Before proving in section 5 the correctness of this algorithm, i. e. basically the fact that it always terminates, we prove in section 4 the accuracy, quality and homeomorphism properties of the resulting mesh. The algorithm has been implemented using the library CGAL [CGAL]. Section 6 provides some implementation details and shows a few examples. Section 7 gives some directions for future work, namely to get rid of the angular restriction on input surface patches.

2 Input, definitions and notations

The domain $\mathcal{O}$ to be meshed is assumed to be a union of 3-dimensional cells whose boundaries are piecewise smooth surfaces, i. e. formed by smooth surface patches joining along smooth curve segments.

More precisely, we define a regular complex as a set of closed manifolds, called faces, such that:

- the boundary of each face is the union of faces of the complex,
- any two faces have disjoint interior.

We consider a 3-dimensional regular complex whose 2-dimensional sub-complex is formed by patches of smooth surfaces and whose 1-dimensional skeleton is formed by smooth curve segments. Each curve segment is assumed to be a compact subset of a smooth closed curve, and each surface patch is assumed to be a compact subset of a smooth closed surface. The smoothness conditions on curves (resp. surfaces) is to be $C^{1,1}$, *i.e.* to be differentiable with a Lipschitz condition on the tangent (resp. normal) field.

The domain $\mathcal{O}$ that we consider is a union of cells, i. e. 3-dimensional faces, in such a regular complex.

We denote by $\mathcal{F}$ the 2-dimensional regular complex which forms the boundaries of the cells in $\mathcal{O}$. The set of faces in $\mathcal{F}$ includes a set $\mathcal{Q}$ of vertices, a set $\mathcal{L}$ of smooth curve segments and a set $\mathcal{S}$ of smooth surface patches, such that $\mathcal{F} = \mathcal{Q} \cup \mathcal{L} \cup \mathcal{S}$.

In the following, $\bigcup \mathcal{F}$ denotes the domain covered by the union of faces in $\mathcal{F}$ and $\mathrm{d}(x, y)$ the Euclidean distance between two points x and y.

We assume that two elements in $\mathcal{F}$ which are neither disjoint nor incident do not form sharp angles. More precisely, we assume the following:

Definition 1 (The angular hypothesis). *There is a distance λ_0 so that, for any pair (F, G) of non disjoint and non incident faces in $\mathcal{F}$, if there is a point z on $F \cap G$ such that $\mathrm{d}(x, z) \leq \lambda_0$ and $\mathrm{d}(y, z) \leq \lambda_0$, then the following inequality holds:*

$$\mathrm{d}(x, y)^2 \geq \mathrm{d}(x, F \cap G)^2 + \mathrm{d}(y, F \cap G)^2. \tag{1}$$

In the special case of linear faces, the angular hypothesis holds when the *projection condition* [She98] holds. The projection condition states that if two elements F and G of $\mathcal{F}$ are neither disjoint nor incident, the orthogonal projection of G on the subspace spanned by F does not intersect the interior of F. For two adjacent planar facets, it means that the dihedral angle must be at least $90°$.

Definition of the local feature size

To describe the sizing field used by the algorithm we need to introduce a notion of *local feature size* (lfs , for short) related to the notion of local feature size used for polyhedra [Rup95, She98] and also to the local feature size introduced [AB99] for smooth surfaces. To account for curvature, we first define a notion of *interrelated points*, (using an idea analogous to the definition of *intertwined* points for anisotropic metric in [LS03]).

Definition 2 (Interrelated points). *Two points x and y of $\mathcal{S}$ are said to be interrelated if:*

- *either they lie on a common face $F \in \mathcal{F}$,*
- *or they lie on non-disjoint faces, F and G, and there exists a point w in the intersection $F \cap G$ such that: $\mathrm{d}(x, w) \leq \lambda_0$ and $\mathrm{d}(y, w) \leq \lambda_0$.*

We first define a feature size $\mathrm{lfs}^{\mathrm{P}}(x)$ analog to the feature size used for a polyhedron. For each point $x \in \mathbb{R}^3$, $\mathrm{lfs}^{\mathrm{P}}(x)$ is the radius of the smallest ball centered at x that contains two non interrelated points of $\bigcup \mathcal{F}$.

We then define a feature size $\mathrm{lfs}_{F_i}(x)$ related to each feature in $\mathcal{L} \cup \mathcal{S}$. Let F_i be a curve segment or surface patch in $\mathcal{F}$. We first define the function $\mathrm{lfs}_{F_i}(x)$ for any point $x \in F_i$ as the distance from x to the medial axis of the smooth curve or the smooth surface including the face F_i. Thus defined, the function $\mathrm{lfs}_{F_i}(x)$ is a Lipschitz function on F_i. Using the technique of Miller, Talmor and Teng [MTT99], we extend it as a Lipschitz function $\mathrm{lfs}_{F_i}(x)$ defined on $\mathbb{R}^3$:

$$\forall x \in \mathbb{R}^3, \ \mathrm{lfs}_{F_i}(x) = \inf \ \{\mathrm{d}(x, x') + \mathrm{lfs}_{F_i}(x') \ : \ x' \in F_i\}.$$

The local feature size $\mathrm{lfs}(x)$ used below is defined as the pointwise minimum:

$$\mathrm{lfs}(x) = \min \left(\mathrm{lfs}^{\mathrm{P}}(x), \min_{F_i \in \mathcal{F}} \mathrm{lfs}_{F_i}(x) \right).$$

Being a pointwise minimum of Lipschitz functions, $\mathrm{lfs}(x)$ is a Lipschitz function.

3 The mesh generation algorithm

The meshing algorithm is based on the notion of Voronoi diagrams, Delaunay triangulations and restricted Delaunay triangulations. Let $\mathcal{P}$ be a set of points and p a point in $\mathcal{P}$. The Voronoi cell $V(p)$ of the point p is the locus of points that are closer to p than to any other point in $\mathcal{P}$. For any subset $\mathcal{T} \subset \mathcal{P}$, the Voronoi face $V(\mathcal{T})$ is the intersection $\bigcap_{p \in \mathcal{T}} V(p)$. The Voronoi diagram $\mathcal{V}(\mathcal{P})$ is the complex formed by the non empty Voronoi faces $V(\mathcal{T})$ for $\mathcal{T} \subset \mathcal{P}$.

We use $\mathcal{D}(\mathcal{P})$ to denote the Delaunay triangulation of $\mathcal{P}$. Let X be a subset of $\mathbb{R}^3$. We call restricted Delaunay triangulation of $\mathcal{P}$ to X, and denote by $\mathcal{D}_X(\mathcal{P})$, the subcomplex of $\mathcal{D}(\mathcal{P})$ formed by the faces in $\mathcal{D}(\mathcal{P})$ whose dual Voronoi faces have a non empty intersection with X. Thus a triangle pqr of $\mathcal{D}(\mathcal{P})$ belongs to $\mathcal{D}_X(\mathcal{P})$ iff the dual Voronoi edge $V(p, q, r)$ intersects X and an edge pq of $\mathcal{D}(\mathcal{P})$ belongs to $\mathcal{D}_X(\mathcal{P})$ iff the dual Voronoi facet $V(p, q)$ intersects X.

The algorithm is a Delaunay refinement algorithm that iteratively builds a set of sample points $\mathcal{P}$ and maintains the Delaunay triangulation $\mathcal{D}(\mathcal{P})$, its restriction $\mathcal{D}_{|\mathcal{O}}(\mathcal{P})$ to the domain $\mathcal{O}$ and the restrictions $\mathcal{D}_{|S_k}(\mathcal{P})$ and $\mathcal{D}_{|L_j}(\mathcal{P})$ to every facet S_k of $\mathcal{S}$ and every edge L_j of $\mathcal{L}$. At the end of the refinement process, the tetrahedra in $\mathcal{D}_{|\mathcal{O}}(\mathcal{P})$ form the final mesh and the subcomplexes $\mathcal{D}_{|S_k}(\mathcal{P})$ and $\mathcal{D}_{|L_j}(\mathcal{P})$ are accurate approximation of respectively S_k and L_j.

The refinement rules applied by the algorithm to reach this goal use the, hereafter defined, notion of encroachment for restricted Delaunay facets and edges.

Let L_j be an edge of $\mathcal{L}$. For each edge qr of the restricted Delaunay triangulation $\mathcal{D}_{|L_j}(\mathcal{P})$, there is at least one ball, centered on L_j, whose bounding sphere passes through q and r and with no point of $\mathcal{P}$ in its interior. Such a ball is centered on a point of the non empty intersection $L_j \cap V(q,r)$ and called here after a *restricted Delaunay ball*. The edge qr of $\mathcal{D}_{|L_j}(\mathcal{P})$ is said to be *encroached* by a point x if x is in the interior of a restricted Delaunay ball of qr.

Likewise, for each triangle qrs of the restricted Delaunay triangulation $\mathcal{D}_{|S_k}(\mathcal{P})$, there is at least one ball, centered on the patch S_k, whose bounding sphere passes through q, r and s and with no point of $\mathcal{P}$ in its interior. Such a ball is called a restricted Delaunay ball (or a *surface Delaunay ball* in this case). The triangle qrs of $\mathcal{D}_{|S_k}(\mathcal{P})$ is said to be encroached by a point x if x is in the interior of a surface Delaunay ball of qrs.

The algorithm takes as input

- the piecewise smooth complex $\mathcal{F}$ describing the boundary of the volume to be meshed.
- A sizing field $\sigma(x)$ defined over the domain to be meshed. The sizing field is assumed to be a Lipschitz function such that for any point $x \in \mathcal{F}$, $\sigma(x) \leq \mathrm{lfs}(x)$.
- Some shape criteria in the form of upper bounds β_3 and β_2 for the radius-edge ratio of respectively the tetrahedra in the mesh and the boundary facets.
- Two parameters α_1 and α_2 such that $\alpha_1 \leq \alpha_2 \leq 1$ whose values will be discussed in section 5.

The algorithm begins with a set of sample points $\mathcal{P}_0$ including $\mathcal{Q}$, at least two points on each segment of $\mathcal{L}$ and at least three points on each patch of $\mathcal{S}$.

At each step, a new sample point is added to the current set $\mathcal{P}$. The new point is added according to the following rules, where rule R_i is applied only when no rule R_j with $j < i$ can be applied. Those rules issue calls to sub-procedures, respectively called `refine-edge`, `refine-facet-or-edge`, and `refine-tet-facet-or-edge`, which are described below.

R1 If, for some L_j of $\mathcal{L}$, there is an edge e of $\mathcal{D}_{|L_j}(\mathcal{P})$ whose endpoints do not both belong to L_j, call `refine-edge`(e).

R2 If, for some L_j of $\mathcal{L}$, there is an edge e of $\mathcal{D}_{|L_j}(\mathcal{P})$ with a restricted Delaunay ball $B(c_e, r_e)$ that does not satisfy $r_e \leq \alpha_1 \sigma(c_e)$, call `refine-edge`$(e)$.

R3 If, for some S_k of $\mathcal{S}$, there is a facet f of $\mathcal{D}_{|S_k}(\mathcal{P})$ whose vertices do not all belong to S_k, call `refine-facet-or-edge`(f).

R4 If, for some S_k of $\mathcal{S}$, there is a facet f of $\mathcal{D}_{|S_k}(\mathcal{P})$, and a restricted Delaunay ball $B(c_f, r_f)$ of f with radius-edge ratio ρ_f, such that

R4.1 either the size criteria, $r_f \leq \alpha_2 \sigma(c_f)$, is not met,

R4.2 or the shape criteria, $\rho_f \leq \beta_2$, is not met,
call `refine-facet-or-edge(f)`.

R5 If there is some tetrahedron t in $\mathcal{D}_{|\mathcal{O}}(\mathcal{P})$, with Delaunay ball $B(c_t, r_t)$ and radius edge ratio ρ_t, such that

R5.1 either the size criteria, $r_t \leq \sigma(c_t)$, is not met,

R5.2 or the shape criteria $\rho_t \leq \beta_3$ is not met,
call `refine-tet-facet-or-edge(t)`.

refine-edge. The procedure `refine-edge(e)` is called for an edge e of the restricted Delaunay triangulation $\mathcal{D}_{|L_j}(\mathcal{P})$ of some edge L_j in $\mathcal{L}$. The procedure inserts in $\mathcal{P}$ the center c_e of the restricted Delaunay ball $B(c_e, r_e)$ of e with largest radius r_e.

refine-facet-or-edge. The procedure `refine-facet-or-edge(f)` is called for a facet f of the restricted Delaunay triangulation $\mathcal{D}_{|S_k}(\mathcal{P})$ of some facet S_k in $\mathcal{S}$. The procedure considers the center c_f of the restricted Delaunay ball $B(c_f, r_f)$ of f with largest radius r_f and performs the following:

- if c_f encroaches some edge e in $\cup_{L_j \in \mathcal{L}} \mathcal{D}_{|L_j}(\mathcal{P})$, call `refine-edge(e)`,
- else add c_f in $\mathcal{P}$.

refine-tet-facet-or-edge. The procedure `refine-tet-facet-or-edge(t)` is called for a cell t of $\mathcal{D}_{|\mathcal{O}}(\mathcal{P})$. It considers the circumcenter c_t of t and performs the following:

- if c_t encroaches some edge e in $\cup_{L_j \in \mathcal{L}} \mathcal{D}_{|L_j}(\mathcal{P})$, call `refine-edge(e)`,
- else if c_t encroaches some facet f in $\cup_{S_k \in \mathcal{L}} \mathcal{D}_{|S_k}(\mathcal{P})$,
    call `refine-facet-or-edge(f)`,
- else add c_t in $\mathcal{P}$.

4 Output Mesh

At the end of the refinement process, the tetrahedra in $\mathcal{D}_{|\mathcal{O}}(\mathcal{P})$ form the final mesh and the features of $\mathcal{L}$ and $\mathcal{S}$ are approximated by their respective restricted Delaunay triangulation. In this section we assume that the refinement process terminates, and we prove that after termination, each connected component O_i of the domain O is represented by a submesh formed with well sized and well shaped tetrahedra and that the boundary of this submesh is an accurate and homeomorphic approximation of bd O_i.

Theorem 1. *If the meshing algorithm terminates, the output mesh* $\mathcal{D}_{|\mathcal{O}}(\mathcal{P})$ *has the following properties.*

Size and shape. *The tetrahedra in* $\mathcal{D}_{|\mathcal{O}}(\mathcal{P})$ *conform to the input sizing field and are well shaped (meaning that their radius-edge ratio is bounded by* β_3*).*

Homeomorphism. *There is an homeomorphism between* $\bigcup \mathcal{F}$ *and* $\mathcal{D}_{\bigcup \mathcal{F}}(\mathcal{P})$ *such that each face F of $\mathcal{F}$ is mapped to its restricted Delaunay triangulation* $\mathcal{D}_F(\mathcal{P})$*. Furthermore, for each connected component O_i of the domain*

O, the boundary $\operatorname{bd} O_i$ of O_i is mapped to the boundary $\operatorname{bd} \mathcal{D}_{O_i}(\mathcal{P})$ of the submesh $\mathcal{D}_{O_i}(\mathcal{P})$.

Hausdorff distance. For each face (curve segment or surface patch) F in $\mathcal{F}$, the Hausdorff distance between the restricted Delaunay triangulation $\mathcal{D}_F(\mathcal{P})$ and F is bounded.

Proof. The first point is a direct consequence of rules R5.2 and R5.1. The rest of this section is devoted respectively to the proof of the homeomorphism properties and to the proof of Hausdorff distance.

The extended closed ball property.

To prove the homeomorphism between $\bigcup \mathcal{F}$ and $\mathcal{D}_{\bigcup \mathcal{F}}(\mathcal{P})$ we make use of the Edelsbrunner and Shah theorem [ES97]. In fact, because the union $\bigcup \mathcal{F}$ is not assumed to be a manifold topological space, we make use of the version of Edelsbrunner and Shah theorem for non manifold topological spaces. This theorem is based on an extended version of the closed balled property recalled here for completeness.

Definition 3 (Extended closed ball property). *A CW complex is a regular complex whose faces are topological balls. A set of point $\mathcal{P}$ is said to have the extended closed ball property with respect to a topological space X of $\mathbb{R}^d$ if there is a CW complex $\mathcal{R}$ with $X = \bigcup \mathcal{R}$ and such that, for any subset $T \subseteq \mathcal{P}$ whose Voronoi face $V(T) = \bigcap_{p \in T} V(p)$ has a non empty intersection with X, the following holds.*

1. *There is a CW subcomplex $\mathcal{R}_T \subset \mathcal{R}$ such that $\bigcup \mathcal{R}_T = V(T) \cap X$.*
2. *Let $\mathcal{R}_T^0$ be the subset of faces in $\mathcal{R}_T$ whose interior is included in the interior of $V(T)$. There is a unique face F_T of $\mathcal{R}_T$ which is included in all the faces $\mathcal{R}_T^0$.*
3. *If F_T has dimension j, $F_T \cap \operatorname{bd} V(T)$ is a $j-1$-sphere.*
4. *For each face $F \in \mathcal{R}_T^0 \setminus \{F_T\}$ with dimension k, $F \cap \operatorname{bd} V(T)$ is a $k-1$ ball.*

Furthermore, $\mathcal{P}$ is said to have the extended generic intersection property for X if for every subset $T \subseteq \mathcal{P}$ and every face $F' \in \mathcal{R}_T \setminus \mathcal{R}_T^0$ there is a face $F \in \mathcal{R}_T^0$ such that $F' \subseteq F$.

Theorem 2 ([ES97]). *If X is a topological space and $\mathcal{P}$ a finite point set that has the extended generic intersection property and the extended closed ball property for X, X and $\mathcal{D}_X(\mathcal{P})$ are homeomorphic.*

In the following, we consider the final sampling $\mathcal{P}$ produced by the meshing algorithm and we show that $\mathcal{P}$ has the extended closed ball property with respect to $\bigcup \mathcal{F}$. For this, we need a CW complex $\mathcal{R}$ whose domain coincides with $\bigcup \mathcal{F}$. We define $\mathcal{R}$ as $\mathcal{R} = \{V(T) \cap F : T \subseteq \mathcal{P}, F \in \mathcal{F}\}$. and our first goal is therefore to prove that each face in this complex is a topological ball.

Surface sampling

The following lemmas are borrowed from the recently developed surface sampling theory [ACDL02, BO05]. They are hereafter adapted to our setting where the faces $S_k \in \mathcal{S}$ are patches of smooth closed surfaces.

Lemma 1 (Topological lemma). *[ACDL02] For any point $x \in \bigcup \mathcal{F}$, any ball $B(x,r)$ centered on x and with radius $r \leq \mathrm{lfs}(x)$ intersects any face F of $\mathcal{F}$ including x according to a topological ball.*

Lemma 2 (Long distance lemma). *[Dey06] Let x be a point in a face S_k of $\mathcal{S}$. If a line l through x makes a small angle $(l, l(x)) \leq \eta$ with the line $l(x)$ normal to S_k at x and intersects S_k in some other point y, $\mathrm{d}(x,y) \geq 2\mathrm{lfs}(x)\cos(\eta)$.*

Lemma 3 (Chord angle lemma). *[ACDL02] For any two points x and y of S_k with $\mathrm{d}(x,y) \leq \eta\,\mathrm{lfs}(x)$, with $\eta \leq 2$, the angle between xy and the tangent plane of S_k at x, is at most $\arcsin\frac{\eta}{2}$.*

Lemma 4 (Normal variation lemma). *[ACDL02] Let x and y be two points of S_k with $\mathrm{d}(x,y) \leq \eta\min(\mathrm{lfs}(x),\mathrm{lfs}(y))$, with $\eta \leq 2$. Let $n(x)$ and $n(y)$ be the normal vectors to S_k at x and y respectively. Assume that $n(x)$ and $n(y)$ are oriented consistently, for instance toward the exterior of the smooth closed surface including S_k. Then the angle $(n(x), n(y))$ is at most $2\arcsin\frac{\eta}{2}$.*

Lemma 5 (Facet normal lemma). *[ACDL02] Let p,q,r be three points of S_k, such that the circumradius of triangle pqr is at most $\eta\,\mathrm{lfs}(p)$. The angle between the line l_f normal to triangle pqr and the line $l(p)$ normals to S_k in p is at most $\arcsin(\eta\sqrt{3}) + \arcsin\frac{\eta}{1-\eta}$.*

At the end of the algorithm, the sampling $\mathcal{P}$ is such that for any patch S_k of $\mathcal{S}$, the subset $\mathcal{P} \cap S_k$ is a *loose α_2-sample* of S_k. This means that any restricted Delaunay ball $B(c,r)$, circumscribed to a face of the restricted Delaunay triangulation $\mathcal{D}_{|S_k}(\mathcal{P} \cap S_k)$, has its radius r bounded by $\alpha_2\mathrm{lfs}(c)$. It is proved in [BO06] that any loose ε-sample of a patch S_k is a ε'-sample of S_k for $\varepsilon' = \varepsilon(1 + O(\varepsilon))$. The following lemma are proved in [Boi06] for closed smooth surfaces, but their proof can be easily adapted in the case of surface patches.

Lemma 6 (Projection lemma). *[Boi06] Let $\mathcal{P}_k$ be a loose ε-sample of the smooth surface patch S_k for $\varepsilon < 0.24$. Any pair f and f' of two facets of $\mathcal{D}_{|S_k}(\mathcal{P}_k)$ sharing a common vertex p, have non overlapping orthogonal projections onto the tangent plane at p i.e. the projections of the relative interiors of f and f' do not intersect.*

Lemma 7 (Small cylinder lemma). *[Boi06] Let $\mathcal{P}_k$ be a loose ε-sample of the smooth surface patch S_k. For any point $p \in \mathcal{P}_k$, if $V_{\mathcal{P}_k}$ denotes the cell* of p in the diagram $\mathcal{V}(\mathcal{P}_k)$, the intersection $V_{\mathcal{P}_k}(p) \cap S_k$ is contained in a small cylinder with axis $l(p)$, height $h = O(\varepsilon^2)\mathrm{lfs}(p)$ and radius $O(\varepsilon\,\mathrm{lfs}(p))$, where $l(p)$ is the line normal to S_k at p.

The same result holds if the function lfs is replaced by any Lipschitz sizing field σ with $\sigma(x) \le \mathrm{lfs}(x)$, both in the definition of loose ε-sample and in the description of the small cylinder.

Proof of the homeomorphism properties

The following lemmas are related to the final sampling $\mathcal{P}$. They assume that the sizing field $\sigma(x)$ is less than $\mathrm{lfs}(x)$ for any point $x \in \mathcal{F}$ and that the constant α_1 and α_2 used in the algorithm are small enough.

Lemma 8.   • *A facet $V(p, q)$ of $\mathcal{V}(\mathcal{P})$ intersects a curve segment L_j of $\mathcal{L}$ iff p and q are consecutive vertices on L_j. Any facet in $\mathcal{V}(\mathcal{P})$ intersects at most one curve segment in $\mathcal{L}$. If non empty, the intersection $V(p, q) \cap L_j$ is a single point, i. e. a 0-dimensional topological ball.*
  • *For any cell $V(p) \in \mathcal{P}$, the intersection $V(p) \cap L_j$ is empty if $p \notin L_j$ and a single curve segment, i.e a 1-dimensional topological ball, otherwise.*

Proof. Proof omitted in this abstract.

Lemma 9.   • *An edge $V(p, q, r)$ of $\mathcal{V}(\mathcal{P})$ may intersect the surface patch S_k only if p, q and r belong to S_k. An edge $V(p, q, r)$ intersects at most one surface patch of $\mathcal{S}$. If non empty, the intersection $V(p, q, r) \cap S_k$ is a single point.*
  • *A facet $V(p, q)$ of $\mathcal{V}(\mathcal{P})$ may intersect the surface patch S_k only if p and q belong to S_k. If non empty, the intersection $V(p, q) \cap S_k$ is a single curve segment, i.e a 1-dimensional topological ball .*
  • *For any cell $V(p) \in \mathcal{P}$, the intersection $V(p) \cap S_k$ is empty if $p \notin S_k$ and a topological disk if $p \in S_k$.*

Proof. Proof omitted in this abstract.

Lemma 10. *The set $\mathcal{R} = \{V(\mathcal{T}) \cap F : \mathcal{T} \subset \mathcal{P}, F \in \mathcal{F}\}$ forms a CW complex and $\mathcal{P}$ has the extended closed ball property for $\bigcup \mathcal{F}$.*

Proof. Lemma 8 and 9 show that each element in $\mathcal{R}$ is a topological ball. Obviously, the boundary of each face in $\mathcal{R}$ is a union of faces in $\mathcal{R}$ and the intersection of two faces in $\mathcal{R}$ is either empty or a face in $\mathcal{R}$. Therefore $\mathcal{R}$ is a CW complex. For any subset $\mathcal{T} \in \mathcal{P}$, $\mathcal{R}_\mathcal{T} = \{V(\mathcal{T}) \cap F : F \in \mathcal{F}\}$ and $\mathcal{R}_\mathcal{T}^0$ is the subset of $\mathcal{R}_\mathcal{T}$ obtained with the faces $F \in \mathcal{F}$ whose interior intersects the interior of $V(\mathcal{T})$. Therefore condition 1 in definition 3 is satisfied and we next show that conditions 2, 3 and 4 are also satisfied.

Let $V(p, q, r)$ be an edge of $\mathcal{V}(\mathcal{P})$. For $\mathcal{T} = \{p, q, r\}$, $\mathcal{R}_\mathcal{T}$ is empty except if p, q, r belong to the same facet $S_k \in \mathcal{S}$. In this last case S_k is the unique face of $\mathcal{F}$ intersected by $V(p,q,r)$ so that $\mathcal{R}_T^0$ reduces to $\{F_T = V(p,q,r) \cap S_k\}$, thus conditions 2 and 4 are trivial and condition 3 is satisfied because $F_T \cap$ bd $V(p,q,r)$ is the empty set which is a 0-dimensional sphere.

Let $V(p,q)$ be a facet of $\mathcal{V}(\mathcal{P})$ and assume that $V(p,q) \cap \mathcal{F}$ is not empty. It results from lemmas 8 and 9 that, either $V(p,q)$ intersects a single surface patch S_k of $\mathcal{S}$ and no curve segment in $\mathcal{L}$ or $V(p,q)$ intersects a single segment L_j of $\mathcal{F}$ and all the patches of $\mathcal{S}$ incident to L_j.

In the first case $\mathcal{R}_T^0$ reduces to $\{F_T = V(p,q) \cap S_k\}$, and conditions 2 and 4 are trivial. Condition 3 is satisfied because $V(p,q)$ does not intersect the boundary of S_k and therefore, bd $V(p,q) \cap F_T = $ bd $V(p,q) \cap S_k = $ bd $(V(p,q) \cap S_k)$, which, owing to lemma 9, is 1-dimensional sphere.

In the second case, $\mathcal{R}_T^0 = \{V(p,q) \cap L_i\} \cup \{V(p,q) \cap S_k : S_k \in \mathcal{S}, L_i \subset S_k\}$, and condition 2 holds with $F_T = V(p,q) \cap L_j$. Furthermore, bd $V(p,q) \cap F_T$ is empty so that conditions 3 holds. For any $F = V(p,q) \cap S_k \in \mathcal{R}_T^0$, bd $V(p,q) \cap F = $ bd $(V(p,q) \cap S_k) - V(p,q) \cap L_j$ which is a single point (from lemmas 8 and 9), so that condition 4 holds.

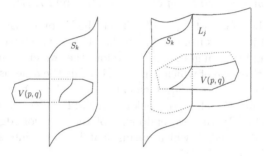

Fig. 1. Intersection with a Voronoi facet $V(p,q)$. Left $V(p,q)$ intersects no curved segment L_j. Right $V(p,q)$ intersects a single curved segment L_j.

Let $V(p)$ be a cell of $\mathcal{V}(\mathcal{P})$. If $p \notin \mathcal{F}$, $V(p)$ intersects no face of $\mathcal{F}$ and $\mathcal{R}_T$ is empty.

If p belongs to some facet S_k of $\mathcal{S}$ but to no edge of $\mathcal{L}$, S_k is the only face in $\mathcal{F}$ intersected by $V(p)$. Thus, $\mathcal{R}_T^0$ reduces to $\{F_T = V(p) \cap S_k\}$ and conditions 2 and 4 are trivial. Condition 3 is satisfied because $V(p)$ does not intersect bd S_k and therefore bd $V(p) \cap F_T = $ bd $V(p) \cap S_k = $ bd $((V(p) \cap S_k)$, which, owing to lemma 9 and is a 1-dimensional topological sphere.

If p belongs to some edge L_i but is not a vertex of $\mathcal{F}$, $V(p)$ intersects L_i and all facets S_k containing L_i. Therefore, $\mathcal{R}_T^0 = \{V(p) \cap L_i\} \cap \{V(p) \cap S_k : S_k \in \mathcal{S}, L_i \subset S_k\}$ and condition 2 holds with $F_T = V(p) \cap L_j$. The intersection bd $V(p) \cap F_T$ is bd $V(p) \cap L_i$, which is two points, i. e. a 0-sphere, from lemma 8. For any $F = V(p) \cap S_k$ in $\mathcal{R}_T^0$, bd $V(p) \cap F = $ bd $(V(p) \cap S_k) - V(p) \cap L_i$ which is from lemmas 9 and 8 a 1-dimensional topological sphere minus a 1-dimensional topological ball, i. e. a 1-dimensional topological ball.

The case where p is a vertex of $\mathcal{F}$ is omitted in this abstract.

$\square$

As a conclusion, for small enough values of the constant α_1 and α_2 used by the algorithm, the final sample set $\mathcal{P}$ has the extended closed ball property for $\bigcup\mathcal{F}$. Adding the reasonable assertion that $\mathcal{P}$ has the extended generic intersection property, we conclude by Theorem 2 that $\bigcup\mathcal{F}$ and $\mathcal{D}_{\bigcup\mathcal{F}}(\mathcal{P})$ are homeomorphic. Moreover, because the proof of Theorem 2 construct the isomorphism step by step between each face $V(\mathcal{T}) \cap F \in \mathcal{R}$ and the corresponding face in $\mathcal{D}_F(\mathcal{P})$ in non decreasing order of dimension, the resulting isomorphism is such that each face F of $\mathcal{F}$ is mapped to its restricted Delaunay triangulation $\mathcal{D}_F(\mathcal{P})$.

The boundary $\mathrm{bd}\, O_i$ of each connected component O_i of the domain O is a union of faces $F \in \mathcal{F}$, each of which is homeomorphic to its restricted Delaunay triangulation $\mathcal{D}_F(\mathcal{P})$. Thus $\mathrm{bd}\, O_i$ is homeomorphic to the restricted Delaunay triangulation $\mathcal{D}_{\mathrm{bd}\, O_i}(\mathcal{P})$, which, from the following lemma, is just $\mathrm{bd}\, \mathcal{D}_{O_i}(\mathcal{P})$.

Lemma 11. *For each cell O_i of $\mathcal{O}$, $\mathcal{D}_{\mathrm{bd}\, O_i}(\mathcal{P}) = \mathrm{bd}\, \mathcal{D}_{O_i}(\mathcal{P})$.*

Proof. Let pqr be a facet of $\mathcal{D}_{\mathrm{bd}\, O_i}(\mathcal{P})$. Then the Voronoi edge $V(p, q, r)$ intersects $\mathrm{bd}\, O_i$ and it results from lemma 9 that $V(p, q, r)$ intersects at most one facet of $\mathrm{bd}\, O_i$ and in a single point. Therefore the two endpoints of $V(p, q, r)$ are Voronoi vertices one of which is inside O_i while the other is outside, which means that one of the tetrahedra incident to pqr belongs to $\mathcal{D}_{O_i}(\mathcal{P})$ while the other does not and therefore pqr belongs to $\mathrm{bd}\, \mathcal{D}_{O_i}(\mathcal{P})$. Reciprocally if pqr belongs to $\mathrm{bd}\, \mathcal{D}_{O_i}(\mathcal{P})$, the two endpoints of the Voronoi edge $V(p, q, r)$ are on a different side of $\mathrm{bd}\, O_i$ which means that $V(p, q, r)$ intersects $\mathrm{bd}\, O_i$ and therefore belongs to $\mathcal{D}_{\mathrm{bd}\, O_i}(\mathcal{P})$. $\square$

Hausdorff distance

We prove here that the mesh generation algorithm provides a control on the Hausdorff distance between F and the approximating linear complex $\mathcal{D}_F(\mathcal{P})$ through the sizing field σ.

Let us first consider a curve segment L_j in $\mathcal{L}$. For each edge $e = pq$ in $\mathcal{D}_{L_j}(\mathcal{P})$, both edge e and the portion $L_j(p, q)$ of L_j joining p to q are included in the restricted Delaunay $B(c_e, r_e)$ circumscribed to e. The Hausdorff distance between $L_j(p, q)$ and e is therefore less than r_e which, from rule R2, is less $\alpha_1\sigma(c_e)$ and the Hausdorff distance between L_j and $\mathcal{D}_{L_j}(\mathcal{P})$ is less than $2\alpha_1 \max_{x \in L_j} \sigma(x)$.

Let us then consider a surface patch S_k. Each triangle pqr in $\mathcal{D}_{S_k}(\mathcal{P})$ is included in its restricted Delaunay ball $B(c, r)$ with radius $r \leq \alpha_2\sigma(c)$ and therefore each point of pqr is at distance less than $\alpha_2\sigma(c)$ from S_k. From rule R4.1, and the small cylinder lemma (Lemma 7), we know that each point x in S_k is at distance $O(\alpha_2)\sigma(p)$ from its closest sample point p. The Hausdorff distance between S_k and $\mathcal{D}_{S_k}(\mathcal{P})$ is less than $O(\alpha_2) \max_{x \in S_k} \sigma(x)$.

5 Termination

This section proves that the refinement algorithm in section 3 terminates, provided that the constants α_1 and α_2, β_2 and β_3 involved in refinement rules are judiciously chosen. The proof of termination is, based on a volume argument. This requires to prove a lower bound for the distance between any two vertices inserted in the mesh.

For each vertex p in the mesh, we denote by $r(p)$ the insertion radius of p, that is the length of the shortest edge incident to vertex p right after the insertion of p, and $\delta(p)$ the distance from p to $\mathcal{F} \setminus \mathcal{F}_p$ where $\mathcal{F}_p$ is the set of features in $\mathcal{F}$ that contain the point p.

Recall that the algorithm depends on four parameters α_1, α_2, β_2 and β_3, and on a sizing field σ. Constants α_1 and α_2 are assumed to be small enough and such that $0 \leq \alpha_1 \leq \alpha_2 \leq 1$. The sizing field σ is assumed to be a Lipschitz function smaller than lfs(x) on $\mathcal{F}$. Let μ_0 be the maximum of σ over $\mathcal{F}$, and σ_0 the minimum of σ over $\mathcal{O}$.

Lemma 12. *For some suitable values of α_1, α_2, β_2 and β_3, there are constants η_2 and η_3 such that $\alpha_1 \leq \eta_3 \leq \eta_2 \leq 1$ and such that the following invariants are satisfied during the execution of the algorithm.*

$$\forall p \in \mathcal{P}, \quad r(p) \geq \alpha_1 \sigma_0 \tag{2}$$

$$\delta(p) \geq \begin{cases} \alpha_1 \sigma_0 & \text{if } p \in \mathcal{L} \\ \frac{\alpha_1 \sigma_0}{\eta_2} & \text{if } p \in \mathcal{S} \\ \frac{\alpha_1 \sigma_0}{\eta_3} & \text{if } p \in \mathcal{P} \setminus \mathcal{S} \end{cases} \tag{3}$$

Proof. (Sketch) The proof is an induction. Invariants (2) and (3) are satisfied by the set $\mathcal{Q}$ and by the set $\mathcal{P}_0$ of initial vertices. We prove that invariants (2) and (3) are still satisfied after the application of any of the refinement rules R1-R5 if the following values are set:

$$\begin{aligned} \alpha_1 &= \frac{1}{(\sqrt{2}+2)\nu_0, (\nu_0+1)} & \alpha_2 &= \frac{1}{\nu_0+1}, \\ \beta_2 &= (\sqrt{2}+2)\nu_0, & \beta_3 &= (\sqrt{2}+2)\nu_0 \, (\nu_0+1), \\ \frac{1}{\eta_2} &= (\sqrt{2}+1)\nu_0, & \frac{1}{\eta_3} &= (\sqrt{2}+2)\nu_0, \end{aligned} \tag{4}$$
$$\text{where } \nu_0 = \frac{2\mu_0}{\sigma_0}.$$

For lack of space and as an example, we only give here the proof for Rule R3 and omit the other cases.

Let us therefore assume that a new vertex c_f is added to $\mathcal{P}$ by application of Rule R3. The vertex c_f belongs to a surface patch S_k and is the center of a Delaunay ball $B(c_f, r_f)$, circumscribing a facet f of $\mathcal{D}_{|S_k}$. At least one of the vertices of f vertices, say p, does not belong to S_k. The insertion radius of c_f is r_f and, $r_f = \mathrm{d}(c_f, p) \geq \mathrm{d}(p, S_k)$. By induction hypothesis, $\mathrm{d}(p, S_k) \geq \alpha_1 \sigma_0$, and invariant (2) is satisfied.

To check invariant (3), we prove hereafter than $\mathrm{d}(c_f, \mathrm{bd}\, S_k) \geq \frac{\alpha_1 \sigma_0}{\eta_2}$, and use the following claim to prove that the same bound holds for $\delta(c_f)$.

Claim. For any point x in a face $F \in \mathcal{F}$, $\delta(x) \geq \min\left(\mathrm{d}(x, \mathrm{bd}\, F), \mathrm{lfs}(x)\right)$.

Proof. Omitted in this abstract.

Let y be the point of $\mathrm{bd}\, S_k$ closest to c_f, and let q be the sample point in $\mathcal{P} \cap \mathrm{bd}\, S_k$ closest to y. Then $\mathrm{d}(c_f, \mathrm{bd}\, S_k) = \mathrm{d}(c_f, y) \geq \mathrm{d}(c_f, q) - \mathrm{d}(y, q)$. The distance $\mathrm{d}(c_f, q)$ is at least r_f. The distance $\mathrm{d}(y, q)$ is at most $2\alpha_1 \mu_0$ because, when Rule R3 is applied, the curve segments of $\mathcal{L}$ are covered by restricted Delaunay balls whose radii are at most $\alpha_1 \mu_0$. Therefore

$$\mathrm{d}(c_f, \mathrm{bd}\, S_k) \geq r_f - 2\alpha_1 \mu_0. \tag{5}$$

- If vertex p does not belongs to $\mathcal{F}$, $r_f = \mathrm{d}(c_f, p) \geq \frac{\alpha_1 \sigma_0}{\eta_3}$ by induction hypothesis, and $\mathrm{d}(c_f, \mathrm{bd}\, S_k) \geq \frac{\alpha_1 \sigma_0}{\eta_3} - 2\alpha_1 \mu_0$ which is at least $\frac{\alpha_1 \sigma_0}{\eta_2}$ with the choices in (4).
- If the vertex p lies on $\mathcal{F}$ but is not interrelated to c_f, $r_f = \mathrm{d}(c_f, p) \geq \mathrm{lfs}(c_f) \geq \sigma_0$, and $\mathrm{d}(c_f, \mathrm{bd}\, S_k) \geq \sigma_0 - 2\alpha_1 \mu_0$ which is more than $\frac{\alpha_1 \sigma_0}{\eta_2}$ with the choices in (4).
- If p lies on $F_i \in \mathcal{F}$ and is interrelated to c_f, we have by angular hypothesis, $r_f^2 = \mathrm{d}(c_f, p)^2 \geq \mathrm{d}(c_f, S_k \cap F_i)^2 + \mathrm{d}(p, S_k \cap F_i)^2$. Thus $\mathrm{d}(c_f, S_k \cap F_i) \leq r_f$, which proves that sample point p cannot lie in a curve segment of $\mathcal{L}$. Indeed if F_i was some curve segment in $\mathcal{L}$, $S_k \cap F_i$ would be a vertex in $\mathcal{Q}$ included in $B(c_f, r_f)$ which contradicts the fact that $B(c_f, r_f)$ is a Delaunay ball. Therefore p belongs to the interior of some surface patch $S_i \in \mathcal{S}$ and by induction hypothesis $\delta(p) \geq \frac{\alpha_1 \sigma_0}{\eta_2}$ which implies that:

$$r_f^2 \geq \mathrm{d}(c_f, \mathrm{bd}\, S_k)^2 + \left(\frac{\alpha_1 \sigma_0}{\eta_2}\right)^2. \tag{6}$$

Equations (5) and (6) and the value chosen for η_2 in (4) implies that $\mathrm{d}(c_f, \mathrm{bd}\, S_k) \geq \frac{\alpha_1 \sigma_0}{\eta_2}$.

□

Lemma 12 provides a constant lower bound, $\alpha_1 \sigma_0$, for the insertion radius of any vertex in the mesh and therefore for the length of any edge in the final mesh. A standard volume argument then ensures that the Delaunay refinement algorithm of section 3 terminates.

6 Implementation and results

The algorithm has been implemented in C++, using the library CGAL [CGAL], which provided us with an efficient and flexible implementation of the three-dimensional Delaunay triangulation. Once the Delaunay refinement process described above is over, a sliver exudation [CDE+00] step is performed. This step does not move or add any vertex but it modifies the mesh by switching the triangulation into a weighted Delaunay triangulation with carefully chosen weights. As in [ORY05] the weight of each vertex in the mesh is chosen in turn so as to maximize the smallest dihedral angle of any tetrahedron incident to that vertex while preserving in the mesh any facet that belongs to the restricted triangulation of an input surface patch.

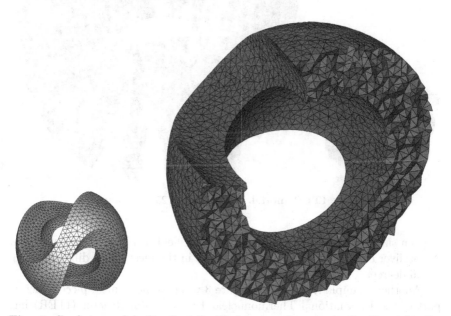

Fig. 2. Sculpt model. On the left: the input surface mesh. On the right: the output mesh (blue: facets of the surface mesh, red: tetrahedra of the volume mesh that intersect a given plan). The mesh counts 22923 tetrahedra.

Our mesh generation algorithm interacts with the input curve segments and surface patches through an oracle that is able to detect and compute intersections between planar triangles and curve segments and between straight segments and surface patches. Currently, we have only one implementation of such an oracle which handles input curve segments and surface patches described as respectively as polylines and triangular meshes. Thus, with respect to input features our algorithm currently act as a remesher but this is not a limitation of the method.

The figure 2 shows a mesh generated by our algorithm. The input surface of the figure 2 is made of six curved edges, and four surface patches The input surface was given by a surface mesh. We have considered that an edge of the mesh is a sub-segment of a curved segment of the surface when the normal deviation at the edge is greater than 60 degree. The curved segments and surfaces patches of the input surface are nicely approximated, as well as the normals (the result surface of figure 2 has been drawn without any OpenGL smoothing). Each element of the result mesh has a Delaunay ball smaller than

Fig. 3. ITER model. A mesh with 72578 tetrahedra.

a given sizing field. In the figure 2, the sizing field has been chosen uniform. After sliver exudation, the worst tetrahedra in the mesh has a dihedral angle of 1.6 degree.

Another example is shown on Figure 3 where a mechanical piece which is part of the International Thermonuclear Experimental Reactor (ITER) has been meshed.

7 Conclusion and future work

The algorithm provided in this paper is able to mesh volumes bounded by piecewise smooth surfaces. The output mesh has guaranteed quality and its granularity adapts to a user defined sizing field. The boundary surfaces and their sharp 1-dimensional features are accurately and homeomorphically represented in the mesh. The main drawback of the algorithm is the restriction imposed on dihedral angles made by tangent planes on singular points. Small angles are known to trigger an ever looping of the Delaunay refinement algorithm. The main idea to handle this problem is to define a protected zone around sharp features where the Delaunay refinement is restricted to prevent looping. This strategy assumes that the mesh already includes restricted Delaunay submeshes homeomorphic to the input surface patches and curved segments. This could be achieved using a strategy analog to the strategy proposed to conform Delaunay triangulation [MMG00, CCY04]. Another promising way to protect sharp features which is proposed by [CDR07] is to use weighted Delaunay triangulation with weighted points on sharp features.

References

[AB99]   N. Amenta and M. Bern. Surface reconstruction by Voronoi filtering. *Discrete Comput. Geom.*, 22(4):481–504, 1999.

[ACDL02]   N. Amenta, S. Choi, TK Dey, and N. Leekha. A simple algorithm for homeomorphic surface reconstruction. Internat. *J. Comput. Geom. & Applications*, 12:125–141, 2002.

[AUGA05]   P. Alliez, G. Ucelli, C. Gotsman, and M. Attene. Recent advances in remeshing of surfaces. *Part of the state-of-the-art report of the AIM@ SHAPE EU network*, 2005.

[BO05]   J.D. Boissonnat and S. Oudot. Provably good sampling and meshing of surfaces. *Graphical Models*, 67(5):405–451, 2005.

[BO06]   J.D. Boissonnat and S. Oudot. Provably good sampling and meshing of Lipschitz surfaces. *Proceedings of the twenty-second annual symposium on Computational geometry*, pages 337–346, 2006.

[BOG02]   C. Boivin and C. Ollivier-Gooch. Guaranteed-quality triangular mesh generation for domains with curved boundaries. *International Journal for Numerical Methods in Engineering*, 55(10):1185–1213, 2002.

[Boi06]   Jean-Daniel Boissonnat. *Voronoi Diagrams, Triangulations and Surfaces*, chapter 5. Inria, 2006.

[CCY04]   D. Cohen-Steiner, É. Colin de Verdière, and M. Yvinec. Conforming Delaunay triangulations in 3d. *Computational Geometry: Theory and Applications*, pages 217–233, 2004.

[CD03]   S.-W. Cheng and T. K. Dey. Quality meshing with weighted Delaunay refinement. *SIAM J. Comput.*, 33(1):69–93, 2003.

[CDE+00]   S.-W. Cheng, T. K. Dey, H. Edelsbrunner, M. A. Facello, and S.-H. Teng. Sliver exudation. *J. ACM*, 47(5):883–904, 2000.

[CDR07]   S.W. Cheng, T.K. Dey, and E.A. Ramos. Delaunay Refinement for Piecewise Smooth Complexes. *Proc. 18th Annu. ACM-SIAM Sympos. Discrete Algorithms*, pages 1096–1105, 2007.

[CDRR04]   S.-W. Cheng, T. K. Dey, E. A. Ramos, and T. Ray. Quality meshing for polyhedra with small angles. In *SCG '04: Proceedings of the twentieth annual symposium on Computational geometry*, pages 290–299. ACM Press, 2004.

[CDRR05]   S.-W. Cheng, T. K. Dey, E. A. Ramos, and T. Ray. Weighted Delaunay refinement for polyhedra with small angles. In *Proceedings 14th International Meshing Roundtable, IMR2005*, 2005.

[CGAL]   CGAL, Computational Geometry Algorithms Library. http://www.cgal.org.

[Che89]   L. P. Chew. Guaranteed-quality triangular meshes. Technical Report TR-89-983, Dept. Comput. Sci., Cornell Univ., Ithaca, NY, April 1989.

[Che93]   L.P. Chew. Guaranteed-quality mesh generation for curved surfaces. *Proceedings of the ninth annual symposium on Computational geometry*, pages 274–280, 1993.

[CP03]   S.-W. Cheng and S.-H. Poon. Graded conforming Delaunay tetrahedralization with bounded radius-edge ratio. In *SODA '03: Proceedings of the fourteenth annual ACM-SIAM symposium on discrete algorithms*, pages 295–304. Society for Industrial and Applied Mathematics, 2003.

[Dey06]   T.K. Dey. *Curve and Surface Reconstruction: Algorithms with Mathematical Analysis*. Cambridge University Press, 2006.

[ES97]   H. Edelsbrunner and N.R. Shah. Triangulating Topological Spaces. *International Journal of Computational Geometry and Applications*, 7(4):365–378, 1997.

[FBG96]   P. J. Frey, H. Borouchaki, and P.-L. George. Delaunay tetrahedrization using an advancing-front approach. In *Proc. 5th International Meshing Roundtable*, pages 31–43, 1996.

[FG00]   P.J. Frey and P.L. George. *Mesh Generation: Application to Finite Elements*. Kogan Page, 2000.

[GHS90]   P.-L. George, F. Hecht, and E. Saltel. Fully automatic mesh generator for 3d domains of any shape. *Impact of Computing in Science and Engineering*, 2:187–218, 1990.

[GHS91]   P.-L. George, F. Hecht, and E. Saltel. Automatic mesh generator with specified boundary. *Computer Methods in Applied Mechanics and Engineering*, 92:269–288, 1991.

[LC87]   W.E. Lorensen and H.E. Cline. Marching cubes: A high resolution 3D surface construction algorithm. *Proceedings of the 14th annual conference on Computer graphics and interactive techniques*, pages 163–169, 1987.

[LS03]   Francois Labelle and Jonathan Richard Shewchuk. Anisotropic voronoi diagrams and guaranteed-quality anisotropic mesh generation. In *SCG '03: Proceedings of the nineteenth annual symposium on Computational geometry*, pages 191–200, New York, NY, USA, 2003. ACM Press.

[LT01]   X.-Y. Li and S.-H. Teng. Generating well-shaped Delaunay meshes in 3d. In *SODA'01: Proceedings of the twelfth annual ACM-SIAM symposium on Discrete algorithms*, pages 28–37. Society for Industrial and Applied Mathematics, 2001.

[MMG00]   M. Murphy, D.M. Mount, and C.W. Gable. A point-placement strategy for conforming Delaunay tetrahedralization. *Proceedings of the eleventh annual ACM-SIAM symposium on Discrete algorithms*, pages 67–74, 2000.

[MTT99]   G.L. Miller, D. Talmor, and S.H. Teng. Data Generation for Geometric Algorithms on Non-Uniform Distributions. *International Journal of Computational Geometry and Applications*, 9(6):577, 1999.

[ORY05]   Steve Oudot, Laurent Rineau, and Mariette Yvinec. Meshing volumes bounded by smooth surfaces. In *Proc. 14th International Meshing Roundtable*, pages 203–219, 2005.

[Rup95]   J. Ruppert. A Delaunay refinement algorithm for quality 2-dimensional mesh generation. *J. Algorithms*, 18:548–585, 1995.

[She98]   J. R. Shewchuk. Tetrahedral mesh generation by Delaunay refinement. In *Proc. 14th Annu. ACM Sympos. Comput. Geom.*, pages 86–95, 1998.

[She02]   J. R. Shewchuk. Delaunay refinement algorithms for triangular mesh generation. *Computational Geometry: Theory and Applications*, 22:21–74, 2002.

5A.2

Large Out-of-Core Tetrahedral Meshing

Aurélien Alleaume, Laurent Francez, Mark Loriot, and Nathan Maman

Distene SAS, Pôle Teratec – Bard-1, Domaine du Grand Rué,
91680 Bruyères-le-Chatel, FRANCE
{`aurelien.alleaume,laurent.francez,mark.loriot,nathan.maman`}@distene.com

Summary. We present an effective strongly uncoupled method, which, given a closed and watertight surface mesh, generates its Delaunay constrained tetrahedrisation into sub-domains. Uncoupled means here that once chosen, a sub-domain interface will not be changed anymore. This method aims at preserving in the final tetrahedral mesh the properties of the mesh which would have been produced by the original sequential algorithm, namely the constrained input surface mesh and the quality and properties of the volume mesh. To achieve these objectives, our method reorders internal vertices insertion (applying the sequential constrained Delaunay kernel) such that the data can then be fully decoupled in sub-domains without introducing more constraints. Moreover, the interfaces are carefully chosen such that the load is roughly the same in all parts, and such that the interfaces separating the sub-domains are "invisible" in the final mesh. Successfully applied to out-of-core large scale mesh generation, this method allowed us to generate in double precision a 210 million tet mesh on a 4GB core memory single processor machine, and a 680 million tet mesh for a very complex geometry.

Introduction

In order to represent accurately complex geometries and to capture fine physical phenomena, high fidelity computational simulations involve more and more quality and density constraints on meshes. Generating large and high quality meshes can be achieved using a parallel approach, in order to gain performance in both memory consumption and CPU time.

A fast and robust sequential mesher like [7] has typical scalable speeds of about 5 million cells/minute on industrial cases in sequential mode. With such a mesher, the size of a mesh is not a problem as long as it fits into core memory. This tends to move the main focus to core memory limitation rather than pure meshing speed. Indeed, hardware and manufacturing limitations make non-distributed large core memory machines very expensive therefore justifies research of out-of-core or distributed parallel alternative methods for the generation of very large computational meshes.

Much effort must also be put on getting meshing properties as close as possible to the ones obtained with classical methods, such as density, number of cells, and quality of cells, but also reliability, so as to limit as much as possible the side effects of using these alternative methods.

Our method targets therefore **memory usage distribution** (rather than speed gain, which will eventually be addressed afterwards), a **full sequential code re-use** to ensure maximum reliability and the **same mesh properties** as the sequentially generated mesh.

To reach these targets, the *keys* of our method are :

1. The **internal edges division process reordering**, in order to **localize data** and to achieve **sub-domain decoupling**.
2. Built up of sub-domain interfaces such that they **do not interfere** with further insertions of vertices through the **constrained Delaunay kernel**.

Following the classification of parallel meshing techniques found in [1], the method presented here is a discrete domain decomposition method for Delaunay constrained mesh generation. Larwood et al. ([10]) propose to decompose the input surface mesh by creating interfaces with two-dimensional Delaunay triangulations. To make their method more robust and effective they also reject some cutting regions depending on angle considerations. Said et al. ([12]) decompose the initial coarse mesh by re-meshing an interface made of coarse tetrahedra faces.

What we propose goes beyond these approaches : in order to decompose the initial coarse mesh, we create the sub-domain interfaces through a three-dimensional Delaunay tetrahedrisation. This allows us to avoid imposing new faces into the mesh, by simply "freezing" some tetrahedra faces to create an interface. This generates interfaces which have a minimal impact in the resulting mesh, as if they had been produced by an *a posteriori* mesh partitioning method applied to the assembled mesh.

The method is based on a recursive bisection. An important feature of our approach is that each bisection determines the final interface between the two newly created sub-domains, very much in the way it would have been produced in the original sequential core-memory mesher. In other words, this method addresses the constraints on the sub-domains interfaces only during their construction and not afterwards, thus allowing a highly uncoupled algorithm.

1 Distributed mesh generation

From sequential to distributed mesh generation

Given a boundary surface mesh, we compute its coarse Delaunay constrained volume mesh. A state of the art implementation of the involved algorithms comes from INRIA [7]. This is the sequential whole-in-core basis we use for our distributed meshing method.

Following a recursive bisection approach, this tetrahedral mesh is then recursively split into two parts until an arbitrary number ns of sub-domains is reached. Each sub-domain is then saturated to its final density using the original whole-in-core sequential algorithm.

The work flow of the method is summarized in figure 1.

The kernel of our method has two main steps:

    1. bisection of a sub-domain mesh S;

    2. topology information management between sub-domains.

To be efficient, the first step involves more or less complex sub-steps while the second mainly requires careful design and implementation.

We will describe them with more details in the following sections. An illustrated example can be found in section (2.1)

1.1 Bisection

This bisection step does not target an optimal mesh decomposition. Thus, for sake of implementation simplicity and to assess the interest of the overall method, we chose to cut the mesh around a bisection plane. Our method can however be generalised to other kinds of decomposition (see [2]). The only prerequisits are :

1. The ability to locate a point relative to the cutting surface, i.e. to find on which side it is.
2. The ability to tell if an edge intersects a given width area around the cutting surface or not.

Note that we use here a mesh partitioning approach only in order to compute a bisection plane which will be used for local mesh generation,

Choosing a good bisection plane

An easy way to choose the cutting plane is to compute the inertia axis of the sub-domain mesh (see [3]). In other words, for a sub-domain S, we compute the inertia matrix

$$M = (\tau_{i,j})_{i,j=1...3} \quad \text{with } \tau_{i,j} = \sum_{x \in G} (x_i - g_i)(x_j - g_j)$$

where $g = (g_i)_{i=1...3}$ is the gravity center of S and G is the set of gravity centers of tetrahedra in S.

The inertia axis is given by an eigenvector associated to the highest singular value of M[1]. This method is very fast and uses the locality of data, i.e.,

---

[1]It can be simply computed by a successive power method.

the more the domain gets divided, the more the inertia axis captures the mesh topology. More information on this subject can be found in [5].

This inertia axis minimises the sub-domain spread (in some sense) from itself. So to minimise the size of the interface, the cutting plane is chosen orthogonal to it with the following considerations in mind:

1. the number of sub-domains may not be a power of two;
   To be able to generate a final number of sub-domains which is not necessarily a power of two, we partition this sub-domain with a ratio not equal to one half.
2. a good load balance may be critical to get a mesh with the smallest number of sub-domains possible and is almost required for most parallel applications that may use such partition.
   To achieve a good load balancing in the subsequent bisection steps, care should be taken that two coarse tetrahedra at this stage will not hold the same number of tetrahedra in the final mesh. Evaluating the target mesh metric and computing tetrahedral volume in that metric gives an acceptable *a priori* final load evaluation, hence a more accurate load is used to select the cutting plane position. For a tetrahedron T we compute:

$$\int_T d\lambda = \iiint_T \frac{1}{h^3(x,y,z)} \, dx \, dy \, dz$$

where λ is an approximation of the metric in S and $h(x, y, z)$ approximates the desired size at x, y, z.

This step will really make a difference in terms of performance and first shot distribution on a non uniform mesh[2].

Refining the mesh around the bisection plane

In this step, vertices are inserted around the cutting plane following a volume constrained Delaunay kernel (see [5]):

Let T_i be the Delaunay tetrahedrisation of a set of points $(P_k)_{k \in \{1..i\}}$. Then, T_{i+1}, the result of the insertion of point $P = P_{i+1}$ in T_i can be written

$$T_{i+1} = T_i - C_P + B_P$$

where :

- C_P is the set of elements whose circumsphere contains P. In order not to lose faces given in the input surface mesh this set can be slightly restricted (and then corrected) such that no required face is an internal face of C_P.

---

[2]Our method excludes over-decomposition techniques that often compensate load imbalance consequences by computing more sub-domains than desired and mapping several ones on each processor.

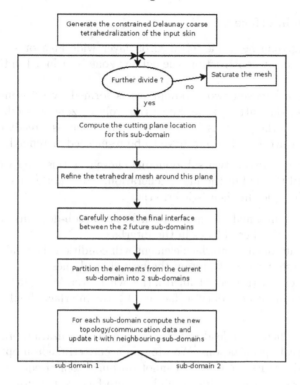

Fig. 1. Method work flow

- $\mathcal{B}_P$ is the set of elements formed by joining P with the external faces of the elements in $\mathcal{C}_P$.

To allow graded mesh generation the inserted vertices are computed by dividing the existing coarse edges that intersect a given area around the bisection plane.

Let the refined zone be the area around the cutting plane which is refined to its final vertex saturation.

The vertices insertion is conducted such that the refined zone is composed of a layer at least two tetrahedron wide on each side of the cutting plane. This will leave some degree of freedom in the next step for choosing the interface faces inside this area.

After this point insertion step, the refined zone is almost meshed as it would have been in the fully sequential method. With very large and fine meshes, this unbalanced vertex insertion often triggers precision issues in the constrained Delaunay sequential mesher. This is especially true when it comes to cavity computation and points localization. Some sequential algorithms have been slightly tuned to address these issues better.

Choosing the interface

Let the cutting zone be a one-tetrahedron layer wide area on each side of the cutting plane. By construction, the cutting zone is included in the refined zone.

The interface between two sub-domains is formed by selecting faces of tetrahedra from the cutting zone. Since this volume zone was refined to its final saturation in the previous step, the selected surface is already consistent with the overall size and does not have to be re-meshed or refined.

The interface between two sub-domains chosen at this step will remain unchanged until the end of the mesh generation. This choice takes therefore into account the following heuristic criteria:

1. As we target minimal communication between sub-domains, we should minimize the number of faces in the interface.
2. As the quality of each interface element will condition the final quality of the final mesh, we need to retain the best shaped ones.
3. To avoid over constrained tetrahedra (i.e. with several faces in sub-domain interfaces or input boundary surface mesh), the interface should be as flat as possible.

These simple heuristics lead to a multi-criteria optimization problem. We solve it with local optimizations. Working on a per tetrahedron optimization basis quickly shows its limits as it cannot optimize some frequent patterns. If we use instead all tetrahedra around an interface vertex, we are able to achieve better interface smoothing.

This method can lower the interface size by approximately thirty percent while retaining the best shaped faces.

1.2 About constrained Delaunay meshing

Let $\mathcal{I}$ be the set of all the faces in the interface build in the previous step and $\mathcal{Z}$ be the set of all the elements from the cutting zone with a face in $\mathcal{I}$.

To ensure that our method generates a constrained Delaunay mesh of the input surface (*i.e.* that interfaces do not introduce more constraints regarding the Delaunay property), a sufficient condition is that for any point P that will be inserted in the future, no interface face is an internal face of the cavity of P, $\mathcal{C}_P$.

A slightly stronger but simpler condition is that for any point P that will be inserted in the future :

$$\mathcal{C}_P \cap \mathcal{Z} = \emptyset$$

This is because if an internal face of $\mathcal{C}_P$ is in the interface, then at least one of its adjacent tetrahedra is in $\mathcal{C}_P \cap \mathcal{Z}$.

Since no point will be inserted further in the refined zone, a sufficient condition is that the circumsphere of all the elements in the cutting zone is included in the refined zone.

This condition is not explicitly verified near the input (constrained) surface mesh when the refined zone is built but the practice shows that it is not really needed, as the extra layers of tetrahedra in the refined zone around the cutting zone create an area large enough.

1.3 Topology information management

Topology information is kept by each sub-domain to allow the merge of all the sub-domains into a single sequential mesh, or the generation of the distributed local communication information which would be required by a parallel solver.

This step does not involve any specific technique but requires careful implementation to assert its correctness and good performance. Basically, after its bisection, each sub-domain sends the topology update to its neighbours. In parallel mode, this step would become a synchronization point.

1.4 Memory consumption and method limits

Memory consumption has two local maxima: the first bisection and the last sub-domain meshing step. The latter can be lowered simply by increasing the desired number of sub-domains whereas the first maximum fixes the limit of the largest achievable mesh on a given machine for this method.

For example, we were able to generate with our method more than 211 million cells in double precision (64 bit) on a 4GB machine (see section 2.3).

2 Applications

2.1 A Simple illustrated application

The first example illustrates how the method works for a decomposition into two sub-domains. The case is a car embedded entirely into a large volume (courtesy of PSA Group).

We deliberately chose here to put the cutting plane on the front of the car, close to but without intersecting it. This is often a difficulty for distributed mesh generation methods which only use surface mesh information.

The figure 2 shows the initial coarse volume mesh before and after it gets refined around the bisection plane. Then figure 3 shows the resulting sub-domain interface with or without faces choice optimisation. Last, figure 4 compares the sequential mesh with the one generated by our method in the previously described configuration. It shows that our interface construction enables us to generate a mesh which is almost identical to the sequential one in terms of density. Moreover, on this final mesh, one cannot identify where the cutting plane lies (see [11] for the planar case analysis in which the interface lines can be seen in the resulting mesh *i.e.* it does not preserve the likewise-sequential property we maintain here; see also [8] for tetrahedral meshes in which the inital coarse mesh surfaces can be seen).

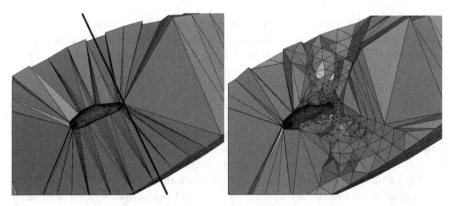

Fig. 2. Initial coarse mesh (on the left, the black line gives the location of the cutting plane) and mesh refined around the cutting plane (right). View is a cut orthogonal to the bisection plane.

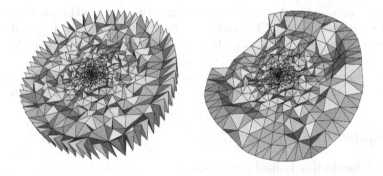

Fig. 3. Natural (left) and optimized (right) sub-domain interface

2.2 Sequential versus out-of-core meshing comparison

In this section we present a more complete comparison of the sequential mesh and the one generated with a 4 domain decomposition.

For this purpose we consider 5 industrial or non trivial test cases. They are illustrated on figures 5, 6, 7, 8, and 9.

The most interesting one is the one on figure 9 representing a cube with 1,000 balls of various size randomly placed in the volume (such geometries are used for the simulation of concrete, where the balls model the aggregates, and the volume between the spheres models the cement). This is a typically challenging case for a mesher because of the numerous randomly placed mesh constraints generated by the balls, and even more challenging in a distributed paradigm such as the one used by out-of-core or parallel meshers.

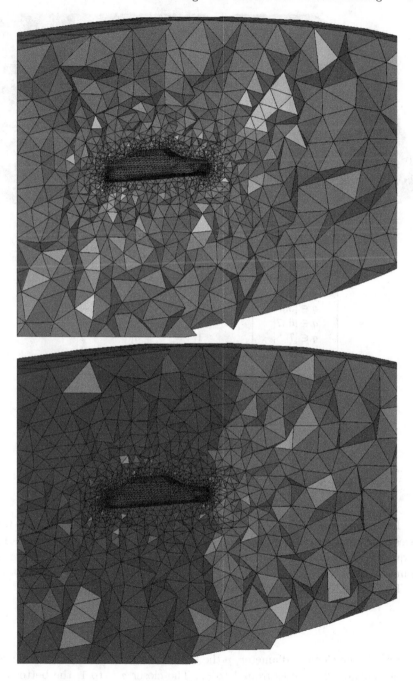

Fig. 4. Comparison of sequential (above) and out-of-core (below) meshes ; colors identify the two sub-domains on the latter. View is a cut orthogonal to the bisection plane.

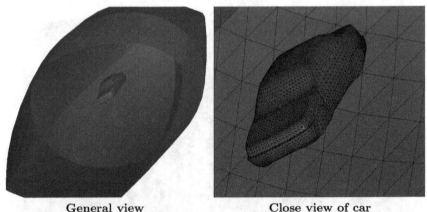

General view   Close view of car

	Sequential	Out-of-core	Difference
$q \in [1..2[$	107,790	84,757	-23,033
$q \in [2..3[$	8,993	6,539	-2,454
$q \in [3..4[$	648	626	-22
$q \in [4..5[$	103	129	26
$q \in [5..6[$	30	26	-4
$q \in [6..7[$	9	9	0
$q \in [7..8[$	1	3	2
$q \in [8..9[$	3	2	-1
$q \in [9..10[$	4	3	-1
$q \in [10..12.37[$	4		
$q \in [10..12.59[$		2	-2
Worst quality	12.37	12.59	
Number of cells	117,585	92,096	-25,489

Quality comparison

Fig. 5. Case #1, Peugeot Car (courtesy of PSA Group)

The tables below the figures expose the quality diagrams of the 5 test cases. For each case, they give the distribution of the tetrahedra in terms of quality, the worst tetrahedron quality and the total number of generated elements.

The element quality we consider here is the finite element quality

$$q = \alpha \frac{h}{\rho}$$

where: h is the element diameter, ρ the inradius, and α a normalisation coefficient. This quality varies from 1 to ∞. The closer q is to 1, the better the element is.

Geometry		Volume mesh cutting plane	
	Sequential	Out-of-core	Difference
$q \in [1..2[$	2,560,869	2,507,554	-53,315
$q \in [2..3[$	43,911	45,461	1,550
$q \in [3..4[$	406	420	14
$q \in [4..5[$	93	86	-7
$q \in [5..6[$	34	25	-9
$q \in [6..101.8[$	22	25	3
Worst quality	101.8	101.8	
Number of cells	2,605,335	2,553,571	-51,764

Quality comparison

Fig. 6. Case #2, Cylinder Head (courtesy of IFP)

These tables show that the meshes generated by our method with a 4 subdomains decomposition are very close to the sequentially generated ones in terms of size and quality comparison.

All these comparisons demonstrate our method's ability to produce meshes very close to their sequential counterpart (when it exists).

2.3 Large scale out-of-core meshing

For the last application, we were interested in testing our method for the generation of very large meshes, such as the ones one will eventually use to simulate with great accuracy the behaviour of a given concrete mixture under stress. Ultimately, these simulations will use tomographic experimental information for the distribution and shape of the numerous aggregates, making it very important to be able to mesh volumes with a very large number of constrained polygonal surface cells of various shapes.

These geometries are very challenging for several reasons: the large number, the varying sizes, the distribution and proximity of the aggregates impose very strong constraints firstly to the sequential volume mesher used, and secondly to our method. The level of details required also induces very large meshes, making them practically impossible to generate with classical methods.

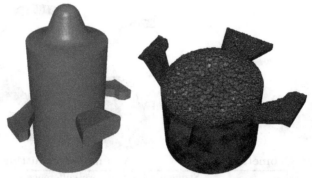

Geometry	Volume mesh cutting plane	

	Sequential	Out-of-core	Difference
$q \in [1..2[$	1,502,094	1,401,389	-100,705
$q \in [2..3[$	22,826	23,518	692
$q \in [3..4.597[$	18		
$q \in [3..4.051[$		17	-1
Worst quality	4.597	4.051	
Number of cells	1,524,938	1,424,924	-100,014

Quality comparison

Fig. 7. Case #3, 2 Stroke Engine

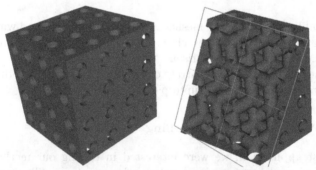

General view	General cut view	

	Sequential	Out-of-core	Difference
$q \in [1..2[$	10,613,508	10,358,106	-255,402
$q \in [2..3[$	149,971	100,248	-49,723
$q \in [3..7.495[$	117		
$q \in [3..6.717[$		67	-50
Worst quality	7.495	6.717	
Number of cells	10,763,596	10,458,421	-305,175

Quality comparison

Fig. 8. Case #4, Non Trivial Artificial Geometry

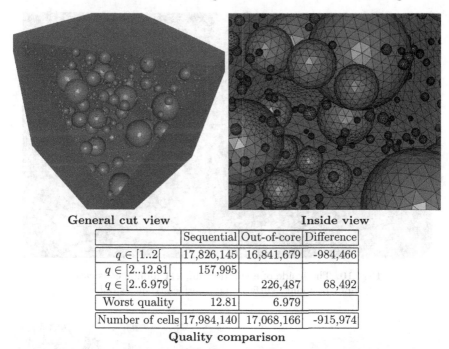

General cut view		Inside view	

	Sequential	Out-of-core	Difference
$q \in [1..2[$	17,826,145	16,841,679	-984,466
$q \in [2..12.81[$	157,995		
$q \in [2..6.979[$		226,487	68,492
Worst quality	12.81	6.979	
Number of cells	17,984,140	17,068,166	-915,974

Quality comparison

Fig. 9. Case #5, Geometry for the modelling of concrete (courtesy of CEA-DEN)

For this purpose, we used two geometries and corresponding surface meshes.

Refined Mesh of Case #5

We refined the input surface mesh of the previous figure 9 case (using the surface remesher Yams[4]) and successfully computed out-of-core sub-domain decomposition using respectively 16, 32 and 64 sub-domains.

The resulting meshes contain approximately 211 million tetrahedra (the inside of the 1,000 included balls was not meshed because the cement behaviour is targetted here). The figure 10 shows the boundary surface mesh of a sub-domain for the 16 sub-domain generation. The computation was carried out in out-of-core mode with our approach on a 4GB Opteron@2.4GHz machine in approximately 5 hours. Generating this mesh with the whole-in-core double precision sequential mesher would have required approximately 60GB of memory.

The memory used at the first bisection was 4GB. At the final sub-domain meshing step it was 3.6GB (respectively 1.8GB) for a decomposition in 16 (respectively 32) sub-domains.

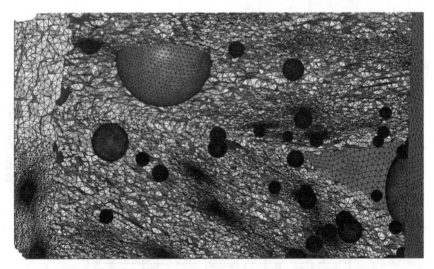

Fig. 10. The inside of a sub-domain, 211 million cells case
(input boundary surface and interface triangles only)

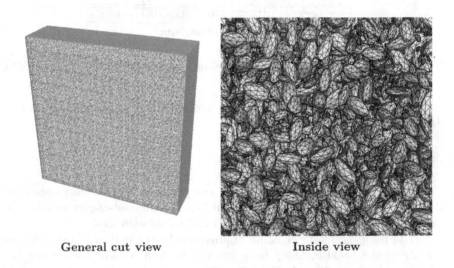

General cut view   Inside view

Fig. 11. Case #6, Geometry for the modelling of concrete (courtesy of C-S)

We can also note that we successfully tested our method on the case #5 geometry with partitions as large as 4096 sub-domains.

Case #6: 500,000 polyhedral cells concrete model

The last application aims at demonstrating our method's robustness through the very high complexity of the geometry : 500,000 polyhedral cells of various shapes and sizes, randomly placed in a rectangular box.

The geometry is shown on figure 11 and provided courtesy of C-S. The input surface mesh contains 40.5 million triangles and 21.2 million vertices. This kind of geometry with a very large input boundary surface defining a rather small volume is one of the worst possible situations regarding memory consumption of our method (see section 1.4), the initial coarse mesh already being very dense.

The resulting volume mesh generated in out-of-core mode contains approximately 680 million tetrahedra (the inside of the 500,000 polyhedral cells was meshed in this case).

The computation of this case was carried out in out-of-core mode on a 64GB Bull Novascale system in double precision for 8 sub-domains. It took approximately 3 days to compute, with almost a third of it spent reading and writing the files, and required 60GB of memory. Generating this mesh with our whole-in-core double precision sequential mesher, would have required approximately 140GB of memory. This is a typical example of extreme geometry hard to mesh in sequential, and even more challenging in parallel because of the constraint imposed by the surface mesh, which must be preserved exactly in the volume mesh. It is also particularly hard for our method, because of the fact that a large amount of the final vertices are already present in the input mesh.

Conclusion and future works

We presented in this article a strongly uncoupled domain decomposition method that relies heavily on an existing sequential mesher with very little modifications. It is able to generate distributed meshes in which parallel interfaces are almost the same as if they came from an *a posteriori* partition of the full sequential mesh.

This method was successfully applied to generate, in an out-of-core mode, large and good-quality meshes on limited core-memory resources, thus increasing the memory capacity by an order of magnitude.

The out-of-core method presented here can be easily adapted to implement a fully parallel message-passing approach, allowing the generation of very large tetrahedral meshes on distributed memory parallel systems more quickly.

It will also offer the advantage of providing an overall behaviour very similar to the original whole-in-core sequential algorithm.

Some adaptation will be made to the vertex generation and insertion, and the cutting zone mesh optimisation by vertex moving to get a consistently better mesh.

Acknowledgement. This work has been partially funded by the Pôle System@tic's IOLS project and achieved with Distene/INRIA tools TetMesh-GHS3D and Yams. The authors would like to thank Institut Français du Pétrole (IFP), Communications et Systèmes (C-S), and Commissariat à l'Énergie Atomique (CEA-DEN) for the test cases, and Bull S.A. for providing access to a Novascale system.

References

1. Chrisochoides, N.P. (2005) A survey of parallel mesh generation methods. Tech. Rep. BrownSC-2005-09, Brown University. Also appears as a chapter in *Numerical Solution of Partial Differential Equations on Parallel Computers* Eds. Bruaset, A.M., Bjørstad, P.E., Tveito, A., Springer Verlag.
2. Farhat, C. and Lesoinne, M. Automatic Partitioning of Unstructured Meshes for the Parallel Solution of Problems in Computational Mechanics *International Journal for the Numerical Methods in Engineering*, 36, pp.745-764
3. Fezoui, L., Loriot, F., Loriot, M., and Regere, J. A 2-D Finite Volume / Finite Element Euler Solver on a MIMD Parallel Machine *Proceedings of "High Performance Computing II" Montpellier, Oct. 1991, M. Durand and F. El Dabaghi Ed., Elsevier Science Publishers, North-Holland, (1991)*.
4. Frey, P.J. (2000) About surface remeshing. *Proceedings of 9th International Meshing Roundtable*, pp. 123-136.
5. Frey, P.J., and George, P.L. (2000) Mesh Generation, application to finite element. Hermes, Paris.
6. Galtier, J., and George, P.L. (1996) Prepartitioning as a way to mesh subdomains in parallel. *5th International Meshing Roundtable*, Sandia National Laboratories, pp. 107–122.
7. George, P.L., Hecht, F., and Saltel, E. (1991) Automatic mesh generator with specified boundary, *Computer Methods in Applied Mechanics and Engineering*, 92:269–288
8. Ito, Y., Shih, A.M., Erukala, A., Soni, B.K., Chernikov, A., Chrisochoides, N., and Nakahashi, K. (2006), Parallel Unstructured Mesh Generation by an Advancing Front Method, *Mathematics and Computers in Simulation*, (accepted).
9. Ivanov, E.G., Andra, H. and Kudryavtsev, A.N. (2006) Domain decomposition approach for automatic parallel generation of tetrahedral grids, *Computational methods in applied mathematics* , Vol.6, No.2, pp.178-193.
10. Larwood, B.G., Weatherill, N.P., Hassan, O., Morgan, K. (2003) Domain decomposition approach for parallel unstructured mesh generation, *International Journal for Numerical Methods in Engineering*, Vol. 58, Issue 2 , pp.177-188
11. Linardakis, L., and Chrisochoides, N. (2007) Graded Delaunay Decoupling Method for Parallel Guaranteed Quality Planar Mesh Generation, *SIAM Journal on Scientific Computing*, 27, submitted December 2006.
12. Said, R., Weatherill, N.P., Morgan, K., Verhoeven, N.A. (1999) Distributed parallel Delaunay mesh generation. *Computer Methods in Applied Mechanics and Engineering*, 177(1-2), pp.109-125.

5A.3

A Practical Delaunay Meshing Algorithm for a Large Class of Domains*

Siu-Wing Cheng[1], Tamal K. Dey[2], and Joshua A. Levine[2]

[1] Dept. of CSE, HKUST, Hong Kong. Email: scheng@cse.ust.hk
[2] Dept. of CSE, Ohio State University, Ohio, USA. Email: tamaldey,levinej@cse.ohio-state.edu

Summary. Recently a Delaunay refinement algorithm has been proposed that can mesh domains as general as piecewise smooth complexes [7]. This class includes polyhedra, smooth and piecewise smooth surfaces, volumes enclosed by them, and above all non-manifolds. In contrast to previous approaches, the algorithm does not impose any restriction on the input angles. Although this algorithm has a provable guarantee about topology, certain steps are too expensive to make it practical.

In this paper we introduce a novel modification of the algorithm to make it implementable in practice. In particular, we replace four tests of the original algorithm with only a single test that is easy to implement. The algorithm has the following guarantees. The output mesh restricted to each manifold element in the complex is a manifold with proper incidence relations. More importantly, with increasing level of refinement which can be controlled by an input parameter, the output mesh becomes homeomorphic to the input while preserving all input features. Implementation results on a disparate array of input domains are presented to corroborate our claims.

1 Introduction

Delaunay refinement is recognized as a versatile tool for meshing geometric domains. Historically, it was developed to mesh polygonal domains in two dimensions [10, 18] and later extended to polyhedral domains in three dimensions [19, 15]. Since the output mesh of a polyhedral domain can conform exactly to the input, topology preservation did not become an issue in these works. For curved domains in two dimensions, topology consideration still remains a mild issue [3]. However, in three dimensions, faithful maintenance of topology becomes a foremost issue. Recently a few works [5, 6, 9] have used Chew's furthest point strategy [11] with the sampling theory [2] to mesh smooth surfaces and volumes enclosed by them [16] with topological guarantees.

*Research supported by NSF, USA (CCF-0430735 and CCF-0635008) and RGC, Hong Kong, China (HKUST 6181/04E).

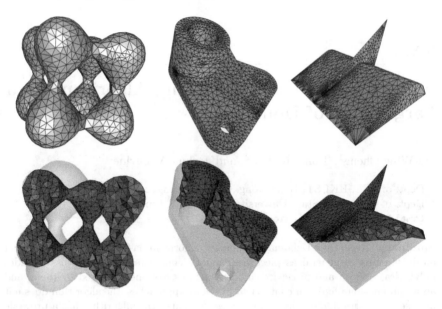

Fig. 1. Meshed PSCs, METABALL (Smooth), PART (Manifold PSC), and WEDGE (Non-manifold, PSC with small angles).

It is well recognized that non-smooth domains pose an added level of difficulty for Delaunay refinement. Boissonnat and Oudot [4] alleviated this problem for a class of surfaces that are only mildly non-smooth as they forbid non-smooth regions subtending small angles. The menace of small angles already appeared in Delaunay meshing of polyhedral domains [9, 17, 20]. Special actions seemed necessary to handle small angles. In a recent work Cheng, Dey, and Ramos [7] succeeded in tackling the problem of non-smoothness with arbitrarily small angles in the input. Drawing upon the idea of protecting small angle regions with balls [9], they protect the curves where different surface patches meet. A novelty of the algorithm is that these protecting balls are turned into weighted points and a Delaunay refinement is run using weighted Delaunay triangulations. The refinement is triggered by violations of some topological conditions introduced by Edelsbrunner and Shah [14] to ensure topology preservations. This algorithm, in theory, enables one to mesh a large class of geometric domains called piecewise smooth complex (PSC). This class includes smooth and non-smooth surfaces both with and without boundaries, volumes enclosed by them, and most importantly non-manifolds. In fact, it is the first algorithm that can compute Delaunay meshes for such a large class of domains with theoretical guarantees. However, the major shortcoming of the algorithm is that it involves expensive computations at each refinement stage making it quite hard for implementation. In this paper we design an algorithm drawing upon the ideas of [7] which is more practical and show its implementation results on a vast array of disparate domains, see Figure 1 for some examples. Due to space constraints, we skip the proofs of theoretical guarantees which will appear elsewhere.

The original algorithm in [7] inserts points in the domain iteratively after the protection phase with four types of violations, namely (i) a Voronoi edge intersects the domain multiple times, (ii) normals on the curves and surface patches vary beyond a threshold within Voronoi cells, (iii) a Delaunay edge in the restricted triangulation connects vertices across different patches, and (iv) the restricted Delaunay triangles incident to points in a patch do not make a topological disk. We replace these four tests with a single one that only checks for topological disk violations and inserts points in each such case. Once intersections of surface patches with Voronoi edges are determined, this test is purely combinatorial making it easily implementable. Obviously, one cannot hope that the output will have the same topology as the input where refinement is triggered by only this single violation. However, with this 'conservative approach' we are able to guarantee that the output restricted to each stratum of the input is a manifold with vertices on that stratum. Furthermore, if the refinement is carried up to a resolution where the triangles are sufficiently small, the output becomes homeomorphic to the input while preserving the features as well. The main observation is that, in practice, this resolution point is achieved quite fast using a resolution parameter in the algorithm. The protection phase requires some involved computations with curve and surface normals. However, these computations are done only once before the refinement steps. Furthermore, the properties of protecting balls required for the insertion phase to work properly can be satisfied in practice by using sufficiently small balls around the non-smooth edges.

2 Notations and Definitions.

2.1 Domain

Throughout this paper, we assume a generic intersection property that a k-manifold $\sigma \subset \mathbb{R}^3$, $0 \leqslant k \leqslant 3$, and a j-manifold $\sigma' \subset \mathbb{R}^3$, $0 \leqslant j \leqslant 3$, intersect (if at all) in a $(k + j - 3)$-manifold if $\sigma \not\subset \sigma'$ and $\sigma' \not\subset \sigma$. We will use both geometric and topological versions of closed balls. A *geometric* closed ball centered at point $x \in \mathbb{R}^3$ with radius $r > 0$, is denoted as $B(x,r)$. We use $\text{int}\,\mathbb{X}$ and $\text{bd}\,\mathbb{X}$ to denote the interior and boundary of a topological space $\mathbb{X}$, respectively.

The input domain $\mathcal{D}$ is a piecewise smooth complex (PSC) where each element is a compact smooth (C^2) k-manifold, $0 \leqslant k \leqslant 3$, possibly with boundary. Further, for $0 \leqslant k \leqslant 3$, each element is contained in a smooth k-manifold without boundary. We use $\mathcal{D}_k$ to denote the kth stratum, i.e., the subset of all k-dimensional elements. $\mathcal{D}_0$ is a set of *vertices*; $\mathcal{D}_1$ is a set of curves called *1-faces*; $\mathcal{D}_2$ is a set of surface patches called *2-faces*; $\mathcal{D}_3$ is a set of volumes called *3-faces*. For $1 \leqslant k \leqslant 2$, we use $\mathcal{D}_{\leqslant k}$ to denote $\mathcal{D}_0 \cup \ldots \cup \mathcal{D}_k$.

The domain $\mathcal{D}$ satisfies the usual proper requirements for being a complex: (i) interiors of the elements are pairwise disjoint and for any $\sigma \in \mathcal{D}$, $\mathrm{bd}\,\sigma \subset \mathcal{D}$; (ii) for any $\sigma, \sigma' \in \mathcal{D}$, either $\sigma \cap \sigma' = \emptyset$ or $\sigma \cap \sigma'$ is a union of elements in $\mathcal{D}$. We use $|\mathcal{D}|$ to denote the underlying space of $\mathcal{D}$. For $0 \leqslant k \leqslant 3$, we also use $|\mathcal{D}_k|$ to denote the underlying space of $\mathcal{D}_k$.

The definition of $\mathcal{D}$ includes smooth surfaces with or without boundaries, piecewise smooth surfaces including polyhedral surfaces, non-manifold surfaces, and volumes enclosed by them. Figure 1 shows some example inputs that can be handled by our algorithm.

For any point x on a 2-face σ, we use $n_\sigma(x)$ to denote a unit outward normal to the surface of σ at x. For any point x on a 1-face σ, $n_\sigma(x)$ denotes a unit oriented tangent to the curve of σ at x.

2.2 Complexes

Our meshing algorithm generates sample points on the input some of which are weighted. A weighted point p with weight w_p is represented with a ball $\hat{p} = B(p, w_p)$. The squared weighted distance of any point $x \in \mathbb{R}^3$ from $\hat{p}$ is given by $\|x-p\|^2 - w_p^2$. One can define a Voronoi diagram and its dual Delaunay triangulation for a weighted point set just like their Euclidean counterparts by replacing Euclidean distances with weighted distances. For a weighted point set $S \subset \mathbb{R}^3$, let $\mathrm{Vor}\,S$ and $\mathrm{Del}\,S$ denote the weighted Voronoi and Delaunay diagrams of S respectively. Each diagram is a cell complex where each k-face is a k-polytope in $\mathrm{Vor}\,S$ and is a k-simplex in $\mathrm{Del}\,S$. Each k-face in $\mathrm{Vor}\,S$ is dual to a $(3-k)$-face in $\mathrm{Del}\,S$ and vice versa. For a simplex $\xi \in \mathrm{Del}S$ we use V_ξ to denote its dual Voronoi face.

Delaunay refinement for curved domains relies upon restricted Delaunay triangulations [5, 6, 9]. These consist of Delaunay simplices whose Voronoi duals intersect the domain. We introduce some notations for these structures. Let S be a point set sampled from $|\mathcal{D}|$. For any sub-collection $\mathbb{X} \subset \mathcal{D}$ we define $\mathrm{Del}\,S|_\mathbb{X}$ to be the Delaunay subcomplex restricted to $\mathbb{X}$, i.e., each simplex $\xi \in \mathrm{Del}\,S|_\mathbb{X}$ is the dual of a Voronoi face V_ξ that intersects $\mathbb{X}$ in non-empty set. By above definition, for any $\sigma \in \mathcal{D}$, $\mathrm{Del}\,S|_\sigma$ denotes the Delaunay subcomplex restricted to the element σ and

$$\mathrm{Del}\,S|_{\mathcal{D}_i} = \bigcup_{\sigma \in \mathcal{D}_i} \mathrm{Del}\,S|_\sigma, \quad \mathrm{Del}\,S|_\mathcal{D} = \bigcup_{\sigma \in \mathcal{D}} \mathrm{Del}\,S|_\sigma.$$

In case of surfaces restricted Delaunay complexes were considered in previous works because they become topologically equivalent (homeomorphic) when sampled set is sufficiently dense. It turns out that even for PSCs, a similar result holds [7]. However, it is computationally very difficult to determine when the sample is sufficiently dense. To bypass this difficulty we sample the domain at a resolution specified by the user. We certify that output mesh restricted to each manifold element is a manifold and when the resolution parameter is small enough, it is homeomorphic too. Empirically we observe that homeomorphism is achieved quite early in the refinement process.

The restricted complexes as defined above may contain some superfluous simplices that can be eliminated safely due to a dimensional reason. For example, if the element is a 2-manifold, we can eliminate restricted edges that do not have any restricted triangles incident to it. Similarly, for 1-manifolds, we can eliminate restricted triangles. This motivates defining special sub-complexes of restricted complexes. For $\sigma \in \mathcal{D}_i$, let $\mathrm{Skl}^i S|_\sigma$ denote the i-dimensional subcomplex of the restricted Delaunay complex $\mathrm{Del}\, S|_\sigma$, that is,

$$\mathrm{Skl}^i S|_\sigma = closure\{t \mid t \in \mathrm{Del}\, S|_\sigma \text{ is an } i\text{-simplex}\}.$$

Intuitively, $\mathrm{Skl}^i S|_\sigma$ is a i-dimensional complex without any hanging lower dimensional simplices. For example, in Figure 2 the dark edge connecting between upper and lower part of σ is eliminated in $\mathrm{Skl}^2 S|_\sigma$.

We extend the definition to strata:

$$\mathrm{Skl}^i S|_{\mathcal{D}_i} = \bigcup_{\sigma \in \mathcal{D}_i} \mathrm{Skl}^i S|_\sigma.$$

Notice that computation of $\mathrm{Skl}^i S|_{\mathcal{D}_i}$ is easier than $\mathrm{Del}\, S|_{\mathcal{D}_i}$ since the former one involves computations of intersections only between $(3-i)$-dimensional Voronoi faces with i-faces in $\mathcal{D}$. In fact, because of our special protections of $\mathcal{D}_1$, the only computation we need to determine $\mathrm{Skl}^1 S|_{\mathcal{D}_1}$ and $\mathrm{Skl}^2 S|_{\mathcal{D}_2}$ is Voronoi edge-surface intersections.

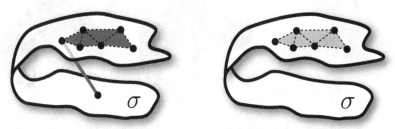

Fig. 2. (left): $\mathrm{Del}\, S|_\sigma$ and (right):$\mathrm{Skl}^i S|_\sigma$.

2.3 Overview

As mentioned earlier, Delaunay meshing of PSCs faces difficulty with small input angles that may be subtended at the input curves and vertices. To overcome this difficulty, we protect all elements in $\mathcal{D}_{\leqslant 1}$ with balls that satisfy certain properties. These balls are turned into weighted points for the next stage of the refinement. Weighted Voronoi diagrams and weighted Delaunay triangulations enter into the picture because of these weighted points. The properties of the protecting balls make sure that the curves in $\mathcal{D}_1$ remain meshed properly throughout the algorithm. In particular, *adjacent* points along any curve in $\mathcal{D}_1$ remain connected with restricted Delaunay edges.

After protection, we insert points iteratively outside the protected regions to mesh 2-faces. This insertion is triggered by a disk condition which essentially imposes that the triangles around a point on a 2-face form a topological disk. After 2-faces are meshed, 3-faces (volumes) are meshed with an usual circumcenter insertion procedure for refining tetrahedra. We show that each inserted point maintains a lower bound on its distances to all existing points. Therefore, the refinement must terminate. At termination the restricted complex $\bigcup_i \mathrm{Skl}^i\, S|_{\mathcal{D}_i}$ is output which has following properties:

Preserved features: All curves in $\mathcal{D}_1$ are meshed homeomorphically with restricted Delaunay edges whose vertices lie on the curves. This preserves non-smooth features or user defined features in the output, see Figure 3.

Faithful topology: All surface patches and volumes in $\mathcal{D}_{\leqslant 3}$ are meshed with a piecewise linear manifold. Furthermore, the algorithm accepts a resolution parameter λ so that it refines the Delaunay triangulations until the restricted triangles have 'size' less than λ. We show that when λ is sufficiently small, the output restricted complex becomes homeomorphic to input $|\mathcal{D}|$.

Fig. 3. Features on ANCHOR are preserved in both surface(middle) and volume(right) meshing.

2.4 Disk conditions

In a mesh of a 2-manifold, the triangles incident to a vertex should form a topological disk. Therefore, one can turn this into a condition for sampling 2-manifolds in the input PSC. Our refinement condition applied to only a single 2-manifold is as simple as this. However, since a PSC may have several 2-manifolds, potentially forming even non-manifolds, one needs to incorporate some more conditions into the disk condition. Let p be a point on a 2-face σ. Let $\mathrm{Umb}_{\mathcal{D}}(p)$ and $\mathrm{Umb}_{\sigma}(p)$ be the set of triangles in $\mathrm{Skl}^2 S|_{\mathcal{D}_2}$ and $\mathrm{Skl}^2 S|_{\sigma}$ respectively which are incident to p. The following disk condition is used for refinement. Once the restricted Delaunay triangles are collected, this check is only combinatorial.

Disk_Conditions(p) : (i) $\mathrm{Umb}_{\mathcal{D}}(p) = \bigcup_{\sigma \ni p} \mathrm{Umb}_{\sigma}(p)$, (ii) for each $\sigma \in \mathcal{D}_2$ containing p, underlying space of $\mathrm{Umb}_{\sigma}(p)$ is a 2-disk which has all vertices in σ. Point p is in the interior of this 2-disk if and only if $p \in \mathrm{int}\,\sigma$. Also, if p is in $\mathrm{bd}\,\sigma$, it is not connected to any other point on $\mathcal{D}_1$ which is not adjacent to it.

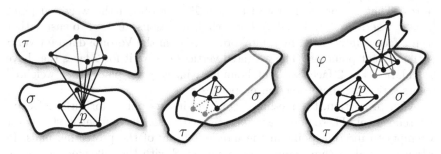

Fig. 4. (left):point $p \in \sigma$ has a disk in σ and another disk in $\tau \neq \sigma$ violating condition (i) (middle): point $p \in \sigma$ has a topological disk but some of its vertices (lightly shaded) belong to τ violating condition (ii), (right): Points p and q satisfy disk condition. Point p, an interior point in σ, lies in the interior of its disk in σ. Point q, a boundary point, has three disks for each of the three 2-faces.

3 Protection

The neighborhoods of the curves and vertices in $\mathcal{D}_{\leqslant 1}$ are regions of potential problems for Delaunay refinements. First, if the elements incident to these curves and vertices make small angles at the points of incidences, usual Delaunay refinement may not terminate. Second, these curves and vertices represent 'features' in the input which should be preserved in the output for many applications. Usual Delaunay refinement may destroy these features [4, 13] or may be made to preserve them for a restricted class of inputs [21]. To overcome these problems we protect elements in $\mathcal{D}_{\leqslant 1}$ with balls and then turn them into weighted points for further meshing following a technique proposed in [7]. The protecting balls should have the following properties.

PROTECTION PROPERTIES: Let $\omega \leqslant 0.076$ be a positive constant and $\mathcal{B}_p$ denote the protecting ball of a point p.

1. Any two adjacent balls on a 1-face must overlap significantly without containing each other's centers.
2. No three balls have a common intersection.
3. Let $p \in \sigma$ be the center of a protecting ball. Further, let $B = B(p, R)$ be a ball with radius R and center p where $R \leqslant c\,\text{radius}(\mathcal{B}_p)$ for some $c \leqslant 8$.
   (a) For $\tau = \sigma$ or any 2-face incident to σ, $\angle n_\tau(p), n_\tau(z) \leqslant 2\omega$ for any $z \in B \cap \tau$. The same result holds for the surfaces of the 2-faces incident to σ.
   (b) B intersects σ in a single open curve and any 2-face incident to σ in a topological disk. The same result holds for the surfaces of the 2-faces incident to σ.

After computing the protecting balls, we turn each of them into a weighted vertex. That is, for each protecting ball $\mathcal{B}_p$, we obtain the weighted point (p, w_p), where $w_p = \text{radius}(\mathcal{B}_p)$. For technical reasons we need to ensure that each 2-face is intersected by some Voronoi edge in the Voronoi diagram $\text{Vor}\,S$ of the current point set. The weighted vertices ensure it for 2-faces that have boundaries. For 2-faces without boundary, initially we place three weighted points satisfying the protection properties.

After protection, the meshing algorithm inserts points for further Delaunay refinement. These points are not weighted. Also, the refinement step never attempts to insert a point in the interior of any of the protecting balls. In essence, our algorithm maintains a point set S with the following two properties : (i) S contains all weighted points placed in protection step, and (ii) other points in S are unweighted and they lie outside the protecting balls. We call such a point set *admissible*.

The following Lemma proved in [7] is an important consequence of the protection properties.

Lemma 1. *Let S be an admissible point set. Let p and q be adjacent weighted vertices on a 1-face σ. Let σ_{pq} denote the curve segment between p and q. V_{pq} is the only Voronoi facet in $\text{Vor}\,S$ that intersects σ_{pq}, and V_{pq} intersects σ_{pq} exactly once.*

4 Meshing PSC

For any triangle $t \in \text{Skl}^2\,S|_\sigma$, define $\text{size}(t, \sigma)$ to be the maximum weighted distance between the vertices of t and points where dual Voronoi edge V_t intersects σ. Notice that if all vertices of t are unweighted, the maximum weighted distance is just the maximum Euclidean distance.

When we mesh volumes, we use the standard technique of inserting circumcenters of tetrahedra that have radius-edge ratio (denoted $\rho()$) greater than a threshold, $\rho_0 \geqslant 1$. If the insertion of the circumcenter threatens to delete any triangle in $\text{Skl}^2\,S|_{\mathcal{D}_2}$, the circumcenter is not inserted. In this case we say that the triangle is encroached by the circumcenter. Essentially, this strategy allows refining most of the tetrahedra except the ones near boundary.

4.1 Algorithm.

The following pseudo-code summarizes our algorithm.

DelPSC $(\mathcal{D}, \lambda, \rho_0)$

1. PROTECTION. Protect elements in $\mathcal{D}_{\leqslant 1}$ with weighted points. Insert three weighted points in each element of $\mathcal{D}_2$ that has no boundary. Let S be the current admissible point set.
2. MESH2COMPLEX.
   a) Let (p, σ) be any tuple where $p \in S \cap \sigma$ and $\sigma \in \mathcal{D}_2$. If **Disk_Conditions**(p) is violated, find the triangle $t \in \mathrm{Umb}_{\mathcal{D}}(p)$ that maximizes $\mathrm{size}(t, \sigma)$ over all σ containing p and insert $x \in V_t|_\sigma$ that realizes $\mathrm{size}(t, \sigma)$ into S. Go to step 2(c).
   b) If $\mathrm{size}(t, \sigma) > \lambda$ for some tuple (t, σ), where $t \in \mathrm{Skl}^2 \, S|_\sigma$, insert $x \in V_t|_{\mathcal{D}}$ that realizes $\mathrm{size}(t, \sigma)$ into S.
   c) Update Del S and Vor S.
   d) If S has grown in the last execution of step 2, repeat step 2.
3. MESH3COMPLEX. For any tuple (t, σ) where t is a tetrahedron in $\mathrm{Skl}^3 \, S|_\sigma$
   a) If $\rho(t) > \rho_0$ insert the center of the Delaunay ball (orthoball) of t into S if it does not encroach any triangle in $\mathrm{Skl}^2 \, S|_{\mathcal{D}}$.
   b) Update Del S and Vor S.
   c) If S has grown in the last execution of step 3, repeat step 3.
4. Return $\bigcup_i \mathrm{Skl}^i \, S|_{\mathcal{D}_i}$.

4.2 Protection computations

To satisfy the protection properties we compute two quantities at the points where balls are centered.

First, we compute the ω-deviation at a point $x \in \sigma$ defined as follows. If $\sigma \in \mathcal{D}_{\leqslant 1}$, for $\omega > 0$, let $\sigma_{x,\omega} = \{y \in \sigma : \angle n_\sigma(x), n_\sigma(y) = \omega\}$. If $\sigma \in \mathcal{D}_2$, define $\sigma_{x,\omega}$ analogously but varying y over the surface of σ. The distance between x and $\sigma_{x,\omega}$ is the ω-deviation radius of x in σ. It is ∞ if $\sigma_{x,\omega} = \emptyset$. Let d_x be the minimum of the ω-deviation radius of x over all σ containing x. By construction, $\angle n_\sigma(x), n_\sigma(y) = \omega$ for some $y \in \sigma$ such that $\|x - y\| = d_x$.

Second, for any 1- or 2-face σ containing x, we compute the tangential contact points between σ and any sphere centered at x. Select the tangential contact point nearest to x (over all 1- and 2-faces containing x). Let d'_x be the distance between x and this nearest tangential contact point.

It is not hard to prove that, for any $r < \min\{d_x, d'_x\}$, $B(x, r) \cap \sigma$ is a closed ball of dimension $\dim(\sigma)$. Also, since $r < d_x$, the normal deviation property 3(a) is satisfied. To satisfy property 1 and 2, we take a fraction of the minimum of d_x and d'_x to determine the size of the ball at x. Let $r_x = \frac{\omega}{8} \min\{d_x, d'_x\}$.

For each curve σ with endpoints, say u and v, we first compute the balls $\mathcal{B}_u$ and $\mathcal{B}_v$ with radii r_u and r_v. Then, starting from, say $\mathcal{B}_u$, we march along

Fig. 5. Protection in action on CASTING. We placed weighted vertices on all elements of $\mathcal{D}_1$ which protect these elements when meshed.

the curve placing the centers of the balls till we reach $\mathcal{B}_v$. These centers can be chosen using a procedure described in [7].

We compute the intersection points $x_0 = \mathcal{B}_v \cap \sigma$ and $x_1 = \mathcal{B}_u \cap \sigma$. The protecting ball at x_1 is $\mathcal{B}_{x_1} = B(x_1, r_{x_1})$. The protecting ball at x_0 is constructed last. We march from $\mathcal{B}_{x_1}$ toward x_0 to construct more protecting balls. For $k \geqslant 2$, let $\mathcal{B}_{x_{k-1}}$ be the last protecting ball placed and let $\mathcal{B}_p$ be the last protecting ball placed before $\mathcal{B}_{x_{k-1}}$. We compute the two intersection points between σ and the boundary of $B(x_{k-1}, \frac{6}{5}r_{x_{k-1}})$. Among these two points let x_k be the point such that $\angle px_{k-1}x_k > \pi/2$. One can show that x_k is well-defined and x_k lies between x_{k-1} and v along σ. Define

$$ r_k = \max \begin{cases} \frac{1}{2}\|x_{k-1} - x_k\| \\ \min_{0 \leqslant j \leqslant k} r_{x_k}/8 + \|x_k - x_j\|/8. \end{cases} $$

If $B(x_k, r_k) \cap B(x_0, r_{x_0}) = \emptyset$, the protecting ball at x_k is

$$ \mathcal{B}_{x_k} = B(x_k, r_k). $$

Figure 6 shows an example of the construction of $\mathcal{B}_{x_k}$. We force $r_k \geqslant \frac{1}{2}\|x_{k-1} - x_k\|$ so that $\mathcal{B}_{x_k}$ overlaps significantly with $\mathcal{B}_{x_{k-1}}$. This is desirable because the protecting balls are supposed to cover σ in the end.

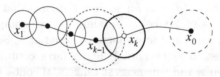

Fig. 6. The two dashed circles denote $B(x_{k-1}, \frac{6}{5}r_{k-1})$ and $B(x_0, r_{x_0})$. The bold circle denotes $\mathcal{B}_{x_k}$.

We continue to march toward x_0 and construct protecting balls until the candidate ball $B(x_m, r_m)$ that we want to put down overlaps with $B(x_0, r_{x_0})$. In this case, we reject x_m and $B(x_m, r_m)$ and compute the intersection points between σ and the bisector plane of x_{m-1} and x_0. Let y_m be the intersection point that lies between x_{m-1} and x_0 along σ. Finally, the protecting ball at y_m is $\mathcal{B}_{y_m} = B(y_m, R)$ and the protecting ball at x_0 is $\mathcal{B}_{x_0} = B(x_0, R)$, where $R = \frac{2}{3}\|x_{m-1} - y_m\| = \frac{2}{3}\|y_m - x_0\|$. It can be shown that the constructed balls satisfy all protection properties.

Lemma 2. *The protecting balls computed by the above described procedure satisfy the protection properties.*

5 Analysis.

The analysis of DelPSC establishes two main facts: (i) the algorithm terminates, (ii) at termination the output mesh satisfies properties T1-T3 as stated later. It will be essential to prove that DelPSC maintains an admissible point set S throughout its execution.

Lemma 3. DelPSC *never attempts to insert a point in any protecting ball.*

Proof. In MESH2COMPLEX, points that are intersection of Voronoi edges and $|\mathcal{D}|$ are inserted. Since no three protecting balls intersect, all points on a Voronoi edge have positive distance from all vertices, weighted or not. This means no point of any Voronoi edge lies inside a protecting ball. Therefore, the inserted points in MESH2COMPLEX must lie outside all protecting balls. For the same reason the orthocenter inserted in MESH3COMPLEX cannot lie in any of the protecting balls.   ⌑

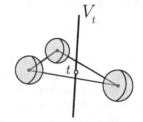

Fig. 7. Inserted points are outside of protecting balls.

5.1 Termination.

We apply the standard argument that there is a lower bound on the distance between each point inserted by DelPSC and all existing points. Then the compactness of $\mathcal{D}$ allows the standard packing argument to claim termination.

The next lemma is the key to proving termination. This result says that if dual Voronoi edges of all restricted triangles incident to a point p in a 2-face σ have nearby intersections with σ, the connected component of σ containing p within V_p satisfies some nice properties. These properties allow one to argue that restricted triangles incident to p will form a disk eventually.

Lemma 4. *Let $p \in S$ be a point on a 2-face σ. Let $\bar{\sigma}$ be the connected component in $V_p|_\sigma$ containing p. There exists a constant $\lambda > 0$ so that following* *holds:*

If some edge of V_p intersects σ and size$(t, \sigma) < \lambda$ for each triangle $t \in$ Skl^2 $S|_{\mathcal{D}_2}$ incident to p, then

(i) there is no 2-face τ where $p \notin \tau$ and τ intersects a Voronoi edge in V_p.
(ii) $\bar{\sigma} = V_p \cap B \cap \sigma$ where $B = B(p, 2\lambda)$ if p is unweighted and $B = B(p, 2\text{radius}(\mathcal{B}_p) + 2\lambda)$ otherwise;
(iii) $\bar{\sigma}$ is a 2-disk;
(iv) any edge of V_p intersects $\bar{\sigma}$ at most once;
(v) any facet of V_p intersects $\bar{\sigma}$ in an empty set or an open curve.

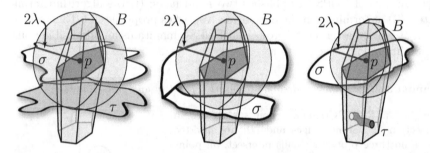

Fig. 8. (left): Within ball B, V_p intersects σ and τ both of which intersect some edge of V_p. This is not possible according to Lemma 4(i), (middle): also not possible since there is another component of σ within $B \cap V_p$ other than $\bar{\sigma}$, (right): Within B, σ intersects V_p in a topological disk. It is possible that there is a different component (τ) which does not intersect any Voronoi edge and hence does not contribute any dual restricted triangle incident to p.

Because of Lemma 4, the restricted triangles incident to a point p on a 2-face σ form a topological disk when λ is sufficiently small (Figure 8). Also, this topological disk cannot include a point from a 2-face other than σ since then the distance between p and that point will be large enough to contradict the resolution level determined by λ.

Theorem 1. DelPSC *terminates.*

Topology Preservation

The output of DelPSC satisfies certain topological properties. Property T1 ensures feature preservation (Figure 9(left)). Property T2 ensures each manifold element is approximated with a manifold and incidence structure among them is preserved (Figure 9(middle)). Property T3 ensures topological equivalence between input and output when resolution parameter is sufficiently small (Figure 9(right) and Figure 10).

(T1) For each $\sigma \in \mathcal{D}_1$, $\mathrm{Skl}^1 \, S|_\sigma$ is homeomorphic to σ and two vertices are joined by an edge in $\mathrm{Skl}^1 \, S|_\sigma$ if and only if these two vertices are adjacent on σ.

(T2) For $0 \leqslant i \leqslant 2$ and $\sigma \in \mathcal{D}_i$, $\mathrm{Skl}^i \, S|_\sigma$ is a i-manifold with vertices only in σ. Further, $\mathrm{bd} \, \mathrm{Skl}^i \, S|_\sigma = \mathrm{Skl}^{i-1} \, S|_{\mathrm{bd}\,\sigma}$. For $i = 3$, the statement is true if the set $\mathrm{Skl}^i \, S|_\sigma$ is not empty at the end of Mesh2Complex.

(T3) There exists a $\lambda > 0$ so that the output mesh of $\mathrm{DELPSC}(\mathcal{D},\lambda)$ is homeomorphic to $\mathcal{D}$. Further, this homeomorphism respects stratification with vertex restrictions, that is, for $0 \leqslant i \leqslant 3$, $\mathrm{Skl}^i \, S|_\sigma$ is homeomorphic to $\sigma \in \mathcal{D}_i$ where $\mathrm{bd} \, \mathrm{Skl}^i \, S|_\sigma = \mathrm{Skl}^{i-1} \, S|_{\mathrm{bd}\,\sigma}$ and vertices of $\mathrm{Skl}^i \, S|_\sigma$ lie in σ.

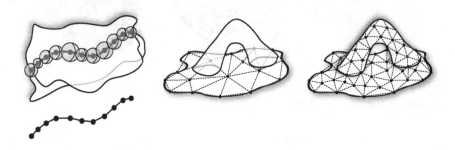

Fig. 9. (left): adjacent points on curves in $\mathcal{D}_1$ are joined by restricted edges, (middle): a surface patch is meshed with a manifold though topology is not fully recovered, (right): topology is fully recovered

The proof of T1 follows immediately from Lemma 1. One requires some nontrivial analysis to prove T2 which we skip here. To prove T3 we need a result of Edelsbrunner and Shah [14].

A CW-complex $\mathcal{R}$ is a collection of closed (topological) balls whose interiors are pairwise disjoint and whose boundaries are union of other closed balls in $\mathcal{R}$. A finite set $S \subset |\mathcal{D}|$ has the extended TBP for $\mathcal{D}$ if there is a CW-complex $\mathcal{R}$ with $|\mathcal{R}| = |\mathcal{D}|$ that satisfies the following conditions for each Voronoi face $F \in \mathrm{Vor}\, S$ intersecting $|\mathcal{D}|$:

(C1) The restricted Voronoi face $F \cap |\mathcal{D}|$ is the underlying space of a CW-complex $\mathcal{R}' \subseteq \mathcal{R}$.

(C2) The closed balls in $\mathcal{R}'$ are incident to a unique closed ball $b_F \in \mathcal{R}'$.

(C3) If b_F is a j-ball, then $b_F \cap \mathrm{bd}\, F$ is a $(j-1)$-sphere.

(C4) Each ℓ-ball in $\mathcal{R}'$, except b_F, intersects $\mathrm{bd}\, F$ in a $(\ell-1)$-ball.

Figure 11 shows two examples of a Voronoi facet F that satisfy the above conditions.

A result of Edelsbrunner and Shah [14] says that if S has the extended TBP for $\mathcal{D}$, the underlying space of $\mathrm{Del}\, S|_\mathcal{D}$ is homeomorphic to $|\mathcal{D}|$. Of course, to apply this result we would require a CW-complex with underlying space as

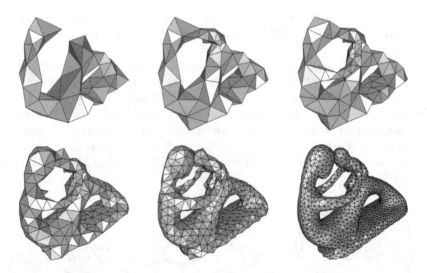

Fig. 10. Output on FERTILITY at different levels of λ. As λ is reduced, eventually the correct topology is achieved.

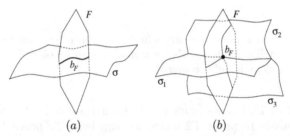

Fig. 11. F is a Voronoi facet. In (a), F intersects a 2-face in a closed topological interval (1-ball) which is b_F. Here b_F intersects bd F at two points, a 0-sphere. In (b), F intersects the 1-face in a single point which is b_F, and for $1 \leqslant i \leqslant 3$, $F \cap \sigma_i$ are closed topological 1-balls incident to b_F. Here $b_F \cap$ bd $F = \emptyset$, a -1-sphere.

$|\mathcal{D}|$. We see that when λ is sufficiently small, Vor $S|_{\mathcal{D}}$ provides such a CW-complex when our algorithm terminates. It can be shown that the following two properties P1, P2 imply Edelsbrunner-Shah conditions C1-C4. Let F be a k-face of Vor S.

- (P1) If F intersects an element $\sigma \in \mathcal{D}_j \subseteq \mathcal{D}$, the intersection is a closed $(k + j - 3)$-ball.
- (P2) There is a unique lowest dimensional element $\sigma_F \in \mathcal{D}$ so that F intersects σ_F and only elements that are incident to σ_F.

One can show that when dual Voronoi edges of all restricted triangles intersect the surface patches within sufficiently small distance from their vertices (that is, size(t, σ) small), properties P1 and P2 hold. The refinement step 2(b) in DelPSC achieve the required condition when λ is sufficiently small. Also, when P1 and P2 hold, Del $S|_{\mathcal{D}}$ equals $\bigcup_i \text{Skl}^i S|_{\mathcal{D}_i}$, the output of DelPSC.

Theorem 2. *The output of* DelPSC *satisfies T1, T2, and T3.*

6 Results

We have implemented the DelPSC algorithm with the aid of the CGAL library for maintaining a weighted Delaunay triangulation. With this implementation we have experimented on a variety of different shapes with varied levels of smoothness, including piecewise-linear (PLC), piecewise-smooth (PSC), and smooth. Our examples incorporate both manifold and non-manifold shapes. Three of these examples also have sharp angles. The different datasets shown in the images are summarized in Table 1. We show the time to mesh each model with the input parameters $\rho_0 = 1.4$ and λ as 10% of the minimum dimension of the bounding box.

Dataset	Smoothness	Manifold	Sharp	Time to Mesh	# of vertices
SERATED	PLC	Yes	Yes	94.4 s	13047
ANCHOR	PSC	Yes	No	43.9 s	7939
CASTING	PSC	Yes	No	170.9 s	19810
PART	PSC	Yes	No	22.2 s	3026
PIN-HEAD	PSC	Yes	Yes	90.1 s	13958
SATURN	PSC	No	No	4.8 s	1340
SWIRL	PSC	No	No	86.7 s	9288
WEDGE	PSC	No	Yes	9.2 s	2980
FERTILITY	Smooth	Yes	No	57.7 s	9113
METABALL	Smooth	Yes	No	12.1 s	3288
HAND	Smooth	No	No	24.4 s	4872

Table 1. Datasets.

The input to DelPSC is a polygonal mesh which represents a PSC. We first mark those edges which are non-manifold or have an inner dihedral angle less than a user-specified parameter and then collect them together to form the complex $\mathcal{D}_1$. From $\mathcal{D}_1$ we mark the input polygons into elements of $\mathcal{D}_2$. An octree is built to bucket these input polygons for quick intersection checks with dual Voronoi edges. The next step is to create the protecting balls for elements of $\mathcal{D}_{\leqslant 1}$ as described in section 4. We finally pass all of this information to the Delaunay mesher described as steps 2 and 3.

We show a variety of the output (both surface and volume meshing) for each input models. Figures 12-17 show additional results. In each figure we show the input model, the output surface mesh with the protected elements

highlighted, and the volume mesh output (if it exists). In particular, the FER-
TILITY and METABALL models (Figures 10 and 1) are taken to be a smooth
manifolds, so no input curves are protected. The SWIRL and HAND models
(Figures 15 and 17) both have no enclosed volumes, so they do not have a
volume mesh associated with them.

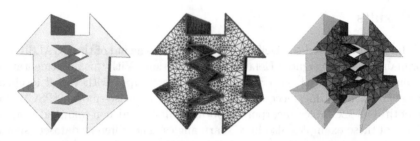

Fig. 12. SERATED: surface (middle) and volume (right) mesh.

Fig. 13. PIN-HEAD: surface (middle) and volume (right) mesh.

Fig. 14. SATURN: Non-manifold, a surface attached to the equator of a ball.

7 Conclusions

We have presented a practical algorithm to mesh a wide variety of geometric
domains with Delaunay refinement technique. The output mesh maintains a
manifold property and captures the topology of the input with a sufficient

Fig. 15. SWIRL: three surface patches meeting along a curve

Fig. 16. GUIDE: surface (middle) and volume (right) mesh.

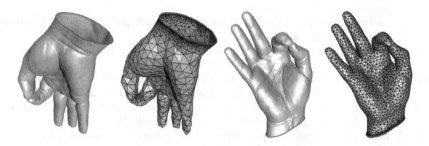

Fig. 17. HAND: manifold with boundary (two levels of refinement).

level of refinement. An interesting aspect of the algorithm is that the input features are preserved in the output.

A number of experimental results validate our claims. It can handle arbitrarily small input angles. When applied to volumes, the algorithm guarantees bounded radius-edge ratio for most of the tetrahedra except near boundary. It can be easily extended to guarantee bounded aspect ratio for most triangles except the ones near non-smooth elements. Furthermore, optimization based techniques can be used to improve qualities of the elements [1].

The question of time and size complexity of the algorithm remains open [12]. Right now the resolution parameter allows uniform refinement. Can it be extended to adaptive refinement? We plan to address these issues in future research.

References

1. P. Alliez, G. Ucelli, C. Goldman, and M. Attene. Recent advances in remeshing of surfaces. State-of-the-art report of AIMSHAPE EU network, 2007.
2. N. Amenta and M. Bern. Surface reconstruction by Voronoi filtering. *Discr. Comput. Geom.*, 22 (1999), 481–504.
3. C. Boivin and C. Ollivier-Gooch. Guaranteed-quality triangular mesh generation for domains with curved boundaries. *Intl. J. Numer. Methods in Engineer.*, 55 (2002), 1185–1213.
4. J.-D. Boissonnat and S. Oudot. Provably good sampling and meshing of Lipschitz surfaces. *Proc. 22nd Ann. Sympos. Comput. Geom.*, 2006, 337–346.
5. J.-D. Boissonnat and S. Oudot. Provably good surface sampling and meshing of surfaces. *Graphical Models*, 67 (2005), 405–451.
6. H.-L. Cheng, T. K. Dey, H. Edelsbrunner, and J. Sullivan. Dynamic skin triangulation. *Discrete Comput. Geom.*, 25 (2001), 525–568.
7. S.-W. Cheng, T. K. Dey, and E. A. Ramos. Delaunay refinement for piecewise smooth complexes. *Proc. 18th Annu. ACM-SIAM Sympos. Discrete Algorithms* (2007), 1096–1105.
8. S.-W. Cheng, T. K. Dey, E. A. Ramos and T. Ray. Quality meshing for polyhedra with small angles. *Internat. J. Comput. Geom. Appl.*, 15 (2005), 421–461.
9. S.-W. Cheng, T. K. Dey, E. A. Ramos and T. Ray. Sampling and meshing a surface with guaranteed topology and geometry. *Proc. 20th Annu. Sympos. Comput. Geom.*, 2004, 280–289.
10. L. P. Chew. Guaranteed-quality triangular meshes. Report TR-98-983, Comput. Sci. Dept., Cornell Univ., Ithaca, New York, 1989.
11. L. P. Chew. Guaranteed-quality mesh generation for curved surfaces. *Proc. 9th Annu. Sympos. Comput. Geom.*, 1993, 274–280.
12. S. Har-Peled and A. Üngör. A time optimal Delaunay refinement algorithm in two dimensions. *Proc. Ann. Sympos. Comput. Geom.*, 2005, 228–236.
13. T. K. Dey, G. Li, and T. Ray. Polygonal surface remeshing with Delaunay refinement. *Proc. 14th Internat. Meshing Roundtable*, 2005, 343–361.
14. H. Edelsbrunner and N. Shah. Triangulating topological spaces. *Internat. J. Comput. Geom. Appl.*, 7 (1997), 365–378.
15. G. L. Miller, D. Talmor, S.-H. Teng and N. Walkington. A Delaunay based numerical method for three dimensions: generation, formulation, and partition. *Proc. 27th Annu. ACM Sympos. Theory Comput.* (1995), 683–692.
16. S. Oudot, L. Rineau, and M. Yvinec. Meshing volumes bounded by smooth surfaces. *Proc. 14th Internat. Meshing Roundtable* (2005), 203–219.
17. S. Pav and N. Walkington. Robust three dimensional Delaunay refinement. *13th Internat. Meshing Roundtable* 2004.
18. J. Ruppert. A Delaunay refinement algorithm for quality 2-dimensional mesh generation. *J. Algorithms*, 18 (1995), 548–585.
19. J. R. Shewchuk. Tetrahedral mesh generation by Delaunay refinement. *Proc. 14th Annu. Sympos. Comput. Geom.* 1998, 86–95.
20. J. R. Shewchuk. Mesh generation for domains with small angles. *Proc. 16th Annu. Sympos. Comput. Geom.*, 2000, 1–10.
21. H. Shi and K. Gärtner. Meshing piecewise linear complexes by constrained Delaunay tetrahedralizations. *Proc. 14th Internat. Meshing Roundtable* (2005), 147–164.

5A.4

Efficient Delaunay Mesh Generation from Sampled Scalar Functions

Samrat Goswami[1], Andrew Gillette[2], and Chandrajit Bajaj[3]

[1] Institute for Computational and Engineering Sciences, University of Texas at Austin samrat@ices.utexas.edu
[2] Department of Mathematics, University of Texas at Austin agillette@math.utexas.edu
[3] Department of Computer Sciences and Institute for Computational and Engineering Sciences, University of Texas at Austin bajaj@cs.utexas.edu

Abstract: Many modern research areas face the challenge of meshing level sets of sampled scalar functions. While many algorithms focus on ensuring geometric qualities of the output mesh, recent attention has been paid to building topologically accurate Delaunay conforming meshes of any level set from such volumetric data.

In this paper, we present an algorithm which constructs a surface mesh homeomorphic to the true level set of the sampled scalar function. The presented algorithm also produces a tetrahedral volumetric mesh of good quality, both interior and exterior to the level set. The meshing scheme presented substantially improves over the existing algorithms in terms of efficiency. Finally, we show that when the unknown sampled scalar function, for which the level set is to be meshed, is approximated by a specific class of interpolant, the algorithm can be simplified by taking into account the nature of the interpolation scheme so as to circumvent some of the critical computations which tend to produce numerical instability.

1 Problem and Motivation

A wide variety of science and engineering applications rely on accurate level set triangulation. This is especially true for multiscale models in biology, such as macromolecular structures extracted from reconstructed single particle cryo-EM (Electron Microscopy), cell-processes and cell-organelles extracted from TEM (Tomographic Electron Microscopy), and even trabecular bone models extracted from SR-CT (Synchrotron Radiation Micro-Computed Tomography) imaging. Computational analysis of these models for estimation of nano, micro, or mesoscopic structural properties depends on the mesh representation of the contour components respecting their topological features.

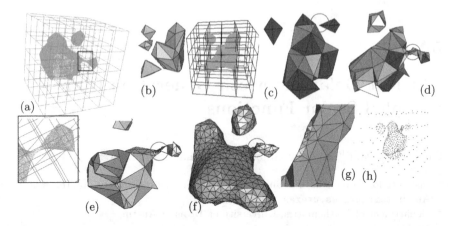

Fig. 1. Various stages of our algorithm. (a) A rectilinear grid with sample values of an unknown function at the grid points. Within cells, the function is approximated with a trilinear interpolant. For the purpose of visualization only, we collect a set of points (green) on the surface and display them. A narrow region of the surface is magnified below. (b) Another view of the data (right) and the same view of the mesh generated by Marching Cubes [22]. Note that the mesh is disconnected in the thin region. (c) The mesh generated by the restricted Delaunay triangulation of only edge and grid points. Blue facets have a grid point as at least one of their vertices. This point set is still not sufficient to produce a Delaunay-conforming mesh. (d) The mesh generated by addition of sample points of Σ. The topology is now recovered (Property I). (e,f) Geometrical refinement for progressively smaller value of ϵ. (g) Even in the magnified portion of the thin region, the triangles approximate the geometry nicely (Property II). (h) All the points involved in construction of the mesh including grid points (blue), edge points (green), and new points added by the algorithm (red). Observe that in order to recover the topology and reduce the geometric error in the approximation, many surface sample points are added to the point set.

Our goal is to find an algorithm to solve the following problem. The input to the algorithm is a rectilinear sampling of a bounded domain of an unknown scalar function F. The rectilinear grid need not be uniform; it may be adaptive as in the case of an octtree. The user then specifies a local interpolant to generate a level set approximation for any isovalue v; we use Σ to denote this level set of the interpolating function. Since the function F is unknown, we must assume that the local interpolant produces a good approximation of the function F within each cell of the grid. Our algorithm is general enough to use any local interpolant, however, in our experience, a trilinear interpolant is the most natural choice.

Our goal is construct a mesh in an efficient manner such that the following properties hold:

I **Topological Guarantee:** M is homeomorphic to Σ.

II **Geometrical Guarantee:** The Hausdorff distance from M to Σ is within a user-specified bound ϵ.

III **Delaunay Conformity:** M is a subcomplex of the Delaunay triangulation of the vertex set of M.

IV **Adaptivity:** The user can decimate part of the volumetric data and still preserve properties I, II, and III.

Once a surface M is generated that is Delaunay conforming, it is possible to improve the mesh quality by applying any Delaunay refinement algorithm. We detail such an algorithm in Section 4.1. Figure 1 visually illustrates a toy data set (a), the failure of a typical isocontouring method, in this case Marching Cubes, in reconstructing the level set (b), generation of the correct topology by our algorithm (c-d), geometric refinement (e-g), and the final Delaunay-conforming surface sample (h).

At this point, we emphasize the novelty of our approach. Although there exist algorithms which can be applied to solve the problem as stated, such algorithms are devised in a more general setting and thus do not exploit the natural structure of the volumetric data. Our approach has a number of unique advantages over its predecessors. As part of the algorithm, we collect some of the grid points around Σ and use them to build a Delaunay conforming mesh efficiently. Once the surface mesh is created and forced to conform to the Delaunay triangulation of the point set, these grid points can either be removed or used to construct an interior or exterior tetrahedral volume mesh.

Additionally, as a result of noise in the input data or a poor choice of the isovalue v, there may exist topological anomalies in the surface Σ. In mesh processing literature, such anomalies have been referred to as "topological noise" [26, 4]. The term "noise" indicates undesirable features of small geometric size that prevent the mesh from being used for further processing. Methods have been developed to remove such artifacts provided that the point sample of Σ is sufficiently dense near the anomalies; our method guarantees this density. Therefore, the output mesh M can be applied without any further refinement to any point processing algorithm that uses prior Delaunay-based reconstruction of geometry to detect and selectively remove these topological features.

Finally, algorithms involving volumetric data and meshes often become computationally demanding due to the large size of data sets. The adaptivity of the algorithm (property IV) provides one way to ease calculations by downsampling less important regions of the volumetric data.

2 Prior Work

Mesh generation techniques have received significant attention in the past two decades. Two works in particular, one by Chew [12] and one by Edelsbrunner and Shah [15], have spawned important and relevant results in the field. We will address the prominent successors of each of these works and compare the relative advantage of our approach.

Chew provided one of the first meshing algorithms for curved surfaces with provably good geometry in [12], although the algorithm as stated in the paper could not guarantee topological correctness. Boissonnat and Oudot showed how Chew's algorithm can be applied to produce a dense sample of Σ and subsequently mesh it. Oudot, Rineau and Yvinec recently improved that method by including sliver exudation, that is, the process of removing tetrahedra with very small dihedral angles from a mesh [24]. Alliez, Cohen-Steiner, Yvinec and Desbrun have also adapted the method to produce nicely graded meshes, i.e. meshes where the tetrahedra vary in size gradually based on their distance to the surface [1].

Separately, Edelsbrunner and Shah established a criterion called the *closed ball property* for ensuring that a mesh is Delaunay conforming [15]. We explain the closed ball property in Section 3. Recent work by Cheng, Dey, Ramos, and Ray uses this property to provide a method for constructing Delaunay conforming meshes that avoids the need to estimate local feature size [9]. This is a significant development as approximating local feature size is computationally expensive and not always numerically robust. Cheng, Dey and Ramos have extended this strategy for piecewise smooth complexes [10]. Very recently, Dey and Levine [14] have given a two phase algorithm to mesh isosurfaces from imaging data.

All of these approaches to mesh generation are useful, however, they all rely on an oracle to know whether an arbitrary ray intersects the surface Σ. The implementation of such an oracle becomes computationally prohibitive when applied to piecewise interpolated surfaces. For example, a common data set size is 100^3 vertices, meaning there exist 100^3 functions in the piecewise decomposition. Thus, a single ray may pass through over a hundred separate function domains, making intersection calculations expensive. As we detail in Section 4.2, a major advantage of the algorithm presented in this paper is that we take advantage of the original rectilinear scaffolding from the data to substantially reduce the computational overhead required.

Our work also improves upon existing methods for isosurface construction. Many well-known techniques exist for isosurfacing including marching cubes [22], active snakes, dual contouring [28], and higher order interpolation [6]. A variety of approaches have also been developed to provide hierarchical isosurface extraction and interior meshing [27, 17, 18]. Shewchuk and Labelle recently provided a straightforward isosurfacing and stuffing algorithm with good quality tetrahedra [19]. Relatively few works, however, take into account the effect of a trilinear interpolant within each grid cell [23, 11, 21]. Attali and Lachaud gave an algorithm for construction of Delaunay conforming isosurfaces [2]. Their method, however, relies on a specific rule established by Lauchaud in [20] that does not accommodate interpolation within grid cells.

Further, none of these techniques generalize easily to data that is sampled adaptively or to arbitrary interpolants.

3 Background

The Voronoi and Delaunay diagrams of a point set P, denoted Vor P and Del P respectively, play an important role in the computations involved in this paper. Due to page limitations, we do not go into the detail of their construction and refer the reader to any standard computational geometry textbook, e.g. [13].

Given a point set P chosen from the same space in which Σ is embedded, the *restricted Delaunay triangulation of* P, denoted Del $P|_\Sigma$, is defined to be the set of Delaunay objects of Del P whose dual Voronoi objects have non-zero intersection with Σ. If P is chosen in a way that respects the structure and local feature size of Σ, Del $P|_\Sigma$ will be a mesh with the desired properties.

Edelsbrunner and Shah gave a sufficient criterion called the *closed ball property* [15], sometimes referred to as the *topological ball property*, for Del $P|_\Sigma$ to be homeomorphic to Σ. A Voronoi object V of dimension k satisfies the closed ball property if $V \cap \Sigma = \emptyset$ or $V \cap \Sigma$ is homeomorphic to a closed ball of dimension $k - 1$. Accordingly, a point set P is said to satisfy the closed ball property if every Voronoi object of Vor P satisfies the closed ball property. Using the notion of *transversality* as defined in [16], their criterion can now be stated precisely.

Theorem 1. *[15] If Σ intersects each Voronoi object of Vor P transversally and Vor P satisfies the closed ball property, then Del $P|_\Sigma$ is homeomorphic to Σ.*

We will show in Section 4 that our algorithm produces a mesh satisfying the closed ball property. This ensures that the mesh is Delaunay conforming (property III from Section 1) and, by the theorem, that the mesh is homeomorphic to Σ (property I).

4 Algorithm

In this section, we describe the algorithm, analyze its efficiency and present some simplifying results for a specific choice of local interpolant. Figure 2 shows an overview of the process in two dimensions.

4.1 Algorithm Description

Our algorithm is motivated primarily by the work of Cheng et al. in [9] who build a Delaunay conforming approximation of the level set of any general

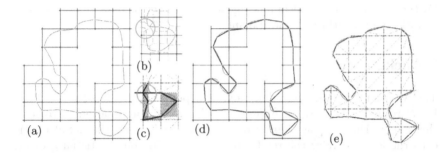

Fig. 2. A 2D example using bilinear interpolation of Σ demonstrating the importance of the closed ball property. (a) Grid points (dark blue) and edge points (light blue) relevant to our algorithm are shown. Note that in the 3D case we form a layer of grid points twice as thick. (b) A portion of the Voronoi diagram is shown in red and a location where the closed ball property is violated is circled. (c) Since the closed ball property is violated, the restricted Delaunay diagram (black) has incorrect topology in the circled region. (d) Red points are inserted where the closed ball property is violated and the restricted Delaunay graph is formed (black). (e) By including those grid points interior to Σ, we efficiently produce an interior mesh that does not alter the Delaunay conforming surface mesh.

implict function. Since our problem is focused on locally interpolated functions, we take advantage of the natural scaffolding of the input grid to substantially improve the computational efficiency of the algorithm. Moreover, we also show that once the Delaunay conforming surface mesh is extracted, it is quite straightforward to build a tetrahedral volumetric mesh of good quality. We call the algorithm for extracting a surface mesh DELSURFMESH and the extension to build a Delaunay tetrahedral interior or exterior mesh DELVOLMESH.

In [9], the authors start with a small sample of points lying on the surface to be meshed. They keep refining the mesh until its vertex set satisfies the closed ball property, thereby providing a Delaunay conforming mesh homeomorphic to Σ. To ensure that a point set P satisfies the closed ball property, it is necessary to check the intersection of Σ with each Voronoi edge, facet, and cell of Vor P. While checking intersections is a computationally expensive task for a general implicit function, it is even more burdensome for the case of a piecewise locally interpolated function as given in our problem. For example, to determine if a certain Voronoi edge intersects Σ more than once, it is necessary to search all voxels containing any subset of Σ which are stabbed by the Voronoi edge, as Voronoi edges may be incident upon many voxels. Matters become worse for Voronoi facets or cells which may touch large regions of the domain.

It is in regards to this difficulty that our approach becomes significant. We exploit the "gridded" nature of the input data set to put $O(1)$ bounds on our intersection calculations. To start, we construct an initial sampling by computing all the points where a grid edge intersects Σ. These points serve as the initial sampling of Σ. However, if we compute the Voronoi diagram of these E points alone, the Voronoi cells can intersect an arbitrary number of voxels, meaning we will still have trouble verifying and enforcing the closed ball property. To circumvent this problem, we compute a protective layer of grid points near Σ which we denote G. The selection of G traps Voronoi cells of E points into a few voxels. As the algorithm progresses, it adds more points to the existing samples and the nature of the insertion process ensures that the Voronoi cells of these new points are also trapped in a constant number of voxels. We derive bounds on the size of Voronoi cells of the initial samples and the new points for uniform and non-uniform rectilinear gridding (Octtree) in Section 4.2.

We now give the pseudocode of the algorithm DELSURFMESH and describe the specifics of the steps subsequently (Figure 3).

DELSURFMESH(Σ)
  1   Compute the point set E sampling Σ.
  2   Compute the protective layer G of grid points.
  3   Compute the Voronoi and Delaunay diagrams of $E \cup G$.
  4   Insert new sample points (N) repeatedly until Vor $(E \cup G \cup N)$ satisfies
        closed ball property.
  5   Output the Restricted Delaunay triangulation Del $(E \cup G \cup N)|_{\Sigma}$.

Fig. 3. Pseudo-code of the DELSURFMESH algorithm.

We have already described how we choose the initial set of points on (E) and near (G) the surface Σ. The next task is to ensure that the closed ball property holds for the set of points. For a general interpolant within every grid cell, we employ the method given in [9]. For completeness, we briefly describe how the closed ball property can be violated and what measures are to be taken. This process is thus divided into three sub-steps CBP_VE for Voronoi edges, CBP_VF for Voronoi facets, and CBP_VC for Voronoi cells. As we show in Section 4.3, one can simplify this process considerably further if the typical trilinear interpolant is used, thereby improving the robustness and efficiency of the algorithm.

- CBP_VE: A Voronoi edge VE violates the closed ball property if it intersects Σ in more than one point. If this occurs, the intersection point which is farthest from the Delaunay triangle dual to VE is inserted into the triangulation.
- CBP_VF: A Voronoi facet VF violates the closed ball property if it intersects Σ in more than one component or if the intersection includes a closed loop in the interior of VF. In either case, the intersection point farthest from the Delaunay edge dual to VF is inserted into the triangulation.
- CBP_VC: A Voronoi cell VC violates the closed ball property if it intersects Σ in more than one component, if the intersection includes an

isolated component of Σ inside VC, or if the intersection includes a surface of positive genus with one or more disks removed.

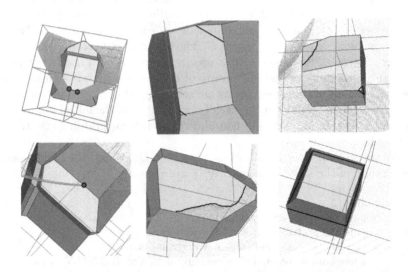

Fig. 4. Top row shows three samples scenarios where the closed ball property is violated for a Voronoi Edge (left), Voronoi Facet (middle) and Voronoi Cell (right). The bottom row shows the cases where the closed ball property is satisfied for Voronoi objects of the corresponding dimension (in the top row).

Note, the above properties are to be checked in the order given. Once a violation is detected, a new point is inserted into the existing Delaunay triangulation, the triangulation is updated, and the processes must begin again. Figure 4 shows different situations that can arise in this context.

Mesh Refinement

Although the topology of M is now correct, it may be possible that the geometry of M is not approximated sufficiently for an application purpose. Hence, we allow a user input ϵ and refine M as follows. For each Voronoi edge VE that intersects Σ, we compute the unique point $p \in VE \cap \Sigma$ and the circumcenter c of the dual Delaunay face to VE. If the distance between p and c is more than ϵ, we add p to the vertex set and regenerate the restricted Delaunay mesh. Every Voronoi edge dual to a restricted Delaunay facet is a normal approximation to Σ locally meaning the distance between p and c is an upper bound for the Hausdorff distance from Σ to the restricted Delaunay mesh. Hence, this process will yield a mesh satisfying property II.

In our current implementation, we maintain the 3D Vor/Del diagram of the point set throughout the process. Very recently, it was shown by Dey and Levine that this is not necessary [14]; one can recover the geometry while manipulating only the 2D mesh data structure of the surface, as long as there are enough points sampling a topologically correct approximation of Σ.

Tetrahedral Meshing of Interior/Exterior

At this stage, we have an accurate mesh approximation of the level set both topologically and geometrically. Since the mesh is already embedded in a Delaunay mesh that includes some grid points, we already have a tetrahedral mesh of both the interior and exterior of Σ. In order to improve the quality of the mesh elements, we use the algorithm DELVOLMESH defined as follows.

Without loss of generality, we describe how DELVOLMESH is used to generate a tetrahedral mesh of the interior of Σ. The input to DELVOLMESH is Del $(G \cup E \cup N)$, the volumetric mesh generated from DELSURFMESH. Note that Del $(G \cup E \cup N)$ has the output of DELSURF MESH, Del $(G \cup E \cup N)|_\Sigma$, as a subcomplex. We form a set G' of all grid vertices of the original rectilinear scaffolding which have function values less than the isovalue and do not belong to G. These points are distributed through the interior of Σ evenly (or evenly relative to an adaptive gridding) and we add them to the Delaunay mesh of the volume.

Here, our protective layer of grid points is crucially important. As we add the points of G', some triangles of the Delaunay mesh will necessarily change but the Delaunay triangles among surface points (E and N vertices) will be unaffected. This is a direct consequence of the fact that a point of G' is, by construction of G, closer to points of G than to points of $E \cup N$. Therefore, Del $(E \cup N \cup G \cup G')$ will still have M as the restricted Delaunay diagram. By throwing out the points of G exterior to Σ, we are left with a tetrahedral mesh of the volume with good quality tetrahedra and a Delaunay conforming surface mesh. The 2D analogue of DELVOLMESH is shown in Figure 2 (d) and (e).

4.2 Efficiency

Since our algorithm uses the natural structure of the rectilinear input data to construct Voronoi and Delaunay diagrams, we are able to provide two important results that reduce the computational burden. We state and discuss the significance of each one.

Theorem 2. *There is an $O(1)$ bound on the number of voxels that a Voronoi cell of an E or N point may intersect. In the case of uniform rectilinear gridding with voxels that are cubes, this bound is four voxels.*

Proof. We prove the following lemma for the simplest case in 2D which will be the crux of the argument for the proof of the more general cases.

Lemma 1. *For data given on an equally spaced 2D grid, the Voronoi cell of an E point is bounded within two pixels.*

Proof. Let $e \in E$ lie on the edge ε be between grid points $g_1, g_2 \in G$ and let $VC(e)$ be the (2D) Voronoi cell defined by e. There exist two lines perpendicular to ε, one passing through the midpoint of g_1 and e and one passing through the midpoint of g_2 and e. By definition, $VC(e)$ is necessarily contained between these two lines which bounds $VC(e)$ to a single column of pixels.

Now consider one of of the two pixels containing both g_1 and g_2. Denote its other vertices as g_3 and g_4. Let s denote the side length of a pixel. Then both $||e - g_3|| > s/2$ and $||e - g_4|| < s/2$. For any point f on the edge between g_3 and g_4, observe that $||f - g_3|| \leq s/2$ or $||f - g_4|| \leq s/2$. Therefore, each point on the edge between g_3 and g_4 is closer to one of those grid points than it is to e, and hence cannot belong to $VC(e)$. This bounds $VC(e)$ to the two pixels containing ε, proving the lemma. A picture version of this proof appears in Figure 5a.   □

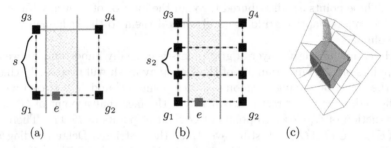

Fig. 5. (a) A picture proof of Lemma 1. The Voronoi cell of the edge point e must lie in the shaded region bounded by the solid lines (green). By symmetry, this restricts the Voronoi cell to a two pixel range. (b) A picture proof of Theorem 2 in the general 2D case. Here, each pixel has dimensions s_1 by s_2 with $\lceil s_1/s_2 \rceil = 3$. By symmetry, the Voronoi cell of e is restricted to a six pixel range. (c) A Voronoi cell of an edge point in 3D, visibly contained within four voxels.

Continuing with the 2D case and the same notation, suppose that each pixel is a rectangle with side lengths s_1 and s_2. Suppose ε has length s_1. As in the proof of the lemma, we can immediately restrict $VC(e)$ to a column of pixels. Consider only the range of $2\lceil s_1/s_2 \rceil$ pixels closest to ε. This range contains the two squares of side length s_1 with ε as one edge. Therefore, we may repeat the analysis in the proof of the lemma using the corners of this range instead of the corners of the pixel. Hence, $VC(e)$ is bounded within $2\lceil s_1/s_2 \rceil$ pixels. Figure 5b shows a simple example.

Now consider the 3D case where each voxel has dimensions s_1, s_2, and s_3 and again assume $e \in E$ lies on the edge ε with side length s_1. There are two planes perpendicular to ε which bounds $VC(e)$ to a flat grid of voxels. In either of the two rectilinear dimensions of this grid, we apply the same analysis as above to conclude that $VC(e)$ intersects at most $2(\lceil s_1/s_2 \rceil + \lceil s_1/s_3 \rceil)$ voxels. Note that this constant reduces to 4 when $s_1 = s_2 = s_3$.

The proof for points of type N is similar. Finally, we note that as we add more E and N points to the diagram, the Voronoi cells can only get smaller.

Therefore, the presence of other E or N points in the diagram is irrelevant to the bound.   □

Theorem 2 has significant implications for solving the ray intersection problem discussed in Section 2. A priori, Voronoi cells may touch an arbitrary number of voxels which cause an explosion in computational time when checking if the closed ball property is satisfied. With Theorem 2, the computational time can be bounded in advance.

Notably, the actual bound depends on the gridding scheme of the input, not the choice of interpolant. Accordingly, the width of the protective layer of grid points collected during the algorithm depends on the gridding scheme as well. For the case of adaptive gridding in an octtree construction, the user must require the "level difference" between two adjacent cells to be no more than some fixed number k. Given k, a loose bound on the number of voxels a Voronoi cell of an E or N point may intersect is $4k^2$, as an edge is incident on at most k^2 cells in each of the other two orthogonal direction. This bound may be improved to 4 for E points and $3k + 1$ for N points by considering limiting cases and using the convexity of Voronoi cells.

We have inserted the G points in order to decrease computations required to check the closed ball property, however we have increased the complexity of the Voronoi and Delaunay triangulations themselves. By Theorem 3 below, the new edges and facets formed by the addition of these G points do not add any significant burden to the closed ball property confirmation process. Further, these facets and edges are used in the algorithm DELVOLMESH described in Section 4.1.

Theorem 3. *Let VF be a Voronoi facet formed between two grid points and VE a Voronoi edge formed among three grid points at any point in the algorithm. Then $VF \cap \Sigma = \emptyset$ and $VE \cap \Sigma = \emptyset$.*

Proof. It suffices to prove the theorem for E points, since the addition of N points only makes existing Voronoi elements smaller. First we treat the 2D case. Consider a pixel with grid points g_1, g_2, g_3, g_4 and two edge points p and q. (If the pixel has more (four) or fewer (zero) edge points, the theorem is vacuous.) The edge points may occur on adjacent or opposite edges of the pixel, as shown in Figures 6 a and b, respectively. We consider the adjacent case first and suppose that the edges on which p and q lie intersect at g_4. Let R denote the region bounded by the rectangle formed with p and q as opposite corners (the shaded yellow regions of Figure 6). Note that $\Sigma \subset R$ since we have chosen a bilinear interpolant. Hence, it suffices to show that the Voronoi edges formed between g_1, g_2 and between g_1, g_3 do not intersect R. If x is a point on the Voronoi edge between g_1, g_2 then by definition $||x - g_1|| = ||x - g_2||$ and $||x - z|| \geq ||x - g_2||$ for all $z \in G \cup E - \{g_1, g_2\}$. However, if $x \in R$ then $||x - q|| < ||x - g_2||$, based on the labelling of Figure 6a. Hence, the Voronoi edge between g_1, g_2 cannot intersect R. Similarly, the Voronoi edge between g_1, g_3 cannot intersect R, proving this case. The case where p and q lie on opposite edges is similar.

The proof for the 3D case is analagous to the 2D case. It suffices to prove the claim for a closed Voronoi facet between two grid points g_1, g_2 as this includes the relevant Voronoi edges. Consider just one of the voxels into which this facet extends. In each of the rectinear directions away from the g_1, g_2 edge, there exists a closest point of E. Taking these two E points and the g_1, g_2 edge, we uniquely define a box. Using the definitions of the Voronoi diagram construction, we can similarly show that Σ lies outside of the box and the Vornoi facet lies inside it, thereby proving the theorem. $\qquad\Box$

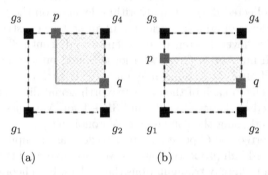

(a)   (b)

Fig. 6. Proof of Theorem 3 in the 2D case. We show that Σ lies entirely inside the shaded yellow region, while the Voronoi edges formed between two grid points necessarily lie outside of it.

4.3 Simplifying Results

Thus far, we have addressed how the inclusion of the grid points G reduces computational overhead. We now examine how particular choices of the interpolant can further ease the calculations. First we consider the process CBP_VF. If $VF \cap \Sigma \neq \emptyset$, then the intersection is a compact 1-manifold by the transversality assumption. All compact 1-manifolds are homeomorphic to a finite collection of line segments and circles and we distinguish between the case of no circles and at least one circle. If no circles exist, the closed ball property is violated if and only if more than two edges of the facet intersect Σ which is easy to check.

If at least one circle occurs in the intersection of Σ and a Voronoi facet, calculations can become more subtle. In the case of trilinear interpolation, these types of intersections do arise in two fashions. First, there exist configurations of relative function values that produce a tunnel topology inside a single voxel, for example, Case 13.5.2 as defined in [21]. In some cases, a Voronoi facet will pass through the entire tunnel, causing an interior loop in its intersection with Σ. In this case, we can detect the topology of the cell by the function values and add the "shoulder points" as described in [21]. The shoulder points are positioned so that the Voronoi diagram breaks these interior loops on Voronoi facets, thereby easing calculations. Alternatively, it may occur that the interior loop of a Voronoi facet passes through multiple voxels, which still provides some difficulties. We conjecture that this phenomenon happens only if the interpolated function is non smooth at a voxel edge or vertex and are working to simplify this problem.

Now we turn to CBP_VC. For this process, use of the trilinear interpolant provides much stronger results to simplify calculations. First, if Σ intersects a Voronoi cell, the intersection manifold must have a boundary component by the following Theorem.

Theorem 4. *If each facet of a Voronoi cell VC has empty intersection with Σ and Σ is defined by a trilinear interpolant, then $VC \cap \Sigma = \emptyset$. That is, there cannot be a component of Σ entirely inside a single Voronoi cell.*

Proof. The trilinear interpolation method precludes the existence of isolated surface components within a single voxel. Therefore, every component of Σ has a point of G in its interior and hence is sampled by at least six points of E, corresponding to the six rectilinear directions. Thus, each component of Σ intersects at least six Voronoi cells. A similar proof holds if the data is not given on a rectilinear grid. $\square$

Therefore, the process of checking the closed ball property for a Voronoi cell is reduced to checking if the intersection along the facets of a cell is a single closed curve (or empty). A priori, the possibility remains that the intersection surface is a manifold of positive genus with a disc removed. Detection of such cases has been discussed in detail in [9]. However, if our algorithm is applied in the common case of trilinear interpolation with voxels that are cubes, these difficult cases can also be excluded, leading to a much simpler algorithm.

Theorem 5. *Let VC be a Voronoi cell whose facets and edges satisfy the closed ball property and let Σ be defined by a trilinear interpolant on a uniform grid with voxels that are cubes. Then the closed ball property is satisfied for VC if and only if $VC \cap \Sigma$ has a single boundary component.*

Proof. It suffices to exclude the minimum case where $VC \cap \Sigma$ is homeomorphic to a torus with a disc removed. To generate such a surface by trilinear interpolation on a grid of cubes requires more than four voxels since the intersection of Σ with a voxel face cannot contain closed loops. By Theorem 2, the surface therefore intersects more than one Voronoi cell. $\square$

5 Implementation and Results

We have used CGAL [8] to build and maintain the Voronoi and Delaunay diagram of the point set. We also needed to compute the intersection of a ray with the surface Σ inside certain voxels; for this task, we have used another publicly available library called SYNAPS [25].

As described in the problem statement in Section 1, only the parameter ϵ is needed to run the algorithm on a given data set. This input dictates the desired upper bound on the Hausdorff distance between the surface approximation M and the interpolated surface Σ. Given this parameter, we reduce the amount of computation needed by snapping some of the points of E in the following manner. If there exist points $g \in G$ and $e_0, e_1, e_2 \in E$ such that the e_i are on grid edges incident to g and $d(g, e_i) < \epsilon$ for $i = 0, 1, 2$, we snap the three e_i points onto the common point g. Since all e_i points are within ϵ distance of the grid point and Σ passes through each e_i, the closet point of Σ to g cannot be more than ϵ away, which is within the user specified limit of tolerance.

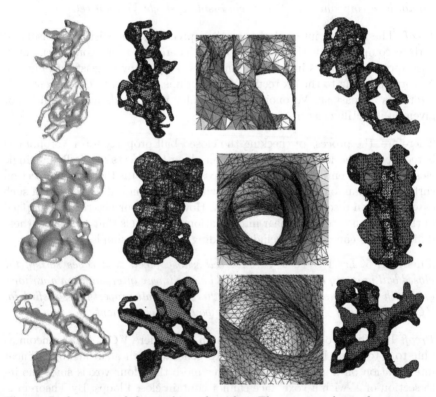

Fig. 7. Performance of the meshing algorithm. The top row shows the geometry, surface mesh, a closeup on the surface mesh and a cut-away of the volumetric tetrahedral mesh for 1CID. The second and the third rows show the same for 1MAG and BONE, respectively.

Although we have reduced the complexity of the ray-surface intersection problem as describe in Section 4, we still have to compute it for a constant number of voxels for every Voronoi edge. This was accomplished by parameterizing the Voronoi edge segment between two Voronoi vertices p_0 and p_1 with a real number t. Inside every candidate voxel, we then detect the intersection points of the segment with the interpolating polynomial function using the Algebraic solver available in the library SYNAPS. If an intersection point lies inside the corresponding voxel, we consider it as a valid intersection.

The performance of the algorithm on biological entities at various scales is shown in Figure 7.

The 1CID data set was obtained via blurring the coordinates of the atoms available from the Protein Data Bank (PDB) [7]. The resulting 3D scalar volume represents the electron density of the protein molecule T Cell Surface Glycoprotein CD4 . The goal in this case is to extract a well-sampled Delaunay conforming surface mesh so that one can analyze the secondary structure using the unstable manifolds of the index 1 and index 2 saddle points of a suitable distance function [5].

1MAG, shown in the second row of Figure 7, is the PDB entry of the ion channel Gramicidin A, obtained from soil bacteria Bacillus brevis. The molecular surface has many tunnels as shown in the left most sub-figure in the second row. However, only the tunnel in the middle is topologically significant as it is important to preserve the property of the ion channel. Once the isosurface has been extracted, it is therefore necessary to remove all the other tunnels and preserve the main tunnel. The algorithm for such removal of topological features has been described in [4] which relies on a Delaunay conforming isosurface in order to detect and remove unwanted topological features.

Finally, BONE data is shown in the third row of Figure 7. This data set is provided by our collaborator from the University of Rome. The goal here is to mesh the internal bone structure for stress-strain analysis.

The size of the 1CID volume data set is 128^3. It took 1 second to build the initial triangulation, 45 seconds to enforce the closed ball property, and 35 seconds to recover the geometry for $\epsilon = 0.00001$. The size of the 1MAG dataset is 128^3. It took 1.5 seconds to build the initial triangulation, 20 seconds to enforce the closed ball property, and 28 seconds to recover the geometry for $\epsilon = 0.00001$. The size of the BONE dataset is 64^3. It took 5 seconds to build the initial triangulation, 65 seconds to enforce the closed ball property, and 89 seconds to recover the geometry for $\epsilon = 0.0001$. Note, the time taken to mesh does not depend on the size of the volume data set since we were able to enforce that the ray-surface intersection calculation is of constant complexity. However, as the level set passes through more voxels, the number of points $(E \cup G)$ increases and that increases the time complexity of the algorithm.

6 Conclusion

We have presented an improvement over traditional level set meshing approaches. In particular, we claimed that the isosurfaces extracted via well-known approaches such as marching cubes or dual contouring are not suitable as they not only fail to capture the topology in a provable manner but also do not produce sufficient samples for the extracted approximation to be embedded in the Delaunay triangulation of the sampled set of points. On the other hand, there are elegant approaches for meshing general implicit surfaces, yet these algorithms suffer from the computational overhead of computing the intersection of a ray and the level set. Typically, the level set of a sampled scalar function is approximated via a number of piecewise interpolating functions and therefore finding the intersection of a ray with the level set requires a large number of checks. In light of this, we have presented an algorithm which not only overcomes the sampling issue of traditional isosurfacing techniques but also adopts a well-known provable algorithm [9] for generating a good sample of the isosurface efficiently. We have also shown that for the commonly used trilinear interpolant, many of the difficult cases that arise in implementation of the algorithm can be avoided, thereby improving numerical robustness.

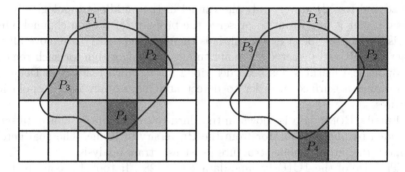

Fig. 8. Data may be partitioned to run the algorithm in parallel. In this 2D example, the voxels indicated by the P_i are to be processed. One can send data from each P_i voxel to a separate processor, along with the collar of protective voxels around it indicated by the same color shading. Two iterations are shown.

Scope

As noted earlier, the scope of this algorithm is immense. First, we have not assumed any particular interpolation scheme in order to prove that we need to check only a few voxels when detecting ray-surface intersection. Therefore, it is possible to extend the algorithm to any higher order interpolant, for example cubic A-patches [3]. Second, adaptive sampling of the scalar function poses no problem to this algorithm as explained previously. Finally, since we exploit the local nature of the interpolated surface Σ, the algorithm can easily be parallelized which is especially important for the large data sets often used in the multiscale biological models we have discussed. Figure 8 indicates how a data set might be partitioned to speed up computational time the analogous 2D case. The different colored pixels marked with P_i can be processed concurrently on separate processors as the Voronoi cells of the

E and N points in those pixels are guaranteed to be inside the protective collar of surrounding pixels. We note that not all pixels can be processed concurrently as the protective collars must be non-overlapping. However, by iterating the process, the most difficult computations can be done in parallel.

Limitations

One limitation of our algorithm is that it tends to generate more samples due to the layer of G points that we use to restrict the search space. Therefore it is necessary to apply adaptive sampling of the scalar function such that bigger voxels are automatically created in regions of less detail and smaller voxels in regions of high detail. By the results of Section 4.2, the efficiency of the algorithm remains the same.

Future Work

The extracted mesh typically has a "gridding" artifact because of the influence of the grid. Further, sometimes the uniformity of the grid causes it to be over-sampled. Therefore, ideally, one would decimate the grid so that the G vertices lie close to the medial axis after the isovalue is known. We plan to develop a scheme for such an optimal placement of the grid vertices. Also, the efficiency of the core computation requires a scaffolding structure. Hence, once the Delaunay conforming surface mesh is extracted, it is not clear how the surface can be decimated (if needed) by throwing away some of the grid vertices, while still efficiently checking that the resulting (decimated) point sample satisfies the closed ball property.

Acknowledgement: This research was supported in part by NSF grants IIS-0325550, CNS-0540033 and NIH contracts P20-RR020647, R01-9M074258, R01-GM07308.

References

1. P. Alliez, D. Cohen-Steiner, M. Yvinec, and M. Desbrun. Variational tetrahedral meshing. In *SIGGRAPH 2005*, pages 617–625, 2005.
2. D. Attali and J.-O. Lachaud. Delaunay conforming iso-surface, skeleton extraction and noise removal. *Comp. Geom.: Theory and Appl.*, 19:175–189, 2001.
3. C. Bajaj, J. Chen, and G. Xu. Modeling with cubic A-patches. *ACM Transactions on Graphics*, 14(2):103–133, 1995.
4. C. Bajaj, A. Gillette, and S. Goswami. Topology based selection and curation of level sets. In *TopoInVis 2007*, Accepted.
5. C. Bajaj and S. Goswami. Automatic fold and structural motif elucidation from 3d em maps of macromolecules. In *ICVGIP 2006*, pages 264–275, 2006.
6. C. Bajaj, G. Xu, and Q. Zhang. Smooth surface constructions via a higher-order level-set method. In *Proc. of CAD/Graphics 2007*, Accepted.
7. H. M. Berman, J. Westbrook, Z. Feng, G. Gilliland, T. Bhat, H. Weissig, I. Shindyalov, and P. Bourne. The Protein Data Bank. *Nucleic Acids Research*, pages 235–242, 2000.

8. CGAL Consortium. CGAL: Computational Geometry Algorithms Library. *http://www.cgal.org*.

9. S.-W. Cheng, T. Dey, E. Ramos, and T. Ray. Sampling and meshing a surface with guaranteed topology and geometry. *SCG '04: Proc. of the 20th Annual Symposium on Computational Geometry*, pages 280–289, 2004.

10. S.-W. Cheng, T. K. Dey, and E. A. Ramos. Delaunay refinement for piecewise smooth complexes. In *SODA*, pages 1096–1105, 2007.

11. E. Chernyaev. Marching cubes 33: Construction of topologically correct isosurfaces. *Technical Report. CERN CN/95-17*, 1995.

12. L. Chew. Guaranteed-quality mesh generation for curved surfaces. In *Proc. SoCG '93*, pages 274–280, 1993.

13. M. de Berg, M. van Kreveld, M. Overmars, and O. Schwarzkopf. *Computational Geometry: Algorithms and Applications*. Springer-Verlag, Berlin, 1997.

14. T. Dey and J. Levine. Delaunay meshing of isosurfaces. In *Proc. Shape Modeling International [to appear]*, 2007.

15. H. Edelsbrunner and N. Shah. Triangulating topological spaces. *Intl. Journal of Comput. Geom. and Appl.*, 7:365–378, 1997.

16. V. Guillemin and A. Pollack. *Differential Topology*. Prentice-Hall Inc., Englewood Cliffs, New Jersey, 1974.

17. I. Guskov, A. Khodakovsky, P. Schroder, and W. Sweldens. Hybrid meshes: multiresolution using regular and irregular refinement. In *SCG '02: Proc. of the 18th Annual Symposium on Computational Geometry*, pages 264–272, 2002.

18. K. Hormann, U. Labsik, M. Meister, and G. Greiner. Hierarchical extraction of iso-surfaces with semi-regular meshes. In *SMA '02: Proc. of 7th ACM Symposium on Solid Modeling and Applications*, pages 53–58, 2002.

19. F. Labelle and J. Shewchuk. Isosurface stuffing: Fast tetrahedral meshes with good dihedral angles. In *SIGGRAPH (to appear)*, 2007.

20. J.-O. Lachaud. Topologically defined iso-surfaces. In *DCGA '96: Proc. 6th Intl. Workshop on Discr. Geom. for Comp. Imagery*, pages 245–256, 1996.

21. A. Lopes and K. Brodlie. Improving the robustness and accuracy of the marching cubes algorithm for isosurfacing. In *IEEE Transactions on Visualization and Computer Graphics*, volume 9, pages 16–29, 2003.

22. W. Lorensen and H. Cline. Marching cubes: A high resolution 3d surface construction algorithm. In *ACM SIGGRAPH '87*, pages 163–169, 1987.

23. B. K. Natarajan. On generating topologically consistent isosurfaces from uniform samples. *The Visual Computer*, 11:52–62, 1994.

24. S. Oudot, L. Rineau, and M. Yvinec. Meshing volumes bounded by smooth surfaces. In *Proc. 14th Intl. Meshing Roundtable*, pages 203–219, 2005.

25. G. D. Reis, B. Mourrain, R. Rouillier, and P. Trébuchet. An environment for symbolic and numeric computation. In *Proc. Internat. Conf. on Mathematical Software*, pages 239–249, 2002.

26. Z. Wood, H. Hoppe, M. Desbrun, and P. Schroder. Removing excess topology from isosurfaces. *ACM Transactions on Graphics*, 23(2):190–208, April 2004.

27. Z. Wood, P. Schroder, D. Breen, and M. Desbrun. Semi-regular mesh extraction from volumes. In *VIS '00: Proc. of the Conference on Visualization 2000*, pages 275–282. IEEE Computer Society Press, 2000.

28. Y. Zhang, C. Bajaj, and B.-S. Sohn. Adaptive and quality 3d meshing from imaging data. In *Proc. of 8th ACM Symposium on Solid Modeling and Applications*, pages 286–291, June 2003.

Applications and Software

5B.1

Parallel Mesh Adaptation for Highly Evolving Geometries with Application to Solid Propellant Rockets

Damrong Guoy[1], Terry Wilmarth[1], Phillip Alexander[1], Xiangmin Jiao[2], Michael Campbell[1], Eric Shaffer[1], Robert Fiedler[1], William Cochran[1], Pornput Suriyamongkol[1]

[1] Center for Simulation of Advanced Rockets,
   Computational Science and Engineering Program,
   University of Illinois at Urbana-Champaign
[2] College of Computing,
   Georgia Institute of Technology

Summary. We describe our parallel 3-D surface and volume mesh modification strategy for large-scale simulation of physical systems with dynamically changing domain boundaries. Key components include an accurate, robust, and efficient surface propagation scheme, frequent mesh smoothing without topology changes, infrequent remeshing at regular intervals or when triggered by declining mesh quality, a novel hybrid geometric partitioner, accurate and conservative solution transfer to the new mesh, and a high degree of automation. We apply these techniques to simulations of internal gas flows in firing solid propellant rocket motors, as various geometrical features in the initially complex propellant configuration change dramatically due to burn-back. Smoothing and remeshing ensure that mesh quality remains high throughout these simulations without dominating the run time.

1 Introduction

Many physical systems involve material interfaces with dynamically changing geometries. Examples include the gas in the vicinity of parachutes, airfoils, helicopter blades, turbine engines, and burning propellants in rockets. Several approaches to discretizing the gas (fluid) and solid (structural) domains in problems of this type may be adopted (embedded boundary method, immersed boundary method, Chimera overset structured grids, ghost fluid method, etc.), but since an accurate representation of the behavior of the solution near the interface is often vitally important to the evolution of the solution throughout the entire fluid domain, body-fitted grids seem to be the most appropriate choice. However, maintaining adequate mesh quality for a body-fitted grid as the domain boundary deforms significantly with time can be a very challenging problem, particularly in large-scale 3-D parallel simulations. In this paper, we describe our multi-tiered, dynamic, parallel mesh modification strategy and demonstrate its effectiveness for two real-world rocket problems. It should be noted that our meshing strategy is broadly applicable to many types of problems with highly evolving domains.

In simulations of physical systems with deforming geometry, the location of the material interface is often part of the solution (e.g., parachutes), rather than a surface whose motion is prescribed as a boundary condition (e.g., a rigid sphere moving through the air). While it may be possible to describe the initial configuration of a parachute using a CAD model, once the chute begins to deform due to its interaction with the air, the only available description of its geometry is the deformed surface mesh of the chute. Surface propagation driven by a discretized velocity field is in general a very challenging problem, which we address using the Face-Offsetting Method.

Generating a new mesh at advanced simulation times requires the ability to derive (locally) a smooth representation from the discrete surface mesh [1]. When a new mesh is needed, we perform surface and volume mesh generation using a package from Simmetrix, Inc. Partitioning the new mesh for parallel execution of the simulation may be accomplished using METIS or ParMETIS, but especially for very large meshes, our recently developed and integrated hybrid geometric partitioner provides superior performance and robustness.

Unfortunately, generating a new mesh and transferring the numerical solution from the old mesh to the new one is both time-consuming and introduces interpolation errors. Therefore, in our simulations we delay remeshing by frequently applying Mesquite from Sandia National Laboratories to smooth and maintain mesh quality without changing mesh topology. Our parallel implementation applies a serial version of Mesquite in a concurrent manner. Interpolation of the solution is not necessary when the mesh is smoothed, allowing us to smooth every few time steps without decreasing solution accuracy.

These remeshing, data transfer, partitioning, and smoothing capabilities have all recently been integrated in the *Rocstar* rocket simulation package to enable us to perform large-scale simulations of physical systems with highly evolving domain boundaries without greatly increasing the wall clock time (compared to simulations of problems with fixed domains). We present results for two rockets with complex changing internal geometries. Experience with these real-world applications has helped us find better criteria for determining when to remesh.

Related work: A number of researchers have worked on mesh generation for dynamically changing geometries. Here we mention several of the methodologies most closely related to our approach. Baker [2] proposed to perform smoothing, coarsening, and refinement at each time step during the geometric evolution. The paper demonstrated effectiveness of the meshing technique without a physics solver. Folwell et al. [3] showed that mesh smoothing can delay abort-time of a computational electromagnetic code TAU3P. Wan et al. [4] successfully applied local mesh modification to metal forming simulations and compared the results with remeshing. Cardoze et al. [5] used bezier-based curved elements with refinement, coarsening, and edge smoothing for dynamic blood flow simulation in two dimensions. Dheeravongkit and Shimada [6] performed pre-analysis to adjust the input mesh for the anticipated deformation. The method was demonstrated successfully for forging simulations in two dimensions. Giraud-Moreau et al. [7] described a geometrical error estimate for adaptive refinement and coarsening of triangular and quadrilateral surface meshes. It was applied to a thin-sheet metal forming simulation using ABAQUS. Compared to the previous works, we focus on large-scale parallel 3-D simulations with emphasis on software integration of various meshing tools.

2 System Integration Overview

The *Rocstar*[8] simulation suite consists of a large number of dynamically loaded software modules coupled together to solve fluid/structure interaction problems including moving and reacting interfaces. Each module encapsulates a physics solver or some service needed by the simulation. Figure 1 depicts the *Rocstar* architecture. This parallel code uses MPI (or AMPI[9]) and typically runs on hundreds to several thousand processors in a batch environment. All module interactions, including the communication of data and function calls, are mediated by a general integration interface. The integration framework is designed to allow each module to retain its own data structures, representation, and parallelism.

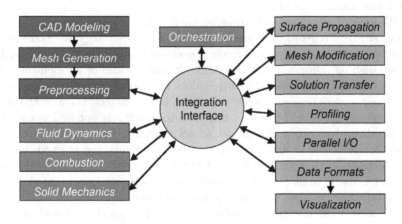

Fig. 1. *Rocstar* simulation component overview.

Our multi-tiered approach to mesh modification involves the steps shown in Figure 2. First, the surface is propagated according to the solution of the physical system and the surface mesh is smoothed to maintain quality. Next, we smooth the volume mesh to maintain its quality, and update the solution field variables. Eventually, the change in geometry becomes so severe that the smoothing operations cannot maintain acceptable mesh quality, and global remeshing is triggered. More details on surface propagation, smoothing, and remeshing are presented in subsequent sections of this paper.

---

1. Propagate surface according to physical solution
2. Smooth surface mesh
3. Smooth volume mesh
4. Physics solvers update solution
5. If mesh and solution are acceptable then continue (Step 1)
6. If mesh or solution quality is too low, trigger remeshing (Step 7)
7. Global remesh of geometry at last good checkpoint
8. Transfer solution from old mesh to new and generate new checkpoint
9. Restart simulation from new checkpoint and continue (Step 1)

Fig. 2. Integrated surface propagation, solution update, and meshing procedures.

When triggered, the remeshing module reads the simulation's last restart disk files (checkpoint) and writes a full set of simulation restart files. This allows our remeshing software to access the mesh and solution data of any domain in an identical manner regardless of the source, and also ensures that we remesh from a known good geometry and solution. Thus, the result of a remeshing phase (trigger, remesh, transfer solution, generate restart files) looks like a normal full simulation restart. This is done in such a way that the batch job does not terminate.

We have explored several strategies for deciding when to remesh. Our first simple strategy is geometry-based triggering. In this strategy, we examine the mesh after each time step and make a decision based on some mesh quality metric such as minimum dihedral angle or aspect ratio of an element. This strategy is easily implemented since it does not rely on any solution data.

In our experience, geometric triggering is insufficient because the mesh may become distorted in such a way that its quality is not bad per se, but the physics solver can no longer get an accurate solution in some part of the mesh. A trivial way to deal with this is to trigger remeshing at fixed intervals short enough so that the mesh is not expected to deform significantly. This strategy is easily implemented and works well for maintaining a good quality mesh, but may be unnecessarily expensive in terms of wall clock time.

We also allow our physics solvers to initiate remeshing based on their own internal decision-making process, whatever that may be. One of our fluid solvers triggers remeshing based on a threshold for the minimum internal timestep. This solver ensures that its explicit time step does not exceed the Courant limit (the shortest time for a signal to cross any cell). This strategy is slightly better than geometry-based triggering, since the local fluid solution is given some weight in the decision-making process.

We believe that the best possible remeshing triggering strategy would be one based on an analysis of the solution accuracy in each domain. This would require internal solver error estimators which are currently under development. We plan to add this capability in the near future.

3 Surface Propagation

Dynamic moving surfaces arise in many scientific and engineering applications, such as crystal growth, dendritic solidification, microchip fabrication, multiphase flows, combustion and biomedical applications. Because singularities and topological changes may develop during the evolution of the surface, propagating it numerically poses daunting challenges in developing accurate, efficient, and robust techniques.

To address the problem, we have developed a novel method for surface propagation, called the *face-offsetting method* (*FOM*) [10]. FOM propagates an explicit surface mesh without requiring a volume mesh, but unlike traditional Lagrangian methods, FOM propagates the faces of the mesh and then reconstructs the vertices from the faces by performing an eigenvalue analysis locally at each vertex to resolve the normal motion (for obtaining surface geometry) and tangential motion (for maintaining mesh quality) simultaneously. This new approach enables us to obtain accurate physical solutions for both advective and wavefrontal types of motion, even at singularities, and to ensure the integrity of the surface as it evolves.

The steps of the face-offsetting method are outlined in Figure 3. The first step "offsets" each individual face by integrating the motion of its vertices or quadrature points, and hence the method is named "face offsetting." After offsetting, the faces may no longer be connected to each other, and the second step reconstructs each vertex to be the "intersection," in a least squares sense, of the offsets of its incident faces. The displacement d_i of the ith vertex is then estimated as the vector from the vertex to the intersection, which essentially propagates the vertex to the point that minimizes a weighted sum of the squared distances to the offsets of its incident faces. This computation results in a linear system $\mathbf{Ad} = \mathbf{b}$ at each vertex, where $\mathbf{A}$ is a 3×3 matrix and $\mathbf{d}$ and $\mathbf{b}$ are 3 vectors. Robustness of the computation is achieved through an eigenvalue analysis of the matrix $\mathbf{A}$.

We then update the tangential motion of the vertices to maintain good mesh quality (step 4). More specifically, we restrict the motion to be within the null space of the matrix $\mathbf{A}$, as it tends to introduce minimal errors. Redistributing the vertices in this manner is referred to as *null-space smoothing* and was proposed in [12]. To maintain a mesh that is free of self-intersections, we next determine the maximum propagation time step $\alpha \Delta t$ that prevents

1. Propagate each face by time-integration of motion at its vertices or quadrature points
2. Reconstruct vertices to estimate displacement $\mathbf{d}_i$ at each vertex $\mathbf{v}_i$
3. For wavefrontal motion, correct vertex displacement $\mathbf{d}_i$ based on Huygens' principle[11]
4. Compute tangential motion $\mathbf{t}_i$ for each vertex $\mathbf{v}_i$ to maintain mesh quality
5. Compute maximum time step $\alpha\Delta t$ that prevents mesh self-intersection for $\alpha \leq 1$ and move each vertex by $\alpha(\mathbf{d}_i + \mathbf{t}_i)$

Fig. 3. Outline of the face-offsetting method.

mesh folding for $\alpha \leq 1$, and then move each vertex $\mathbf{v}_i$ to $\mathbf{v}_i + \alpha(\mathbf{d}_i + \mathbf{t}_i)$ (step 5).

If the motion is wavefrontal, such as in burning or etching, the estimated displacement from step 2 yields a poor approximation, and therefore an additional step is needed to correct the vertex displacements. This is done by adjusting the face offsets at expansions based on Huygens' principle[11] (step 3), so that the solution erodes expanding features. The effectiveness of the face offsetting method can be seen in Figures 10, 12, and 13.

4 Parallel Mesh Smoothing

We perform mesh optimization in *Rocmop*, a module which improves the quality of unstructured volume meshes through nodal repositioning, a process referred to as mesh smoothing. The purpose of smoothing is to delay degradation of mesh quality (and therefore the need for remeshing) due to the geometric evolution of the domain boundary as the simulation progresses. Because our fluid and structural dynamics equation solvers are formulated on moving meshes (i.e., Arbitrary Lagrangian-Eulerian [ALE] formulation), no interpolation of the solution is required when the mesh is smoothed. Instead, the equations of motion include the contribution to the change in the solution due to mesh motion. This allows us to smooth the mesh as often as desired without decreasing solution accuracy.

Nodes on the domain boundary move (at a nominal rate of about 1 cm/s in solid propellant rockets) due to the physics of the problem. Our fluid dynamics solver relies on *Rocmop* to move interior nodes, in particular those close to the surface, in order to avoid rapid generation of highly skewed elements there. *Rocmop* must be called at least every few time steps to maintain the same mesh quality as a run with smoothing performed every fluid time step. The main limitation appears to be stability of the numerical solution: if smoothing is not called often enough, the smoother makes more drastic changes to the nodal locations when it finally is invoked, which can cause the numerical solution to blow up.

An important feature of any module that may be called many times during a simulation is efficiency. In an early implementation of *Rocmop*, we noticed that more computational time was spent in mesh smoothing than in the fluid solver. After careful performance tuning, if called once every 3 fluid steps, *Rocmop* now uses around 30 percent of the total wall clock time. One lesson learned was that the execution time to evaluate the mesh quality measure (maximum dihedral angle) is significant compared to one time step of the physics solver. We wanted to check at every call whether this measure was below some specified threshold, and to perform additional smoothing if improvement was needed, but that turned out to be quite impractical. We now normally avoid computing the quality measure altogether.

Rocmop calls Mesquite [13, 14, 15], a powerful sequential mesh-smoothing package from Sandia National Laboratories. We apply Mesquite concurrently to each mesh partition independently, and then use a simple averaging scheme to realign the nodes shared by neighboring partitions. We were concerned that this averaging would degrade mesh quality, and were computing mesh quality metrics at substantial computational cost to ensure that mesh quality remained high. This turned out to cost far more than it was worth. We saw little improvement when we repeated the process of smoothing the mesh partitions independently and then averaging the coordinates of partition boundary nodes.

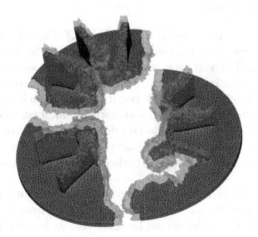

Fig. 4. Three mesh partitions with their ghost cells.

Note that Mesquite does not move nodes on surface patches or partition boundaries. As mentioned in Section 3, the surface mesh is smoothed frequently by our surface propagation module. Unlike most surface patches, partition boundaries do not often have a very regular geometry, especially when using a tetrahedral mesh. We have found that it is extremely important to allow Mesquite to relocate nodes on partition boundaries freely, as if they were interior nodes. We accomplish this by passing to Mesquite the real elements (local to a partition) plus the *ghost* cells (non-local elements) that share at least one node with a real element on the partition. The ghost nodes are used in the fluid solver's parallel implementation to store local copies of quantities needed from neighboring partitions. Figure 4 shows an example of three mesh partitions with their ghost cells. Utilizing these expanded mesh partitions, Mesquite can move the nodes on the partition boundary to their optimal locations.

5 Remeshing and Solution Data Transfer

We have developed a suite of remeshing and solution transfer tools called *Rocrem* to carry out a complete on-line remeshing process to improve the surface and volume quality of 3-D unstructured tetrahedral meshes. This approach is utilized when other methods, such as smoothing and local mesh repair, fail to improve mesh quality sufficiently for the continued use of the mesh (i.e. by a physics solver). *Rocrem* performs two key tasks: the generation of a better quality mesh and the transfer of solution data from the old mesh to the new mesh.

5.1 Surface Remeshing

Because of the use of commercial off-the-shelf (COTS) software for the mesh generation component of *Rocrem*, the first step is the serialization of the partitioned mesh. The surface mesh, like the physical domain boundary that it represents, is organized into *patches*, each having a particular boundary condition (BC). Each surface mesh partition may include regions ("*panes*") from various patches. The panes on each partition are first "stitched" together into a single surface mesh with patch information preserved by including it with the data for each triangular element. These surface partitions are then sent to the rank zero processor, and stitched together to form the serial surface mesh.

This serial mesh is then imported to Simmetrix's Simulation Modelling Suite (SMS). The SMS constructs a *discrete model* from the surface mesh. The patches and BCs are preserved by specifying each patch to Simmetrix as a *model face*, and assigning each triangular element to the appropriate model face.

At this point, mesh sizing prescribed by user-specified parameters is imposed, ranging from absolute sizing to fine control of the sizing relative to the discrete model. A new surface mesh is then generated according to the discrete model and these sizing parameters.

There are a few special situations that must be handled at the surface remeshing stage that are of particular importance in simulations with evolving mesh geometries. One is the disappearance of geometric features, and the other is the far more complicated disappearance of entire patches. The techniques for addressing these situations are notable in that they expand the functionality of our remeshing tools (beyond merely improving the quality of a mesh) into the realm of deviating slightly from the discrete model to properly continue the mesh evolution during a simulation. We describe each of these situations below.

Disappearing Geometric Features: Geometric feature information is determined by the SMS, and during surface remeshing those features are maintained by default. However, as the mesh evolves, certain features may become smaller and might eventually disappear, while others will arise and expand. As a feature disappears, its size will first become smaller than that of the edge sizing desired in the discretization. This will cause the appearance of poor quality elements, whose edges have a very high aspect ratio. These elements will invariably result in poor quality elements in the volume mesh. To remove such features and thereby recover higher quality in both the surface and volume meshes, it is necessary to deviate slightly from the discrete model that was determined from the original surface mesh. This is done by carefully selecting the parameters to the surface mesh modification function provided by the SMS. In particular, the SMS is instructed to remove poor quality elements that fall below a specified aspect-ratio threshold. This approach allows for the removal of a variety of features at the appropriate times during simulations with evolving geometries, such as the star-shaped region in some solid propellant rockets (see Section 7).

Disappearing Patches: A disappearing patch is a special case of a disappearing geometric feature. It arises when a geometric feature is shrinking and about to disappear, but the entire feature itself is a patch that was specified to the SMS discrete model as being a model face. While disappearing features on the same model face can be removed as described in the previous section, a model face that has a width of a single layer of very thin, poor quality elements cannot be removed without violating the model.

To handle this case, the mesh is examined to determine which patch is disappearing, and then the patch is reassigned to a neighboring patch. This allows for the removal of the thin elements using the same approach as in the previous section. The neighboring patch effectively takes over the space occupied by the reassigned patch.

In our rocket simulations, this situation typically arises when a burning surface neighboring an inhibited (non-burning) surface propagates in such as way as to eliminate the inhibited surface. When the inhibited surface is reduced to a thin ring and the solver cannot continue due to the poor element quality in that area, the inhibited surface elements are reassigned to the encroaching burning surface. Figure 5 illustrates this situation.

(a) 52.3s   (b) 87.6s   (c) 98.8s

Fig. 5. A geometric feature that is an entire patch in the RSRM joint-slot disappears over time.

Our current implementation requires user intervention in determining to which neighboring patch the disappearing patch should be reassigned. We are in the process of developing a means to detect this situation and automatically reassign the patch.

5.2 Volume Remeshing

Once the surface mesh is obtained, a volume mesh is generated. This typically results in a mesh with a coarser discretization deep in the interior. To control the interior mesh sizing and its gradient from the surface to deep in the interior, we set sizing parameters on the entire mesh and perform mesh adaptation using the SMS. The resulting mesh is smoothed and optimized, and the final product is then partitioned and ghost layers are added to it by ParFUM to facilitate communication and smoothing. For partitioning, we make use of either METIS- or ParMETIS-based[16, 17] partitioning provided by ParFUM[18], or our novel geometric partitioner (see Section 6). The new mesh is then ready for the solution transfer process.

5.3 Solution Data Transfer

We perform parallel solution data transfer using a conservative volume-weighted averaging strategy. Conservative volume-data transfer has been previously investigated in the literature, for example [19]. Our strategy begins by extruding the surface of the old mesh just enough to encapsulate the boundary of the new mesh. Cells generated by extrusion are assigned solution values from the cells whose faces were extruded to create them. Next, the new mesh is superimposed on the old mesh, and we determine which elements of the two meshes intersect. This is accomplished by using the parallel CHARM++ Collision Detection library[20]. This library can be configured to determine lists of potentially colliding elements quite efficiently, carrying the associated element data to the new mesh in the process. These lists each contain entries for both source (old mesh) and destination (new mesh) elements. Each pair of source and destination elements is examined to determine their overlapping volume. The solution on the old element is then weighted by the percentage of the new element's volume that is overlapped, and this portion of the solution is accumulated on the new element. Preliminary tests verify that the solution transfer scheme is conservative to machine precision, and the accuracy is acceptable for this first order simulation. Further work will be devoted to detailed error quantification.

5.4 Future Directions in Remeshing

Efforts to parallelize the bulk of the remeshing phase are on-going. Approaches are selected based on how much the mesh quality has degraded. For example, a mesh with high surface quality but low volume quality is improved by remeshing the volume of each partition in parallel. This approach relies on the use of the geometric partitioner discussed in Section 6 to create well-formed partitions that the SMS can accept as valid meshes. This approach is complete and full integration of the new partitioner is underway.

Another approach we are taking to relieve the remeshing bottleneck addresses the problem of poor surface mesh quality. There is no mechanism to pass a mesh to the SMS in parallel and improve the surface. This phase will proceed as it currently does in *Rocrem*: the mesh is serialized, the surface is improved, and a new volume mesh is generated. However, one time-consuming phase involves the adaptation of the new volume mesh to a desired sizing. We intend to partition the mesh and perform this expensive step in parallel.

6 Parallel Hybrid Mesh Partitioner

As mesh-based simulations become larger and larger, the need for scalable, parallel mesh partitioners increases. Generally, mesh partitioners fall into two categories: topological and geometric. Each category has advantages and disadvantages relative to the other [21]. For instance, geometric partitioners can create disjoint partitions on non-convex domains. On the other hand, topological partitioners can create partitions with many neighboring partitions.

We seek to combine the advantages of each by developing a hybrid partitioner. The idea is to compute a coarse topological partitioning of an irregular mesh, segmenting large portions of the mesh that a geometric partitioner would have trouble partitioning into simpler domains. This coarse partitioning is then refined by a geometric partitioner.

At present, we have implemented in parallel a nested dissection geometric partitioner [22]. It has proven scalable for meshes with tens of millions of cells on tens of thousands of compute nodes.

A novel topological partitioner is under development. The partitioner finds separators in the medial surface of the domain. The medial surface is computed in parallel with only a simple partitioning (Figure 6). Once computed, the medial surface is walked to find edges that bound large features. These features

Fig. 6. Medial surface computed on 8 processors.

are broken off of the mesh and partitioned geometrically. Preliminary results of the hybrid partitioner show regular partitions with a bounded number of neighbors and no disjoint partitions (Figure 7).

Fig. 7. Hybrid partition example.

7 Results

In this section, we present *Rocstar* simulations of two different rockets whose propellant is of significant geometrical complexity, a two-meter long test motor called StarAft and the 30-meter long Space Shuttle Reusable Solid Rocket Motor (RSRM).

Figure 8 shows the outer surface of the fluid domain in the StarAft rocket at four different simulation times during the test firing. The colors indicate the mesh partitions produced by METIS. The initial domain has a 6-pointed star-shaped cross section that tapers to a cylindrical profile near the head end (left hand side of Figure 8) and an inert nozzle at the aft end. As the propellant burns back to the cylindrical case, the fluid domain expands to fill the new fluid volume. When the propellant has burned completely, the fluid volume is over four times larger than it was initially. The number of

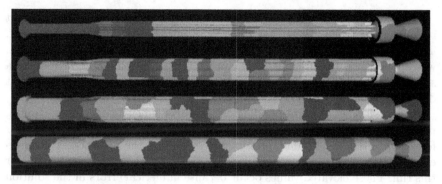

Fig. 8. Internal fluid domain of StarAft at (from top) 0.1, 3.4, 6.8, and 10.2 seconds.

tetrahedral elements increases from about 500K to as many as 1.7M. These computations were performed on up to 96 CPUs on various platforms. We used the Slow-timescale Acceleration technique [23] to speed up the burn-back rate by a factor Z from 64 to 1000. This reduces the number of time steps and mesh smoothings (performed every fluid time step) by a factor of Z, and therefore reduces by Z the ratio of the wall clock time for these operations compared to remeshing, whose invocation frequency and wall clock time are not affected by the acceleration technique. For $Z = 64$, the error in the fluid solution introduced by the acceleration is less than 2 percent; the percentages of wall clock time usage are: fluid solver 36, smoothing 41, and remeshing 23.

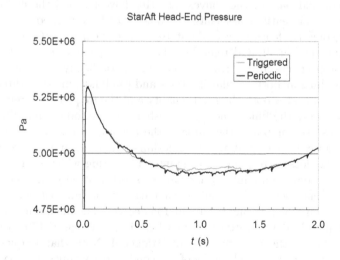

Fig. 9. StarAft pressure history for two different remeshing criteria.

Figure 9 shows the gas pressure at a particular location in the StarAft rocket for two simulations that differ only in their criterion for remeshing. The curve labeled "Periodic" (black) derives from a run in which remeshing was applied at 0.1 s intervals. One can see that each remeshing introduces a small (less than 0.3 percent amplitude) transient in the pressure history that dies away in about 0.02 s.

In the simulation corresponding to the curve labeled "Triggered" (pink), remeshing was triggered only when the time step computed by the fluid solver fell below a threshold value, presumably indicating the presence of sliver elements. Note the elevated pressures for times between 0.6 s and 1.3 s for this run. These elevated values occur because numerical errors in the solution increase as the mesh expands along with the fluid domain, reducing the spatial resolution. Whenever a new mesh is generated, additional interior nodes are inserted automatically to help preserve the initial local mesh spacing. Thus, the more frequent remeshings which occur in the periodically remeshed run better preserve both solution accuracy and mesh quality, although at a somewhat higher computational cost. In practice, we have found that periodic remeshing can also help prevent occasional simulation crashes due to large nodal displacements induced by the mesh smoother while attempting to improve a mesh of marginal quality.

Figure 10 shows the grid near the aft end of the StarAft rocket at four different times. The colors again indicate the partitioning pattern. The image sequence shows how the geometric features evolve. By 3.4 s, the concave star grooves have become sharp ridges, and the convex star tips have become several times wider. By 6.8 s, the annular surface at the aft end of the star has vanished, and part of the convex star tips have reached the cylindrical case. By 10.2 s, all features associated with the star have burned away.

When geometric features are about to change, nearby surface triangles often become very small, and remeshing is more frequently triggered by the fluid time step criterion. For this reason, our "periodically" remeshed runs enforce remeshing at both regular intervals and the fluid time step threshold.

Figure 11 indicates both the mesh quality and the frequency of remeshing in a run with only the fluid time step threshold (no periodic remeshing). At 0.1 s intervals, we computed the value of the maximum dihedral angle. The black dots represent times at which remeshing occurred, and line segments connecting the dots correspond to time intervals during which no remeshing was triggered. Some remeshing periods are so short that only one data point was used for that period. We can see that each remeshing yielded a mesh with a maximum dihedral angle of about 155 degrees. The maximum dihedral angle subsequently deteriorated to somewhere in the range of 170 to nearly 180 degrees before the next remesh was triggered. Note that frequency of remeshing varied from several tenths of a second at early times, to very often for times between about 6 to 8 seconds, and then several tenths of a second again towards burn out. This observation agrees with the fact that geometric features changed frequently between 6 s and 8 s, as mentioned above. Note

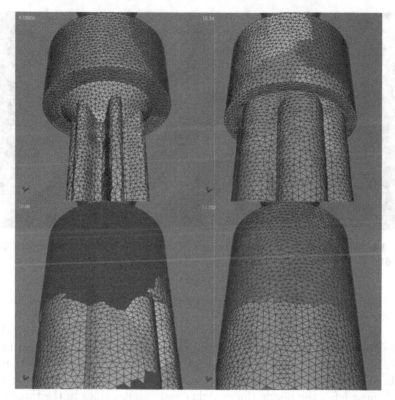

Fig. 10. Mesh of StarAft near the aft end at 0.1, 3.4, 6.8, and 10.2 seconds.

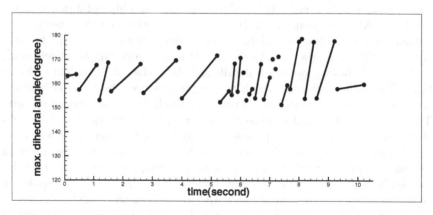

Fig. 11. Maximum dihedral angle of StarAft. Data points are connected together for each interval between remeshing.

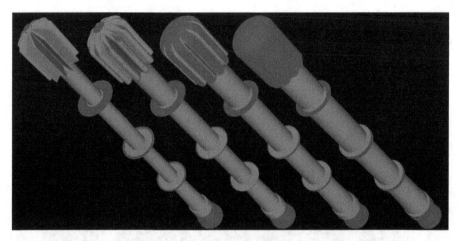

Fig. 12. Burning (red) and non-burning surfaces in the RSRM at 0.1 s, 22 s, 50 s, and 61 s.

that the remeshing interval was also very short near 4 s, when the star grooves became sharp ridges.

Next, we simulated near-complete propellant burn-out in the RSRM on up to 256 processors. Figure 12 illustrates the progression of propellant burnback. Shortly after ignition (leftmost image), burning propellant covers most of the surface of the fluid domain, except for a hole for the igniter at the head end (upper left), the nozzle (lower right) and annular rings in the three joint slots where a coating inhibits burning. At 22 s, the star tips first reach the rocket case, which is essentially cylindrical with an ellipsoidal dome at the head end. At subsequent times, the star tips broaden and then merge. The burning surface in the joint slots expand radially until they reach the case, and then slide along the rocket axis. By 111 s, the propellant is entirely gone.

Figure 13 shows geometrical details and the surface mesh at four different times. The color scale corresponds to the gas pressure. Shortly after ignition, at 0.2 s, the pressure is still low, since the propellant has just begun to burn. There are several triangles across the concave grooves in the star-shaped region, and the convex star tips are thin. By 7.1 s, the concave grooves are just beginning to merge. The merging of features occurs over the course of several remeshings. During each remeshing, all triangles below a certain size are eliminated, but this process may remove only part of the corresponding geometrical feature; the rest of the feature remains in the model until the remaining triangles become small enough to be eliminated during a subsequent remeshing. By 15.1 s, the concave grooves have become ridges, and the wider bumps at the bases of the star tips are about to merge. By 23.1 s, the bumps have merged and the star tips are flattening where they have reached the case.

The mesh in this simulation had 1.7 M tetrahedra initially, and 4.1 M by 23.1 s. Remeshing was invoked every 0.5 s, and also triggered whenever

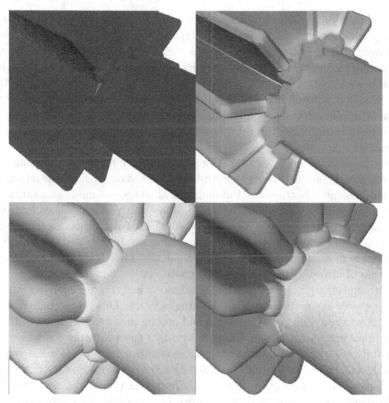

Fig. 13. Aft end of the star-shaped region in the RSRM. Simulation times are (clockwise from upper left) 0.1 s, 7.1 s, 15.1 s, and 23.1 s.

the fluid time step dropped below 1 microsecond. There were roughly 50,000 fluid steps between remeshings. The Slow-timescale Acceleration factor Z was 64. Each entire remeshing procedure takes about 46 minutes and produces a mesh with a maximum dihedral angle near 155 degrees. For a run that is not accelerated ($Z = 1$), in which we smooth the mesh (approximately 3 M elements) every 3 fluid time steps, the percentages of wall clock time usage are: fluid solver 70, smoothing 27, and remeshing 3. Accelerating the run by a large factor essentially eliminates the fluid and smoothing times (even if we smooth every fluid step), and therefore serial remeshing limits the maximum possible speedup (through Slow-timescale Acceleration) to a factor of about 33.

8 Conclusion and Discussion

The strategy for modifying body-fitted meshes described in this paper enables our integrated simulation package to obtain accurate numerical solutions for many classes of physical problems with moving domain boundaries. For the solid propellant rockets studied here, the nominal burn rate is quite slow compared to the Courant limit on the explicit time step that can be used in our unstructured-mesh fluid dynamics solver. As a result, even though we perform some remeshing stages using a sequential commercial package, these operations do not dominate the total simulation wall clock time. In contrast, frequent application of a sequential mesh smoother in a parallel manner is essential to minimizing the total computational cost of both smoothing and remeshing.

For problems with more rapid domain boundary motions, including those we produced by accelerating the burn-back times in our rockets, serial remeshing can become a bottleneck. Parallelizing all stages of mesh generation is a much more difficult problem than smoothing mesh partitions concurrently. Although there are parallel versions of some meshing tools, including Simmetrix's, we believe that a parallel remeshing strategy based on concurrent application of serial packages has important advantages, including ease of integration of alternative serial meshing tools.

Our experience with large-scale simulations with highly evolving domains suggests some future directions in meshing research. One research area is software integration with physics solvers and among meshing packages. Although many meshing packages can successfully solve the problems for which they were designed, integrating diverse packages into a large simulation remains difficult and time consuming. Various obstacles prevented us from using the native parallel versions of Mesquite and Simmetrix (which did not exist when we developed Rocstar in its present form). For example, Mesquite has an API that conforms to the TSTT specification, but Simmetrix does not. Another obstacle involves the discrete geometry of partitioned meshes. The mesh in a large-scale simulation is already partitioned when remeshing is needed, and the domain geometry can be defined only by a collection of surface patches distributed among a large number of processors. The surface patches carry additional attributes, such as boundary conditions, whose values must remain associated with the appropriate faces when the domain is remeshed and repartitioned.

Additional research is needed to improve the efficiency of mesh smoothing, which for evolving geometries can consume a significant fraction of the run time, even when performed concurrently. Our simulations would benefit from the use of a mesh quality measure that is both quick to evaluate and takes into account the element size distribution of the existing mesh. The smoothed mesh should be as similar as possible to the existing mesh to help prevent sudden large motions of nodes that could cause solution instability in the physics solvers. This constraint and the fact that the existing mesh is usually of reasonably good quality might be used to improve the rate of convergence to the optimal smoothed mesh configuration.

Acknowledgements

This work is supported by the Center for Simulation of Advanced Rockets (CSAR) at the University of Illinois at Urbana-Champaign. The CSAR research program is supported by the US Department of Energy through the University of California under subcontract B523819. The authors thank John Norris for the visualization software *Rocketeer*, Mu Sun for generating pictures and movies from simulation results, Mark Brandyberry for helping with initial mesh generation, Gengbin Zheng for the orchestration module *Rocman*, and Professor Michael Heath for general direction of the project.

References

1. Owen S J, White D R (2003) Mesh-based geometry. Int. J. Numer. Meth. Engrg. 58:375–395.
2. Baker T J (2001) Mesh movement and metamorphosis. In Proc. 10th Int. Meshing Roundtable, Newport Beach, California, pp. 387–395.
3. Folwell N, Knupp P, Brewer M (2003) Increasing TAU3P abort-time via mesh quality improvement. In Proc. 12th Int. Meshing Roundtable, Santa Fe, New Mexico, pp. 379–390.
4. Wan J, Kocak S, Shephard M S (2004) Automated adaptive forming simulations. In Proc. 13th Int. Meshing Roundtable, Williamsburg, Virginia, pp. 323–334.
5. Cardoze D E, Miller G L, Olah M, Phillips T (2004) A bezier-based moving mesh framework for simulation with elastic membranes. In Proc. 14th Int. Meshing Roundtable, San Diego, California, pp. 71–79.
6. Dheeravongkit A, Shimada K (2004) Inverse pre-deformation of finite element mesh for large deformation analysis. In Proc. 14th Int. Meshing Roundtable, San Diego, California, pp. 81–94.
7. Giraud-Moreau L, Borouchaki H, Cherouat A (2006) A remeshing procedure for numerical simulation of forming processess in three dimensions. In Proc. 16th Int. Meshing Roundtable, Birmingham, Alabama, pp. 127–144.
8. Dick W, Fiedler R A, Heath M T (2006) Building rocstar: Simulation science for solid propellant rocket motors. In 42nd AIAA/ASME/SAE/ASEE Joint Propulsion Conference and Exhibit, AIAA-2006-4590.
9. Huang C, Lawlor O, Kalé L V (2003) Adaptive MPI. In Proceedings of the 16th International Workshop on Languages and Compilers for Parallel Computing (LCPC 2003), LNCS 2958, College Station, Texas, pp. 306–322.
10. Jiao X (2007) Face offsetting: a unified framework for explicit moving interfaces. J. Comput. Phys. 220:612–625.
11. Baker B, Copson E (1950) The Mathematical Theory of Huygens' Principle. Clarendon Press, Oxford.
12. Jiao X, Alexander P (2005) Parallel feature-preserving mesh smoothing. In Computational Science and Its Applications - ICCSA 2005, vol. 3483/2005 of *Lecture Notes in Computer Science*, pp. 1180–1189.
13. Knupp P (2006) Mesh quality improvement for SciDAC applications. Journal of Physics: Conference Series 46:458–462.

14. Brewer M, Diachin L, Knupp P, Leurent T, Melander D (2003) The Mesquite mesh quality improvement toolkit. In Proc. 12th Int. Meshing Roundtable, Santa Fe, New Mexico, pp. 239–250.

15. Freitag L, Knupp P, Leurent T, Melander D (2002) MESQUITE design: Issues in the development of a mesh quality improvement toolkit. In Proc. 8th Int. Conf. Numer. Grid Gen. in Comput. Field Sim., Honolulu, Hawaii, pp. 159–168.

16. Karypis G, Kumar V (1998) Multilevel k-way partitioning scheme for irregular graphs. Journal of Parallel and Distributed Computing 48:96 – 129.

17. Karypis G, Kumar V (1997) A coarse-grain parallel formulation of multilevel k-way graph partitioning algorithm. In Proc. of the 8th SIAM conference on Parallel Processing for Scientific Computing.

18. Lawlor O, Chakravorty S, Wilmarth T, Choudhury N, Dooley I, Zheng G, Kale L (2006) Parfum: A parallel framework for unstructured meshes for scalable dynamic physics applications. Engineering with Computers 22 3-4:215–235.

19. Grandy J (1999) Conservative remapping and region overlays by intersecting arbitrary polyhedra. J. Comput. Phys. 148 2:433–466.

20. Lawlor O S, Kalé L V (2002) A voxel-based parallel collision detection algorithm. In Proc. Int. Conf. in Supercomputing, ACM Press, pp. 285–293.

21. Hendrickson B (1998) Graph partitioning and parallel solvers: Has the emperor no clothes? (extended abstract). In IRREGULAR '98: Proc. 5th Int. Symp. on Solving Irregularly Structured Problems in Parallel, Springer-Verlag, London, UK, ISBN 3-540-64809-7, pp. 218–225.

22. Vidwans A (1999) A Framework for Grid Solvers. Ph.D. thesis, Univeristy of Illinois, Urbana-Champaign, Urbana, IL.

23. Haselbacher A, Najjar F M, Massa L, Moser R (2006) Enabling three-dimensional unsteady srm burn-out computations by slow-time acceleration. In 42nd AIAA/ASME/SAE/ASEE Joint Propulsion Conference and Exhibit, AIAA-2006-4591.

New Applications of the **Verdict** Library for Standardized Mesh Verification
Pre, Post, and End-to-End Processing

Philippe P. Pébay[1]*, David Thompson[1]*, Jason Shepherd[2]*, Patrick Knupp[2]*, Curtis Lisle[3‡], Vincent A. Magnotta[4‡], and Nicole M. Grosland[4‡]

[1] Sandia National Laboratories, P.O. Box 969, Livermore CA 94551, U.S.A.
   {pppebay,dcthomp}@sandia.gov
[2] Sandia National Laboratories, P.O. Box 5800, Albuquerque NM 87185, U.S.A.
   {jfsheph,pknupp}@sandia.gov
[3] Knowledgevis, Llc, 1905 South Blvd, Maitland, FL 32751, U.S.A.
   clisle@ieee.org
[4] The University of Iowa, Iowa City, IA 52242, U.S.A.
   {vincent-magnotta,nicole-grosland}@uiowa.edu

Summary. Verdict is a collection of subroutines for evaluating the geometric qualities of triangles, quadrilaterals, tetrahedra, and hexahedra using a variety of functions. A quality is a real number assigned to one of these shapes depending on its particular vertex coordinates. These functions are used to evaluate the input to finite element, finite volume, boundary element, and other types of solvers that approximate the solution to partial differential equations defined over regions of space. This article describes the most recent version of Verdict and provides a summary of the main properties of the quality functions offered by the library. It finally demonstrates the versatility and applicability of Verdict by illustrating its use in several scientific applications that pertain to pre, post, and end-to-end processing.

1 Introduction

Verdict is a library for evaluating the geometric qualities of regions of space. A region of space can be, for example, a finite element or a volume associated with a finite volume mesh. This paper presents the design of the library and its application to several problems of interest. This introduction briefly

* These authors were supported by the United States Department of Energy, Office of Defense Programs. Sandia is a multiprogram laboratory operated by Sandia Corporation, a Lockheed-Martin Company, for the United States Department of Energy under contract DE-AC04-94-AL85000.
‡ These authors gratefully acknowledge the financial support provided by NIH Awards EB005973, EB001501, and U54 EB005149.

defines quality functions and discusses the history and motivation behind the creation of Verdict. The paper continues with the practical aspects of obtaining and using the library. Finally, several applications currently using Verdict are reviewed and followed with a list of quality functions provided by the library. The goals of this paper are to increase the visibility Verdict by illustrating its use in several scientific applications, and to propose it as a reference implementation for geometric qualities.

1.1 Quality Functions

In general, one can state that

Definition 1. *A* quality function *is a function that maps any region of space – possibly along with a parametrization defined over it – to a non-negative real number.*

Since a quality function maps an entire region of space to a single real number, it cannot completely describe the shape of its corresponding region. For example, the set of triangular shapes (*i.e.*, equivalence classes of triangles that are identical up to similarity) can be described as a 3-dimensional space. Therefore, some information is necessarily lost when a triangular shape is mapped to the real line. Thus, most quality functions are used to identify a single type of problem with a region's shape. Moreover, it is widely recognized that acceptable quality of regions depends on the particular problem of interest and its solution. For example, equilateral triangles do not always constitute the best region shape for every finite element simulation using triangular elements. The quality functions presently in Verdict do not take the PDE nor the solution into account when evaluating region quality. Therefore, when the desired optimal shapes can be specified by means of an appropriate geometric transform, regions should be transformed accordingly before quality is evaluated. This ensures, in this case, that distances are computed in the correct anisotropic metric [14]. Nevertheless, in applications which possess physics that are nearly homogeneous and isotropic, Verdict can also be used directly, independent of an application, for detecting inverted and other problematic regions.

Each quality function may take on any value on the real number line, but a typical use of quality functions is to focus on subsets of this range that are of interest for the application at hand. In particular, if one is interested in filtering out bad elements, as illustrated by applications in §3, a quality function with range $[1, \infty[$, where the value 1 is attained at and only at optimal geometric shapes, and tends to ∞ where regions are degenerate, may be advantageous. Therefore, we say that quality functions that have this property are *filterable*.

1.2 Assessing Quality

Verdict groups quality functions by the topological definition of the region on which they operate. The topological definition of a region is related to the number and type of geometric discontinuities on its boundary. For instance, regions with 8 boundary corners, 12 boundary edges, and 6 boundary faces (hexahedra) will have one corresponding set of quality functions. This is to be expected since the number and type of degenerate configurations vary with the topological definition. It is usually these degenerate boundary configurations where filterable quality functions should tend to $+\infty$. Examples where filterable quality functions should tend to $+\infty$ include crossed quadrilaterals, triangles with edges of vastly different lengths, and regions with coincident corner vertices. Some quality functions defined on regions of a given topology place additional caveats on their use. For instance, many quadrilateral quality functions are intended for planar quadrilaterals, and thus the numbers they may yield for non-planar regions are not guaranteed to be meaningful.

1.3 History of Verdict

Verdict has its roots in VERDE[5], a simple program to read ExodusII [33] meshes, and analyze them for possible problems, in particular in terms of finite element quality. VERDE, in turn, has roots in the CUBIT[6] mesh generation code. Many of the original quadrilateral and hexahedral quality functions, for example, were first coded in CUBIT based on the papers of [32] and later transferred over to VERDE. After a period of independent development, it was realized that VERDE and CUBIT did not yield the same results when assessing mesh quality. Consequently, Verdict was created so that both applications could share the same code and produce consistent results. Verdict was later extended to include quality functions for simplicial regions and the algebraic quality functions [16]. Verdict was initially licensed under the LGPL.

Meanwhile, the Visualization Tool Kit (VTK)[7] did not support general purpose mesh quality assessment. Due to the need for such a tool and the incompatibility between licenses, P. Pébay and D. Thompson generalized the vtkMeshQuality[8] class in 2004 to compute the quality of triangles, quadrilaterals, tetrahedra, and hexahedra using a variety of quality functions.

Following a change in Verdict's licensing scheme, from LGPL to BSD-style, it was decided in late 2006 to use Verdict in VTK for the same reasons that Verdict was initially created, by moving all quality functions from vtkMeshQuality to Verdict while retaining the best implementation when the same quality function was implemented in both software packages, using

---

[5] http://www.cs.sandia.gov/capabilities/VerdeMeshVerificationSuite/
[6] http://cubit.sandia.gov/
[7] http://www.vtk.org/
[8] http://www.vtk.org/doc/nightly/html/classvtkMeshQuality.html

vtkMeshQuality as a wrapper around Verdict, and resolving naming inconsistencies and redundancies. It is important to former Verdict users to note
that the latter action has resulted in changes to Verdict's API, although efforts have been made to preserve backwards-compatibility as often as possible
(*cf.* [39] for a summary of these changes).

2 Practicalities

The main goals of the library are (1) correctness of the implementation; (2) order invariance; and (3) computational efficiency. Correctness of the implementation is assessed empirically as Verdict has undergone an extensive amount of
testing and debugging, and users are invited to report bugs and other problems they may encounter. Note that the quality functions in Verdict are all
checked for overflow as follows: given a double-precision quality value q, if
$q > 0$, then $q \leftarrow \min(q, D_{\max})$; otherwise $q \leftarrow \max(q, -D_{\max})$. An algorithm
for computing quality is *order invariant* if the same result (to within machine
truncation error) is returned regardless of the order in which nodes are specified, as long as the nodes specify the same region[9]. Where applicable, the
quality functions in Verdict are verified against theory for order invariance.
When multiple quality functions are requested for the same geometric region,
an efficient implementation will only compute intermediate results used by
different quality functions one time.

2.1 Obtaining Verdict

The Verdict repository now resides at Kitware, Inc. and is publicly available.
A formal release has not yet been made since the repository has been moved
and so Verdict source code must be obtained from Kitware's CVS server at
www.vtk.org. If you intend to build VTK, you need not obtain or compile
Verdict separately since it is included with VTK.

Verdict can be built and installed on most systems, including Linux, Mac
OS X, and even Windows systems. For configuration, compilation, and instal-
lation details, please refer to [39].

2.2 Application Programming Interface (API)

Verdict was designed with a C interface so that it can be used in a variety
of applications. It provides a number of quality functions, implemented as
routines whose input consists of a list of coordinates corresponding to a list
of nodes or vertices within a region and of the cardinality of this list. Note
that this interface does not allow to pass the parametrizations mentioned in

---

[9] A corollary to this is that regions with different arrangements, such as inverted
finite elements, may take on different values.

Definition 1, but the implementation may be expanded in this order, *e.g.*, to accommodate higher order elements. Each available quality function has a corresponding routine; for instance, the hexahedron "Condition Number" is:

```
double v_hex_condition(int num_nodes,
                       double node_coordinates[][3])
```

and it may be used as follows:

```
double coords[8][3];
...
double condition_value = v_hex_condition(8, coords);
```

If a region's quality must be assessed with multiple functions, it is less computationally expensive to take advantage of common expressions: *e.g.*, the hexahedron "Jacobian" and "shape" quality functions both use the Jacobian matrix. Therefore, to improve computational efficiency, one function for each type of region computes multiple quality functions at the same time:

```
double v_hex_quality(int num_nodes,
                     double node_coordinates[][3],
                     unsigned int request_flag,
                     struct HexMetricVals* quality_vals)
```

e.g., to obtain together "Jacobian" and "shape" of a hexahedron:

```
double coords[8][3];
HexMetricVals vals;
double jacobian_value;
double shape_value;
int request = V_HEX_JACOBIAN | V_HEX_SHAPE;
...
v_hex_quality(8, coords, request, &vals);
double jacobian_value = vals.jacobian;
double shape_value = vals.shape;
```

2.3 Available Quality Functions

Verdict provides a total of 78 quality functions: triangles (13), quadrilaterals (23), tetrahedra (16), hexahedra (23), pyramids (1), wedges (1), and knives (1), which are now summarized in tabulated form. In principle, one should be able to reduce the 78 quality functions to a list of at most a dozen quality functions that correspond to basic geometric properties. However, the quality functions available in Verdict represent a collection of quality functions that have accumulated over the years by various practitioners at various times. Some quality functions may therefore be redundant.

A brief summary of the available qualities and their main properties is given in the Appendix; *cf.* Verdict's Reference Manual [39] for the explicit definitions of these functions. The summary provides two tables for each type of region, with the exception of pyramids, wedges, and knives which are lumped into the same tables. The first table first indicates the function's dimension, in the sense of the associated units: L (resp. A) denotes dimensions of length (resp. angle). When a quality function has a repeated dimension unit, an exponent is used to show the count: *e.g.*, volume has dimension L^3. While the precise units of length depend on the input coordinates, angles are always reported in degrees. Note that filterable quality functions are dimensionless (denoted 1), but the converse is not true. Second, a bibliographic reference where the function is defined and discussed is provided, if available; otherwise, the formula is one that is traditionally used but not readily available in the literature. Last, the corresponding Verdict function name is indicated. For each type of region, the second table presents information on the acceptable, normal, and full ranges of values taken on. The acceptable ranges are subjective and are provided as a suggestion for people unfamiliar with the meaning of the value of a particular quality function. The ranges were selected based on visual appearance – poor-looking regions are considered to have unacceptable quality. The acceptable ranges should not be taken as authoritative and they should be used with care for quality verification of application meshes. The normal column describes the range of values taken on by all valid, non-degenerate regions. The full range column includes the values taken on by invalid and/or degenerate regions. These tables also contain an entry for the reference value that the quality takes on for the corresponding optimal element. For triangular, quadrilateral, tetrahedral, and hexahedral shapes, this is respectively an equilateral triangle, a square, a regular tetrahedron, and a cube. For filterable quality functions, this value is 1.

3 Applications

Geometric quality is important in simulations. Typically an initial quality must be met and be maintained as the simulation progresses. When the simulation domain is deformed during the course of a simulation, this can require mesh relaxation or re-meshing. When the quality is not maintained, simulations can diverge. This makes quality important during pre-processing (for achieving the initial quality desired), during simulation (for maintaining the quality), and during post-processing (for inspecting the results to verify the impact of geometric quality on the results). In this section, five applications that build on Verdict to evaluate mesh quality in all phases of mesh generation and analysis are used to illustrate the efficiency and versatility of this library at providing support for geometric quality assessment in a variety of contexts.

3.1 Pre-processing: **VERDE**

VERDE is a pre-processing tool that uses Verdict for verifying the quality of finite element models. Utilizing Verdict, VERDE provides a wide range of

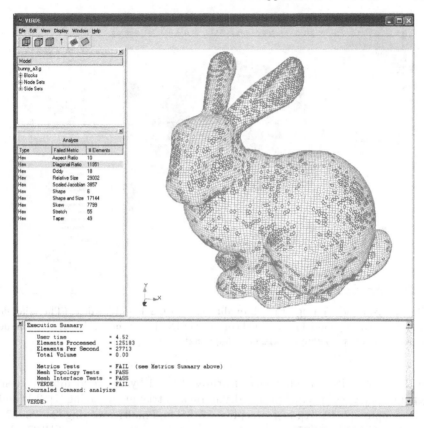

Fig. 1. Example filtering of mesh quality using VERDE and Verdict.

element qualities, including state of the art algebraic quality functions for calculating model topology, interface continuity, and locating mesh connectivity problems. It features a cross-platform graphical user interface and a graphical and numeric output useful for evaluating the quality of a finite element model, and was also designed to be a testbed for additional mesh quality research. Fig. 1 shows an example mesh loaded in the VERDE environment with highlighted elements filtered using mesh qualities calculated in Verdict.

3.2 Pre-processing: CUBIT

CUBIT[11] is a pre-processing tool for model and mesh generation that uses Verdict for verifying the mesh quality of finite element models. CUBIT is a full-featured software toolkit for geometric model generation and robust generation of 2-D and 3-D finite element meshes. The main development goal for CUBIT is to dramatically reduce the time required to generate meshes, particularly large hex meshes of complicated, interlocking assemblies. CUBIT incorporates a solid-model based preprocessor for generating meshes that conform

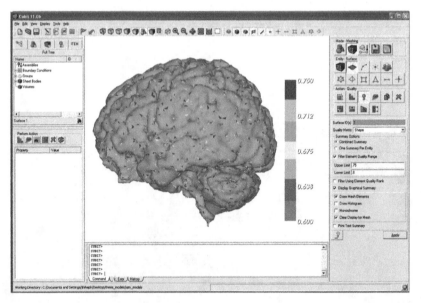

Fig. 2. Example filtering of mesh quality using CUBIT and Verdict. (The triangle mesh for the brain model is provided courtesy of INRIA by the AIM@SHAPE Shape Repository (http://shapes.aim-at-shape.net/index.php).)

to the geometric and topological features defined by the solid model. Mesh generation algorithms include quadrilateral and triangular paving [7], 2-D and 3-D mapping [9, 42, 41], hex sweeping [17, 35] and multi-sweeping [24, 23, 37], tetrahedral meshing [25], and various special purpose primitives. CUBIT also contains many algorithms for controlling and automating much of the meshing process, such as automatic scheme selection [43], interval matching [27], sweep grouping and sweep verification [37, 26], and includes state-of-the-art smoothing algorithms [8, 19, 20, 18, 21]. Fig. 2 shows a mesh loaded in the CUBIT environment with highlighted elements filtered using Verdict quality functions.

3.3 Post-processing: ParaView

ParaView[10] is a post-processing visualization tool that uses Verdict to provide mesh quality inspection. More specifically, the vtkMeshQuality class provides an interface to Verdict, whose functionalities are exposed as a filter in ParaView. vtkMeshQuality computes quality average, minimum, maximum, and variance over the entire mesh for each type of region, and stores these statistics in the output mesh's FieldData. In addition, the per-cell qualities are added to the mesh cell's data in an array named Quality to allow for

---

[10] http://www.paraview.org/

visualization and/or further processing as shown in Fig. 3. This figure demonstrates techniques we have found useful during post-processing. We are often interested in spatial trends in quality and their correlation with other variables such as simulation error estimates; a logarithmic scale with a user-adjustable color range helps to make these correlations along with volume rendering regions using a nearly transparent palette (resp. opaque) for good (resp. poor) quality values. This is especially true when the quality of interest is related to skew or torsion of elements. However, when quality functions meant to detect coincident vertices, slivers, or other disparities in scale are used, thresholding is more apropos since these elements are likely to take up very little screen space and thus require close-up inspection without nearby elements obscuring the rendering.

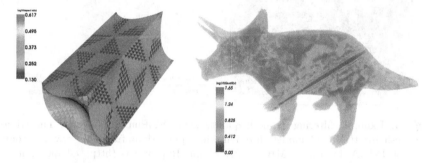

Fig. 3. Surface (left) and volume (right) renderings with ParaView of the per-region base-10 logarithms of the aspect ratios for two tetrahedral meshes.

3.4 End-to-end-processing: SCIRun

SCIRun [34] is a scientific programming environment which provides for the interactive construction, debugging and steering of large scale computer simulations. A visual programming interface allows scientific computations to be composed, executed, controlled, and tuned via a visual programming interface. Users may extend the set of scientific computations provided. In order to make the data flow paradigm applicable to large simulations, ways have been identified to avoid excessive memory use. Most importantly, SCIRun has been applied to solving large scale problems in computational biomedical imaging, [2, 4, 6] simulations [1] and visualization [3], explosives and fire simulations [31], and fusion reactions [5]. Verdict has been added to this environment to provide mesh quality verification and inspection. Fig. 4 shows an example mesh loaded in the SCIRun environment with highlighted regions filtered using mesh qualities calculated in Verdict.

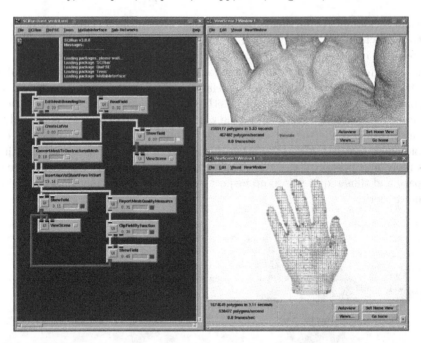

Fig. 4. Example filtering of mesh quality using SCIRun and Verdict. The triangle mesh for the hand model shown in the upper right frame is provided courtesy of INRIA by the AIM@SHAPE Shape Repository (http://shapes.aim-at-shap.net/index.php). The hexahedral mesh with mesh quality filtering shown in the lower right frame was generated by Jason Shepherd.

3.5 IA-FEMesh

IA-FEMesh[11] is a meshing tool designed to generate hexahedral finite element models appropriate for surface contact analyses in orthopedic applications. The mesh definitions initiate with segmented (manual or automated) medical image datasets, from which a surface representation of the region or regions of interest is generated. Two main meshing algorithms have been developed to generate a hexahedral mesh bounded by the triangulated isosurface representation. The first technique maps a block, or blocks, of elements onto the surface definition [38]. This algorithm has been used to mesh structures ranging from the relatively cylindrical phalanx bones of the hand to the complex geometries of the spinal vertebrae. Secondly, a mapped meshing technique has also been implemented. The objective is to map a predefined mesh of high quality (template) directly onto a new (subject-specific) bony surface definition, thereby yielding a similar mesh with minimal user intervention.

Regardless of the meshing technique, a check of the resulting mesh quality is imperative; consequently, a stand alone mesh quality viewer has been inte-

---

[11] www.ccad.uiowa.edu/mimx

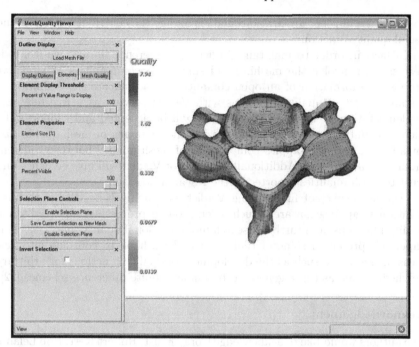

Fig. 5. Evaluation of the mesh quality for a cervical vertebra using IA-FEMesh and Verdict.

grated into the IA-FEMesh program suite. The objective is to make the mesh generation process more efficient by providing rapid, visual feedback to the user (Fig. 5). Our mesh quality application utilizes Verdict to analyze element quality according to several functions including: volume, Jacobian, scaled Jacobian, edge-ratio, and the Frobenius aspect. The tool supports the ability to display qualities and to interact with the mesh in real time using VTK. Since datasets generally contain a large number of elements, interactive tools have been developed that allow the operator to isolate and examine both individual elements and element groupings. A Laplacian smoothing algorithm has also been implemented to help improve quality in regions exhibiting elements of sub par quality. This work is currently being integrated into Slicer3, the core tool of the National Alliance for Medical Image Computing (NA-MIC).

4 Conclusions

This paper has outlined several applications that use Verdict for a variety of tasks in the context of simulations. We believe its adoption is evidence of an easy-to-use programmatic interface and an attention to speed and accuracy of calculation. Verdict is now distributed under a modified BSD license which makes it available to both open- and closed-source projects. We hope this will foster its use in other applications and encourage the contribution of new and useful quality functions.

In fact, in order to maintain Verdict as a reference library, we feel it is necessary to involve the meshing and simulation communities at large; we propose the formation of an open committee to adjudicate the definition of a standard set of quality functions with Verdict serving as a free reference implementation. It is hoped that Verdict will be accepted and supported by the community as a tool for standardizing the reporting of mesh quality and for allowing apples-to-apples comparisons of mesh quality between different mesh generation tools. Additionally, the use of Verdict will give credence and trust to mesh qualities reported in the research literature.

Another aspect of maintaining Verdict as a living code base is to push it into new application such as end-to-end processing, non-traditional quality functions, and further research into traditional quality functions. This paper has presented some of that work and we hope to continue into new areas in the future such as the development of quality functions for arbitrary polyhedra, such as those generated by mimetic finite difference schemes [22].

Acknowledgement

The authors would like to acknowledge Corey Ernst, Robert Kerr, Rob Leland, Ray Meyers, Karl Merkley, Scott Mitchell, Will Schroeder, Greg Sjaardema, Clinton Stimpson, Tim Tautges, and David White.

References

1. BioFEM: Bioelectric field finite element method software. Scientific Computing and Imaging Institute (SCI). URL www.sci.utah.edu/cibc/software/index.html.
2. BioImage: Volumetric image analysis and visualization. Scientific Computing and Imaging Institute (SCI). URL www.sci.utah.edu/cibc/software/index.html.
3. BioPSE: Bioelectric problem solving environment for modeling, simulation, image processing, and visualization of bioelectric fields. Scientific Computing and Imaging Institute (SCI). URL www.sci.utah.edu/cibc/software/index.html.
4. BioTensor: A SCIRun Power App for processing and visualizing diffusion tensor images. Scientific Computing and Imaging Institute (SCI). URL www.sci.utah.edu/cibc/software/index.html.
5. FusionView: Problem solving environment for MHD visualization. Scientific Computing and Imaging Institute (SCI). URL software.sci.utah.edu/fusion.html.
6. Seg3D: Volumetric image segmentation and visualization. Scientific Computing and Imaging Institute (SCI). URL www.sci.utah.edu/cibc/software/index.html.
7. Ted D. Blacker. Paving: A new approach to automated quadrilateral mesh generation. *Intl. J. for Numerical Methods in Engineering*, 32:811–847, 1991.

8. Michael Brewer, Lori Freitag-Diachin, Patrick Knupp, Thomas Leurent, and Darryl J. Melander. The MESQUITE mesh quality improvement toolkit. In *Proc. 12th Intl. Meshing Roundtable*, pages 239–250. Sandia National Laboratories, September 2003.

9. W. A. Cook and W. R. Oakes. Mapping methods for generating three-dimensional meshes. *Computers In Mechanical Engineering*, CIME Research Supplement:67–72, August 1982.

10. MSC Software Corporation. *Finite Element Modeling*, volume 3 of *MSC.PATRAN Reference Manual*. MSC.Software, Los Angeles, CA, 2003.

11. The CUBIT geometry and mesh generation toolkit, Sandia National Laboratories. URL `cubit.sandia.gov`.

12. V. N. Parthasarathy et al. A comparison of tetrahedron quality measures. *Finite Elem. Anal. Des.*, 15:255–261, 1993.

13. Inc. Fluent. *Fluent FIMESH User's Manual*. Fluent, Inc.

14. P.J. Frey and P.-L. George. *Mesh Generation*. Hermes Science Publishing, Oxford & Paris, 2000.

15. P. Knupp. Achieving finite element mesh quality via optimization of the Jacobian matrix norm and associated quantities. *Intl. J. Numer. Meth. Engng.*, 48: 1165–1185, 2000.

16. P. Knupp. Algebraic mesh quality metrics for unstructured initial meshes. *Finite Elements in Analysis and Design*, 39:217–241, 2003.

17. Patrick Knupp. Next-generation sweep tool: A method for generating all-hex meshes on two-and-one-half dimensional geometries. In *Proc. 7th Intl. Meshing Roundtable*, pages 505–513. Sandia National Laboratories, October 1998.

18. Patrick Knupp and Scott A. Mitchell. Integration of mesh optimization with 3D all-hex mesh generation, LDRD subcase 3504340000, final report. SAND 99-2852, October 1999.

19. Patrick M. Knupp. Winslow smoothing on two-dimensional unstructured meshes. In *Proc. 7th Intl. Meshing Roundtable*, pages 449–457. Sandia National Laboratories, 1998.

20. Patrick M. Knupp. Hexahedral and tetrahedral mesh shape optimization. *Intl. J. for Numerical Methods in Engineering*, 58(1):319–332, 2003.

21. Patrick M. Knupp. Hexahedral mesh untangling and algebraic mesh quality metrics. In *Proc. 9th Intl. Meshing Roundtable*, pages 173–183. Sandia National Laboratories, October 2000.

22. Yuri Kuznetsov, Konstantin Lipnikov, and Mikhail Shashkov. Mimetic finite difference method on polygonal meshes. Technical Report LA-UR-03-7608, Los Alamos National Laboratories, 2003.

23. Mingwu Lai, Steven E. Benzley, Gregory D. Sjaardema, and Timothy J. Tautges. A multiple source and target sweeping method for generating all-hexahedral finite element meshes. In *Proc. 5th Intl. Meshing Roundtable*, pages 217–228. Sandia National Laboratories, October 1996.

24. Mingwu Lai, Steven E. Benzley, and David R. White. Automated hexahedral mesh generation by generalized multiple source to multiple target sweeping. *Intl. J. for Numerical Methods in Engineering*, 49(1):261–275, September 2000.

25. Mark Loriot. TetMesh-GHS3D v3.1 the fast, reliable, high quality tetrahedral mesh generator and optimiser. URL `www.simulog.fr/mesh/tetmesh3p1d-wp.pdf`.

26. Scott A. Mitchell. Choosing corners of rectangles for mapped meshing. In *13th Annual Symposium on Computational Geometry*, pages 87–93. ACM Press, June 1997.

27. Scott A. Mitchell. High fidelity interval assignment. In *Proc. 6th Intl. Meshing Roundtable*, pages 33–44. Sandia National Laboratories, October 1997.

28. A. Oddy, J. Goldak, M. McDill, and M. Bibby. A distortion metric for isoparametric finite elements. *Trans. CSME*, 38-CSME-32, 1988. Accession No. 2161.

29. P. P. Pébay. Planar quadrangle quality measures. *Engineering with Computers*, 20(2):157–173, 2004. URL dx.doi.org/10.1007/s00366-004-0280-8.

30. P. P. Pébay and T. J. Baker. Analysis of triangle quality measures. *AMS Mathematics of Computation*, 72(244):1817–1839, 2003. URL dx.doi.org/10.1090/S0025-5718-03-01485-6.

31. Combustion research group of University of Utah. C-Safe - Center for the Simulation of Accidental Fires and Explosions. URL www.csafe.utah.edu/.

32. J. Robinson. CRE method of element testing and the Jacobian shape parameters. *Eng. Comput.*, 4, 1987.

33. Larry A. Schoof and Victor R. Yarberry. EXODUS II: A finite element data model. Technical Report SAND92-2137, Sandia National Laboratories, 1992.

34. Scirun: A scientific computing problem solving environment. Scientific Computing and Imaging Institute (SCI). URL software.sci.utah.edu/scirun.html, 2007..

35. Michael A. Scott, Matthew N. Earp, Steven E. Benzley, and Michael B. Stephenson. Adaptive sweeping techniques. In *Proc. 14th Intl. Meshing Roundtable*, pages 417–432. Sandia National Laboratories, September 2005.

36. SDRC. *SDRC/IDEAS Simulation: Finite Element Modeling – User's Guide*. SDRC.

37. Jason F. Shepherd, Scott A. Mitchell, Patrick Knupp, and David R. White. Methods for multisweep automation. In *Proc. 9th Intl. Meshing Roundtable*, pages 77–87. Sandia National Laboratories, October 2000.

38. K.H. Shivanna, B.D. Adams, V.A. Magnotta, and N.M. Grosland. Towards automating patient-specific finite element model development. In K. Miller and D. Poulikakos, editors, *Computational Biomechanics For Medicine 2006*, pages 85–93. Medical Image Computing and Computer Assisted Intervention Conference, 2006.

39. C. Simpson, C. D. Ernst, P. Knupp, P. P. Pébay, and D. C. Thompson. *The Verdict Library Reference Manual*. Sandia National Laboratories, April 2007. URL www.vtk.org/Wiki/images/6/6b/VerdictManual-revA.pdf.

40. L. M. Taylor and D. P. Flanagan. Pronto3D - a three dimensional transient solid dynamics program. Technical Report SAND87-1912, Sandia National Laboratories, 1989.

41. David R. White. *Automatic Quadrilateral and Hexahedral Meshing of Pseudo-Cartesian Geometries Using Virtual Subdivision*. Published Master's Thesis, Brigham Young University, June 1996.

42. David R. White, Mingwu Lai, Steven E. Benzley, and Gregory D. Sjaardema. Automated hexahedral mesh generation by virtual decomposition. In *Proc. 4th Intl. Meshing Roundtable*, pages 165–176. Sandia National Laboratories, October 1995.

43. David R. White and Timothy J. Tautges. Automatic scheme selection for toolkit hex meshing. *Intl. J. for Numerical Methods in Engineering*, 49(1):127–144, September 2000.

Appendix: Summary of **Verdict** Quality Functions

Quality name	Dim.	Reference	Verdict function
area	L^2	–	v_tri_area
aspect Frobenius	1	[30]	v_tri_aspect_frobenius
aspect ratio	1	[30]	v_tri_aspect_ratio
condition	1	[15, 16]	v_tri_condition
distortion	1	[36] (adapt.)	v_tri_distortion
edge ratio	1	[30]	v_tri_edge_ratio
maximum included angle	A^1	–	v_tri_maximum_angle
minimum included angle	A^1	[30]	v_tri_minimum_angle
radius ratio	1	[30]	v_tri_radius_ratio
relative size squared	1	[16]	v_tri_relative_size_squared
scaled Jacobian	1	[15]	v_tri_scaled_jacobian
shape and size	1	[16]	v_tri_shape_and_size
relative size squared	1	[16]	v_tri_shape

Table 1. Verdict triangle quality functions.

Quality name	Dim.	Reference	Verdict function
area	L^2	–	v_quad_area
aspect ratio	1	[29]	v_quad_aspect_ratio
condition	1	[15]	v_quad_condition
distortion	1	[36]	v_quad_distortion
edge ratio	1	[29]	v_quad_edge_ratio
Jacobian	L^2	[15]	v_quad_jacobian
maximum aspect frobenius	1	[29]	v_quad_max_aspect_frobenius
maximum included angle	A^1	–	v_quad_maximum_angle
maximum edge ratio	1	[32]	v_quad_max_edge_ratio
mean aspect frobenius	1	[29]	v_quad_med_aspect_frobenius
minimum included angle	A^1	–	v_quad_minimum_angle
Oddy	1	[28]	v_quad_oddy
radius ratio	1	[29]	v_quad_radius_ratio
relative size squared	1	[16]	v_quad_relative_size_squared
scaled Jacobian	1	[15]	v_quad_scaled_jacobian
shape and size	1	[16]	v_quad_shape_and_size
shape	1	[16]	v_quad_shape
shear and size	1	[16]	v_quad_shear_and_size
shear	1	[16]	v_quad_shear
skew	1	[32] (adap.)	v_quad_skew
stretch	1	[13]	v_quad_stretch
taper	1	[32] (adap.)	v_quad_taper
warpage	1	–	v_quad_warpage

Table 2. Verdict quadrilateral quality functions.

Quality name	Dim.	Reference	Verdict function
aspect β	1	[12]	v_tet_aspect_beta
aspect Frobenius	1	[15]	v_tet_aspect_frobenius
aspect γ	1	[12]	v_tet_aspect_gamma
aspect ratio	1	[14]	v_tet_aspect_ratio
collapse ratio	1	[10]	v_tet_collapse_ratio
condition	1	[15]	v_tet_condition
distortion	1	[36] (adap.)	v_tet_distortion
edge ratio	1	–	v_tet_edge_ratio
Jacobian	L^3	[16]	v_tet_jacobian
minimum dihedral angle	A^1	–	v_tet_minimum_angle
radius ratio	1	[12]	v_tet_radius_ratio
relative size squared	1	[16]	v_tet_relative_size_squared
scaled Jacobian	1	[15]	v_tet_scaled_jacobian
shape and size	1	[16]	v_tet_shape_and_size
shape	1	[16]	v_tet_shape
volume	L^3	[12]	v_tet_volume

Table 3. Verdict tetrahedron quality functions.

Quality name	Dim.	Reference	Verdict function
diagonal	1	–	v_hex_diagonal
dimension	L^1	[40] (adap.)	v_hex_dimension
distortion	L^3	[36] (adap.)	v_hex_distortion
edge ratio	1	–	v_hex_edge_ratio
Jacobian	L^3	[15]	v_hex_jacobian
maximum aspect frobenius	1	[15]	v_hex_max_aspect_frobenius
maximum edge ratio	1	[40] (adap.)	v_hex_max_edge_ratio
mean aspect frobenius	1	–	v_hex_med_aspect_frobenius
Oddy	1	[28] (adap.)	v_hex_oddy
relative size squared	1	[16]	v_hex_relative_size_squared
scaled Jacobian	1	[15]	v_hex_scaled_jacobian
shape and size	1	[16]	v_hex_shape_and_size
shape	1	[16]	v_hex_shape
shear and size	1	[16]	v_hex_shear_and_size
shear	1	[16]	v_hex_shear
skew	1	[40] (adap.)	v_hex_skew
stretch	1	[13] (adap.)	v_hex_stretch
taper	1	[40] (adap.)	v_hex_taper
volume	L^3	–	v_hex_volume

Table 4. Verdict hexahedron quality functions.

Quality name	Dim.	Reference	Verdict function
pyramid volume	L^3	–	v_pyramid_volume
wedge volume	L^3	–	v_wedge_volume
knife volume	L^3	–	v_knife_volume

Table 5. Other Verdict quality functions.

Quality name	Accpt.	Normal	Full	Ideal
area	$[\,0, D_{\max}]$	$[\,0, D_{\max}]$	$[\,0, D_{\max}]$	$\frac{\sqrt{3}}{4}$
aspect Frobenius	$[\,1,\ 1.3]$	$[\,1, D_{\max}]$	$[\,1, D_{\max}]$	1
aspect ratio	$[\,1,\ 1.3]$	$[\,1, D_{\max}]$	$[\,1, D_{\max}]$	1
condition	$[\,1,\ 1.3]$	$[\,1, D_{\max}]$	$[\,1, D_{\max}]$	1
distortion	$[\,0.5,\ 1]$	$[\,0,\ 1]$	$[-D_{\max}, D_{\max}]$	1
edge ratio	$[\,1,\ 1.3]$	$[\,1, D_{\max}]$	$[\,1, D_{\max}]$	1
maximum included angle	$[\,60°,\ 90°]$	$[\,60°,\ 180°]$	$[\,0°,\ 180°]$	60°
minimum included angle	$[\,30°,\ 60°]$	$[\,0°,\ 60°]$	$[\,0°,\ 360°]$	60°
radius ratio	$[\,1,\ 3]$	$[\,1, D_{\max}]$	$[\,1, D_{\max}]$	1
relative size squared	$[\,0.25,\ 1]$	$[\,0,\ 1]$	$[\,0,\ 1]$	N/A
scaled Jacobian	$[\,0.2,\ \frac{2\sqrt{3}}{3}]$	$[-\frac{2\sqrt{3}}{3},\ \frac{2\sqrt{3}}{3}]$	$[-D_{\max}, D_{\max}]$	1
shape and size	$[\,0.2,\ 1]$	$[\,0,\ 1]$	$[\,0,\ 1]$	N/A
relative size squared	$[\,0.25,\ 1]$	$[\,0,\ 1]$	$[\,0,\ 1]$	1

Table 6. Properties of Verdict triangle quality functions.

Quality name	Accpt.	Normal	Full	Ideal
area	$[\,0, D_{\max}]$	$[\,0, D_{\max}]$	$[-D_{\max}, D_{\max}]$	1
aspect ratio	$[\,1,\ 1.3]$	$[\,1, D_{\max}]$	$[\,1, D_{\max}]$	1
condition	$[\,1,\ 4]$	$[\,1, D_{\max}]$	$[\,1, D_{\max}]$	1
distortion	$[\,0.5,\ 1]$	$[\,0,\ 1]$	$[-D_{\max}, D_{\max}]$	1
edge ratio	$[\,1,\ 1.3]$	$[\,1, D_{\max}]$	$[\,1, D_{\max}]$	1
Jacobian	$[\,0, D_{\max}]$	$[\,0, D_{\max}]$	$[-D_{\max}, D_{\max}]$	1
maximum aspect frobenius	$[\,1,\ 1.3]$	$[\,1, D_{\max}]$	$[\,1, D_{\max}]$	1
maximum included angle	$[\,90°,\ 135°]$	$[\,90°,\ 360°]$	$[\,0°,\ 360°]$	90°
maximum edge ratio	$[\,1,\ 1.3]$	$[\,1, D_{\max}]$	$[\,1, D_{\max}]$	1
mean aspect frobenius	$[\,1,\ 1.3]$	$[\,1, D_{\max}]$	$[\,1, D_{\max}]$	1
minimum included angle	$[\,45°,\ 90°]$	$[\,0°,\ 90°]$	$[\,0°,\ 360°]$	90°
Oddy	$[\,0,\ \frac{1}{2}]$	$[\,0, D_{\max}]$	$[\,0, D_{\max}]$	0
radius ratio	$[\,1,\ 1.3]$	$[\,1, D_{\max}]$	$[\,1, D_{\max}]$	1
relative size squared	$[\,0.3,\ 1]$	$[\,0,\ 1]$	$[\,0,\ 1]$	N/A
scaled Jacobian	$[\,0.2,\ 1]$	$[-1,\ 1]$	$[-1,\ 1]$	1
shape and size	$[\,0.2,\ 1]$	$[\,0,\ 1]$	$[\,0,\ 1]$	N/A
shape	$[\,0.2,\ 1]$	$[\,0,\ 1]$	$[\,0,\ 1]$	1
shear and size	$[\,0.2,\ 1]$	$[\,0,\ 1]$	$[\,0,\ 1]$	N/A
shear	$[\,0.3,\ 1]$	$[\,0,\ 1]$	$[\,0,\ 1]$	1
skew	$[\,0.5,\ 1]$	$[\,0,\ 1]$	$[\,0,\ 1]$	1
stretch	$[\,0.25,\ 1]$	$[\,0,\ 1]$	$[\,0, D_{\max}]$	1
taper	$[\,0,\ 0.7]$	$[\,0, D_{\max}]$	$[\,0, D_{\max}]$	0
warpage	$[\,0,\ 0.7]$	$[\,0,\ 2]$	$[\,0, D_{\max}]$	0

Table 7. Properties of Verdict quadrilateral quality functions.

Quality name	Accpt.	Normal	Full	Ideal
aspect β	[1, 3]	[1, D_{max}]	[1, D_{max}]	1
aspect Frobenius	[1, 1.3]	[1, D_{max}]	[1, D_{max}]	1
aspect γ	[1, 3]	[1, D_{max}]	[1, D_{max}]	1
aspect ratio	[1, 3]	[1, D_{max}]	[1, D_{max}]	1
collapse ratio	[0.1, D_{max}]	[0, D_{max}]	[0, D_{max}]	$\frac{\sqrt{6}}{3}$
condition	[1, 3]	[1, D_{max}]	[1, D_{max}]	1
distortion	[0.5, 1]	[0, 1]	[$-D_{max}$, D_{max}]	0
edge ratio	[1, 3]	[1, D_{max}]	[1, D_{max}]	1
Jacobian	[0, D_{max}]	[0, D_{max}]	[$-D_{max}$, D_{max}]	$\frac{\sqrt{2}}{2}$
minimum dihedral angle	[40°, α^*]	[0°, α^*]	[0°, 360°]	α^*
radius ratio	[1, 3]	[1, D_{max}]	[1, D_{max}]	1
relative size squared	[0.3, 1]	[0, 1]	[0, 1]	N/A
scaled Jacobian	[0.2, $\frac{\sqrt{2}}{2}$]	[$-\frac{\sqrt{2}}{2}$, $\frac{\sqrt{2}}{2}$]	[$-D_{max}$, D_{max}]	1
shape and size	[0.2, 1]	[0, 1]	[0, 1]	N/A
shape	[0.3, 1]	[0, 1]	[0, 1]	1
volume	[0, D_{max}]	[$-D_{max}$, D_{max}]	[$-D_{max}$, D_{max}]	N/A

Table 8. Properties of Verdict tetrahedron quality functions ($\alpha^* = \frac{180°}{\pi} \arccos \frac{1}{3} \approx 70.53°$).

Quality name	Accpt.	Normal	Full	Ideal
diagonal	[0.65, 1]	[0, 1]	[1, D_{max}]	1
dimension	[1, D_{max}]	[0, D_{max}]	[0, D_{max}]	1
distortion	[0.5, 1]	[0, 1]	[$-D_{max}$, D_{max}]	1
edge ratio	[1, 1.3]	[1, D_{max}]	[1, D_{max}]	1
Jacobian	[0, D_{max}]	[0, D_{max}]	[$-D_{max}$, D_{max}]	1
maximum aspect frobenius	[1, 3]	[1, D_{max}]	[1, D_{max}]	1
maximum edge ratio	[1, 1.3]	[1, D_{max}]	[1, D_{max}]	1
mean aspect frobenius	[1, 3]	[1, D_{max}]	[1, D_{max}]	1
Oddy	[0, $\frac{1}{2}$]	[0, D_{max}]	[0, D_{max}]	0
relative size squared	[0.5, 1]	[0, 1]	[0, 1]	N/A
scaled Jacobian	[0.2, 1]	[-1, 1]	[-1, D_{max}]	1
shape and size	[0.2, 1]	[0, 1]	[0, 1]	N/A
shape	[0.3, 1]	[0, 1]	[0, 1]	1
shear and size	[0.2, 1]	[0, 1]	[0, 1]	N/A
shear	[0.3, 1]	[0, 1]	[0, 1]	1
skew	[0, $\frac{1}{2}$]	[0, 1]	[0, D_{max}]	0
stretch	[0.25, 1]	[0, 1]	[0, D_{max}]	1
taper	[0, $\frac{1}{2}$]	[0, D_{max}]	[0, D_{max}]	0
volume	[0, D_{max}]	[0, D_{max}]	[$-D_{max}$, D_{max}]	N/A

Table 9. Properties of Verdict hexahedron quality functions.

Quality name	Accpt.	Normal	Full	Ideal
pyramid volume	[0, D_{max}]	[$-D_{max}$, D_{max}]	[$-D_{max}$, D_{max}]	N/A
wedge volume	[0, D_{max}]	[$-D_{max}$, D_{max}]	[$-D_{max}$, D_{max}]	N/A
knife volume	[0, D_{max}]	[$-D_{max}$, D_{max}]	[$-D_{max}$, D_{max}]	N/A

Table 10. Properties of other Verdict quality functions.

An Immersive Topology Environment for Meshing

Steven J. Owen[1], Brett W. Clark[1], Darryl J. Melander[1], Michael Brewer[1], Jason F. Shepherd[1], Karl Merkley[2], Corey Ernst[2], and Randy Morris[2]

[1] Sandia National Laboratories [†], Albuquerque, New Mexico
{sjowen,bwclark,djmelan,mbrewer,jfsheph}@sandia.gov
[2] Elemental Technologies, American Fork, Utah
{karl,corey,randy}@elemtech.com

Summary. The Immersive Topology Environment for Meshing (ITEM) is a wizard-like environment, built on top of the CUBIT Geometry and Meshing Toolkit. ITEM is focused on three main objectives: 1) guiding the user through the simulation model preparation workflow; 2) providing the user with intelligent options based upon the current state of the model; and 3) where appropriate, automating as much of the process as possible. To accomplish this, a *diagnostic-solution* approach is taken. Based upon diagnostics of the current state of the model, specific solutions for a variety of common tasks are provided to the user. Some of these tasks include geometry simplification, small feature suppression, resolution of misaligned assembly parts, decomposition for hex meshing, and source and target selection for sweeping. The user may scroll through a list of intelligent solutions for a specific diagnostic and entity, view a graphical preview of each solution and quickly perform the solution to resolve the problem. In many cases, automatic solutions for these tasks can be generated and executed if the user chooses. This paper will discuss the various diagnostics and geometric reasoning algorithms and approaches taken by ITEM to determine solutions for preparing an analysis model.

Key words: geometry simplification, sweep decomposition, design through analysis, hexahedra, sweeping, assembly meshing, imprint/merge, meshing user interface

1 Introduction

At a cursory inspection of the computational simulation process, the creation of a mesh may seem like a relatively trivial task. In most cases, significant energy and thought is put into the numerics for computing the physics of the

[†]Sandia is a multiprogram laboratory operated by Sandia Corporation, a Lockheed Martin Company, for the United States Department of Energy under Contract DE-AC04-94AL85000

system. A mesh may often be thought of as simply a means to represent the geometric domain of the system and its significance frequently diminished. Once the problems to be simulated advance beyond simple academic proto-types of blocks and cylinders, the true magnitude of the meshing problem readily becomes apparent. It is not unusual for the meshing process to take upwards of three-quarters of the entire simulation time. At Sandia National Labs, for instance, a survey[1] of analysts was conducted in 2005 to determine where the bulk of their time was being spent in modeling and simulation.

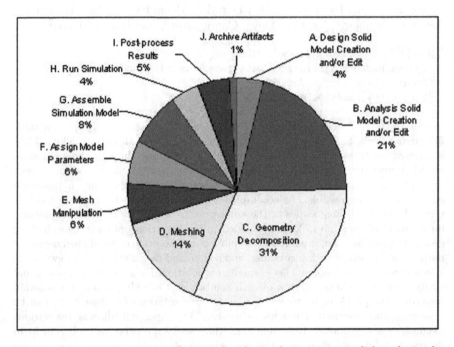

Fig. 1. Approximate percent of time taken by analysts to accomplish tasks in the modeling and simulation process at Sandia National Laboratories

Analysts were asked to quantify the amount of time they spend in each of 10 separate tasks. Analysts were selected from a wide variety of disciplines including modal, linear and non-linear structural, heat transfer, fluid flow and radiation transport. Figure 1 shows a summary of the results of this survey. Of significant note is the relatively large amount of time devoted to building the analysis solid model, geometry decomposition, meshing, and mesh manipula-tion (Tasks B through E). These tasks were reported to take 73% of the total time as compared to just 4% to actually run the simulation. These statistics illustrate where the major bottlenecks remain in the simulation process.

With the current state of the art, these tasks are inherently very user inter-active. Analysts at Sandia have access to almost any state-of-the-art modeling and simulation tools developed within the lab and available commercially. They often choose to use methods that perform better with a hexahedral mesh definition. While tetrahedral methods are adequate for many situations, specific advantages are frequently cited for using hexahedral meshes. State of the art meshing tools, such as CUBIT[3] are developed to support the meshing and geometry needs of the analysts at Sandia.

CUBIT uses a toolbox approach to providing a meshing solution. Incorporating state-of-the-art algorithms for quadrilateral, triangle, tetrahedral and hexahedral meshing it tries to address a diverse range of mesh generation needs from across many disciplines throughout the Laboratories. As a solid-model based system it allows the user the flexibility of importing existing CAD models from commercial tools such as Pro/Engineer[4] and Solidworks[5], but also includes many tools for generating a solid model directly within CUBIT. Models that have been developed in commercial solid modeling tools are rarely created with simulation in mind and must frequently be simplified. Geometric translation errors introduced as a result of incompatible modeling standards between commercial tools further complicate the model preparation and can be time consuming to address. Geometry decomposition, another time-consuming task, is also often needed to provide suitable topology for hexahedral meshing algorithms.

CUBIT, first developed at Sandia in the early 1990s as a research platform for new geometry and meshing research, has become the most used tool by Sandia's engineers for generating meshes for simulation on a day-to-day basis. It is also available world-wide through a government or academic use license as well as for commercial distribution. With CUBIT's many sophisticated and technically advanced tools developed for a wide range of application areas, it can be an unwieldy endeavor to become proficient enough with the software to quickly generate a mesh from a complex geometry. As a result, the *Immersive Topology Environment for Meshing* (ITEM) was developed.

With the ultimate goal of reducing the time to generate a mesh for simulation, ITEM has been developed within the CUBIT Geometry and Meshing Toolkit to take advantage of its extensive tool suite. Built on top of these tools it attempts to improve the user experience by accomplishing three main objectives:

1. Guiding the user through the workflow
2. Providing the user with smart options
3. Automating geometry and meshing tasks

---

[3]http://cubit.sandia.gov
[4]http://www.proengineer.com
[5]http://www.solidworks.com/

1.1 Guiding the user through the workflow

In software of any complexity where usage may be occasional or infrequent, the overhead of learning the new tool to a point of proficiency may be daunting. Given a solid model that may have been designed for manufacturing purposes, an analyst may be faced with generating a mesh. They may not be working with CUBIT on a daily basis, but would like to take advantage of the powerful tools provided by the software.

To address this, ITEM provides a wizard-like environment that steps the user through the geometry and meshing process. For someone unfamiliar with the software, it provides an interactive, step-by-step set of tools for accomplishing the major tasks in the process. For those more familiar with the tools, it serves as a reminder of the major tasks, but is flexible enough to accommodate a more iterative approach, allowing them to jump between major tasks easily. Currently restricting the workflow to models requiring three-dimensional, solid elements, ITEM uses the following steps:

1. Define the Geometric Model: Import a CAD model or create geometry within the CUBIT environment.
2. Set up the model: Define basic information such as element shape, volumes to be meshed and element sizes or budgets.
3. Prepare the geometry: Detect and remove unwanted geometric features on the CAD model, resolve problems with conformal assemblies and identify and provide suggestions to make the geometry sweepable.
4. Meshing: Perform the meshing operation and provide feedback if it is unsuccessful.
5. Validate the mesh: Check element quality and perform mesh improvement operations.
6. Apply boundary conditions regions: Define regions where boundary conditions may be applied using nodeset, sideset and block definitions.
7. Export the mesh: Define a target analysis code format and export the mesh.

1.2 Providing the user with smart options

Solid models used for analysis may have a huge variety of different characteristics that could prevent them from being easily meshed. Questions such as, "What are the problems associated with my model?", "What are the current roadblocks to generating a mesh on this model?" and "What should I do to resolve the problems?", are constantly being asked by the analysts as they work with models. Without an extensive knowledge of the tools and algorithms, it may be difficult to answer these questions effectively.

ITEM addresses this issue by providing smart options to the user. Based on the current state of the model, it will automatically run diagnostics and determine potential solutions that the user may consider. For example, where unwanted small features may exist in the model, ITEM will direct the user to these features and provide a range of geometric solutions to the problem. Scrolling through the solutions provides a preview of the expected result. The user can then select the solution that seems most appropriate and execute the solution to change or simplify the geometry. This *diagnostic-solution* approach is the basis for the ITEM design and is the common mode of user interaction while in this environment. This contrasts with the more traditional *hunt-and-guess* approach of providing the user with an array of buttons and icons from which they may choose and guessing what may result. On the other hand, ITEM serves in effect as an expert providing guidance to the user as they proceed through the geometry and meshing process.

To illustrate the *diagnostic-solution* approach, Figure 2 shows an example of one of the panels in the ITEM environment. In this panel a diagnostic is run to determine what volumes are not yet meshable based on the criteria for sweep meshing[19]. After selecting one of the volumes, a set of potential operations for decomposing the model is computed and presented to the user as a solution list. Selecting or browsing the solution list will preview the decomposition, shown to the right of the panel. Once satisfied with one of the solutions, the user may quickly perform the displayed operation, which in turn will update the diagnostics and display a new set of volumes for consideration.

1.3 Automating geometry and meshing tasks

With all of the advanced research and development that has gone into the meshing and geometry problem, a push-button solution for any arbitrary solid model may seem like the ideal objective of any meshing tool. Although for many cases this would be the best solution, for others it may not even be desirable. A push-button solution assumes a certain amount of trust in the geometric reasoning the software chooses to provide. This may be more trust than an occasional user who is tasked with a high consequence simulation may be willing to give. Even if the user is willing to accept full automation, in many cases, the geometric complexity of the model may be beyond the capability of current algorithms to adequately resolve.

Alternatively, once users are familiar with the characteristics of the solutions that the software provides, they may not be concerned with examining and intervening on every detail of the model creation process. Instead, in the interest of increasing efficiency, they may want the fastest solution possible. As a result, the idea of providing the option for full user automation while still maintaining the alternative for user control is a central objective of ITEM.

For various characteristic geometric problems that are encountered in a solid model, ITEM can determine from the potential geometric solutions, which may be most applicable and apply that solution without any user intervention. For many configurations of geometry, a completely automated solution may be available. For others, ITEM may be able to automate only

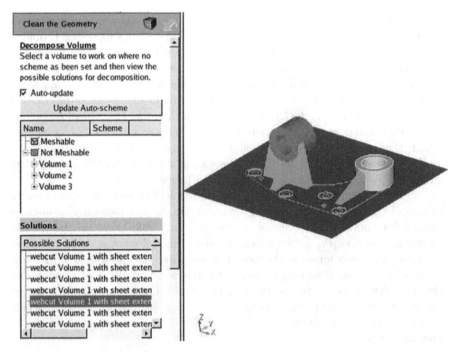

Fig. 2. Example of a GUI panel in ITEM illustrating the diagnostic-solution approach

a portion of the process. Where an adequate solution cannot be determined automatically, the smart options described above are available to help guide the user. As new advances in geometric reasoning and advanced meshing algorithms are developed, ITEM will incorporate these into the solutions for automation.

Although ITEM utilizes a variety of common meshing algorithms for meshing surfaces and solids, a full description of these methods is beyond the scope of this document. Instead, ITEM has primarily been designed to be *meshing algorithm independent*. The diagnostics and solutions proposed in ITEM are developed with the objective of being able to successfully utilize one or more of CUBIT's meshing schemes.

The remainder of this document highlights some of the key aspects of the analysis model preparation process and describes how they are addressed within the context of ITEM. These aspects focus primarily on preparing the geometric model for meshing, with some focus on final mesh quality. They include removing small features, resolving problems in a conformal assembly, building a sweepable topology, and improving mesh quality. For each of these aspects of model preparation, a description of the typical problems encountered will be defined along with proposed diagnotics that can detect these situations. Once a problem has been detected, the basic logic for determining a list of solutions to address the specific problems are outlined.

2 Geometry Clean Up

Meshing packages have the challenge of dealing with a host of geometry problems. Many of these problems can be generalized as file translation issues. Typically, the geometry used in a meshing package has not been created there but in one of many CAD packages. Exporting these files out of CAD and into a neutral file format (IGES, STEP, SAT) accepted by the meshing software can introduce misrepresentations in the geometry. If the CAD and meshing packages do not support the same file formats, a second translation may be necessary, possibly introducing even more problems.

Another complication caused by file translation is that of tolerances. Some CAD packages see two points as coincident if they are within 1e-3 units, while others use 1e-6. If the meshing software's tolerance is finer than the CAD package's, this disparity in tolerance can cause subsequent geometry modification operations in the meshing package to inadvertently create *sliver* features, which tend to be difficult and tedious to deal with. This tolerance problem also causes misalignment issues between adjacent volumes of assemblies, hindering the sharing of coincident geometry in order to produce a conformal mesh.

Modeling errors caused by the user in the CAD package is another problem that the meshing package has to correct. In the CAD package, the user may not create the geometry correctly, or there may simply be very detailed components causing some parts to overlap, or introduce small gaps between parts that should touch. Many times these problems are detected in the meshing package at a point when it is not feasible to simply go back into the CAD system and fix the problem, so the meshing package must be capable of correcting it.

Several approaches for addressing the geometry cleanup problem have been proposed in the literature [2, 3, 4, 5]. They typically provide operations that are automatically applied to the geometry once one or more topology problems have been identified. While very effective in many cases, they generally lack the ability for the user to have control over the resolution of these CAD issues while still maintaining the option for automation. The proposed environment provides tools to both diagnose these common issues and to provide a list of solutions from which the user may select that will correct the problems. The specific diagnostics and solutions for dealing with small features, whether created from geometry misrepresentations or inadvertently from imprinting are first addressed.

2.1 Small Feature Detection and Removal

The small feature removal area of ITEM focuses on identifying and removing small features in the model that will either inhibit meshing or force excessive mesh resolution near the small feature. Small features may result from translating models from one format to another or may be intentional design features. Regardless of the origin small features must often be removed in order to generate a high quality mesh.

ITEM will recognize small features that fall in four classifications: 1) small curves, 2) small surfaces, 3) narrow surfaces, and 4) surfaces with narrow regions. These operations may involve either real, virtual or a combination of both types of operations to remove these features. A virtual operation is one in which does not modify the CAD model, but rather modifies an overlay topology on the original CAD model. Real operations, on the other hand directly modify the CAD model. Where real operations are provided by the solid modeling kernel upon which CUBIT is built, virtual operations are provided by CUBIT's CGM [6] module and are implemented independently of the solid modeling kernel. The following describes the diagnostics for finding each of the four classifications of small features and the methods for removing them.

Small Curves

Diagnostic: Small curves are found by simply comparing each curve length in the model to a user-specified characteristic small curve size. A default ϵ is automatically calculated as 10 percent of the user specified mesh size, but can be overridden by the user.

Solutions: ITEM provides three different solutions for eliminating small curves from the model. The first solution uses a virtual operation to composite surfaces. Two surfaces near the small curve can often be composited together to eliminate the small curve as shown in Figure 3(a)

The second solution for eliminating small curves is the collapse curve operation. This operation combines partitioning and compositing of surfaces near the small curve to generate a topology that is similar to pinching the two ends of the curve together into a single point. The partitioning can be done either as a real or virtual operation. Figure 3(b) illustrates the collapse curve operation.

The third solution for eliminating small curves is the remove topology operation. This operation can be thought of as cutting out an area around the small curve and then reconstructing the surfaces and curves in the cut-out region so that the small curves no longer exist. [7] provides a detailed description of the remove topology operation. This operation has more impact on the actual geometry of the model because it redefines surfaces and curves in the vicinity of a small curve. The reconstruction of curves and surfaces is done

using real operations followed by composites to remove extra topology introduced during the operation. Figure 3(c) shows the results using the remove
topology operation.

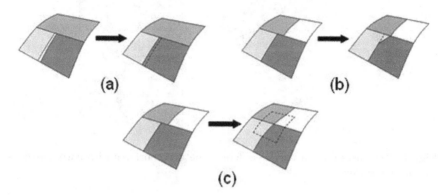

(a)

(b)

(c)

Fig. 3. Three operators used for removing small curves (a) composite; (b) collpase
curve; (c) remove topology

Small and Narrow Surfaces

ITEM also addresses the problem of small and narrow surfaces. Both are dealt
with in a similar manner and are described here.

Diagnostic: Small surfaces are found by comparing the surface area with a
characteristic *small area*. The characteristic small area is defined simply as
the characteristic small curve length squared or ϵ^2.

Narrow surfaces are distinguished from *surfaces with narrow regions* by
the characteristic that the latter can be split such that the narrow region
is separated from the rest of the surface. Narrow surfaces are themselves a
narrow region and no further splits can be done to separate the narrow region.
Figure 4 shows examples of each. ITEM provides the option to split off the
narrow regions, subdividing the surface so the narrow surfaces can be dealt
with independently.

Narrow regions/surfaces are also recognized using the characteristic value
of ϵ. The distance, d_i from the endpoints of each curve in the surface to the
other curves in the surface are computed and compared to ϵ. When $d_i < \epsilon$
other points on the curve are sampled to identify the beginning and end of
the narrow region. If the narrow region encompasses the entire surface, the
surface is classified as a narrow surface. If the region contains only a portion
of the surface, it is classified as a surface with a narrow region

Solutions: ITEM provides four different solutions for eliminating small and
narrow surfaces from the model. The first solution uses the regularize operation. Regularize is a real operation provided by the solid modeling kernel

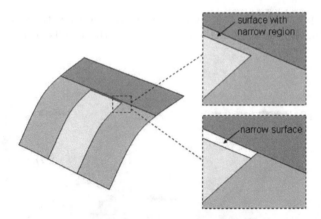

Fig. 4. Two cases illustrating the difference between surfaces with narrow regions and narrow surfaces

that removes unnecessary/redundant topology in the model. In many cases a small/narrow surface's normal may be the same as a surface next to it and therefore the curve between them is not necessary and can be regularized out. An example of regularizing a small/narrow surface out is shown in Figure 5.

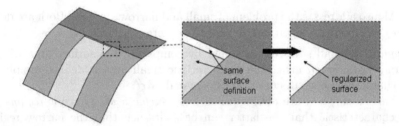

Fig. 5. When the small surface's underlying geometric definition is the same as a neighbor the curve between them can be regularized out.

The second solution for removing small/narrow surfaces uses the remove operation. Remove is also a real operation provided by the solid modeling kernel. However, it differs from regularize in that it doesn't require the neighboring surface(s) to have the same geometric definition. Instead the remove operation removes the specified surface from the model and then attempts to extend the existing adjacent surface definitions by computing new trimming curves at their intersections to close the volume. An example of using the remove solution is shown in Figure 6.

The third solution for removing small/narrow surfaces uses the virtual composite operation to composite the small surface with one of its neighbors.

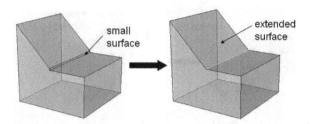

Fig. 6. The remove operation extends an adjacent surface to remove a small surface

This is very similar to the use of composites for removing small curves. An example is shown in Figure 7.

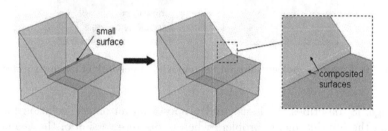

Fig. 7. Composite solution for removing a narrow surface

The final solution for removing small/narrow surfaces uses the remove topology operation[7]. The remove topology operation behaves the same as when used for removing small curves in that it cuts out the area of the model around the small/narrow surface and replaces it with a simplified topology. In the case of a small surface where all of the curves on the surface are smaller than the characteristic small curve length, the small surface is replaced by a single vertex. In the case of a narrow surface where the surface is longer than the characteristic small curve length in one of its directions, the surface is replaced with a curve. The remove topology operation can be thought of as a local dimensional reduction to simplify the topology. The remove topology operation can also be used to remove networks of small/narrow surfaces in a similar fashion. Examples of using the remove topology solution to remove small/narrow surfaces are shown in Figures 8 and Figure 9.

2.2 Resolving Problems with Conformal Assemblies

Where more than a single geometric volume is to be modeled, a variety of common problems may arise that must be resolved prior to mesh generation. These are typically a result of misaligned volumes defined in the CAD package or problems arising from the imprint and merge operations in the meshing

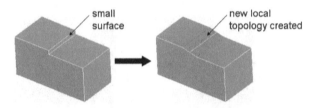

Fig. 8. Remove topology solution for removing a narrow surface

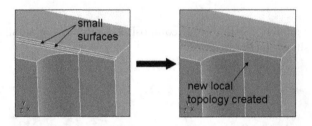

Fig. 9. Remove topology solution for removing a network of narrow surfaces

package. [8] describes the issues and proposes an automatic solution for resolving the imprint/merge problem where a discrete version of the geometry is used. ITEM addresses some of the same problems by allowing the option for user interaction as well as full automation using the CAD geometry representation. Two main diagnostics to detect potential problems are utilized: the misalignment check and overlapping surfaces check. Associated with both of these are solutions that are specific to the entity and from which the user may preview and select to resolve the problem.

Resolving Misaligned Volumes

Diagnostics: The *near coincident vertex check* or *misalignment check* is used to diagnose possible misalignments between adjacent volumes. This diagnostic is performed prior to the imprint operation in order to reduce the sliver surfaces and other anomalies which can occur as a result of imprinting misaligned volumes. With this diagnostic, the distance between pairs of vertices on different volumes are measured and flagged when they are just beyond the merge tolerance. The merge tolerance, T, is the maximum distance at which the geometry kernel will consider the vertices the same entity. A secondary tolerance T_s is defined where $T_s > T$ which is used for determining which pairs of vertices may also be considered for merging. Pairs of vertices whose distance d is $T < d > T_s$ are presented to the user, indicating areas in the model that may need to be realigned. Although not yet implemented at this writing, the misalignment check should also detect small distances between vertices and curves on adjacent volumes.

Solutions: When pairs of vertices are found that are slightly out of tolerance, the current solution is to move one of the surfaces containing one vertex of the pair to another surface containing the other vertex in the pair. Moving or extending a surface is known as *tweaking*. Solutions for determining which surfaces to tweak are generated as follows:

- Given that **vertex A** and **vertex B** are slightly outside of tolerance **T** by a distance δ as shown in Figure 11.
- Gather all surfaces that contain **vertex A**. Call this group of surfaces **Group A**.
- Gather all surfaces that contain **vertex B**. Call this group of surfaces **Group B**.
- For each surface in **Group A**, extend it out twice its size. Call this surface **extended A**
  - See if **extended A** overlaps within a distance > **T** and < δ to any surface in **Group B**.
  - If such an overlap pair is found, present two mutually exclusive solutions:
    - tweak **surface A** to **surface B**
    - tweak **surface B** to **surface A**

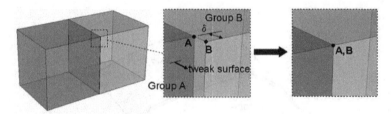

Fig. 10. Example of a solution generated to correct misaligned volumes using the tweak operator

The result of this procedure will be a list of possible solutions that will be presented to the users. They can then graphically preview the solutions and select the one that is most appropriate to correct the problem.

Correcting Merge Problems

The merge operation is usually performed immediately following imprinting and is also subject to occasional tolerance problems. In spite of correcting misalignments in the volume, the geometry kernel may still miss merging surfaces that may occupy the same space on adjacent volumes. If volumes in an assembly are not correctly merged, the subsequent meshes generated on the volumes will not be conformal. As a result, it is vital that all merging issues be resolved prior to meshing. The proposed environment provides a diagnostic and several solutions for addressing these issues.

Diagnostic: An overlapping surface check is performed to diagnose the failed sharing of topology between adjacent volumes. In contrast to the misalignment check, the check for overlapping surfaces is performed after the imprinting and merging operations. The overlapping surface check will measure the distance between surfaces on neighboring volumes to ensure that they are greater than the merge tolerance apart. Pairs of surfaces that failed to merge and that are closer than the merge tolerance are flagged and displayed to the user as potential problems.

Solutions: If imprinting and merging has been performed and a subsequent overlapping surface check finds overlapping surface pairs, the user may be offered three different options for correcting the problem: force merge, tolerant imprint of vertex locations and tolerant imprint of curves.

If the topology for both surfaces in the pair is identical, the force merge operation can generally be utilized. The merge operation will remove one of the surface definitions in order to share a common surface between two adjacent volumes. Normally this is done only after topology and geometry have been determined to be identical, however the force merge will bypass the geometry criteria and perform the merge. Figure 11 shows a simple example where the bounding vertices are identical but the surface definitions are slightly different so that the merge operation fails. Force merge in this case would be an ideal choice.

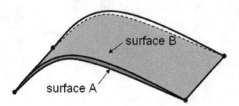

Fig. 11. Example where the merge operation will fail, but force merge will be successful

The force merge operation is presented as a solution where a pair of overlapping surfaces are detected and if any of the following criteria are satisfied:

- All curves of both surfaces are merged
- All vertices between the two surfaces are merged and all the curves are coincident to within 1% of their length or 0.005, whichever is larger
- All the curves of both surfaces are either merged or overlapping and a vertex of any curve of one surface that will imprint onto any other curve of the other surface cannot be identified
- At least one curve of one surface may be imprinted onto the other and if both surfaces have an equal number of curves and vertices, and the overlapping area between the 2 surfaces is more than 99% of the area of each surface. This situation generally prevents generating sliver surfaces

- At least one vertex of **surface** B may be imprinted onto **surface** A, and if both surfaces have equal number of curves and vertices, and the vertex(s) of **surface** B to imprint onto **surface** A lies too *close* to any vertices of **surface** A
- All the curves of both surfaces are either merged or overlapping and no vertices of any curve of **surface** A will imprint onto any other curve of **surface** B

Individual vertices may need to be imprinted in order to accomplish a successful merge. The solution of imprinting a position x,y,z onto surface A or B is presented to the user if the following criteria is met

- Curves between the two surfaces overlap within tolerance, and a vertex of **curve** A lies within tolerance to **curve** B and outside tolerance to any vertex of **curve** B. Tolerance is 0.5% of the length of the smaller of the 2 curves or the merge tolerance (0.0005), whichever is greater.

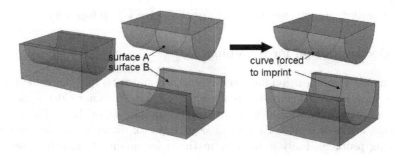

Fig. 12. Curve on **surface** A was not imprinted on **surface** B due to tolerance mismatch. Solution is defined to detect and imprint the curve

In some cases one or more curves may not have been correctly imprinted onto an overlapping surface which may be preventing merging. This may again be the result of a tolerance mismatch in the CAD translation. If this situation is detected a tolerant imprint operation may be performed which will attempt to imprint the curve onto the adjacent volume. Figure 12 shows an example where a curve on **surface** A is forced to imprint onto **surface** B usign tolerant imprint, because it did not imprint during normal imprinting. The solution of a curve of **surface** A to be imprinted onto **surface** B may be presented to the user if all 3 of the following conditions are satisfied:

- there are no vertices to imprint onto the owning volume of either surface
- curve of **surface** A is not overlapping another curve of **surface** B
- curve of **surface** A passes tests to ensure that it is really ON **surface** B

3 Building a Sweepable Topology

The hex meshing problem presents a number of additional challenges to the user that tetrahedral meshing does not. Where a good quality tetrahedral mesh can generally be created once small features and imprint/merge problems have been addressed, the hexahedral meshing problem poses additional topology constraints which must be met. Although progress has been made in automating the hex meshing process, the most robust meshing algorithms still rely on geometric primitives. Mapping [9] and sub-mapping [10] algorithms rely on parametric cubes and sweeping [11, 12] relies on logical extrusions. Most real world geometries do not automatically fit into one of these categories so the topology must be changed to match the criteria for one of these meshing schemes. ITEM addresses the hex meshing topology problem through three primary diagnostic and solution mechanisms.

1. Detecting and suggesting decomposition operations
2. Recognizing nearly sweepable topologies and suggesting source-target pairs
3. Detecting and compositing surfaces to force a sweep topology.

3.1 Decomposition for Sweeping

Automatic decomposition has been researched and tools have been developed which have met with some limited success [13, 14]. Automatic decomposition requires complex feature detection and sub-division algorithms. The decomposition problem is at least on the same order of difficulty as the auto-hex meshing problem. Fully automatic methods for quality hexahedral meshing have been under research and development for many years [15, 16, 17]. However, a method that can reliably generate hexahedral meshes for arbitrary volumes, without user intervention and that will build meshes of an equivalent quality to mapping and sweeping techniques, has yet to be realized. Although fully automatic techniques continue to progress [18], the objective of the proposed environment is to reduce the amount of user intervention required while utilizing the tried and true mapping and sweeping techniques as its underlying meshing engine.

Instead of trying to solve the all-hex meshing problem automatically, the ITEM approach to this problem is to maintain user interaction. The ITEM algorithms determine possible decompositions and suggest these to the user. The user can then make the decision as to whether a particular cut is actually useful. This process helps guide new users by demonstrating the types of decompositions that may be useful. It also aids experienced users by reducing the amount of time required to set up decomposition commands.

Diagnostics: The current diagnostic for determining whether a volume is mappable or sweepable is based upon the autoscheme tool described in [19]. Given a volume, the autoscheme tool will determine if the topology will admit a mapping, sub-mapping or sweeping meshing scheme. For volumes where a scheme cannot be adequately determined, a set of decomposition solutions are generated and presented to the user.

Solutions: The current algorithm for determining possible cut locations is based on the algorithm outlined in [13] and is described here for clarity:

- Find all curves that form a dihedral angle less than an input value (currently 135°)
- Build a graph of these curves to determine connectivity
- Find all curves that form closed loops
- For each closed loop
  - Find the surfaces that bound the closed loop
  - Save the surface
  - Remove the curves in the closed loop from the processing list
- For each remaining curve
  - Find the open loops that terminates at a boundary
  - For each open loop
    - Find the surfaces that bound the open loop
    - Save the surfaces
- For each saved surface
  - Create an extension of the surface
  - Present the extended surface to the user as a possible decomposition location.

This relatively simple algorithm detects many cases that are useful in decomposing a volume. Future work will include determining symmetry, sweep, and cylindrical core decompositions. These additional decomposition options should increase the likelihood of properly decomposing a volume for hexahedral meshing.

Figure 13 shows an example scenario for using this tool. The simple model at the top is analyzed using the above algorithm. This results in several different solutions being offered to the user, three of which are illustrated here. As each of the options is selected, the extended cutting surface is displayed providing rapid feedback to the user as to the utility of the given option. Note that all solutions may not result in a volume that is closer to being successfully hex-meshed. Instead the system relies on some user understanding of the topology required for sweeping. Each time a decomposition solution is selected and performed, additional volumes may be added, which will in turn be analyzed by the autoscheme diagnostic tool. This interactive process continues until the volume is successfully decomposed into a set of volumes which are recognized as either mappable or sweepable.

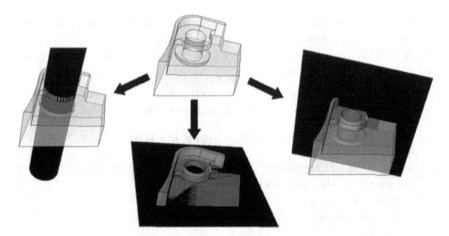

Fig. 13. ITEM decomposition tool shows 3 of the several solutions generated that can be selected to decompose the model for hex meshing

3.2 Recognizing Nearly Sweepable Regions

The purpose of geometry operations such as decomposition is to transform an unmeshable region into one or more meshable regions. However, even the operations suggested by the decomposition tool can degenerate into guesswork if they are not performed with a specific purpose in mind. Without a geometric goal to work toward, it can be difficult to recognize whether a particular operation will be useful.

Incorporated within the proposed ITEM environment are algorithms that are able to detect geometry that is nearly sweepable, but which are not fully sweepable due to some geometric feature or due to incompatible constraints between adjacent sections of geometry. By presenting potential sweeping configurations to the user, ITEM provides suggested goals to work towards, enabling the user to make informed decisions while preparing geometry for meshing.

Unlike the decomposition solutions presented in the previous section, the purpose of recognizing nearly sweepable regions is to show potential alternative source-target pairs for sweeping even when the autoscheme tool does not recognize the topology as strictly sweepable. When combined with the decomposition solutions and the forced sweep capability described later, it provides the user with an additional powerful strategy for building a hexahedral mesh topology.

Diagnostics: In recognizing nearly sweepable regions, the diagnostic tool employed is once again the autoscheme tool described in [19]. Volumes that do not meet the criteria defined for mapping or sweeping are presented to the user. The user may then select from these volume for which potential source-target pairs are computed.

Solutions: The current algorithm for determining possible sweep configurations is an extension of the autoscheme algorithm described in [19]. Instead of rejecting a configuration which does not meet the required sweeping constraints, the sweep suggestion algorithm ignores certain sweeping roadblocks until it has identified a nearly feasible sweeping configuration. The suggestions are presented graphically, as seen in Figure 14(a). In most cases, the source-target pairs presented by the sweep suggestion algorithm are not yet feasible for sweeping given the current topology. The user may use this information for further decomposition or to apply solutions identified by the forced sweepability capability described next. The sweep suggest algorithm also provides the user with alternative feasible sweep direction solutions as shown in Figure 14(b). This is particularly useful when dealing with interconnected volumes where sweep directions are dependent on neighboring volumes.

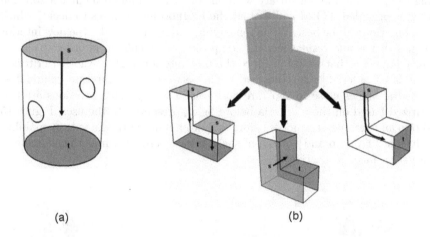

(a)   (b)

Fig. 14. (a) ITEM displays the source and target of a geometry that is nearly sweepable. The region is not currently sweepable due to circular imprints on the side of the cylinder. (b) Alternative feasible sweep directions are also computed

3.3 Forced Sweepability

In some cases, decomposition alone is not sufficient to provide the necessary topology for sweeping. The forced sweepability capability attempts to force a model to have sweepable topology given a set of source and target surfaces. The source-target pairs may have been identified manually by the user, or defined as one the solutions from the sweep suggestion algorithm described above. All of the surfaces between source and target surfaces are referred to as linking surfaces. Linking surfaces must be mappable or submappable in order for the sweeping algorithm to be successful. There are various topology configurations that will prevent linking surfaces from being mappable or submappable.

Diagnostics: The first check that is made is for small curves. Small curves will not necessarily introduce topology that is not mappable or submappable but will often enforce unneeded mesh resolution and will often degrade mesh quality as the mesh size has to transition from small to large. Next, the interior angles of each surface are checked to see if they deviate far from 90° multiples. As the deviation from 90° multiples increases the mapping and submapping algorithms have a harder time classifying corners in the surface. If either of these checks identify potential problems they are flagged and potential solutions are generated.

Solutions: If linking surface problems are identified ITEM will analyze the surface and generate potential solutions for resolving the problem. Compositing the problem linking surface with one of its neighbors is a current solution that is provided. ITEM will look at the neighboring surfaces to decide which combination will be best. When remedying bad interior angles the new interior angles that would result after the composite are calculated in order to choose the composite that would produce the best interior angles. Another criterion that is considered is the dihedral angle between the composite candidates. Dihedral angles close to 180° are desirable. The suggested solutions are prioritized based on these criteria before being presented to the user. Figure 15 shows an example of a model before and after running the forced sweepability solutions. The top and bottom of the cylinder were chosen as the source and target surfaces respectively.

Fig. 15. Non-submappable linking surface topology is composited out to force a sweepable volume topology

4 Mesh Quality

Advancements in the mesh generation algorithms have significantly reduced the amount of quality problems seen in the initially generated mesh. Further, ITEM generally relies on the most robust meshing algorithms available in CUBIT, specifically sweeping for hexahedral mesh generation[12] and the Tetmesh-GHS3D[20] meshing software[6]. However, some problems can still exist, and therefore ITEM has integrated quality diagnostics and solution options.

Diagnostics: After the mesh has been generated, the user may choose to perform element quality checks. ITEM utilizes the Verdict[21] library where a large number of mesh quality metrics have been defined and available as a modular library. If no user preference is specified, ITEM uses the Scaled Jacobian distortion metric to determine element quality. This check will warn users of any elements that are below a default or user-specified threshold, allowing various visualization options for displaying element quality.

Solutions: If the current element quality is unacceptable, ITEM will present several possible mesh improvement solutions. The most promising solutions are provided through ITEM's interface to two smoothers: mean ratio optimization and Laplacian smoothing. These are provided as part of the Mesquite[22] mesh quality improvement tool built within CUBIT. The user has the option of performing these improvements on the entire mesh, subsets of the mesh defined by the element quality groups, or on individual elements. The Laplacian smoothing scheme allows the users to smooth just the interior nodes or to simultaneously smooth both the interior and boundary nodes in an attempt to improve surface element quality.

5 Conclusion

A new approach to presenting the problem of preparing a finite element mesh to an intermittent user of modeling and simulation technology has been proposed. The Immersive Topology Environment for Meshing (ITEM) addresses a wide range of problems and issues commonly encountered during this process. Its intent is to reduce the learning, and re-learning often associated with complex software tools and to ultimately reduce the time to mesh. This is accomplished through a step-by-step wizard-like approach where users may address common problems by using built-in diagnostics and are then presented with specific intelligent solutions to these problems.

Table 1 summarizes the problems addressed by the proposed environment along with associated diagnostics and solutions. Details of each of the diagnostics and solutions are discussed within the body of the paper.

At this writing the proposed ITEM environment is still under development with plans for release shortly. The current set of diagnostics and solutions defined in Table 1 represent a reasonable set of tools for preparing models for analysis, however it is recognized that these tools will be modified, tuned and expanded based on user feedback and experience and as new technology is developed.

---

[6]http://www.distene.com

Problem	Diagnostic	Solutions
Small Curves	Curve length $< \epsilon$	1. composite surfaces 2. collapse curve 3. remove topology
Small Surfaces	Surface area $< \epsilon^2$	1. regularize
Narrow Surfaces	$d_i < \epsilon$ for all curves on surface	2. remove/extend surfaces 3. composite surfaces 4. remove topology
Surfaces with Narrow Regions	$d_i < \epsilon$ for some curves on surface	1. split off narrow region and treat as narrow surface
Misaligned volumes	Near coincident vertex or misalignment check	1. tweak surf A to surf B 2. tweak surf B to surf A
Unmerged surfaces	Overlapping surfaces check	1. force merge 2. imprint vertices 3. imprint curves
Non-sweepable/mappable topology	Autoscheme tool	1. cut locations based upon dihedral angles and connectivity graph
Nearly sweepable	Autoscheme tool + sweep suggestions	1. suggested source/target pairs
Linking surfaces not mappable	Linking surfaces: 1. Curve length $< \epsilon$ 2. Interior angles deviate significantly from $90°$	1. composite surfaces
Poor mesh quality	Quality metric $<$ threshold	Mean Ratio or Laplacian smoothing applied to: 1. entire mesh 2. element quality group 3. individual elements

Table 1. Summary of problems and associated diagnostics and solutions that are addressed with ITEM

Prior to release, extensive user testing will be performed in order to determine the impact that ITEM has on the time to mesh. This will include a series of prescribed models that will be meshed by several intermittent users of meshing software. Metrics will be gathered comparing times to complete the mesh in ITEM compared with previous technology. Understanding the difficulty of accurately measuring the time to mesh, the testing and metrics gathering procedure will attempt to control for factors including learning, user expertise and model complexity. These factors are outlined as follows:

1. *Learning*: Much of the meshing procedure involves a trial and error process of learning a strategy for model cleanup and decomposition. The second time a user attempts to mesh a model, they will very likely be able to be more efficient regardless of which system they use. As a result, a single tester will not attempt the same model more than once regardless of which system they use. The order in which they use ITEM and the previous technology will also be interspersed to randomize the effect of learning one system over the other.

2. *User Expertise*: Depending on how much experience a particular user has with a specific software system will effect how quickly they can complete a task compared to others with much less experience. An attempt will be made to enlist analysts with equivalent experience, however the timed results for any one particular model will be averaged across all users to reduce the effect of user expertise.

3. *Model Complexity*: Very complex models will inherently take longer to prepare and mesh than models of less complexity. Averaging the time taken for all models for any one user should reduce the effect of model complexity.

Since many diverse human factors may be involved, it is clear that any solution to gather metrics to guage improved time to mesh will be flawed in some way. It is however healthy and important to implement these measurements to independently measure the effectiveness of ITEM or any proposed system claiming to reduce the time to mesh. The results will ultimately provide insights and new input for further improvement.

While it is recognized that it is still a work in progress, the main contribution of the current work includes the infrastructure proposed for presenting and managing the model preparation process and its potential impact on reducing the time to generate analysis models.

References

1. Hardwick, Mike (2005) DART System Analysis Presented to Simulation Sciences Seminar, Sandia National Laboratories, Albuquerque, NM June 28, 2005
2. Tautges, Timothy J. (2001), Automatic Detail Reduction for Mesh Generation Applications, *Proceedings, 10th International Meshing Roundtable*, pp.407-418

3. Sheffer, A. (2001), Model simplification for meshing using face clustering, *Computer-Aided Design*, Vol. 33, No. 13, pp. 925-934(10)

4. Butlin, Geoffrey and Clive Stops (1996) CAD Data Repair, *5th International Meshing Roundtable*, pp.7-12

5. Mezentsev, Andrey A. (1999) Methods and Algorithms of Automated CAD Repair For Incremental Surface Meshing, *Proceedings, 8th International Meshing Roundtable*, pp.299-309

6. Tautges, Timothy J. (2000) The Common Geometry Module (CGM): A Generic, Extensible Geometry Interface, *Proceedings, 9th International Meshign Roundtable*, pp. 337-348

7. Clark Brett W. (2007) Removing Small Features with Real Solid Modeling Operations, Submitted to *16th International Meshing Roundtable*

8. White, David R. and Sunil Saigal (2002) Improved Imprint and Merge for Conformal Meshing, *Proceedings, 11th International Meshing Roundtable*, pp.285-296,

9. Cook, W. A. and W. R. Oakes (1982) Mapping methods for generating three-dimensional meshes, *Computers In Mechanical Engineering, CIME Research Supplement*:67-72, August 1982

10. Whiteley, M., D. White, S. Benzley and T. Blacker (1996) Two and Three-Quarter Dimensional Meshing Facilitators, *Engineering with Computers*, Vol 12, pp. 155-167

11. Knupp, Patrick M. (1998) Next-Generation Sweep Tool: A Method For Generating All-Hex Meshes On Two-And-One-Half Dimensional Geomtries, *Proceedings, 7th International Meshing Roundtable*, pp. 505-513

12. Scott, Michael A., Matthew N. Earp, Steven E. Benzley and Michael B. Stephenson (2005) Adaptive Sweeping Techniques, *Proceedings, 14th International Meshing Roundtable*, pp. 417-432

13. Yong Lu, Rajit Gadh, and Timothy J. Tautges (1999) Volume decomposition and feature recognition for hexahedral mesh generation, *Proceedings, 8th International Meshing Roundtable*, pp. 269-280

14. Staten, Matthew L., Steven J. Owen, Ted D. Blacker (2005) Unconstrained Paving & Plastering: A New Idea for All Hexahedral Mesh Generation, *Proceedings, 14th International Meshing Roundtable*, pp.399-416

15. Blacker T.D. and Meyers R.J. (1993) Seams and Wedges in Plastering:A 3D Hexahedral Mesh Generation Algorithm, *Engineering with Computers*, Vol. 2, No. 9, pp. 83-93

16. Folwell, Nathan T. and Scott A. Mitchell (1998) Reliable Whisker Weaving via Curve Contraction, *Proceedings, 7th International Meshing Roundtable*, pp.365-378

17. Price, M.A. and C.G. Armstrong (1995) Hexahedral Mesh Generation by Medial Surface Subdivision: Part I, Solids With Convex Edges, *International Journal for Numerical Methods in Engineering*, Vol. 38, No. 19, pp. 3335-3359

18. Staten, Matthew L., Robert A. Kerr, Steven J. Owen, Ted D. Blacker (2006) Unconstrained Paving and Plastering: Progress Update, *Proceedings, 15th International Meshing Roundtable*, pp.469-486

19. White, David R. and Timothy J. Tautges (2000) Automatic scheme selection for toolkit hex meshing, *International Journal for Numerical Methods in Engineering*, Vol. 49, No. 1, pp. 127-144

20. George, P.L., F. Hecht and E. Saltel (1991) Automatic Mesh Generator with Specified Boundary, *Computer Methods in Applied Mechanics and Engineering*, Vol. 92, pp. 269-288

21. Stimpson, CJ, Ernst, CD, Knupp, P, Pebay; P, and Thompson, D. (2007) The Verdict Geometric Quality Library, *Sandia Report SAND2007-1751*

22. Brewer, Michael, Lori Freitag Diachin, Patrick Knupp, Thomas Leurent and Darryl Melander (2003) The Mesquite Mesh Quality Improvement Toolkit, *Proceedings, 12th International Meshing Roundtable*, pp. 239-250

CUBIT and Seismic Wave Propagation Based Upon the Spectral-element Method: An Advanced Unstructured Mesher for Complex 3D Geological Media

Emanuele Casarotti[1,5], Marco Stupazzini[2], Shiann Jong Lee[3], Dimitri Komatitsch[4], Antonio Piersanti[5], and Jeroen Tromp[6]

[1] Seismological Laboratory, California Institute of Technology. 1200 E. California Blvd., MS 252-21, Pasadena, California, 91125, USA.
emanuele@gps.caltech.edu
[2] Department of Earth- and Environmental Sciences, Geophysics section, Ludwig-Maximilians-Universität. Theresienstrasse 41, 80333, Munich, Germany.
stupa@geophysik.uni-muenchen.de
[3] Institute of Earth Science, Academia Sinica. No.128, Section2, Academia Road, Nankang, Taipei, Taiwan 115. sjlee@earth.sinica.edu.tw
[4] Laboratoire de Modlisation et d'Imagerie en Gosciences UMR 5212, Universit de Pau et des Pays de l'Adour. Btiment IPRA - Avenue de l'Universit, BP 1155, 64013, Pau, France. dimitri.komatitsch@univ-pau.fr
[5] Istituto Nazionale di Geofisica e Vulcanologia. Via di Vigna Murata 605, 00143, Rome, Italy. piersanti@ingv.it
[6] Seismological Laboratory, California Institute of Technology. 1200 E. California Blvd., MS 252-21, Pasadena, California, 91125, USA. jtromp@gps.caltech.edu

Abstract

Unstructured hexahedral mesh generation is a critical part of the modeling process in the Spectral-Element Method (SEM). We present some examples of seismic wave propagation in complex geological models, automatically meshed on a parallel machine based upon CUBIT (Sandia Laboratory, cubit.sandia.gov), an advanced 3D unstructured hexahedral mesh generator that offers new opportunities for seismologist to design, assess, and improve the quality of a mesh in terms of both geometrical and numerical accuracy. The main goal is to provide useful tools for understanding seismic phenomena due to surface topography and subsurface structures such as low wave-speed sedimentary basins. Our examples cover several typical geophysical problems: 1) "layer-cake" volumes with high-resolution topography and complex solid-solid interfaces (such as the Campi Flegrei Caldera Area in Italy), and 2) models with an embedded sedimentary basin (such as the Taipei basin in Taiwan or the Grenoble Valley in France).

1 The spectral-element method in seismology

Wave propagation phenomena can nowadays be studied thanks to many powerful numerical techniques. We have seen rapid advances in computational seismology at global, regional, and local scales thanks to various numerical schemes, such as the finite-difference method (FDM) [15], the finite-element method (FEM) [1], the spectral-element method (SEM) [12], the integral boundary-element method (IBEM) [2], and the Arbitrary High-Order Discontinuous Galerkin method (ADER-DG) [8]. Spurred by the computational power made available by parallel computers, geoscientists and engineers can now accurately compute synthetic seismograms in realistic 3D Earth models. Among the methods previously listed, the SEM has convincingly demonstrated the ability to handle high-resolution simulations of seismic wave propagation in 3D domains.

The purpose of this article is not to give a complete description of the SEM, but a basic introduction to the main properties of the SEM is needed in order to understand the various constraints imposed on the meshing process. The SEM is as a generalization of the FEM based on the use of high-order piecewise polynomial functions. A crucial aspect of the method is its capability to provide an arbitrary increase in accuracy by simply increasing the algebraic degree of these functions (the spectral degree n). From a practical perspective, this operation is completely transparent to the users, who limit themselves to choosing the spectral degree at runtime, leaving to the software the task of building up suitable quadrature points and the relevant degrees of freedom. Obviously, one can also use mesh refinement to improve the accuracy of the numerical solution, thus following the traditional finite-element approach.

Referring to the literature for further details [12], we begin by briefly summarizing the key features of the SEM. We start from the general differential form of the equation of elastodynamics:

$$\rho\partial_t^2\mathbf{s} = \nabla \cdot \mathbf{T} + \mathbf{F}, \qquad (1)$$

where $\mathbf{s}(\mathbf{x}, t)$ denotes the displacement at position $\mathbf{x}$ and time t, $\rho(\mathbf{x})$ is the density distribution, and $\mathbf{F}(\mathbf{x}, t)$ the external body force. The stress tensor $\mathbf{T}$ is related to the strain tensor by the constitutive relation $\mathbf{T} = \mathbf{c} : \nabla\mathbf{s}$, where $\mathbf{c}$ denotes a fourth-order tensor. No particular assumptions are made regarding the structure of $\mathbf{c}$, which describes the (an)elastic properties of the medium (the formulation is general and can incorporate full anisotropy [9] or non-linear constitutive relationships [6]).

For seismological applications, a numerical technique needs to facilitate at least the following: (i) an attenuating medium, (ii) absorption of seismic energy on the fictitious boundaries of the domain in order to mimic a semi-infinite medium (the free-surface condition is a natural condition in the SEM), and finally (iii) seismic sources. In the SEM all these features are readily accommodated [12].

As in the FEM, the dynamic equilibrium problem for the medium can be written in a weak, or variational, form, and, through a suitable discretization procedure that depends on the numerical approach, can be cast as a system of ordinary differential equations with respect to time. Time-marching of this system of equations may be accomplished based upon an explicit second-order finite-difference scheme, which is conditionally stable and must satisfy the well known Courant condition [4].

The key features of the SEM discretization are as follows:

1. Like in the FEM, the model volume Ω is subdivided into a number of non-overlapping elements $\Omega_e, e = 1, \ldots, n_e$, such that $\Omega = \bigcup_{e=1}^{n_e} \Omega_e$.
2. The expansion of any function within the elements is accomplished based upon Lagrange polynomials of suitable degree n constructed from $n + 1$ interpolation nodes.
3. In each element, the interpolation nodes are chosen to be the Gauss-Lobatto-Legendre (GLL) points, i.e., the $n+1$ roots of the first derivatives of the Legendre polynomial of degree n. On such nodes, the displacement, its spatial derivatives, and integrals encountered in the weak formulation are evaluated.
4. The spatial integration is performed based upon GLL quadrature (while most classical FEMs use Gauss quadrature).

Thanks to this numerical strategy, exponential accuracy of the method is ensured and the computational effort minimized, because the resulting mass matrix is exactly diagonal. The latter feature does not occur in so-called hp FEMs, nor in SEMs based upon Chebychev polynomials.

2 Mesh design for Spectral-Element Methods

The first critical ingredient required for a spectral-element simulation is a high-quality mesh appropriate for the 3D model of interest. This process generally requires discouraging expertise in meshing and preprocessing, and is subject to several tough constraints: 1) the number of grid points per shortest desired wavelength, 2) the numerical stability condition, 3) an acceptable distortion of the elements, 4) balancing of numerical cost and available computing resources. A poor-quality mesh can generate numerical problems that lead to an increase in the computational cost, poor (or lack of) convergence of the simulation, or inaccurate results. For example, a geological model often includes a layered volume, and a staircase sampling of the interfaces between the layers can produce fictitious diffractions. Therefore, a good mesh should honor at least the major geological discontinuities of the model.

As noted in Section 1, the SEM is similar to a high-degree FEM, and in fact these methods share the first part of the discretization process. The present paper is focused on this first meshing step, and thus the results are relevant to both FEMs and SEMs. In the SEM, the subsequent step is the evaluation of the model on the GLL integration points [12]. Here we only note that if we use Lagrange polynomials and GLL quadrature, the mass matrix is exactly diagonal, resulting in a dramatic improvement in numerical accuracy and efficiency. With this choice, each element of the mesh contains $(n + 1)^3$ GLL points. Unfortunately, this approach requires that the mesh elements are hexahedra [12]. It is worth mentioning that for 2D problems it is possible to develop a SEM with triangles keeping the diagonal mass matrix but with an higher numerical cost [10]. The fact that we are restricted to hexahedral elements complicates matters significantly. Whereas 3D unstructured tetrahedral meshes can be constructed relatively easily with commercial or non-commercial codes, the creation of 3D unstructured hexahedral meshes remains a challenging and unresolved problem. For complex models, as in the case of realistic geological volumes, generating an all-hexahedral mesh based upon the available meshing algorithms can require weeks or months, even for an expert user [21]. One of the features of the SEM that impacts the creation of the mesh is the polynomial degree n used to discretize the wave field. The following heuristic rule has emerged to select n for an unstructured mesh of a heterogeneous medium: if $n < 4$ the inaccuracies are similar to the standard FEM, while if $n > 10$ the accuracy improves but the numerical cost of the simulation becomes prohibitive. The choice of n is related to the grid spacing Δh: in order to resolve the wave field to a shortest period T_0, the number of points per wavelength λ should be equal or greater than 5, leading to the constraint expressed in Eq. 2. If $n = 4$, then Δh is roughly equal to λ. We note that, for the same accuracy, a classical low-degree FEM requires a higher number of elements. Since the material properties are stored for each GLL point and can vary inside an element, we are able to interpolate the geological interfaces that our mesh is not able to follow. Nevertheless, this is an undesirable staircase sampling of the model, which introduces non-physical diffractions. Therefore, it is necessary that the mesh honors the major seismological contrasts. Furthermore, when a discontinuity is honored, the elements across the interface share some nodes, which will have the properties of the material below the discontinuity in one case, and the proprieties of the material above the discontinuity in the other case. Thus, the actual behavior of the seismic waves at the geological interface is perfectly mimicked in a way that cannot be achieved by an interpolation solely based upon Lagrange polynomials and the GLL quadrature.

Another constraint on the design of the mesh is the stability condition imposed by the adoption of an explicit conditionally stable time scheme. For a given mesh, there is an upper limit on the time step above which the calculations are unstable. We define the Courant stability number of the time scheme as $C = \Delta t (v/\Delta h)_{\max}$, where Δt is the time step and $(v/\Delta h)_{\max}$ denotes the maximum ratio of the compressional-wave speed and the grid spacing. The Courant stability condition ensures that calculations are stable if the Courant number is not higher than an upper limit $C_{\max}$ [4], providing a condition that determines the time step Δt (Eq. 3). Again, an heuristic rule suggests that C_{max} is roughly equal 0.3–0.4 for a deformed and heterogeneous mesh [12].

Like any technique based upon an unstructured deformed mesh, the SEM requires a smooth variation of the Jacobian and an acceptable distortion of the elements. Usually, to produce an acceptable accuracy the maximum equiangle skewness should not be greater than 0.5, although empirical tests show that $s < 0.8$ can sometimes be acceptable (Eq. 4) [12].

To sum up, spectral-element simulations require an unstructured all-hexahedral conforming mesh subject to the following heuristic constraints:

$$\Delta h = v_{min} \, T_0 \, \frac{n+1}{f(n)}, \tag{2}$$

$$\Delta t < C_{max} \frac{v_{min}}{v_{max}} \, T_0 \, \frac{n+1}{f(n)}, \tag{3}$$

$$s < 0.8, \tag{4}$$

where v_{max} and v_{min} are the maximum and minimum seismic wave speeds in the element, C_{max} is the Courant constant discussed above, T_0 is the shortest period that we seek to resolve, and $f(n)$ is an empirical function related to the number of points per wavelength and the polynomial degree n; for $n = 4$ $f(n) = 5$.

The mesh should have a similar number of grid points per wavelength throughout the entire model (i.e., the same numerical resolution everywhere), and since seismic wave speeds vary inside the volume, the mesh should have a variable element density, in accordance with Eq. 1. Wave speeds increase with depth in a typical geological model, so that the mesh should be denser near the surface and close to low wave-speed regions, such as an alluvial basin. In elements located at the top surface, v_{min} is the surface-wave speed, which controls the resolution of the mesh. Inside the volume it is the shear waves that have the slowest speed and thus determine Δh. From Eq. 3, we note that the smaller the grid spacing, the smaller the time step needed to obtain numerical stability, at the expense of computational cost. Therefore, Δh should be carefully evaluated in order to optimize numerical efficiency and accurate geometrical representation of the model. In the current SEM implementation, the time step is constant for the whole model. It is evident that a mesh that coarsens with depth accommodates a larger time step than a mesh with a constant element size throughout the entire volume, leading to a significant saving in computational cost.

Finally, one additional difficulty is the enormous number of elements that a typical seismic wave propagation simulation requires, ranging from hundreds of thousands of elements for local-scale simulations with low-frequency content, to tens of millions of elements for regional-scale simulations or for a high-frequency local-scale simulation. This implies that we need a parallel computational approach not only for the simulation of seismic wave propagation, but also for the mesh generation process.

The package of choice for our simulations is SPECFEM3D (geodynamics.org), which simulates global, regional (continental-scale) and local seismic wave propagation. Effects due to lateral variations in compressional-wave speed, shear-wave speed, density, a 3-D crustal model, ellipticity, topography and bathymetry, the oceans, rotation, and self-gravitation are all included, as well as full 21-parameter anisotropy and lateral variations in attenuation. We also present some SEM results obtained based upon the GeoELSE software package [21]. This SEM implementation is more suited for local- or small-scale simulations of seismic wave propagation (typically the goal is the study of soil-structure interaction problems). The main features of this kernel are the capability of 1) dealing with externally created 3D unstructured meshes, 2) handling the partitioning and load balancing of the computational domain by incorporating the METIS software library [13, 21], 3) implementing complex constitutive behavior [6] and 4) communicating with other codes through a sub-structuring interface based upon a domain reduction method [7].

Mesh examples

In this section we present some examples of unstructured hexahedral meshes that have been developed for simulations with SPECFEM3D.

The SPECFEM3D_GLOBE software package has been created primarily to simulate 3D seismic wave propagation on a global scale. For this purpose, a unique optimized mesh has been generated based upon a cubed sphere decomposition [17]. The details are described in [12]. Here we note that the need to densify the mesh close to the surface has been resolved by introducing doubling layers. The mesh honors topography and bathymetry and the major internal discontinuities, but not the Moho (the bottom of the Earth's crust) and the intra-crustal interfaces, which are interpolated based upon the Lagrange interpolation. This approximation is justified since the shapes of the Moho and the intra-crustal interfaces are insufficiently known to support the increase in elements that a full description of these boundaries would require. The mesh adequately resolves broadband data with periods down to 3.5 seconds on modern computers [12]. As an example, for a simulation executed on the Japanese Earth Simulator (JAMSTEC), the mesh was composed of 206 million hexahedral elements, corresponding to 36.6 billion degrees of freedom.

More critical from the point of view of mesh creation is the application of the SEM to regional or local problems, especially if we are interested in wave propagation with periods below 1 second. The software package SPECFEM3D_BASIN simulates seismic wave propagation in sedimentary basins, attempting to take into account the full complexity of these models. The solver is general and can be used to simulate seismic wave propagation on regional or local scales, but the mesh generator is currently specifically written for Southern California [11]. The mesh that honors the Moho, surface topography, and the deeper parts of the basement of the Los Angeles and Ventura sedimentary basins. It is coarsened twice with depth: first below the low wave-speed layer, and then below the basement. The mechanical proprieties of the volume are stored at each GLL node, taking into account the 3D Southern California seismic model. The mesh contains approximately 700 thousand spectral elements (corresponding to 136 million degrees of freedom). The minimum resolved period of the seismic waves is roughly 2 seconds [11]. The mesh is used routinely to simulate earthquakes with a magnitude greater than 3.5 in Southern California. At periods of about 4 seconds and longer, the resulting simulated wave motion closely matches the data recorded at the stations of the Southern California Seismic Network. Animations of seismic wave propagation are posted in near real-time at `shakemovie.caltech.edu`. If we want to analyze seismic wave propagation at shorter periods, we need to refine the mesh and constrain the mesh to honor the shape of the shallow parts of the basins, which is currently not done. Furthermore, the mesher is designed for this specific region. Unfortunately, generalizing the adopted meshing procedure to other geographical areas requires experience, internal modification of the code, and a significant amount of time.

For example, the SPECFEM3D_BASIN mesh generator was modified by [14] to produce a mesh for the Taipei basin (Taiwan). This work represents the state of the art for the SEM meshing process. The SPECFEM3D_BASIN mesher was modified to take into account the notable topography around Taipei city and the complex geometry of the underlying basin. Three new control surfaces were introduced at various depths and one new doubling layer was added in order to honor the main geological features of the region. The top of the model is controlled by surface topography based upon a detailed Digital Elevation Model. Just below the surface, a buffer layer is used to accommodate mesh distortions induced by the steep topography. This is critical to ensure the accurate and stable accommodation of the free surface. This layer also defines the top of the first mesh doubling, needed to take into account the low seismic wave speeds inside the basin. The shallow sedimentary basin is introduced between the surface topography and the buffer layer. Seismic wave speeds in the basin are quite low, with a sharp contrast between the basin and the surrounding basement. Compared to the Los Angeles basin, the Taipei basin is relatively small and shallow. The meshing strategy of [11] would result in a basement layer honoring only a small portion of the deepest part of the basin, and most of the basin boundaries would be not constrained by the mesh. To overcome these limitations, the nodes were empirically adjusted close to the sedimentary basin to redefine the basin boundary mesh [14]. Several additional empirical criteria were used to define how and where to move the nodes, leading to a stable and accurate mesh implementation. The resulting mesh is composed of 4.5 million spectral elements and 900 million degrees of freedom, and it is adequate for periods down to 1 second. As in the Los Angeles basin example, we note that this meshing strategy is ad-hoc for this particular application. Furthermore, the size of the hexahedra on the surface is controlled by the speed of the surface waves inside the basin, leading to an oversampled mesh in the surrounding regions (i.e., an enormous number of unnecessary elements, with a serious impact on the computational cost of the simulation). Finally, empirical adjustment of the nodes close to the basin basement was an attainable solution for the Taipei basin, but this approach cannot be generalized.

Meshing in parallel

The meshes discussed in the previous section are simply too large to be built on a single processor or even on a large shared memory machine: they require clusters or other parallel computers with distributed memory. The approach in SPECFEM3D is to partition the model into one mesh slice for each available CPU. Mesh generation and the corresponding wave propagation simulation are accomplished in parallel, the latter requiring only one communication phase between shared slice interfaces at each simulation time step. Since the communication map is constant throughout the simulation, it is built once and for all at the end of the mesh generation phase. The challenge is to load-balance the mesh, i.e., to make sure that each mesh slice contains roughly the same number of spectral-elements as any other mesh slice. This further limits the flexibility of the mesher, and adds further complexity to the procedure.

3 New meshing strategies: CUBIT

The aim of this article is to show examples of new approaches to the meshing process by means of the CUBIT mesh generation tool kit from Sandia National Laboratories (CUBIT.sandia.gov), with as an ultimate goal a more general parallel mesh generator for SPECFEM3D_BASIN. From the overview in the previous section it is clear that we are looking for software that has the following attributes: 1) the ability to accommodate general geological models, 2) minimal user intervention, 3) high resolution, 4) conforming, unstructured hexahedral elements, 5) a parallel, load-balanced implementation. The experience gained by the authors for various meshing problems [21] suggests that CUBIT comes very close to fulfilling these requirements. Although it currently lacks parallel capabilities, it incorporates a set of powerful and advanced meshing schemes developed to automatically handle the unstructured meshing problem. However, it is worth noting that meshing a large complex domain, such as any geological model, does not seem feasible with a desirable "single-button" procedure, due to the shape of the interfaces, the wide range of element sizes, and the broad range of possible geological models. A starting point would be to mesh in a parallel and automated way some basic classes of problems that are important in seismology. For example, a layer-cake volume combined with a shallow sedimentary basin.

3.1 Layer-cake geological volumes

The first step in interfacing CUBIT with the SPECFEM3D workflow is to reproduce the mesh that the current mesher creates. Since SPECFEM3D is written in Fortran90, we have developed a Fortran code that takes advantage of the journal scripting capabilities of CUBIT. The adopted strategy is valid for all the geological models that can be described by a layer-cake volume, consisting of a stack of non-degenerate heterogeneous quasi-parallel horizontal layers. The volume is split into rectangular slices of equal size, one slice for each processor following the original MPI map of SPECFEM3D. We use a serial Fortran routine to build a journal file for each slice, creating and storing all the commands that drive the meshing process inside CUBIT. Schematically, we can summarize each journal file in five steps: 1) creation of the geological geometry by means of a bottom-up approach (i.e., vertices → splines → surfaces → volumes) and geometry healing, 2) definition of the meshing intervals along the curves based upon Eqs. 2 and 3, 3) definition of the meshing schemes and meshing, 4) refinement and smoothness, and 5) blocks and sidesets definition and export (Figure 1). One of the main advantages of using CUBIT is that the entire process can be fairly well automated both for mesh creation and quality checking. The user is only asked to provide the shape of the topographic and geological surfaces and a good choice of intervals based upon Eqs. 2 and 3. Furthermore, it is possible to add or subtract geological discontinuities and refinement layers, and to correct mesh details in an easier way than in the current workflow. Unfortunately, some constraints on the meshing scheme are needed. As we will see later, due to the conformity requirements for the whole mesh and at the MPI interfaces (i.e., the surfaces that are shared by two adjacent slices), the vertical surfaces must be structured, while the horizontal surfaces can be meshed with structured or unstructured schemes. The so-constructed volume has a 2.5D logical symmetry, therefore, the sweep algorithm [19] can be directly applied without any further decomposition. As explained in the previous section, generally the profiles of seismic wave speeds demand a mesh coarsening with depth. Since the mesh resulting from the approach described above is structured along the vertical direction, the only way to produce a vertical densification of elements is the refinement algorithm included in CUBIT [5], which produces a transition between 1 to 27 hexahedra, tripling the elements along a linear direction in only one hex sheet 1b. It is more efficient than the refinement used in [11] (i.e., a transition from 1 element to 8 in 2 hex sheets) and it is perfect for a sharp high-contrast in seismic wave speeds. Nevertheless, where the variation of seismic wave speeds is smoother, the higher transition ratio in CUBIT's refinement produces a larger difference between the number of points per wavelength across the mesh (i.e., an unevenly sampled mesh). For this reason, the possibility of introducing a refinement with a lower transition ratio is under development. In this respect, let us mention that a refinement with linear ratio of 2, gping from two elements to four in one sheet has successfully been implemented in SPECFEM3D_GLOBE V4.0. After the creation of the journal files, a second Fortran routine provides the MPI environment needed to build the mesh in parallel. CUBIT is executed on each processor (slice) and it plays back the related script file, meshing its portion of volume. A quality analysis of the skewness is then performed. Since each slice is meshed independently, we carefully check that each corresponding node in the surfaces shared by two adjacent independent slices has the same coordinates. The "refined-structured" scheme imposed in the vertical direction guarantees that the number of elements in each surface is the same; nevertheless, the possibility of a discrepancy in location between two corresponding nodes is not negligible, due to the applied smoothness algorithm and the influence of the background elements. The routine communicates the nodes to the adjacent MPI slice following the map in Figure 1c. If the nodes in the slices show a numerically significant location discrepancy compared to the corresponding neighboring nodes, they are adjusted accordingly. Then, the GLL points are evaluated for each element and the material proprieties are assigned. The GLL node numbering follows the same procedure as in [11]. Consequently, the resulting GLL mesh can be written in order to follow the format required by the SPECFEM_BASIN solver, so that simulation can be performed directly.

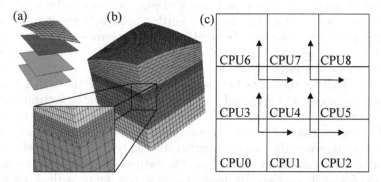

Fig. 1. Basic sketch for the mesh creation of a layer-cake geological model: (a) Step 1 of the strategy: creation of the geological geometry by means of a bottom-up approach. (b) Final result with a detailed zoom showing the tripling refinement layer in the upper part of the model. (c) The routine communicates the nodes to the adjacent MPI slices (up and right) following the map

Benchmark

In order to investigate the impact on the simulations of the coupling between SPECFEM3D and CUBIT in terms of both numerical dispersion and non-physical diffractions, we have performed some benchmarks for seismic wave propagation associated with a moderate earthquake using meshes produced by CUBIT and by the current SPECFEM3D_BASIN mesh generator. Table 1 summarizes the various benchmarks, which have been performed on 64 processors of CITERRA (a high-performance computing cluster in the Division of Geological & Planetary Sciences at Caltech [3]). The meshed volume covers part of the Taipei basin mesh described in Section 2.

Table 1. Summary of performed benchmarks

ID	model	mesh	differences
1	homogenous	regular mesh	no
2	homogenous	1 refinement	no
3	2 heterogeneous layers	1 refinement	no
4	homogenous	regular mesh & topography	negligible
5	2 layer & basin	2 refinement & topography	see Figure 2

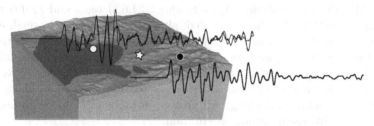

Fig. 2. Taipei basin CUBIT mesh used in the benchmark Test 5. Comparison between two seismograms obtained through two simulations of the seismic wave propagation produced by a M 6 earthquake (star): one simulation is performed with the mesh obtained by the current SPECFEM3D_BASIN mesher (dotted line) and one with the mesh created by CUBIT (solid line). The seismogram are recorded on bedrock (black circle) and soft soil basin deposits.

The goal of these benchmarks is not only to test the MPI communication capabilities of SPECFEM3D_CUBIT, but also to provide information about the impact of the tripling refinement layers and the topography on the quality of the simulation. Tests 1–3 shows that there is no significant impact due to the distorted elements in the refinement layer ($s_{\max} = 0.63$). Test 4 indicates that negligible differences in the seismograms are due to the different spline representation of the topography (Figure 1a). In Test 5 we applied the procedure for the Los Angeles basin [11] to the Taipei basin. The mesh honors the Moho interface, the topography and the deeper parts of the basin. The outcrop and the outline of the basin are interpolated by the Lagrange polynomials. Due to the different transition ratios for the refinement in SPECFEM3D_BASIN and SPECFEM3D_CUBIT, the resolution on the surface is better for SPECFEM3D_CUBIT (70 m vs 80 m). As shown in Figure 1b, if we consider the synthetic seismograms for a station on the bedrock, the two seismograms match almost perfectly, despite the different sampling. However, if we consider a station in the basin, some differences appear due to the interpolation of the discontinuity. Since no analytical solution exists for such a 3D seismological model, it is impossible to determine the seismogram that is closer to the exact solution of the problem, therefore we can only say that the results are globally in agreement for seismic waves with wavelengths greater that the dimension of the basin outcrop. Nevertheless, in order to resolve shorter wavelengths we would need a mesh that would honor the boundary of the basin better.

Campi Flegrei Caldera

The first application of SPECFEM3D_CUBIT is to simulate seismic wave propagation in the Campi Flegrei caldera (Italy). It is an active 13-km wide quaternary volcanic caldera whose history is dominated by two eruptions that produced widespread ash-flow deposits about 34,000 years and 12,000 years ago. Intense activity has been recorded, although it is limited to small eruptions and dramatic uplift and subsidence of the terrain. Particularly noteworthy is the subsidence episode of 1982–1984 and the contemporaneous seismic crisis. The caldera is located just a few kilometers west of the city of Naples. Because of the high population density, it is considered a highly dangerous volcanic area and a challenging civil protection problem faced by authorities and scientists in case of imminent eruption.

Volcanic structures represent a challenge for seismic wave simulations due to the high level of heterogeneity and topographic variations, and the presence of fractures, magmatic intrusions, and low wave-speed regions. For our simulation, we adopted a 3D model for Campi Flegrei [18], composed of 3 layers with a P-wave speed ranging from 2.7 to 6.5 km/s. The interfaces have a resolution of 250 m, while the topography is a based upon a 60 m Digital Elevation Model. The dimensions of the volume are 40 km×40 km×9 km. Both the mesh generation and the simulation are performed on 256 processors of CITERRA. The mesh contains 6.2 million hexahedral elements and 1.2 billion degrees of freedom. The resolution at the surface is 60 m and the mesh is numerically accurate down to the maximum frequency content of the seismic source (6 Hz). The mesh is created in 5 min of user-time: 60% of this time is dedicated to geometry creation, and only 20% to evaluation at the GLL points and assignment of material properties. Figure 3 shows the mesh and synthetic seismograms with a duration of 20 s for the main seismic event of the 1983 caldera unrest.

In summary, we have developed a parallel, semiautomatic mesher for a general geological layer-cake structure, capable of accommodating surface topography and the main internal discontinuities. The user provides only the shapes of the interfaces and the desired numerical resolution (balanced against the time step, i.e., against the numerical cost). The mesher includes some Fortran routines taken from the SPECFEM3D_BASIN solver and which allow the execution of CUBIT in an MPI environment. The main limitation is due to the requirement that each slice has the exact same number of hexahedra. It is likely that in the next version of the solver this limitation will be overcome, opening new possibilities for simulating seismic wave propagation in complex geological volumes. The ultimate goal is to mesh sedimentary basins fairly automatically. Another considerable limitation of this meshing strategy is the use of Fortran as the language to create the journal files, an approach that does

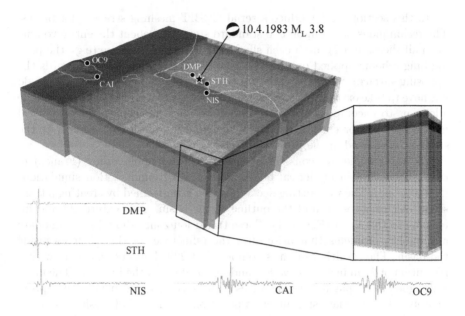

Fig. 3. CUBIT Mesh adopeted for the simulation on Campi Flegrei caldera. The seismograms show 20 s of ground motion associated with the main event of the 1982–84 crisis. The mesh is composed of more than six million elements and is accurate for the propagation of seismic waves with frequencies up to 6 Hz. The close up highlights the single refinement layer.

not provide the possibility to "interact" with the mesh during its creation. A significant step towards major flexibility is using the Python Interface, which is contained in the CUBIT package and provide the user with access to CUBIT commands from a python program. To take advantage of this versatility, we have developed some Python routines wrapping the fortran codes. In this case, the MPI environment is provided by PyMPI, a parallel interpreter of Python [16]. First, we have reproduced the same kinds of meshes for the layer-cake model class that we have presented in this subsection. The flexibility of CU-BIT Python Interface enables us to handle fairly automatically more complex cases, such as sedimentatary basins.

3.2 Sedimentary basins

Our interest in modelling alluvial basins is related to the seismic properties of these regions. Since they involve high wave-speed contrasts between alluvial deposits and bedrock, seismic waves are trapped inside the basin, producing reverberations that are stronger and longer than in the surrounding areas, consequently involving higher seismic hazard.

In this section we introduce a serial CUBIT meshing strategy for basins. The technique is based upon two standard steps: 1) webcut the entire volume in small slices, and 2) mesh each slice with a standard scheme (e.g., the pave meshing scheme applied on one of the surfaces and swept along towards the opposing surface). Since the entire process is executed on one processor, we do not have to take extra care to check the conformity of the mesh. This technique allows us to decompose the basin into hexahedra, honoring the geometrical interfaces, but worsens the quality of the resulting mesh with respect to the strategy developed in the previous section.

As an example, we consider a mesh of the Grenoble valley (France) in the framework of a numerical benchmark of 3D ground motion simulations (EGS06, [22]). The webcutting decomposition is performed by creating a transition belt of slices around the outline of the basin, both externally and internally to the basin (Figure 4). These transition zones should accommodate any quality problems that arise after the refinement of the elements inside the basin. The resulting mesh is composed of 220 thousand elements with a resolution of 21 m inside the valley, and up to 900 m in the bedrock. The mesh is designed to propagate seismic waves accurately up to frequencies of approximately 3 Hz. The last step involves partitioning the global mesh in order to assign the corresponding part of the mesh to each processor for seismic wave propagation. The velocity snapshots shown in Figure 4 are produced by the spectral-element software package GeoELSE [21], which resolves the problem of partitioning and load balancing by incorporating the METIS software library [13]. This task is completely independent of the meshing process. The creation of the mesh requires detailed knowledge of CUBIT and its meshing schemes. In particular, the webcut decomposition is a long and tedious step that can take days. The entire process requires one or two weeks of dedicated time for an experienced user.

This strategy forms the basis of an automated way to handle mesh generation for a sedimentary basin in parallel. The logical scheme of the problem is depicted in Figure 5a. The procedure takes advantage of the Python Interface to have access to CUBIT. The strategy is based upon the idea of creating a so-called partitioner from a coarse mesh on the surface. This is the tool we will use to decompose the model and create the MPI communication map between the slices. To understand how the mesh is built in practice, we show the procedure for a conceptual model in Figure 5. Before proceeding with the actual meshing process, the user should prepare in a general XYZ ASCII format: 1) the topography (labeled S) and the interfaces of the layer-cake volume (one of these is a surface that accommodates the deeper parts of the basin), 2) the internal outline curve (labeled I) of the part of the basin that is honored, and 3) outline curve (labeled B) of the basin, surrounded by 4) a transition curve (labeled T) (Figure 5a). Let us emphasize that the creation of the transition curve is the only intervention required from the user in this fairly automatic process. Furthermore, the user can add to the process some personal commands by means of a standard CUBIT journal file in order to adapt

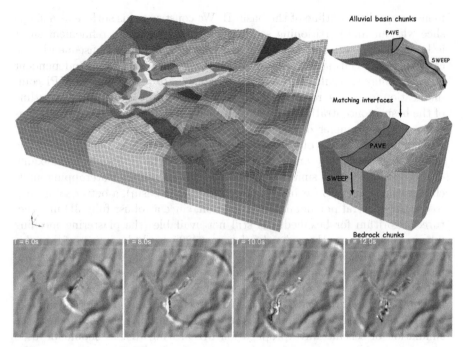

Fig. 4. (a) 3D numerical model used for the simulations of ESG06 "Grenoble Benchmark" with the GeoELSE code. The computational domain is subdivided into small chunks and each one is meshed starting from the alluvial basin down to the bedrock. (b) Snapshots of the velocity component for the Strong case 1, ESG06 "Grenoble Benchmark" [22].

the general case to a specific problem (e.g., the presence of a fault). Curves I and T are projected and imprinted on surface S, which is subsequently partitioned in three surfaces S_I, S_E, and S_T. The first two, S_I and S_E, are meshed with quadrilaterals using the pave scheme. The quad size is defined by the user and is related to the number of available processors. Each quadrilateral that shares at least one side with curves I and T is transformed into a curved element as in 5b. The group of these elements is the partitioner; one processor is assigned to each quad and the MPI map, with all the communication addresses, is created. In the MPI environment provided by PyMPI, each processor webcuts the volume V with the assigned element of the partitioner and keeps only its corresponding part, which it meshes independently following the procedure described for the layer-cake model (Figure 5c). In accordance with the MPI map, it sends (and receives) the position of the nodes on the lateral surface at the neighboring slices, so that a conforming mesh can be assured by the adjustment mentioned above. The hexahedral elements of the mesh that are inside the basin are refined in accordance with the constraints imposed by the wave speeds. Surface S_T is created in order to facilitate horizontal refinements around the outline of the basin B. We construct this surface as a single slice without any partitioning so that the entire transition refinement sheet is built inside a single chunk (i.e., by a single processor). Consequently, we can avoid propagating the information of the refinement to some independent adjacent chunks, resulting in a higher quality mesh and a simpler MPI communication map. The resulting mesh honors the deep sections and the outline of the basin, concentrating horizontal refinement only where it is needed, thus saving a huge number of elements. Nevertheless, with respect to the serial procedure, the outcrop of the basin is generally not honored. This is possible only if the outcropping angle is fairly constant (with respect to the hexahedral dimensions of the simulation). On the contrary, if the outcropping angle changes along curve B (as in the case of the Taipei basin), a better solution is to use the spectral polynomial interpolation, since a robust fully 3D unstructured algorithm for hexahedra is still not available (the plastering meshing scheme that is currently under development [20] should be very useful in this context). Further study is required in order to understand when the interpolation is acceptable. The Python code that drives the meshing process is also applicable if more than one basin is present in the volume. Since each slice is different both in terms of the number of elements and in terms of the number of vertical side surfaces, the mesh is not well balanced. Therefore, before the simulation of seismic wave propagation METIS/ParMETIS should be used to partition the global graph, as is done in the GeoELSE package, thereby producing a load-balanced mesh and minimizing MPI communications.

4 Conclusion

In the coming Petaflops era, the SEM should become a standard tool for the study of seismic wave propagation, both for forward and inverse problems. The more the power provided by computer clusters, the higher the resolution that is available for the simulations. Consequently, the definition of a good geological model and the creation of an all-hexahedral unstructured mesh are critical. While a full 3D unstructured algorithm is still awaited, we have presented some meshing examples for typical seismological problems, such as layer-cake volumes and sedimentary basins. The adopted strategies are valid for classes of models that have the same logical geometrical representation. The procedures work automatically (or with minimal user intervention) embedded in a parallel MPI environment, thereby easily creating unstructured meshes with hundreds of millions of elements.

Acknowledgments

Emanuele Casarotti is supported by Project SP3D, Marie Curie Fellowship UE/INGV n. MOIF-CT-2005-008298. Marco Stupazzini is supported by SPICE, MRTN-CT-2003-504267. The numerical simulations for this research

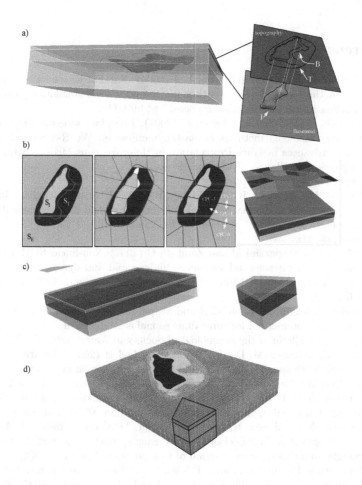

Fig. 5. Mesh of a conceptual model composed of a basin embedded in a layer-cake volume. For visual simplicity, we assume that all the surface are planar. (a) The "ingredients" that the user should provide are: interfaces, outline of the basin, internal outline corresponding to the part of the basin basement honored by the mesh, and a transition curve. (b) The creation of the "partitioner": from left to right, 1) the imprinting of the curves on the topography, 2) quad mesh of S_E and S_I, 3) transformation of the quads into curved elements (we note that S_T is a single slice), 4) the partitioner with respect of the mesh that should be meshed. (c) Each element of the partitioner is assigne to a processor and it is used as tool for the webcut. (d) The resulting mesh obtained reassembling the slices meshed by the independent processors; note the refinement in both the vertical and horizontal directions. The latter (lighter gray) is entirely included in the transition zone.

were performed on Caltechs Division of Geological & Planetary Sciences Dell cluster.

References

1. Bielak, J, Ghattas, O, and Kim, E J (2005) Parallel octree-based finite element method for large-scale earthquake ground motion simulation. Computer Modeling in Engineering and Sciences, 10(2):99–112.
2. Bouchon, M. and Sanchez-Sesma, F. (2007). Boundary integral equations and boundary elements methods in elastodynamics. In: Wu R-S and Maupin V (eds), Advances in Wave Propagation in Heterogeneous Media, Advances in Geophysics, 48 Elsevier - Academic Press.
3. citerra.gps.caltech.edu
4. Courant, R, Friedrichs, K O, and Lewy, H (1928) Über die partiellen differenzgleichungen der mathematischen physik (on the partial difference equations of mathematical physics). Mathematische Annalen, 100:32–74.
5. cubit.sandia.gov
6. Di Prisco C, Stupazzini M and Zambelli C (2007) Non-linear SEM numerical analyses of dry dense sand specimens under rapid and dynamic loading. International Journal for Numerical and Analytical Methods in Geomechanics, 31-6:757–788.
7. Faccioli E, Vanini M, Paolucci R and Stupazzini M (2005) Comment on "Domain reduction method for three-dimensional earthquake modeling in localized regions". Bulletin of the Seismological Society of America, 95:763–769
8. Käser M, Dumbser M, De la Puente J and Igel H (2007) An arbitrary high order discontinuous Galerkin method for elastic waves on unstructured meshes iii: Viscoelastic attenuation. Geophysical Journal International, 168:224–242
9. Komatitsch, D, Barnes, C, and Tromp, J (2000) Simulation of anisotropic wave propagation based upon a spectral element method. 65(4):1251–1260.
10. Komatitsch, D, Martin, R, Tromp, J, Taylor, M A, and Wingate, B A (2001) Wave propagation in 2-D elastic media using a spectral element method with triangles and quadrangles. Journal of Computational Acoustic, 9(2):703–718.
11. Komatitsch, D, Liu, Q, Tromp, J, Süss, P, Stidham, C, and Shaw, J H (2004) Simulations of ground motion in the Los Angeles basin based upon the Spectral-Element method. Bulletin Seismological Society American, 94:187–206.
12. Komatitsch, D, Tsuboi, S, and Tromp, J (2005) The spectral-element method in seismology. In: Levander A. and Nolet G., (eds), Seismic Earth: Array Analysis of Broadband Seismograms, volume 157 of Geophysical Monograph, pages 205–228 American Geophysical Union, Washington DC, USA.
13. glaro.sdtcumn.edu/gkhome/views/metis
14. Lee S J, Chen H W, Liu Q, Komatitsch D, Huang B S, and Tromp J (2007) Mesh generation and strong ground motion simulation in the taipei basin based upon the spectral-element method. submitted.
15. Moczo P, Robertsson J O A, and Eisner L (2007) The finite-difference time-domain method for modeling of seismic wave propagation, in advances in wave propagation in heterogeneous Earth In: Wu R-S and Maupin V (eds), Advances in Wave Propagation in Heterogeneous Media, Advances in Geophysics, 48 Elsevier - Academic Press.

16. pympi.sourceforge.net
17. Ronchi C, Ianoco R and Paolucci P S (1996) The "Cubed Sphere": a new method for the solution of partial differential equations in spherical geometry. J Comput Phys, 124:93–114.
18. Satriano C, Zollo A, Capuano P, Russo G, Vanorio T, Caielli G, Lovisa L and Moretti M (2006) A 3D velociy model for earthquake location in Campi Flegrei area application to the 1982-84 uplift event. In: Zollo A, Capuano P and Corciulo M (eds.), Geophysical Exploration of the Campi Flegrei (Southern Italy) caldera interiors: Data, Methods and Results. DoppiaVoce ed, Naples.
19. Scott M A, Earp M N, Benzley S E and Stephenson M B (2005) Adaptive sweeping techniques. In Proceedings of the 14th International Meshing Roundtable, pages 417–432 Springer.
20. M L Staten , R A Kerr, S.J.Owen and T D Blacker (2006). Unconstrained paving and plastering: Progress update. In: Proceedings, 15th International Meshing Roundtable, pages 469–486. Springer-Verlag.
21. Stupazzini M (2004) A spectral element approach for 3D dynamic soil-structure interaction problems. PhD thesis, Milan University of Technology, Milan, Italy.
22. Stupazzini M. (2006) 3D ground motion simulation of the grenoble valley by geoelse. In: Third International Symposium on the Effects of Surface Geology on Seismic Motion, August 30th - September 1st, Grenoble, France.

Index of Authors & Co-Authors

Index by Affliation